Device Electronics for Integrated Circuits

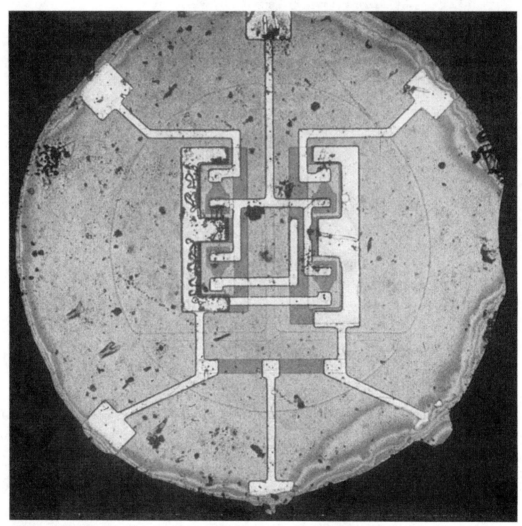

First silicon planar integrated circuit—a bipolar OR gate—built by Robert Noyce and colleagues at Fairchild Semiconductor in 1961. (*Courtesy of Dr. B. E. Deal, Fairchild Semiconductor*)

Device Electronics for Integrated Circuits

Third Edition

Richard S. Muller   and   **Theodore I. Kamins**

Electrical Engineering
and Computer Science
University of California
Berkeley, California

Hewlett-Packard Laboratories
Palo Alto, California, and
Stanford University
Stanford, California

with

Mansun Chan
Hong Kong University of Science & Technology
Clear Water Bay, Kowloon, Hong Kong

John Wiley & Sons

Acquisitions Editor   *Bill Zobrist*
Marketing Manager   *Katherine Hepburn*
Senior Production Editor   *Valerie A. Vargas*
Senior Designer   *Harry Nolan*
Production Management Services   *Suzanne Ingrao of Ingrao Associates*

Photomicrograph of the Pentium 4 microprocessor which contains 42 million transistors; see Fig. 2.2e for a block diagram. (Courtesy of Intel Corporation.)

This book is printed on acid-free paper. ∞

Library of Congress Cataloging-in-Publication Data

Muller, Richard S.
  Device electronics for integrated circuits / Richard S. Muller, Theodore I. Kamins—3rd ed.
    p. cm.
  Includes bibliographical references.
  ISBN 978-0-471-59398-0 (cloth : alk. paper)
  ISBN 0-471-42877-9 (WIE)
    1. Semiconductors. 2. Integrated circuits. I. Kamins, Theodore I. II. Title.

TK7871.85 .M825 2002
621.3815'2–dc21   2002026742
  ISBN 978-0-471-59398-0
  ISBN 0-471-42877-9 (WIE)

10 9 8 7 6 5

To our TEACHERS . . .
who opened new vistas,
and to our STUDENTS . . .
who will see further than we.

Richard S. Muller has worked on silicon integrated circuits since their invention in the early 1960s. After earning a doctorate at the California Institute of Technology in 1962 he joined the forward-looking research and teaching program then being formed at the University of California, Berkeley. Through associations with circuit designers, device specialists, and technologists at the university and through consulting activities at several companies in Silicon Valley, he developed awareness and practical experience in integrated circuits. This experience provided a firm foundation for his creation of a new course having the title later given to this book which he introduced at Cal and which he has taught to multiple generations of Berkeley students. In the late 1970s, Muller began research in the area now known as MEMS and he joined with a colleague in 1986 to found the Berkeley Sensor & Actuator Center. He has been elected to membership in the National Academy of Engineering and as a Fellow of the IEEE; other distinctions include the IEEE Brunetti Award, Stevens Institute of Technology Renaissance Award, Alexander von Humboldt Research Award, Berkeley Citation, and both NATO and Fulbright Fellowships. Dr. Muller presently serves on the Board of Trustees of Stevens Institute of Technology.

Theodore I. Kamins is Principal Scientist in the Quantum Science Research group at Hewlett-Packard Laboratories in Palo Alto, California, where he is conducting research on advanced nanostructured electronic materials and devices. He is also a Consulting Professor in the Electrical Engineering Department at Stanford University. He received his degrees from the University of California, Berkeley. He then joined the Research and Development Laboratory of Fairchild Semiconductor, where he worked with epitaxial and polycrystalline silicon before moving to Hewlett-Packard. In addition to being co-author with R. S. Muller of the textbook "Device Electronics for Integrated Circuits," Ted is author of the book "Polycrystalline Silicon for Integrated Circuits and Displays." He is a Fellow of the IEEE and a Fellow of the Electrochemical Society. He received the 1989 Electronics Division Award of the Electrochemical Society. He has also taught at the University of California, Berkeley, and at Stanford University. He has been an Associate Editor of the IEEE Transactions on Electron Devices and has presented short courses for the University of California, Oxford University, the IEEE, the American Vacuum Society, and SEMI.

Mansun Chan received his Ph.D. degree in 1995 at the University of California at Berkeley before joining the Hong Kong University of Science & Technology. His research interests cover a broad area in the field of silicon devices, ranging from deep submicron process development to device design, characterization, and modeling. He is one of the major contributors to the unified BSIM3 model for SPICE, which has been accepted by most U.S. companies and the Compact Model Council as an industry standard model. In January 2002, Mansun returned to UC Berkeley to co-direct the BSIM project and the development of the next-generation device models for sub-70 nm devices.

Mansun has been teaching device and technology classes at the advanced undergraduate and graduate levels for more than six years at the Hong Kong University of Science & Technology, and recently, at the University of California at Berkeley. He has received a number of appreciation awards for his teaching efforts. In addition, he has given public lectures to the semiconductor industry and presented tutorials at international conferences.

The dramatic impact of integrated circuits (ICs) on engineering and on the broader society continues to grow with each passing year. Recognition of this impact, as much as acknowledgment of the basic inventive genius that gave it birth, led the Nobel Prize Committee to award the Physics prize in 2000* to Jack Kilby for work that he did more than 40 years earlier at Texas Instruments, Inc. In addition to Kilby's inventive concept for the integrated fabrication of an electrical circuit, which was proven on a germanium substrate, the pioneering work by Jean Hoerni, working with a team at Fairchild Semiconductor in California, established the silicon planar process, providing a direct route to today's integrated circuits. In the early 1960s, Robert Noyce joined with colleagues at Fairchild to produce the first planar-processed silicon integrated circuit (the binary OR gate shown in our frontispiece)—a forerunner of hundreds of billions of descendents that control and power our modern digitally processed world.

From the inventive ideas of Kilby, Hoerni, Noyce, and their co-workers has grown a worldwide market in integrated circuits that (in 2000) totaled $177 billion, supporting an end-product market of roughly $1.15 trillion (also in 2000). A token of the societal impact of integrated circuits is also evident in the recognition awarded recently to Andrew Grove, former CEO of Intel, who was named "Man of the Year" in 1997 by TIME Magazine. Dr. Grove is, of course, a pioneer in developing integrated circuits, as well as the author in 1967 of a seminal text that described for the first time the physics, design, and technology of planar-processed integrated circuits ("Physics and Technology of Semiconductor Devices," Wiley, New York). In the newly unfolding third millennium, as we approach a half-century since ICs were born, their influence extends beyond traditional electronics to mechanics, optics, and to chemical and biological integrated systems. In these diverse fields, techniques, design paradigms, materials, and processes developed for ICs are being extended and enriched to advance what has become known as microelectromechanical systems (MEMS).

Today's integrated circuits contain tens-of-millions of active devices on a chip, but the vast majority of chips are still formed of silicon, as was Noyce's 1961 IC. Today, the overwhelming majority of devices are metal-oxide-semiconductor (MOS) field-effect transistors, which displaced the formerly dominant bipolar junction transistors in the 1980s. Technology changes in the IC world over the past two decades have mostly focused on ways to optimize the performance of MOSFETs and to reduce their sizes as much as possible. To place a particular emphasis on MOSFETs as they are now designed and analyzed, the two original authors of "Device Electronics," Richard Muller and Ted Kamins have enlisted the experienced talent of Professor Mansun Chan, a specialist in the field of MOSFETs for integrated circuits and therefore a strong contributor to the material in the final two chapters. These chapters have been substantially revised from earlier editions.

Technology advances have similarly made it appropriate to change considerably the material explaining the IC fabrication process (Chapter 2). The increasing complexity of ICs (and the power of the computers they enable) motivates the use of computer-aided process and device simulation, which we introduce in new sections of this book (in Chapters 2 and 5). The need for higher speed transistors is leading to the adoption of heterojunction bipolar transistors (HBTs), especially when silicon-based HBTs can be integrated along with

---

* Awarded jointly with Z. Alferov and H. Kroemer (for their work in related solid-state-electronics fields—rather than specifically on ICs).

more conventional silicon devices. The concepts of heterojunctions and HBTs are introduced throughout Chapters 4–7, together with the more usual development for homojunctions.

Our goal in this, as in earlier editions of "Device Electronics," is to emphasize those electronic processes that govern the behavior of the solid-state devices used most frequently to build ICs. In each chapter, we present the physics associated with a device and conclude with a detailed device discussion that illustrates the basic physical principles. The text includes more discussion than typically can be covered in a one-semester course, allowing instructors to focus on topics that they choose to emphasize. In an introductory course, much of the material in Chapters 2, 7, and 10 can be omitted. Alternatively, students having a strong solid-state physics background may find Chapter 1 unnecessary. Example problems are incorporated into many discussions and answers to some of the chapter-end problems (those designated by an asterisk *) are printed at the end of the book. We have marked sections that contain more advanced topics and also more challenging problems with a dagger superscript (†) to indicate that they can be omitted without a loss of continuity. In the following, we introduce the book through a series of capsule descriptions of the chapters.

Chapter 1 SEMICONDUCTOR ELECTRONICS. This chapter gives an overview of the physical electronics of semiconductors: energy-band theory, doping principles, free-carrier statistics, and drift and diffusion. Only topics that are needed for later device discussions are presented; physical pictures, rather than mathematical formulations, are emphasized. Integrated Hall-effect magnetic sensing circuits are the subjects of a concluding section on specific IC applications. An appendix provides a review of the use of Gauss' law to solve electrostatic problems, an important technique for basic device analysis.

Chapter 2 SILICON TECHNOLOGY. Silicon crystal growth, wafer preparation, oxidation, film deposition, wafer doping by diffusion and ion implantation, pattern definition and transfer, lead attachment and packaging; in brief, the techniques of the modern planar process are described in this chapter. There is considerable detail in some sections that can be used for reference, rather than being studied linearly. By omitting these sections, the chapter can be covered in a shorter time while still affording the student a broad overview of silicon technology. The concept of computer-aided process modeling is introduced to show how increasingly expensive experiments are augmented by simulation. Processing of compound semiconductors is briefly considered. The device section discusses diffused resistors in ICs.

Chapter 3 METAL-SEMICONDUCTOR CONTACTS. In this chapter the student is introduced to the concept of thermal equilibrium between dissimilar solids. The electronics of Schottky-barrier contacts are described in detail. Optional sections present a rigorous derivation of the current-voltage relationship for the junction, the nature of Schottky ohmic contacts, and surface effects at metal-semiconductor contacts. Practical Schottky diodes are considered in the device section.

Chapter 4 _pn_ JUNCTIONS. The topic of this chapter is the electronics associated with inhomogeneously doped semiconductors. The behavior of _pn_ junctions—both homojunctions and heterojunctions—under reverse bias is described in detail. A section on junction breakdown is included, but may be omitted without loss of continuity. In the concluding section, the junction field-effect transistor is discussed.

Chapter 5 CURRENTS IN _pn_ JUNCTIONS. The continuity equations for free carriers are derived. The concepts of generation and recombination are introduced and then applied to describe the electronics of _pn_ junctions under forward (injecting) bias. A discussion of minority-carrier storage and numerical simulation of diodes provides a foundation for the later development of transistor equivalent circuits. Optional sections include a full description of Shockley-Hall-Read recombination theory and space-charge-region generation and recombination. The device discussion focuses on the uses of junction diodes in ICs and briefly considers light-emitting diodes made from compound semiconductors. After Chapter 5, readers having a strong interest in MOS electronics may, with minor loss of continuity, consider the materials in Chapters 8 through 10 before returning to the bipolar transistor discussions in Chapters 6 and 7.

Chapter 6 BIPOLAR TRANSISTORS I: Basic Operation. The framework for the discussion of bipolar junction transistors in this chapter is the integrated-charge (charge-control) model. The familiar homogeneously doped transistor is treated as a special case of the general theory to convey the physical principles. This approach to bipolar transistors creates a smooth transition into topics such as bias ranges, current gain, and the Ebers-Moll model. The basic concept of the bipolar junction transistor is extended to consider the heterojunction bipolar transistor. The device discussion focuses on properties of planar IC transistors for amplifying and switching circuits.

Chapter 7 BIPOLAR TRANSISTORS II: Limitations and Models. The topics covered in this chapter are oriented toward computer modeling of the transistor. For example, the Early effect is introduced in terms of the *Early voltage*, and effects at low- and high-emitter bias are considered with the use of the integrated-charge model. The same approach is taken to describe the electronics of the base transit time. This topic leads naturally to the charge-control model for the transistor and to examples of its use. The small-signal transistor (hybrid-pi) model is deduced from the charge-control model, and we consider equivalences between the various representations for the transistor. Discussion of the frequency response of bipolar transistors shows the differences between homojunction and heterojunction bipolar transistors. The device discussion focuses on *pnp* transistors for integrated circuits.

Chapter 8 PROPERTIES OF THE OXIDE-SILICON SYSTEM. This chapter provides the necessary additional background in physical electronics that is needed for a discussion of MOS devices. The control exercised by the gate to accumulate, deplete, or invert the semiconductor surface is described, and the concepts underlying the flat-band and threshold voltages of an MOS system are presented. An optional section describes surface effects on *pn* junctions. The chapter concludes with a discussion of MOS capacitors for memories and charge-coupled devices (CCDs).

Chapter 9 MOS FIELD-EFFECT TRANSISTORS I: Physical Effects and Models. The classical theory of MOSFETs is first derived from an approximate charge-control model, and then refined to include short-channel effects, mobility degradation, and velocity saturation. The concept of MOSFET scaling is presented to provide a practical perspective to the rapid development of advanced CMOS structures. A simplified model for circuit design, together with the parameter-extraction technique based on the industry-standard BSIM model, is given to bridge the gap between technology development and practical design.

Chapter 10 MOS FIELD-EFFECT TRANSISTORS II: High-Field Effects. This chapter considers several high-field effects that are extremely important in modern CMOS technology, but often not considered in basic descriptions of MOSFET operation: channel-length modulation, output resistance, substrate current, gate current, MOSFET breakdown, device degradation, and hot-carrier-resistant structures. A device section describes both the design of floating-gate, non-volatile memory elements, which utilize high-field effects, and the integration of memory elements into a gate-array architecture.

1. SEMICONDUCTOR ELECTRONICS

2. SILICON TECHNOLOGY

3. METAL-SEMICONDUCTOR CONTACTS

4. *pn* JUNCTIONS

Device Electronics for Integrated Circuits

SEMICONDUCTOR ELECTRONICS

From everyday experience we know that the electrical properties of materials vary widely. If we measure the current I flowing through a bar of homogeneous material with uniform cross section when a voltage V is applied across it, we can find its resistance $R = V/I$. The resistivity ρ—a basic electrical property of the material comprising the bar—is related to the resistance of the bar by a geometric ratio

$$\rho = R\frac{A}{L} \tag{1.1}$$

where L and A are the length and cross-sectional area of the sample.

The resistivities of common materials used in solid-state devices cover a wide range. An example is the range of resistivities encountered at room temperature for the materials used to fabricate typical silicon integrated circuits. Deposited metal strips, made from very low-resistivity materials, connect elements of the integrated circuit; aluminum and copper, which are most frequently used, have resistivities at room temperature of about 10^{-6} Ω-cm. On the other end of the resistivity scale are insulating materials such as silicon dioxide, which serve to isolate portions of the integrated circuit. The resistivity of silicon dioxide is about 10^{16} Ω-cm—22 orders of magnitude higher than that of aluminum. The resistivity of the plastics often used to encapsulate integrated circuits can be as high as 10^{18} Ω-cm. Thus, a typical integrated circuit can contain materials with resistivities varying over 24 orders of magnitude—an extraordinarily wide range for a common physical property.

For electrical applications, materials are generally classified according to their resistivities. Those with resistivities less than about 10^{-2} Ω-cm are called *conductors*, while those having resistivities greater than about 10^5 Ω-cm are called *insulators*. In the intermediate region is a class of materials called *semiconductors* that have profound electrical importance because their resistivities can be varied by design and precisely controlled. Equally important, they can be made to conduct by one of two types of current carriers. The utilization of these unique properties of semiconductors is the primary topic of this book.

In the first portion of this chapter we consider what gives rise to the enormous range of resistivities in solids. We then briefly review some important concepts from semiconductor physics. To focus on devices, however, our discussion of materials is brief, and we assume that the reader has had some exposure to basic semiconductor and solid-state physics such as that contained in reference 1. The treatment here is merely a review of some of the more important concepts in a form useful for later application. The appendix to this chapter briefly reviews some concepts of electromagnetic theory.

1.1   PHYSICS OF SEMICONDUCTOR MATERIALS

An understanding of the physics of electrons in solids can be achieved by first considering electrons in an isolated atom. By logically perturbing the electrons in this system, we can learn about some important electronic properties of solids.

We therefore begin by considering the allowed energies of an electron influenced only by an isolated atomic core. Then we look at the effect of bringing other atoms near the first atom so that the core of one atom influences the electrons associated with the other atoms, as is true in a solid material. In this manner we can study the effect of the atomic cores in a crystal on the behavior of the associated electrons. We then investigate the effect on the electrons of applying an electric field to the solid material.

Two complementary models are used in the discussion of semiconductor physics: the *energy-band model* and the *crystal-bonding model*. We first consider the energy-band description and bring in the bonding model later to introduce other concepts.

Band Model of Solids

We know that an electron acted on by the Coulomb potential of an atomic nucleus can have only certain allowed energies (Figure 1.1). In particular, the electron can occupy one of a series of energy levels

$$E_n = \frac{-Z^2 m_0 q^4}{8\epsilon_0^2 h^2 n^2} \tag{1.1.1}$$

below a reference energy taken as $E = 0$.

The electron carries charge $-q$ so that q in Equation 1.1.1 is a positive number. This is our convention throughout the book. The other symbols in Equation 1.1.1 are Z, the number

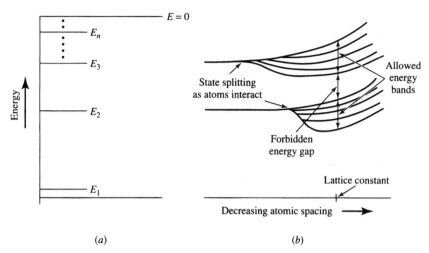

FIGURE 1.1 (*a*) Allowed energy levels of an electron acted on by the Coulomb potential of an atomic nucleus. (*b*) Splitting of energy states into allowed bands separated by a forbidden energy gap as the atomic spacing decreases; the electrical properties of a crystalline material correspond to specific allowed and forbidden energies associated with an atomic separation related to the lattice constant of the crystal.

of protons in the nucleus; m_0, the free electron mass; ϵ_0, the permittivity* of free space; h, Planck's constant; and n, which is a positive integer. For the hydrogen atom with $Z = 1$, the allowed energies are $-2.19 \times 10^{-18}/n^2$ J or $-13.6/n^2$ electron volts (eV) below the zero reference level.** At low temperatures and when more than one electron are associated with the atom, the electrons fill the allowed levels starting with the lowest energies. By the Pauli exclusion principle, at most two electrons (of opposite spins) can occupy any energy level.

Let us now consider the electron in the highest occupied energy level of an atom and neglect the lower filled levels. When two isolated atoms are separated by a large distance, the electron associated with each atom has the energy E_n given by Equation 1.1.1. If two atoms approach one another, however, the atomic core of the first atom exerts a force on the second electron, changing the potential that determines the energy levels of the electron. All allowed energy levels for the electron are consequently modified because of this change in potential.

We must also consider the Pauli exclusion principle when describing a two-atom system. An energy level E_n, which can contain at most two electrons of opposite spin, is associated with each isolated atom so that the total system can contain at most four electrons. When the two atoms are brought together to form one system, however, only two electrons can be associated with the allowed energy level E_n. The allowed energy level E_n of the isolated atoms must split into two levels with slightly different energies in order to retain space for a total of four electrons. Therefore, bringing two atoms close together not only slightly perturbs each energy level of the isolated atom but also splits each of the energy levels of the isolated atoms into two slightly separated energy levels. As the two atoms are brought closer together, stronger interactions are expected and the splitting increases.

---

* Sometimes called the *dielectric constant*.

** The use of the unit electron volt (1 eV $= 1.6 \times 10^{-19}$ J; see the conversion factors on the inside front cover) is convenient in semiconductor physics because it often avoids cumbersome exponents. Although not an *MKS* unit, the electron volt is widely employed and is used extensively in our discussion.

As more atoms are added to form a crystalline structure, the forces encountered by each electron are altered further, and additional changes in the energy levels occur. Again, the Pauli exclusion principle demands that each allowed electron energy level have a slightly different energy so that many distinct, closely spaced energy levels characterize the crystal. Each of the original quantized levels of the isolated atom is split many times, and each resulting group of energy levels contains one level for each atom in the system. When N atoms are included in the system, the original *energy level E_n* splits into N different allowed levels, forming an *energy band*, which may contain at most $2N$ electrons (because of spin degeneracy). In Figure 1.1*b* the splitting of two discrete allowed energy levels of isolated atoms is indicated in a sketch that shows what the allowed energy levels would be if the interatomic distances for a large assemblage of atoms could be varied continuously. Under such conditions, the discrete energies that characterize isolated atoms split into multiple levels as the atomic spacing decreases. When the atomic spacing equals the crystal-lattice spacing (indicated in Figure 1.1*b*), regions of allowed energy levels are typically separated by a *forbidden energy gap* in which electrons cannot exist. An analogue to atomic energy-level splitting exists in mechanical and electrical oscillating systems which have discrete resonant frequencies when they are isolated, but multiple resonance values when a number of similar systems interact.

Because the number of atoms in a crystal is generally large—of the order of 10^{22} cm^{-3}—and the total extent of the energy band is of the order of a few electron volts, the separation between the N different energy levels within each band is much smaller than the thermal energy possessed by an electron at room temperature ($\sim 1/40$ eV), and the electron can easily move between levels. Thus, we can speak of a continuous band of allowed energies containing space for $2N$ electrons. This allowed band is bounded by maximum and minimum energies, and it may be separated from adjacent allowed bands by *forbidden-energy gaps*, as shown in Figure 1.2*a*, or it may overlap other bands. The detailed behavior of the bands (whether they overlap or form gaps and, if so, the size of the gaps) fundamentally determines the electronic properties of a given material. It is the essential feature differentiating conductors, insulators, and semiconductors.

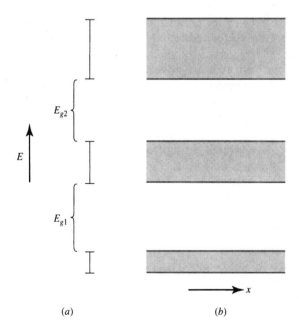

FIGURE 1.2 Broadening of allowed energy levels into allowed energy bands separated by forbidden-energy gaps as more atoms influence each electron in a solid: (*a*) one-dimensional representation; (*b*) two-dimensional diagram in which energy is plotted versus distance.

Although each energy level of the original isolated atom splits into a band composed of 2N levels, the range of allowed energies of each band can be different. The higher energy bands generally span a wider energy range than do those at lower energies. The cause for this difference can be seen by considering the Bohr radius r_n associated with the nth energy level:

$$r_n = \frac{n^2 \epsilon_0 h^2}{Z \pi m_0 q^2} = \frac{n^2}{Z} \times 0.0529 \text{ nm} \tag{1.1.2}$$

For higher energy levels (larger n) the electron is less tightly bound and can wander farther from the atomic core. If the electron is less tightly confined, it comes closer to the adjacent atoms and is more strongly influenced by them. This greater interaction causes a larger change in the energy levels so that the wider energy bands correspond to the higher energy electrons of the isolated atoms.

We represent the energy bands at the equilibrium spacing of the atoms by the one-dimensional picture in Figure 1.2a. The highest allowed level in each band is separated by an energy gap E_g from the lowest allowed level in the next band. It is convenient to extend the band diagram into a two-dimensional picture (Figure 1.2b) where the vertical axis still represents the electron energy while the horizontal axis now represents position in the semiconductor crystal. This representation emphasizes that electrons in the bands are not associated with any of the individual nuclei, but are confined only by the crystal boundaries. This type of diagram is especially useful when we consider combining materials with different energy-band structures into semiconductor devices. In this brief discussion of the basis for energy bands in crystalline solids, we implicitly assumed that each atom is like its neighbor in all respects including orientation; that is, we are considering *perfect crystals*. In practice, a very high level of crystalline order, in which defects are measured in parts per billion or less, is normal for device-quality semiconductor materials.

The formation of energy bands from discrete levels occurs whenever the atoms of any element are brought together to form a solid. However, the different numbers of electrons within the energy bands of different solids strongly influence their electrical properties. For example, consider first an alkali metal composed of N atoms, each with one valence electron in the outer shell. When the atoms are brought close together, an energy band forms from this energy level. In the simplest case this band has space for 2N electrons. The N available electrons then fill the lower half of the energy band (Figure 1.3a),

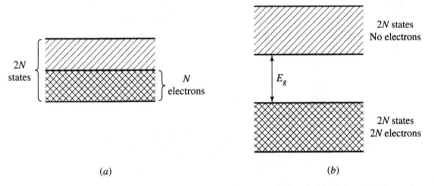

(a)

(b)

FIGURE 1.3 Energy-band diagrams: (a) N electrons filling half of the 2N allowed states, as can occur in a metal. (b) A completely empty band separated by an energy gap E_g from a band whose 2N states are completely filled by 2N electrons, representative of an insulator.

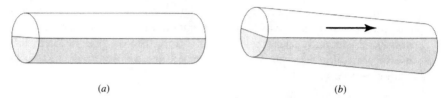

(a)   (b)

FIGURE 1.4 Electron motion in an allowed band is analogous to fluid motion in a glass tube with sealed ends; the fluid can move in a half-filled tube just as electrons can move in a metal.

and there are empty states just above the filled states. The electrons near the top of the filled portion of the band can easily gain small amounts of energy from an applied electric field and move into these empty states. In these states they behave almost as free electrons and can be transported through the crystal by an externally applied electric field. In general, metals are characterized by partially filled energy bands and are, therefore, highly conductive.

Markedly different electrical behavior occurs in materials in which the valence (outermost shell) electrons completely fill an allowed energy band and there is an energy gap to the next higher band. In this case, characteristic of insulators, the closest allowed band above the filled band is completely empty at low temperatures as shown in Figure 1.3b. The lowest-energy empty states are separated from the highest filled states by the energy gap E_g. In insulating material, E_g is generally greater than 5 eV ($\sim$8–9 eV for SiO_2),* much larger than typical thermal or field-imparted energies (tenths of an eV or less). In the idealization that we are considering, there are no electrons close to empty allowed states and, therefore, no electrons can gain small energies from an externally applied field. Consequently, no electrons can carry an electric currrent, and the material is an insulator.

An analogy may be helpful. Consider a horizontal glass tube with sealed ends representing the allowed energy states and fluid in the tube representing the electrons in a solid. In the case analogous to a metal, the tube is partially filled (Figure 1.4). When a force (gravity in this case) is applied by tipping the tube, the fluid can easily move along the tube. In the situation analogous to an insulator, the tube is completely filled with fluid (Figure 1.5). When the filled tube is tipped, the fluid cannot flow because there is no empty volume into which it can move; that is, there are no empty allowed states.

Both electrical insulators and semiconductors have similar band structures. The electrical difference between insulators and semiconductors arises from the size of the forbidden-energy gap and the ability to populate a nearly empty band by adding conductivity-enhancing impurities to a semiconductor. In a semiconductor the energy gap separating the highest band that is filled at absolute zero temperature from the lowest empty band is typically of the order of 1 eV (silicon: 1.1 eV; germanium: 0.7 eV; gallium arsenide: 1.4 eV). In an impurity-free semiconductor the uppermost filled band is populated by electrons that were the valence electrons of the isolated atoms; this band is known as the *valence band.*

The band structure of a semiconductor is shown in Figure 1.6. At any temperature above absolute zero, the valence band is not entirely filled because a small number of electrons possess enough thermal energy to be excited across the forbidden gap

---

* Specifying band gaps in amorphous materials may cause anxiety to theorists, but short-range order in such materials leads to effects that can be interpreted with the aid of an energy-band description.

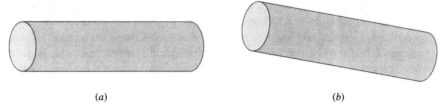

(a)   (b)

FIGURE 1.5 No fluid motion can occur in a completely filled tube with sealed ends.

into the next allowed band. The smaller the energy gap and the higher the temperature, the greater the number of electrons that can jump between bands. The electrons in the upper band can easily gain small amounts of energy and can respond to an applied electric field to produce a current. This band is called the *conduction band* because the electrons that populate it are conductors of electricity. The current density J (current per unit area) flowing in the conduction band can be found by summing the charge $(-q)$ times the net velocity (v_i) of each electron populating the band. The summation is then taken over all electrons in the conduction band in a unit volume of the material.

$$J_{cb} = \frac{I}{A} = \sum_{cb} (-q) v_i \qquad (1.1.3)$$

Since only a small number of electrons exist in this band, however, the current for a given field is considerably less than that in a metal.

Holes

When electrons are excited into the conduction band, empty states are left in the valence band. If an electric field is then applied, nearby electrons can respond to the field by moving into these empty states to produce a current. This current can be expressed by summing the motion of all electrons in the valence band of a unit volume of the material.

$$J_{vb} = \sum_{vb} (-q) v_i \qquad (1.1.4)$$

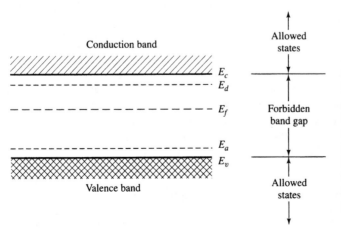

Conduction band

Valence band

E_c
E_d

E_f

E_a
E_v

Allowed
states

Forbidden
band gap

Allowed
states

FIGURE 1.6 Energy-band diagram for a semiconductor showing the lower edge of the conduction band E_c, a donor level E_d within the forbidden gap, the Fermi level E_f, an acceptor level E_a, and the top edge of the valence band E_v.

Because there are many electrons in the valence band, but only a few empty states, it is easier to describe the conduction resulting from electrons interacting with these empty states than to describe the motion of all the electrons in the valence band. Mathematically, we can describe the current in the valence band as the current that would flow if the band were completely filled minus that associated with the missing electrons. Again, summing over the populations per unit volume, we find

$$J_{vb} = \sum_{vb} (-q)v_i = \sum_{Filled\ band} (-q)v_i - \sum_{Empty\ states} (-q)v_i \qquad (1.1.5)$$

since no current can flow in a completely filled band (because no net energy can be imparted to the electrons populating it), the current in the valence band can be written as

$$J_{vb} = 0 - \sum_{Empty\ states} (-q)v_i = \sum_{Empty\ states} qv_i \qquad (1.1.6)$$

where the summation is over the empty states per unit volume. Equation 1.1.6 shows that we can express the motion of charge in the valence band in terms of the vacant states by treating the states as if they were particles with positive charge. These "particles" are called *holes*; they can only be discussed in connection with the energy bands of a solid and cannot exist in free space. Note that energy-band diagrams, such as those shown in Figures 1.3 and 1.6, are drawn for electrons so that the energy of an electron increases as we move upward toward the top of the diagram. However, because of its opposite charge, the energy of a hole increases as we move downward on this same diagram.*

The concept of holes can be illustrated by continuing our analogy of fluids in glass tubes. We start with two sealed tubes—one completely filled, and the other completely empty (Figure 1.7a). When we apply a force by tipping the tubes, no motion can occur (Figure 1.7b). If we transfer a small amount of fluid from the lower tube to the upper tube (Figure 1.7c), however, the fluid in the upper tube can move when the tube is tilted (Figure 1.7d). This flow corresponds to electron conduction in the conduction band of the solid. In the lower tube, a bubble is left because of the fluid we removed. This bubble is analogous to holes in the valence band. It cannot exist outside of the nearly filled tube, just as it is only useful to discuss holes in connection with a nearly filled valence band. When the tube is tilted, the fluid in the tube moves downward but the bubble moves in the opposite direction, as if it has a mass of opposite sign to that of the fluid. Similarly the holes in the valence band move in a direction opposite to that of the electrons, as if they had a charge of the opposite sign. Just as it is easier to describe the motion of the small bubble than that of the large amount of fluid, it is easier to discuss the motion of the few holes rather than the motion of the electrons, which nearly fill the valence band.

Bond Model

The discussion of free holes and electrons in semiconductors can also be phrased in terms of the behavior of completed and broken electronic bonds in a semiconductor crystal. This viewpoint, which is often called the *bond model*, fails to account for important quantum-mechanical constraints on the behavior of electrons in crystals, but it does illustrate several useful qualitative concepts.

---

* As in Figure 1.6 we often simplify the energy-band diagram by showing only the upper bound of the valence band (denoted by E_v) and the lower bound of the conduction band (denoted by E_c) because we are primarily interested in states near these two levels and in the energy gap E_g separating them.

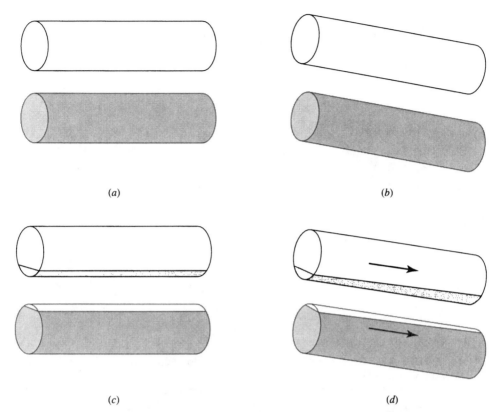

FIGURE 1.7 Fluid analogy for a semiconductor. (a) and (b) No flow can occur in either the completely filled or completely empty tube. (c) and (d) Fluid can move in both tubes if some of it is transferred from the filled tube to the empty one, leaving unfilled volume in the lower tube.

To discuss the bond model, we consider the diamond-type crystal structure that is common to silicon and germanium (Figure 1.8). In the diamond structure, each atom has covalent bonds with its four nearest neighbors. There are two tightly bound electrons associated with each bond—one from each atom. At absolute zero temperature, all electrons are held in these bonds, and therefore none are free to move about the crystal in response to an applied electric field. At higher temperatures, thermal energy breaks some of the bonds and creates nearly free electrons, which can then contribute to the current under the influence of an applied electric field. This current corresponds to the current associated with the conduction band in the energy-band model.

After a bond is broken by thermal energy and the freed electron moves away, an empty bond is left behind. An electron from an adjacent bond can then jump into the vacant bond, leaving a vacant bond behind. The vacant bond, therefore, moves in the opposite direction to the electrons. If a net motion is imparted to the electrons by an applied field, the vacant bond can continue moving in the direction opposite to the electrons as if it had a positive charge.* This vacant bond corresponds to the hole associated with the valence band in the energy-band picture.

A crystal lattice with many similarities to the diamond structure is the *zincblende* lattice which characterizes several important compound semiconductors composed of

---

* A positive charge is associated with the vacant bond because there are insufficient electrons in its vicinity to balance the proton charges on the atomic nuclei.

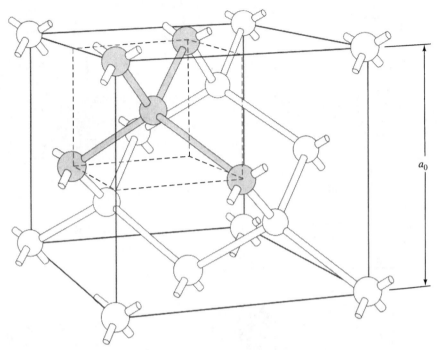

FIGURE 1.8 The diamond-crystal lattice characterized by four covalently bonded atoms. The lattice constant, denoted by a_0, is 0.356, 0.543 and 0.565 nm for diamond, silicon, and germanium, respectively. Nearest neighbors are spaced ($\sqrt{3}a_0/4$) units apart. Of the 18 atoms shown in the figure, only 8 belong to the volume a_0^3. Because the 8 corner atoms are each shared by 8 cubes, they contribute a total of 1 atom; the 6 face atoms are each shared by 2 cubes and thus contribute 3 atoms, and there are 4 atoms inside the cube. The atomic density is therefore $8/a_0^3$, which corresponds to 17.7, 5.00, and 4.43 × 10^{22} cm^{-3}, respectively. (After W. Shockley: *Electrons and Holes in Semiconductors*, Van Nostrand, Princeton, N.J., 1950.)

atoms in the third and fifth columns of the periodic table (called III-V semiconductors). Some III-V semiconductors, particularly gallium arsenide (GaAs) and gallium phosphide (GaP), have important device applications. Many properties of the elemental and compound semiconductors are given in Table 1.3, which appears at the end of this chapter. Table 1.3 also includes data for some insulating materials used in the manufacture of integrated circuits. A second table at the end of the chapter (Table 1.4) contains additional properties of the most important semiconductor, silicon.

Donors and Acceptors

Thus far we have discussed a pure semiconductor material in which each electron excited into the conduction band leaves a vacant state in the valence band. Consequently, the number of negatively charged electrons n in the conduction band equals the number of positively charged holes p in the valence band. Such a material is called an *intrinsic* semiconductor, and the densities of electrons and holes in it (carriers cm^{-3}) are usually subscripted i (that is, n_i and p_i). However, the most important uses of semiconductors arise from the interaction of adjacent semiconductor materials having differing densities of the two types of charge carriers. We can achieve such a structure either by physically joining two materials with different band gaps (as we will discuss in Sec. 4.2) or by varying the number of carriers in one semiconductor material (as we consider here).

The most useful means for controlling the number of carriers in a semiconductor is by incorporating *substitutional impurities*; that is, impurities that occupy lattice sites in place of the atoms of the pure semiconductor. For example, if we replace one silicon atom (four valence electrons) with an impurity atom from group V in the periodic table, such as phosphorus (five valence electrons), then four of the valence electrons from the impurity atom fill bonds between the impurity atom and the adjacent silicon atoms. The fifth electron, however, is not covalently bonded to its neighbors; it is only weakly bound to the impurity atom by the excess positive charge on the nucleus. Only a small amount of energy is required to break this weak bond so its the fifth electron can wander about the crystal and contribute to electrical conduction. Because the substitutional group V impurities donate electrons to the silicon, they are known as *donors*.

In order to estimate the amount of energy needed to break the bond to a donor atom, we consider the net Coulomb potential that the electron experiences because of the core of its parent atom. We assume that the electron is attracted by the single net positive charge of the impurity atom core weakened by the polarization effects of the background of silicon atoms. The energy binding the electron to the core is then

$$E = \frac{m_n^* q^4}{8h^2 \epsilon_0^2 \epsilon_r^2} = \frac{13.6}{\epsilon_r^2} \frac{m_n^*}{m_0} \text{ eV} \tag{1.1.7}$$

where ϵ_r is the relative permittivity of the semiconductor and m_n^* is the effective mass of the electron in the semiconductor conduction band. The use of an effective mass accounts for the influence of the crystal lattice on the motion of an electron. For silicon with $\epsilon_r = 11.7$ and $m_n^* = 0.26 \, m_0$, $E \approx 0.03$ eV, which is only about 3% of the silicon band-gap energy (1.1 eV). (More detailed calculations and measurements indicate that the binding energy for typical donors is somewhat higher: 0.044 eV for phosphorus, 0.049 eV for arsenic, and 0.039 eV for antimony.) The small binding energies make it much easier to break the weak bond connecting the fifth electron to the donor than to break the silicon-silicon bonds.

n-type Semiconductors. According to the energy-band model, it requires only a small amount of energy to excite the electron from the donor atom into the conduction band, while a much greater amount of energy is required to excite an electron from the valence band to the conduction band. Therefore, we can represent the state corresponding to the electron when bound to the donor atom by a level E_d about 0.05 eV below the bottom of the conduction band E_c (Figure 1.6). The density of donors (atoms cm^{-3}) is generally designated by N_d. Thermal energy at temperatures greater than about 150 K is generally sufficient to excite electrons from the donor atoms into the conduction band. Once the electron is excited into the conduction band, a fixed, positively charged atom core is left behind in the crystal lattice. The allowed energy state provided by a donor (*donor level*) is, therefore, neutral when occupied by an electron and positively charged when empty.*

If most impurities are of the donor type, the number of electrons in the conduction band is much greater than the number of holes in the valence band. Electrons are then called the *majority carriers*, and holes are called the *minority carriers*. The material is said to be an *n-type* semiconductor because most of the current is carried by the *negatively charged* electrons. A graph showing the conduction electron concentration versus temperature

---

* Other fields of study, such as chemistry, define a donor differently, possibly leading to some confusion.

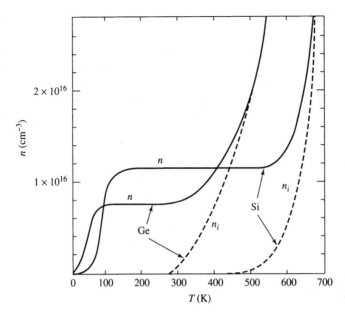

FIGURE 1.9 Electron concentration versus temperature for two n-type doped semiconductors: (*a*) Silicon doped with 1.15×10^{16} arsenic atoms cm^{-3}[1], (*b*) Germanium doped with 7.5×10^{15} arsenic atoms cm^{-3}[2].

for silicon and germanium is sketched in Figure 1.9. Because the hole density is at most equal to n_i, this figure shows clearly that electrons are far more numerous than holes when the temperature is in the range sufficient to ionize the donor atoms (about 150 K) but not adequate to free many electrons from silicon-silicon bonds (about 600 K).

p-type Semiconductors. In an analogous manner, an impurity atom with three valence electrons, such as boron, can replace a silicon atom in the lattice. The three electrons fill three of the four covalent silicon bonds, leaving one bond vacant. If another electron moves to fill this vacant bond from a nearby bond, the vacant bond is moved, carrying with it positive charge and contributing to hole conduction. Just as a small amount of energy was necessary to initiate the conduction process in the case of a donor atom, only a small amount of energy is needed to excite an electron from the valence band into the vacant bond caused by the trivalent impurity. This energy is represented by an energy level E_a slightly above the top of the valence band E_v (Figure 1.6). An impurity that contributes to hole conduction is called an *acceptor impurity* because it leads to vacant bonds, which easily accept electrons. The acceptor concentration (atoms cm^{-3}) is denoted as N_a. If most of the impurities in the solid are acceptors, the material is called a *p-type* semiconductor because most of the conduction is carried by *positively charged* holes. An acceptor level is neutral when empty and negatively charged when occupied by an electron.

Semiconductors in which conduction results primarily from carriers contributed by impurity atoms are said to be *extrinsic*. The donor and acceptor impurity atoms, which are intentionally introduced to change the charge-carrier concentration, are called *dopant* atoms.

In compound semiconductors, such as gallium arsenide, certain group IV impurities can be substituted for either element. Thus, silicon incorporated as an impurity in gallium arsenide contributes holes when it substitutes for arsenic and contributes electrons when it substitutes for gallium. This *amphoteric* doping behavior can be difficult to control and is not found for all group IV impurities; for example, the group IV element tin is incorporated almost exclusively in place of gallium in gallium arsenide and is therefore a useful *n*-type dopant. Impurities from group VI that substitute for arsenic, such as tellurium, selenium, or sulfur, are also used to obtain *n*-type gallium arsenide, whereas group II elements like zinc or cadmium have been used extensively to obtain *p*-type material.

Other impurity atoms or crystalline defects may provide energy states in which electrons are tightly bound so that it takes appreciable energy to excite an electron from a bound state to the conduction band. Such *deep donors* may be represented by energy levels well below the conduction-band edge in contrast to the *shallow donors* previously discussed, which had energy levels only a few times the thermal energy kT below the conduction-band edge. Similarly, *deep acceptors* are located well above the valence-band edge. Because deep levels are not always related to impurity atoms in the same straightforward manner as shallow donors and acceptors, the distinction between the terms donor and acceptor is made on the basis of the possible charge states the level can take. A deep level is called a donor if it is neutral when occupied by an electron and positively charged when empty, while a deep acceptor is neutral when empty and negative when occupied by an electron.

Compensation. The intentional doping of silicon with shallow donor impurities, to make it *n*-type, or with shallow acceptor impurities, to make it *p*-type, is the most important processing step in the fabrication of silicon devices. An especially useful feature of the doping process is that one may *compensate* a doped silicon crystal (for example an *n*-type sample) by subsequently adding the opposite type of dopant impurity (a *p*-type dopant in this example). Reference to Figure 1.6 helps to clarify the process. In Figure 1.6 donor atoms add allowed energy states to the energy-band diagram at E_d, close to the conduction-band energy E_c, whereas acceptor atoms add allowed energy states at E_a, close to the valence-band energy E_v. At typically useful temperatures for silicon devices, each donor atom has lost an electron and each acceptor atom has gained an electron. Because the acceptor atoms provide states at lower energies than those either in the conduction band or at the donor levels, the electrons from the donor levels transfer (or "fall") to the lower-energy acceptor sites as long as any of these remain unfilled. Hence, in a doped semiconductor, the effective dopant concentration is equal to the magnitude of the difference between the donor and acceptor concentrations $|N_d - N_a|$; the semiconductor is *n*-type if N_d exceeds N_a and *p*-type if N_a exceeds N_d. Although in theory one can achieve a zero effective dopant density through compensation (with $N_d = N_a$), such exact control of the dopant concentrations is technically impractical. As we will see in Chapter 2 (where technology is discussed), compensation doping usually involves adding a dopant density that is about an order-of-magnitude higher than the density of dopant that is initially present.

EXAMPLE *Donors and Acceptors*

A silicon crystal is known to contain 10^{-4} atomic percent of arsenic (As) as an impurity. It then receives a uniform doping of 3×10^{16} cm^{-3} phosphorus (P) atoms and a subsequent uniform doping of 10^{18} cm^{-3} boron (B) atoms. A thermal annealing treatment then completely activates all impurities.

(a) What is the conductivity type of this silicon sample?

(b) What is the density of the majority carriers?

Solution   Arsenic is a group V impurity, and acts as a donor. Because silicon has 5×10^{22} atoms cm^{-3} (Table 1.3), 10^{-4} atomic percent implies that the silicon is doped to a concentration of

$$5 \times 10^{22} \times 10^{-6} = 5 \times 10^{16} \text{ As atoms cm}^{-3}$$

The added doping of 3×10^{16} P atoms cm^{-3} increases the donor doping of the crystal to 8×10^{16} cm^{-3}.

Additional doping by B (a group III impurity) converts the silicon from *n*-type to *p*-type because the density of acceptors now exceeds the density of donors. The net acceptor density is, however, less than the density of B atoms owing to the *donor compensation*.

(a) Hence, the silicon is *p*-type.

(b) The density of holes is equal to the net dopant density:

$$p = N_a(B) - [N_d(As) + N_d(P)]$$
$$= 10^{18} - [5 \times 10^{16} + 3 \times 10^{16}]$$
$$= 9.2 \times 10^{17} cm^{-3}$$

■

Thermal-Equilibrium Statistics

Before proceeding to a more detailed discussion of electrical conduction in a semiconductor, we consider three additional concepts: first, the concept of thermal equilibrium; second, the relationship at thermal equilibrium between the majority- and minority-carrier concentrations in a semiconductor; and third, the use of Fermi statistics and the Fermi level to specify the carrier concentrations.

Thermal Equilibrium. We saw that free-carrier densities in semiconductors are related to the populations of allowed states in the conduction and valence bands. The densities depend upon the net energy in the semiconductor. This energy is stored in crystal-lattice vibrations (phonons) as well as in the electrons. Although a semiconductor crystal can be excited by external sources of energy such as incident photoelectric radiation, many situations exist where the total energy is a function only of the crystal temperature. In this case the semiconductor spontaneously (but not instantaneously) reaches a state known as *thermal equilibrium*. Thermal equilibrium is a dynamic situation in which every process is balanced by its inverse process. For example, at thermal equilibrium, if electrons are being excited from a lower energy E_1 to a higher energy E_2, then there must be equal transfer of electrons from the states at E_2 to those at E_1. Likewise, if energy is being transferred into the electron population from the crystal vibrations (phonons), then at thermal equilibrium an equal flow of energy is occurring in the opposite direction. A useful thought picture for thermal equilibrium is that a moving picture taken of any event can be run either backward or forward without the viewer being able to detect any difference. In the following we consider some properties of hole and electron populations in semiconductors at thermal equilibrium.

Mass-Action Law. At most temperatures of interest to us, there is sufficient thermal energy to excite some electrons from the valence band to the conduction band. A dynamic equilibrium exists in which some electrons are constantly being excited into the conduction band while others are losing energy and falling back across the energy gap to the valence band. The excitation of an electron from the valence band to the conduction band corresponds to the *generation* of a hole and an electron, while an electron falling back across the gap corresponds to electron-hole *recombination* because it annihilates both carriers. The generation rate of electron-hole pairs G depends on the temperature T but is, to first order, independent of the number of carriers already present. We therefore write

$$G = f_1(T) \tag{1.1.8}$$

where $f_1(T)$ is a function determined by crystal physics and temperature. The rate of recombination R, on the other hand, depends on the concentration of electrons n in the conduction band and also on the concentration of holes p (empty states) in the valence band, because both species must interact for recombination to occur. We therefore represent the recombination rate as a product of these concentrations as well as other factors that are included in $f_2(T)$:

$$R = npf_2(T) \qquad (1.1.9)$$

At equilibrium the generation rate must equal the recombination rate. Equating G and R in Equations 1.1.8 and 1.1.9, we have

$$npf_2(T) = f_1(T)$$

or

$$np = \frac{f_1(T)}{f_2(T)} = f_3(T) \qquad (1.1.10)$$

Equation 1.1.10 expresses the important result that at thermal equilibrium the product of the hole and electron densities in a given semiconductor is a function only of temperature.

In an intrinsic (i.e., undoped) semiconductor all carriers result from excitation across the forbidden gap. Consequently, $n = p = n_i$, where the subscript i reminds us that we are dealing with intrinsic material. Applying Equation 1.1.10 to intrinsic material, we have

$$n_i p_i = n_i^2 = f_3(T) \qquad (1.1.11)$$

The intrinsic carrier concentration depends on temperature because thermal energy is the source of carrier excitation across the forbidden energy gap. The intrinsic concentration is also a function of the size of the energy gap because fewer electrons can be excited across a larger gap. We will soon be able to show that under most conditions n_i^2 is given by the expression

$$n_i^2 = N_c N_v \exp\left(\frac{-E_g}{kT}\right) \qquad (1.1.12)$$

where N_c and N_v are related to the density of allowed states near the edges of the conduction band and valence band, respectively. Although N_c and N_v vary somewhat with temperature, n_i is much more temperature dependent because of the exponential term in Equation 1.1.12. For silicon with $E_g = 1.1$ eV, n_i doubles for every 8°C increase in temperature near room temperature. Because the intrinsic carrier concentration n_i is constant for a given semiconductor at a fixed temperature, it is useful to replace $f_3(T)$ by n_i^2 in Equation 1.1.10. Therefore the relation

$$np = n_i^2 \qquad (1.1.13)$$

holds for both extrinsic and intrinsic semiconductors; it shows that increasing the number of electrons in a sample by adding donors causes the hole concentration to decrease so that the product np remains constant. This result, often called the *mass-action law*, has its counterpart in the behavior of interacting chemical species, such as the concentrations of hydrogen and hydroxyl ions (H^+ and OH^-) in acidic or basic solutions. As we see from our derivation, the law of mass action is a straightforward consequence of equating generation and recombination, that is, of thermal equilibrium.

In the neutral regions of a semiconductor (i.e., regions free of field gradients), the number of positive charges must be exactly balanced by the number of negative charges.

Positive charges exist on ionized donor atoms and on holes, while negative charges are associated with ionized acceptors and electrons.* If there is charge neutrality in a region where all dopant atoms are ionized,

$$N_d + p = N_a + n \tag{1.1.14}$$

Rewriting Equation 1.1.14 and using the mass-action law (Equation 1.1.13), we obtain the expression

$$n - \frac{n_i^2}{n} = N_d - N_a \tag{1.1.15}$$

which may be solved for the electron concentration n:

$$n = \frac{N_d - N_a}{2} + \left[\left(\frac{N_d - N_a}{2} \right)^2 + n_i^2 \right]^{1/2} \tag{1.1.16}$$

In an n-type semiconductor $N_d > N_a$. From Equation 1.1.16 we see that the electron density depends on the net excess of ionized donors over acceptors. Thus, as we saw in the previous example, a piece of p-type material containing N_a acceptors can be converted into n-type material by adding an excess of donors so that $N_d > N_a$. In Chapter 2 we will see how this conversion is carried out in fabricating silicon integrated circuits.

For silicon at room temperature, n_i is 1.45×10^{10} cm^{-3} while the net donor density in n-type silicon is typically about 10^{15} cm^{-3} or greater: Hence $(N_d - N_a) \gg n_i$ and Equation 1.1.16 reduces to $n \approx (N_d - N_a)$. Consequently, from Equation 1.1.13,

$$p = \frac{n_i^2}{n} \approx \frac{n_i^2}{N_d - N_a} \tag{1.1.17}$$

Thus, for $N_d - N_a = 10^{15}$ cm^{-3} we have $p = 2 \times 10^5$ cm^{-3}, and the minority-carrier concentration is nearly 10 orders of magnitude below the majority-carrier population. In general, the concentration of one type of carrier is many orders of magnitude greater than that of the other in extrinsic semiconductors.

Fermi Level.   The numbers of free carriers (electrons and holes) in any macroscopic piece of semiconductor are relatively large—usually large enough to allow use of the laws of statistical mechanics to determine physical properties.** One important property of electrons in crystals is their distribution at thermal equilibrium among the allowed energy states. Basic considerations of ways to populate allowed energy states with particles subject to the Pauli exclusion principle leads to an energy distribution function for electrons that is called the Fermi-Dirac distribution function. It is denoted by $f_D(E)$ and has the form

$$f_D(E) = \frac{1}{1 + \exp[(E - E_f)/kT]} \tag{1.1.18}$$

where E_f is a reference energy called the *Fermi energy* or *Fermi level*. From Equation 1.1.18, we see that $f_D(E_f)$ always equals $\frac{1}{2}$. The Fermi-Dirac distribution function, often

---

* When we speak of *electrons* in our discussion of devices in subsequent chapters, we generally refer to electrons in the conduction band; exceptions are explicitly noted. The term *holes* always denotes vacant states in the valence band.

** However, in some devices made with submicrometer dimensions, the number of dopant atoms in the active regions is so small that statistical fluctuations in the number of dopant atoms can affect device characteristics.

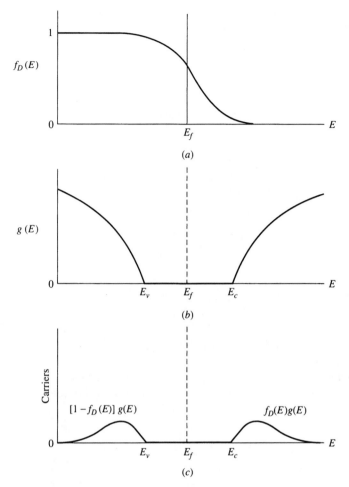

FIGURE 1.10 (a) Fermi-Dirac distribution function describing the probability that an allowed state at energy E is occupied by an electron. (b) The density of allowed states for a semiconductor as a function of energy; note that $g(E)$ is zero in the forbidden gap between E_v and E_c. (c) The product of the distribution function and the density-of-states function.

called simply the *Fermi function*, describes the probability that a state at energy E is filled by an electron. As shown in Figure 1.10a, the Fermi function approaches unity at energies much lower than E_f, indicating that the lower energy states are mostly filled. It is very small at higher energies, indicating that few electrons are found in high-energy states at thermal equilibrium—in agreement with physical intuition. At absolute zero temperature all allowed states below E_f are filled and all states above it are empty. At finite temperatures, the Fermi function does not change so abruptly; there is a small probability that some states above the Fermi level are occupied and some states below it are empty.

The Fermi function represents only a probability of occupancy. It does not contain any information about the states available for occupancy and, therefore, cannot by itself specify the electron population at a given energy. Applying quantum physics to a given system provides information about the density of available states as a function of energy. We denote this function by $g(E)$. A sketch of $g(E)$ for an intrinsic semiconductor is shown in Figure 1.10b. It is zero in the forbidden gap ($E_c > E > E_v$), but it rises sharply within both the valence band ($E < E_v$) and the conduction band ($E > E_c$). The actual distribution of electrons as a function of energy can be found from the product of the density of allowed states $g(E)$ within a small energy interval dE and the probability $f_D(E)$ that these states are filled. The total density of electrons in the conduction band can be obtained by multiplying the density-of-states function $g(E)$ in the conduction band by the Fermi function and integrating over the conduction band:

$$n = \int_{cb} f_D(E)g(E)\, dE \tag{1.1.19}$$

Similarly, the density of holes in the valence band is found by multiplying the density-of-states function in the valence band by the probability $[1 - f_D(E)]$ that these states are empty and integrating over the valence band.

In n-type material that is not too highly doped, only a small fraction of the allowed states in the conduction band are filled. The Fermi function in the conduction band is very small, and the Fermi level is well below the bottom of the conduction band. Then $(E_c - E_f) \gg kT$, and the Fermi function given by Equation 1.1.18 reduces to the mathematically simpler Maxwell-Boltzmann distribution function:

$$f_M(E) = \exp\left[\frac{-(E - E_f)}{kT}\right] \tag{1.1.20}$$

This thermal-equilibrium distribution function can also be derived independently by omitting the limitations imposed by the Pauli exclusion principle; that is, the Boltzmann function applies to the case that any number of electrons can exist in an allowed state. At energies well above the Fermi level, the fraction of available states that are occupied is so small that the exclusion-principle limitation has no practical effect, and Maxwell-Boltzmann statistics are applicable.

Using Equation 1.1.20 in the integration described in Equation 1.1.19 and making several approximations, we express the carrier concentration in the conduction band in terms of the Fermi level by

$$n = N_c \exp\left[-\frac{(E_c - E_f)}{kT}\right] \tag{1.1.21}$$

Similarly, in moderately doped p-type material, the Fermi level is significantly above the top of the valence band, and

$$p = N_v \exp\left[-\frac{(E_f - E_v)}{kT}\right] \tag{1.1.22}$$

where $(E_c - E_f)$ is the energy between the bottom edge of the conduction band and the Fermi level, and $(E_f - E_v)$ is the energy separation from the Fermi level to the top of the valence band. The quantities N_c and N_v, called the effective densities of states at the conduction- and valence-band edges, respectively, are given by the expressions

$$N_c = 2\left(\frac{2\pi m_n^* kT}{h^2}\right)^{3/2} \tag{1.1.23}$$

and

$$N_v = 2\left(\frac{2\pi m_p^* kT}{h^2}\right)^{3/2} \tag{1.1.24}$$

where m_n^* and m_p^* are the effective masses of electrons and holes. These effective masses are related to m^* as introduced in Equation 1.1.7 but differ somewhat from it because of the details of the energy band structure. As we see from Equations 1.1.21 and 1.1.22, the quantities N_c and N_v effectively concentrate all of the distributed conduction- and valence-band states at E_c and E_v. They can be used to calculate thermal-equilibrium densities whenever the Fermi level is a few kT or more removed from a band edge.

Except for slight differences in the values of m^*, all terms in Equations 1.1.23 and 1.1.24 are equal so that $N_c \simeq N_v$. Hence, in an n-doped material for which $n \gg p$, $(E_c - E_f) \ll (E_f - E_v)$; this means that the Fermi level is much closer to the conduction band than it is to the valence band. Similarly, the Fermi level is nearer the valence band than the conduction band in a p-type semiconductor.

In an intrinsic semiconductor $n = p$. Therefore, $(E_c - E_f) \approx (E_f - E_v)$ and the Fermi level is nearly at the middle of the forbidden gap $[E_f = (E_c + E_v)/2]$. We denote this *intrinsic Fermi level* by the symbol E_i. Just as the quantity n_i is useful in relating the carrier concentrations even in an extrinsic semiconductor (Equation 1.1.13), E_i is frequently used as a reference level when discussing extrinsic semiconductors. In particular, because

$$n_i = N_c \exp\left[\frac{-(E_c - E_i)}{kT}\right] = N_v \exp\left[\frac{-(E_i - E_v)}{kT}\right] \tag{1.1.25}$$

the expressions for the carrier concentrations n and p in an extrinsic semiconductor (Equations 1.1.21 and 1.1.22) can be rewritten in terms of the intrinsic carrier concentration and the intrinsic Fermi level:

$$n = n_i \exp\left[\frac{(E_f - E_i)}{kT}\right] \tag{1.1.26}$$

and

$$p = n_i \exp\left[\frac{(E_i - E_f)}{kT}\right] \tag{1.1.27}$$

Thus, the energy separation from the Fermi level to the intrinsic Fermi level is a measure of the departure of the semiconductor from intrinsic material. Because E_f is above E_i in an n-type semiconductor, $n > n_i > p$, as we found before.

When the semiconductor contains a large dopant concentration $[N_d \to N_c \text{ or } N_a \to N_v$ $(\sim 10^{19} \text{ cm}^{-3} \text{ for Si})]$, we can no longer ignore the limitations imposed by the Pauli exclusion principle. That is, the Fermi-Dirac distribution cannot be approximated by the Maxwell-Boltzmann distribution function. Equations 1.1.21–22 and 1.1.26–27 are no longer valid, and more exact expressions must be used or the limited validity of the simplified expressions must be realized. Very highly doped semiconductors ($N_d \gtrsim N_c$ or $N_a \gtrsim N_v$) are called *degenerate* semiconductors because the Fermi level is within the conduction or valence band. Therefore, allowed states for electrons exist very near the Fermi level, just as is the case in metals. Consequently, many of the electronic properties of very highly doped semiconductors *degenerate* into those of metals.

EXAMPLE   *Thermal-Equilibrium Statistics*

Find the equilibrium electron and hole concentrations and the location of the Fermi level (with respect to the intrinsic Fermi level E_i) in silicon at 300 K if the silicon contains $8 \times 10^{16} \text{ cm}^{-3}$ arsenic (As) atoms and $2 \times 10^{16} \text{ cm}^{-3}$ boron (B) atoms.

Solution   Because the donor (As) density exceeds the acceptor (B) density, the crystal is n-type. The net doping concentration is the difference between the donor dopant density (8×10^{16}) and the acceptor dopant density (2×10^{16}) and is $6 \times 10^{16} \text{ cm}^{-3}$.

The electron density equals the net dopant concentration.

$$n = 6 \times 10^{16} \text{ cm}^{-3}$$

The hole density is (from Equation 1.1.13)

$$p = \frac{n_i^2}{n} = 3.5 \times 10^3 \text{ cm}^{-3}$$

From Equation 1.1.26,

$$\begin{aligned}
E_f - E_i &= kT \ln(n/n_i) \\
&= 0.0258 \ln(6 \times 10^{16}/1.45 \times 10^{10}) \\
&= 0.393 \text{ eV}
\end{aligned}$$

Note that the Fermi level can be specified with respect to the conduction band by using Equation 1.1.21:

$$\begin{aligned}
E_c - E_f &= kT \ln(N_c/n) \\
&= 0.0258 \ln(2.8 \times 10^{19}/6 \times 10^{16}) \\
&= 0.159 \text{ eV}
\end{aligned}$$

The sum of these two energies is 0.55 eV, half the bandgap energy of Si.

$$\begin{array}{ll}
\downarrow 0.159 & \\
\rule{5cm}{0.4pt} & E_c \\
\text{-----------------------------------} & E_f \\
\updownarrow 0.393 & \\
\text{---.---.---.---.---.---} & E_i
\end{array}$$

∎

Inhomogeneously Doped Semiconductors.

At thermal equilibrium, electrons are distributed in energy according to the Fermi-Dirac distribution function, which at a given temperature is determined by the Fermi energy (Equation 1.1.18). The Fermi energy, furthermore, must have the same value throughout a system to assure the detailed balance required at thermal equilibrium for electron transfers. This very important requirement is discussed more fully in Secs. 3.1 and 4.1; at this point we consider the effect on the semiconductor energy-band structure of the constancy of the Fermi level throughout a system.

As we found in the example, the Fermi energy is in the middle of the forbidden energy gap in an undoped (intrinsic) semiconductor. For extrinsic (doped) semiconductors, the Fermi level is closer to one of the bands [to the conduction band in a semiconductor doped with donors (n-type material) and to the valence band in a semiconductor doped with acceptors (p-type material)].

Drawing the energy-band diagram for a piece of semiconductor with nonuniform doping illustrates some useful concepts. Consider, for example, silicon in which the doping in the region $x < a$ is N_{d1} donors cm^{-3}. At $x = a$ the doping decreases abruptly to N_{d2} donors cm^{-3}, as shown in Figure 1.11a. When the two regions of the semiconductor are in intimate contact, the entire piece of semiconductor corresponds to a single system of states. An energy-band diagram for this crystal can be drawn by noting first that the Fermi level must be constant at thermal equilibrium and then drawing the conduction and valence band edges around the Fermi level by using other constraints. In the present case we know that the silicon doping establishes the electron concentrations far from the interface at $x = a$. In the region $x \ll a$ $n \approx N_{d1}$, and for $x \gg a$, $n \approx N_{d2}$. The electron density changes between these two values over a small region near the interface plane at $x = a$.

We sketch an energy-band diagram for this inhomogeneously doped system in Figure 1.11b. As described above, we start constructing this diagram by first drawing a constant horizontal line to represent the constant Fermi level. Away from the interface, we can then locate the conduction-band edge in each region where the electron density is equal to the donor density. The energy interval between the Fermi level and the

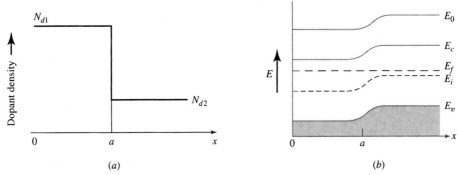

FIGURE 1.11 (a) Dopant density in a silicon crystal. (b) Band diagram at thermal equilibrium for an n-type semiconductor having doping N_{d1} for $0 < x < a$ and N_{d2} for $x > a$ where ($N_{d1} > N_{d2}$).

conduction-band edge is calculated using Equation 1.1.26 to find $E_c - E_f$ at each end. We then draw a smooth transition through the interface plane at $x = a$ to connect the two end regions. The detailed variation of the band edges in the transition region will be discussed in detail in Chapter 4 after further physical theory is developed. Because the energy gap is constant throughout the piece of semiconductor, we can draw the valence-band edge parallel to the conduction-band edge.

As we see in Figure 1.11b, the conduction and valence band edges in the silicon are not constant along x, but rather move to higher energies when the donor density decreases. The energy difference between the conduction and the valence band (the forbidden-gap energy) is a property of the silicon lattice that is not changed by lightly or moderately doping the crystal. (The case of heavy doping is considered later in this section.) The increase in energies associated with the silicon band edges represents an increased potential energy for electrons in the less heavily doped region*. We will return to consider this and other points about Figure 1.11b in Chapter 4. For the present, our emphasis is on the constancy of the Fermi level and on the use of the thermal-equilibrium principle to construct the energy-band diagram shown in Figure 1.11b.

Quasi-Fermi Levels.[†]   We have already found the Fermi level to be a useful concept to explain the behavior of semiconducting materials; we will see many further applications as we extend our discussion to devices. The Fermi level arises from the statistics of an ensemble of electrons at thermal equilibrium, and in fact only for thermal equilibrium is there a fundamental physical definition for the Fermi energy. Often, however, thermal equilibrium is disturbed by excitation such as incident radiation or the application of bias to *pn* junctions. To analyze these nonequilibrium cases, it is useful to introduce two related parameters called *quasi-Fermi levels*.**

We define the quasi-Fermi levels in a manner that preserves the relationship between the intrinsic-carrier density and the electron and hole densities as expressed for thermal equilibrium in Equations 1.1.26 and 1.1.27. Under nonequilibrium conditions similar

---

* Dopants for electrons provide localized positive charges that are attached to donor-atom sites. Higher densities of donors (and hence of positively charged ions) in a region tend to attract electrons to that region and, therefore, lower the potential energy of an electron.

** Some authors also use the term *Imref*, which can be taken to mean "imaginary reference" as well as being *Fermi* spelled backwards, instead of quasi-Fermi level.

equations can only be written if two different quasi-Fermi levels are defined, one for electrons and one for holes.

These conditions are met if we define the quasi-Fermi level for electrons E_{fn} (and its corresponding quasi-Fermi potential $\phi_{fn} = -E_{fn}/q$), and the quasi-Fermi level for holes E_{fp} (and corresponding potential $\phi_{fp} = -E_{fp}/q$), by

$$E_{fn} = E_i + kT \ln(n/n_i) \quad \text{and} \quad \phi_{fn} = \phi_{fi} - \frac{kT}{q} \ln(n/n_i) \tag{1.1.28}$$

and

$$E_{fp} = E_i - kT \ln(p/n_i) \quad \text{and} \quad \phi_{fp} = \phi_{fi} + \frac{kT}{q} \ln(p/n_i) \tag{1.1.29}$$

where ϕ_{fi} is the potential associated with E_i and $\phi_{fi} = -E_i/q$. Under nonequilibrium conditions, the np product is not equal to the thermal equilibrium value n_i^2 but is a function of the separation of the two quasi-Fermi levels. From Equations 1.1.28 and 1.1.29, we can derive

$$np = n_i^2 \exp[(E_{fn} - E_{fp})/kT] \tag{1.1.30}$$

The separation between the two quasi-Fermi levels is, therefore, a measure of the deviation from thermal equilibrium of the semiconductor free-carrier populations, and is identically zero at thermal equilibrium.

The concept of quasi-Fermi levels is especially useful when considering photoconduction, in which excess electrons and holes are generated by light. In general, it is helpful to use quasi-Fermi levels to discuss generation and recombination, as we will see in more detail in Chapter 5.

Photoconduction†

The covalent bonds holding electrons at atomic sites in the lattice can be broken by incident radiant energy (photons) if the photon energy is sufficient. When the bonds are broken, both the freed electrons and the vacant bonds left behind are able to move through the semiconductor crystal and act as current carriers. In terms of the energy-band picture this process of free-carrier production, called *photogeneration*, is equivalent to exciting electrons from the valence band into the conduction band, leaving free holes behind. The required photon energy for photogeneration is thus at least equal to the bandgap energy, and the number of holes created equals the number of generated electrons. The band gap in silicon (1.1 eV) is energetically equivalent to photons in the far infrared portion of the electromagnetic spectrum (1.1 μm wavelength).

The radiation incident on the semiconductor surface is absorbed as it penetrates into the crystal lattice. The amount of energy ΔI absorbed in each small increment of length Δx along the path of the radiation is described by an *absorption coefficient* α:

$$\Delta I = I(x) - I(x + \Delta x) = I(x) \times \alpha \Delta x \tag{1.1.31}$$

where $I(x)$ is the energy reaching the position x. Treating Δx as an infinitesimal quantity, Equation 1.1.31 can be rewritten as a differential equation whose solution is

$$I(x) = I_0 \exp(-\alpha x) \tag{1.1.32}$$

where I_0 is the energy that enters the solid at $x = 0$.

The absorption coefficient is typically a strong function of photon energy, as can be seen in the plot of α versus wavelength (and photon energy)* for silicon shown

---

* Electromagnetic wavelength λ is related to photon energy E by the equation $\lambda = hc/E$, where hc is the product of Planck's constant and the speed of light. For λ in μm, and E in eV, the conversion equation is $\lambda = 1.24/E$.

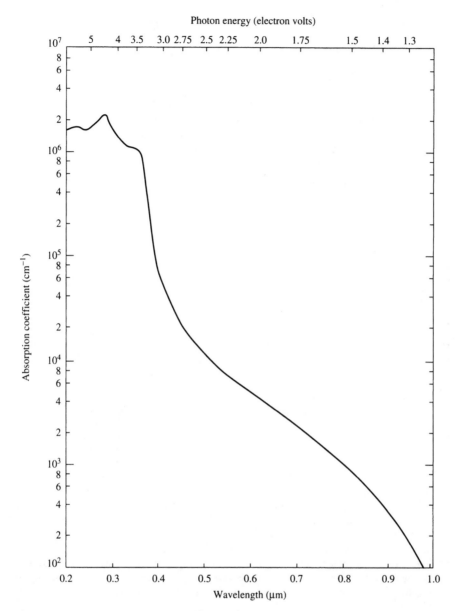

FIGURE 1.12 Absorption coefficient of light in silicon.

in Figure 1.12. High-energy ultraviolet (UV) light is absorbed with a characteristic length (equal to α^{-1}) that is less than 10 nm, while light of 1 μm wavelength (in free space) is not efficiently absorbed and penetrates about 100 μm into silicon before decaying appreciably. Absorption of photons having energies greater than the bandgap is almost entirely due to the generation of holes and electrons. The specific shape of the light-absorption curve is related to the details of the energy-band picture for silicon, but a full discussion of this important topic is better reserved for a fundamental course in solid-state physics.

When photogeneration occurs in silicon, the incident radiation supplies energy which adds to the thermal energy of the crystal. Hence, the silicon is not at thermal equilibrium, and quasi-Fermi levels are appropriate measures for the free-carrier densities.

EXAMPLE *Photogeneration and Quasi-Fermi Levels*

A silicon wafer is doped with 10^{15} cm^{-3} donor atoms.

(a) Find the electron and hole concentrations and the location of the Fermi level with respect to the intrinsic Fermi level.

(b) Light irradiating the wafer leads to a steady-state photogenerated density of electrons and holes equal to 10^{12} cm^{-3}. We assume that the wafer is thin compared to the absorption depth for the light so that the free carriers are generated uniformly throughout its volume. Find the overall electron and hole concentrations in the wafer and calculate the positions of the quasi-Fermi levels for the two carrier types.

(c) Repeat the calculations of (b) under the condition that the light intensity is increased so that the photogeneration produces 10^{18} cm^{-3} electron-hole pairs.

Solution

(a) $n = N_d = 10^{15}$ cm^{-3}

$$p = \frac{n_i^2}{n} = 2.1 \times 10^5 \text{ cm}^{-3}$$

$$E_f - E_i = kT \ln(n/n_i) = 0.29 \text{ eV}$$

(b) $n = 10^{15} + 10^{12} \approx 10^{15}$ cm^{-3}

$$p = 2.1 \times 10^5 + 10^{12} \approx 10^{12} \text{ cm}^{-3}$$

$$E_{fn} - E_i = kT \ln(n/n_i) = 0.29 \text{ eV}$$

$$E_i - E_{fp} = kT \ln(p/n_i) = 0.11 \text{ eV}$$

(c) $n = 10^{15} + 10^{18} \approx 10^{18}$ cm^{-3}

$$p = 2.1 \times 10^5 + 10^{18} \approx 10^{18} \text{ cm}^{-3}$$

$$E_{fn} - E_i = kT \ln(n/n_i) = 0.47 \text{ eV}$$

$$E_i - E_{fp} = kT \ln(p/n_i) = 0.47 \text{ eV}$$

In part (b), photogeneration is shown to change the minority-carrier concentration by seven orders of magnitude without causing any appreciable variation in the majority-carrier density. Consequently, the electron quasi-Fermi level is close to the thermal-equilibrium Fermi level, but the hole quasi-Fermi level is displaced by 0.40 eV. When the light intensity increases as in part (c), both the hole and electron densities are affected, and both quasi-Fermi levels are strongly displaced from the thermal-equilibrium position. The two carrier densities in this case are nearly equal, as in an intrinsic semiconductor at high temperatures. Most instances of photogeneration in doped semiconductors are similar to case (b) in the example; that is, the minority-carrier densities are greatly changed by the incident radiation while the majority-carrier concentrations are essentially unaffected. ■

Heavy Doping.[†]   In much of our discussion, we simplified the statistical expressions for the densities of holes and electrons in semiconductors by assuming that only a small fraction of the available electron states in the conduction band were full and only a small fraction of valence-band states were empty. Under these assumptions, for example, we were able to make approximations for the integral in Equation 1.1.19 to define the "effective density of conduction-band states" N_c and to approximate the Fermi-Dirac statistics for electron density by the simpler Maxwell-Boltzmann statistics (Equation 1.1.21). However, these approximations become invalid when a crystal is doped with impurities at densities that approach N_c. In addition to expressing the free-carrier statistics correctly, other, more basic, effects need to be considered when a semiconductor is heavily doped.

If moderate concentrations of dopant impurities are present (for example, the bulk doping in a silicon wafer), the individual impurity atoms do not interact with one another, and they do not perturb the band structure of the host crystal. For example, a dopant density of 5×10^{15} cm^{-3} represents only about one atom of dopant in 10^7 atoms of silicon. Each of the dopant atoms then adds a discrete allowed donor energy level in the silicon bandgap. If the dopant density is increased sufficiently to become a significant fraction of the silicon-atom density, however, the band structure itself begins to be perturbed.

The most significant perturbation is a reduction in the size of the silicon bandgap. The reduced bandgap energy causes the product of the free-carrier densities p and n to increase. This effect is usually expressed in terms of a value for the pn product in the form

$$pn = n_i^2 \exp(\Delta E_g/kT) = n_{ie}^2 \qquad (1.1.33)$$

where ΔE_g expresses the effective bandgap narrowing caused by heavy doping, and n_{ie} is an effective value of the intrinsic-carrier density. Measurements of bandgap narrowing indicate that this effect is negligible for dopant densities less than 10^{18} cm^{-3}, but at higher dopant densities it can become sizable. Some experimental data showing ΔE_g as a function of the free-electron density n in silicon are plotted in Figure 1.13. Heavy doping

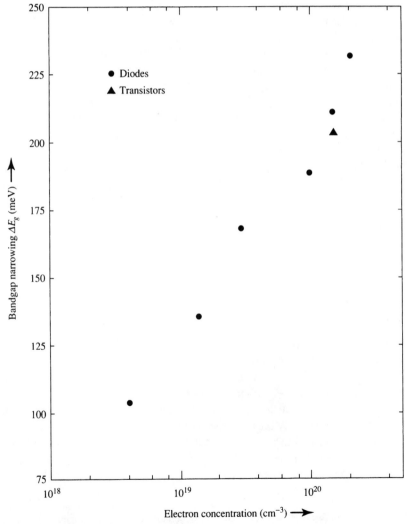

FIGURE 1.13 Energy-gap narrowing ΔE_g as a function of electron concentration. [A. Neugroschel, S. C. Pao, and F. A. Lindholm, IEEE Trans. Electr. Devices, **ED-29**, 894 (May 1982).]

effects begin to be noticeable between 10^{18} and 10^{19} electrons cm^{-3}; at an electron concentration of $10^{19} cm^{-3}$, ΔE_g is more than 10% of the bandgap energy.

A detailed study of the effect of heavy doping on the semiconductor band structure shows that as the dopant densities increase, the energy levels they introduce are no longer distinct, but instead broaden into bands. These *impurity bands* can overlap the adjacent conduction or valence bands so that no energy is required to ionize the dopant atoms and provide free carriers. Therefore, under heavy doping conditions, the formulas derived earlier in this chapter for silicon doping need modification.

The most important device effect of heavy doping is to limit the achievable current gain of bipolar transistors. It can also increase undesired leakage current in both bipolar and MOS transistors.

1.2 FREE CARRIERS IN SEMICONDUCTORS

Our first reference to the electronic properties of solids earlier in this chapter was to the familiar linear relationship that is often found between the current flowing through a sample and the voltage applied across it. This relationship is known as Ohm's law: $V = IR$. Although a thorough derivation of the physics of ohmic conduction can be quite complex, an approximate representation of the process provides adequate background for our purposes. To accomplish this we first develop a picture of the kinetic properties of free electrons without any external fields. We then consider the addition of low to moderate fields, characteristic of many device applications, and finally we discuss the high-field case.

We begin by recalling that electrons (and holes) in semiconductors are almost "free particles" in the sense that they are not associated with any particular lattice site. The influences of crystal forces are incorporated in an *effective mass* that differs somewhat from the free-electron mass. Using the laws of statistical mechanics, we can assert that electrons and holes have the thermal energy associated with classical free particles: $\frac{1}{2}kT$ units of energy per degree of freedom where k is Boltzmann's constant and T is the absolute temperature. This means that electrons in a crystal at a finite temperature are not stationary, but are moving with random velocities. Furthermore, the mean-square thermal velocity v_{th} of the electrons is approximately* related to the temperature by the equation

$$\frac{1}{2}m_n^* v_{th}^2 = \frac{3}{2}kT \qquad (1.2.1)$$

where m_n^* is the effective mass of conduction-band electrons. For silicon $m_n^* = 0.26\, m_0$ (where m_0 is the free electron rest mass), and v_{th} is calculated from Equation 1.2.1 to be $2.3 \times 10^7\ cm\ s^{-1}$ at $T = 300$ K. The electrons may be pictured as moving in random directions through the lattice, colliding among themselves and with the lattice. At thermal equilibrium the motion of the system of electrons is completely random so that the net current in any direction is zero. Collisions with the lattice result in energy transfer between the electrons and the atomic cores that form the lattice. The time interval between collisions averaged over the entire electron population is τ_{cn}, the mean scattering time for electrons. These considerations all apply to the field-free, thermal-equilibrium crystal.

---

* The formulation of Equation 1.2.1 is slightly in error because of improper averaging. However, we are only concerned with the order-of-magnitude of the result. At 300 K, v_{th} is typically taken to be $10^7\ cm\ s^{-1}$ for electrons or holes in silicon.

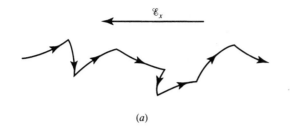

(a)

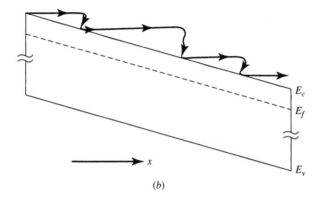

E_c
E_f

x

E_v

(b)

FIGURE 1.14 (a) The motion of an electron in a solid under the influence of an applied field. (b) Energy-band representation of the motion, indicating the loss of energy when the electron undergoes a collision.

Drift Velocity

Let us now apply a small electric field to the lattice. The electrons are accelerated along the field direction during the time between the collisions. Figure 1.14a is a sketch of the motion typical of a crystal electron in response to a small applied field $\mathcal{E}$. Note in the figure that the field-directed motion is a small perturbation on the random thermal velocity. Therefore τ_{cn}, the mean scattering time, is not altered appreciably by the applied field.

In Figure 1.14b electron motion in a small applied field is schematically represented on a band diagram. A constant applied field results in a linear variation in the energy levels in the crystal. Electrons (which move downward on energy-level diagrams) tend to move to the right on the diagram, as is appropriate for a field directed in the negative x direction. The electrons exchange energy when they collide with the lattice and drop toward their thermal-equilibrium positions. If the field is small, the energy exchanged is also small, and the lattice is not appreciably heated by the passage of the current. The slope of the energy bands and the energy losses associated with lattice collisions are exaggerated in Figure 1.14b to show the process schematically. In fact, the energy lost in each collision is much less than the mean thermal energy of an electron. An electron at rest in the conduction band is at the band edge E_c. The kinetic energy of an electron in the conduction band is therefore measured by $(E - E_c)$, and the mean value at thermal equilibrium is just $(E - E_c) = \frac{3}{2}kT$ or about 0.04 eV at 300 K. This is less than 4% of the bandgap energy.

The net carrier velocity in an applied field is called the *drift velocity*, v_d. It can be found by equating the impulse (force × time) applied to an electron during its free flight between collisions with the momentum gained by the electron in the same period. This equality is valid because steady state is reached when all momentum gained between collisions is lost to the lattice in the collisions. The force on an electron is $-q\mathcal{E}$ and the momentum gained is $m_n^* v_d$. Thus (within the statistical limitations mentioned in the footnote to Equation 1.2.1),

$$-q\mathscr{E}\tau_{cn} = m_n^* v_d \tag{1.2.2}$$

or

$$v_d = -\frac{q\mathscr{E}\tau_{cn}}{m_n^*} \tag{1.2.3}$$

Equation 1.2.3 states that the electron drift velocity v_d is proportional to the field with a proportionality factor that depends on the mean scattering time and the effective mass of the nearly free electron. The proportionality factor is an important property of the electron called the *mobility* and is designated by the symbol μ_n.

$$\mu_n = \frac{q\tau_{cn}}{m_n^*} \tag{1.2.4}$$

Because $v_d = -\mu_n\mathscr{E}$, the mobility describes how easily an electron moves in response to an applied field.

From Equation 1.2.3 the current density flowing in the direction of the applied field can be found by summing the product of the charge on each electron times its velocity over all electrons n per unit volume.

$$J_n = \sum_{i=1}^{n} -qv_i = -nqv_d = nq\mu_n\mathscr{E} \tag{1.2.5}$$

Entirely analogous arguments apply to holes. A hole with zero kinetic energy resides at the valence-band edge E_v. The kinetic energy of a hole in the valence band is therefore measured by $(E_v - E)$. If a band edge is tilted, the hole moves upward on an electron energy-band diagram. The hole mobility μ_p is defined as $\mu_p = q\tau_{cp}/m_p^*$. The total current can be written as the sum of the electron and hole currents:

$$J = J_n + J_p = (nq\mu_n + pq\mu_p)\mathscr{E} \tag{1.2.6}$$

The term in parentheses in Equation 1.2.6 is defined as the conductivity σ of the semiconductor:

$$\sigma = q\mu_n n + q\mu_p p \tag{1.2.7}$$

In extrinsic semiconductors, only one of the components in Equation 1.2.7 is generally significant because of the large ratio between the two carrier densities. The resistivity, which is the reciprocal of the conductivity, is shown as a function of the dopant concentration in Figure 1.15 for phosphorus-doped, n-type silicon and for boron-doped, p-type silicon. There are slight variations in the resistivities obtained for different dopant species, particularly in the heavy-doping range. In most practical cases, however, Figure 1.15 can be used for any dopant species.

A property of a solid that is closely related to its conductivity is its *dielectric relaxation time*. The dielectric relaxation time is a measure of the time it takes for charge in a semiconductor to become neutralized by conduction processes. It is small in metals and can be large in semiconductors and insulators. The magnitude of the dielectric relaxation time can be used to obtain qualitative insight. Several device concepts can be readily interpreted according to the relative sizes of the charge transit time through a material and the dielectric relaxation time in the same material. Problem 1.12 provides further introduction to the dielectric relaxation time.

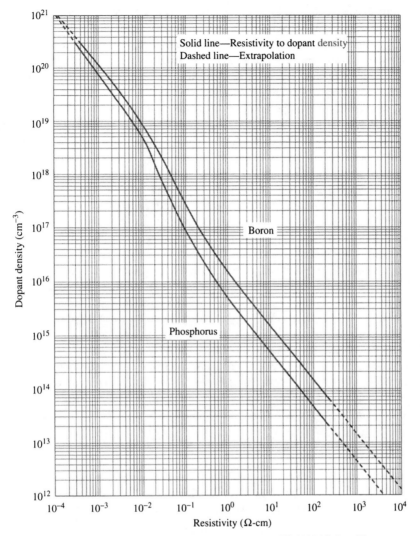

FIGURE 1.15 Dopant density versus resistivity at 23°C (296 K) for silicon doped with phosphorus and with boron. The curves can be used with little error to represent conditions at 300 K. [W. R. Thurber, R. L. Mattis, and Y. M. Liu, National Bureau of Standards Special Publication 400–64, 42 (May 1981).]

Mobility and Scattering

Quantum-mechanical calculations indicate that a perfectly periodic lattice does not scatter free carriers; that is, the carriers do not interchange energy with a stationary, perfect lattice. However, at any temperature above absolute zero the atoms that form the lattice vibrate. These vibrations disturb periodicity and allow energy to be transferred between the carriers and the lattice.* The interactions with lattice vibrations can be viewed as collisions with energetic "particles" called *phonons*. Phonons, like photons, have energies quantized in units of $h\nu$, where ν is the lattice-vibrational frequency and h is Planck's constant. The theories of thermal

---

* Because this energy is supplied to the carriers by the applied field, scattering processes lead to heating of the semiconductor. The dissipation of this heat is often a limiting factor in the size of semiconductor devices. A device must be large enough to avoid heating to temperatures at which it no longer functions.

and electrical conduction can often be simplified when formulated in terms of phonon interactions. For silicon at room temperature and moderate electric fields, the lowest vibrational mode for the lattice corresponds to a phonon energy of 0.063 eV, and the energy of an electron is changed by this amount when it interacts with these lowest-energy phonons. At higher temperatures lattice vibrations are larger, and collisions between electrons and the lattice vibrations are more important. Theoretical analysis indicates that the mobility should decrease with increasing temperature in proportion to T^{-n} with n between 1.5 and 2.5 when lattice scattering dominates. Experimentally, values of n range from 1.66 to 3, with $n = 2.5$ being common.

In addition to lattice vibrations, dopant impurities also cause local distortions in the lattice and scatter free carriers. However, unlike scattering from lattice vibrations, scattering from ionized impurities becomes less significant at higher temperatures. Because the carriers are moving faster at higher temperatures, they remain near the impurity atom for a shorter time and are therefore less effectively scattered. Consequently, when impurity scattering dominates, the mobility increases with increasing temperature. Scattering can also be caused by collisions with unintentional impurities and with crystal defects. These defects can arise from poor control of the quality of the semiconductor, or they can be related to boundaries between grains of a polycrystalline material. An example of the latter is the thin films of polycrystalline silicon used to form portions of many MOS integrated circuits (Chapters 2 and 9). The grain boundaries and defects in polycrystalline material can reduce the mobility to a small fraction of its value in single-crystal material with the same dopant concentration.

Two or more of the scattering processes discussed above can be important at the same time, and their combined effect on mobility must be assessed. To do this we consider the number of particles that are scattered in a time interval dt. The probability that a carrier is scattered in a time dt by process i is dt/τ_i where τ_i is the average time between scattering events resulting from process i. The total probability dt/τ_c that a carrier is scattered in the time interval dt is then the sum of the probabilities of being scattered by each mechanism:

$$\frac{dt}{\tau_c} = \sum_i \frac{dt}{\tau_i} \tag{1.2.8}$$

This is a reasonable way to combine the scattering probabilities because the average scattering time resulting from all of the processes acting simultaneously is less than that resulting from any one process and is dominated by the shortest scattering time. Because the mobility μ equals $q\tau_c/m^*$, we can write

$$\frac{1}{\mu} = \sum_i \frac{1}{\mu_i} \tag{1.2.9}$$

The mobility of a carrier subjected to several different scattering mechanisms can thus be found by combining the reciprocal mobilities corresponding to each type of scattering; the resulting mobility is smaller than that determined by any of the individual scattering mechanisms. Because of the reciprocal relation for combining mobility components (Equation 1.2.9), the overall mobility is dominated by the process for which τ_i is smallest.

We can apply these considerations to the mobilities of electrons and holes in silicon at room temperature which are plotted in Figure 1.16. The figure represents a "best fit" to measured data reported in a number of different sources. In lightly-doped material, the mobility resulting from ionized impurity scattering is higher than that due to lattice scattering. Hence, for silicon having impurity concentrations less than about 10^{15} cm^{-3}, the mobilities both for holes and electrons remain nearly constant as the dopant concentration varies. At higher dopant concentrations, however, scattering by ionized impurities becomes comparable to or greater than that resulting from lattice vibrations, and the total mobility decreases.

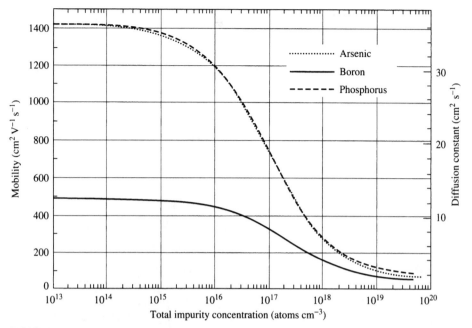

FIGURE 1.16 Electron and hole mobilities in silicon at 300 K as functions of the total dopant concentration. The values plotted are the results of curve fitting measurements from several sources. The mobility curves can be generated using Equation 1.2.10 with the following values of the parameters [3].

Parameter	Arsenic	Phosphorus	Boron
μ_{min}	52.2	68.5	44.9
μ_{max}	1417	1414	470.5
N_{ref}	9.68×10^{16}	9.20×10^{16}	2.23×10^{17}
α	0.680	0.711	0.719

An important practical consequence of the dependence of mobility on total impurity concentration is observed if a semiconductor is converted from one type to the other (p to n or n to p) by compensating the dopant impurity atoms already present. While the carrier densities depend on the difference between the concentrations of the two types of dopant impurities ($N_d - N_a$) (Equations 1.1.16 and 1.1.17), the scattering depends on the sum of the ionized impurity concentrations ($N_d + N_a$). Thus, the mobilities in a compensated semiconductor can be markedly lower than those in an uncompensated material with the same net carrier density.

The equation used in Figure 1.16 to represent the measured electron and hole mobilities in silicon is [3]

$$\mu = \mu_{min} + \frac{\mu_{max} - \mu_{min}}{1 + (N/N_{ref})^\alpha} \qquad (1.2.10)$$

where N is the total dopant concentration in the silicon and the four parameters μ_{max}, μ_{min}, N_{ref}, and α have different values for each dopant species. Values of these parameters for the most common dopants in silicon are included in the caption for Figure 1.16. Table 1.1 gives numerical values of the mobility (as calculated from Equation 1.2.10) at decade values of N.

The dependence of electron mobility on dopant species is seen, from Figure 1.16, to be slight for total impurity concentrations less than 10^{19} cm^{-3}. In the high doping range (for $N > 10^{19}$ cm^{-3}), the mobility of phosphorus-doped silicon is 10 to 20% greater than

TABLE 1.1   Mobilities in Silicon (cm^2 V^{-1}s^{-1})

	Electrons		Holes
N	Arsenic	Phosphorus	Boron
10^{13}	1423	1424	486
10^{14}	1413	1416	485
10^{15}	1367	1374	478
10^{16}	1184	1194	444
10^{17}	731	727	328
10^{18}	285	279	157
10^{19}	108	115	72

that of silicon doped with arsenic. At very high dopant densities (above about 10^{20} cm^{-3}), measured mobilities drop below the minimum values shown in Figure 1.16.

Temperature Dependence.   As briefly discussed above, the different scattering mechanisms that affect free-carrier mobilities have varying dependences on temperature. For example, scattering by ionized impurities becomes less effective as the temperature increases because the faster moving carriers interact less effectively with stationary impurities. However, scattering by lattice vibrations (phonon collisions) becomes more effective at higher temperatures. Because of this, at lower temperatures the mobility characteristically increases with rising temperature (because impurity-scattering predominates), while at higher temperatures the mobility decreases (because phonon collisions dominate). These competing temperature variations lead to a characteristic maximum in the mobility versus temperature relation as seen in the experimental data shown in Figure 1.17. At the peak in mobility, the two temperature variations are balanced, and the mobility has its minimum temperature sensitivity.

For design and analysis, equations for the dependence of mobility on temperature and dopant concentration are useful. Such expressions have been derived empirically for

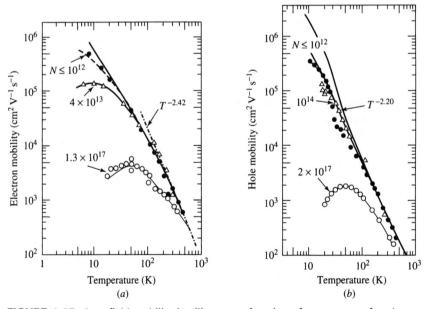

FIGURE 1.17   Low-field mobility in silicon as a function of temperature for electrons (*a*), and for holes (*b*). The solid lines represent the theoretical predictions for pure lattice scattering [5].

silicon and can be written as [4]

$$\mu_n = 88\, T_n^{-0.57} + \frac{1250\, T_n^{-2.33}}{1 + [N/(1.26 \times 10^{17}\, T_n^{2.4})]0.88\, T_n^{-0.146}}$$

and

$$\mu_p = 54.3\, T_n^{-0.57} + \frac{407\, T_n^{-2.23}}{1 + [N/(2.35 \times 10^{17}\, T_n^{2.4})]0.88\, T_n^{-0.146}} \tag{1.2.11}$$

where $T_n = T/300$ with T measured in K (Kelvin), and N is the total dopant density in the silicon. Equation 1.2.11 is useful up to dopant densities of 10^{20} cm^{-3} and for temperatures between 250 and 500 K.

Velocity Limitations. In the simplified treatment presented thus far, we assumed (by taking τ_c to be insensitive to $\mathscr{E}$) that the velocity imparted to the free carriers by the applied field is much less than the random thermal velocity, which we found from Equation 1.2.1 to be approximately 10^7 cm s^{-1} at room temperature in silicon. For electrons in silicon with $\mu_n \approx 1400$ cm^2 V^{-1} s^{-1} the drift velocity at a typical field of 100 V cm^{-1} is roughly 1.5% of the thermal velocity, and the applied field does not appreciably change the total velocity or energy of the carrier. At high fields, however, the drift velocity becomes comparable to the random thermal velocity, and can no longer be considered as a small increment to the thermal motion. The total energy of the carrier then increases significantly as the field increases. When carriers reach energies above the ambient thermal energy, they are often called *hot carriers* and characterized by an effective temperature T_e that rises with increasing field and corresponds to the increased kinetic energy of the carriers [6].

At high electric fields, the energy of the *hot electrons* reaches a critical value at which an additional scattering process (collisions with high energy "optical" phonons) becomes important, and the mobility decreases from its low-field value. Because optical-phonon scattering is very effective in transferring energy from the hot carriers to the lattice, the carriers cannot gain significant additional net energy, and the drift velocity approaches a limiting or *saturation* value v_l or v_{sat} at high fields. Figure 1.18 shows measured values

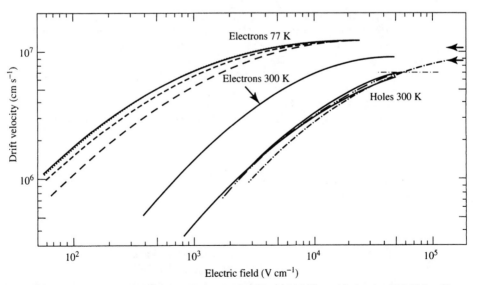

FIGURE 1.18 Drift velocities of electrons (at 77 K and 300 K) and holes (at 300 K) in silicon as functions of the applied field, showing velocity saturation at high fields. The presence of several curves indicates the variation in reported data. An empirical "best fit" to these curves is given in Equation 1.2.12 and Table 1.2 [5].

TABLE 1.2 Parameters for Field Dependence of Drift Velocity

Parameter	Electrons		Holes	
	Expression	at 300 K	Expression	at 300 K
v_l cm s^{-1}	$1.53 \times 10^9 \ T^{-0.87}$	1.07×10^7	$1.62 \times 10^8 \ T^{-0.52}$	8.34×10^6
$\mathscr{E}_c$ V cm^{-1}	$1.01 \ T^{1.55}$	6.91×10^3	$1.24 \ T^{1.68}$	1.45×10^4
β	$2.57 \times 10^{-2} \ T^{0.66}$	1.11	$0.46 \ T^{0.17}$	2.637

of drift velocity for electrons (at 77 K and 300 K) and holes (at 300 K) in silicon as functions of the applied field. At low voltages, the curves are linear, indicating a constant mobility. At fields above a few thousand volts per centimeter, however, there are noticeable deviations from constant mobility. Because fields of this magnitude are frequently present in integrated-circuit devices (equivalent to a few hundred millivolts across a micrometer), velocity saturation must be considered in analyzing many practical devices.

As a useful approximation, the data shown in Figure 1.18 can be modeled empirically by the expression [5]

$$|v_d| = v_l \frac{\mathscr{E}}{\mathscr{E}_c} \left[\frac{1}{1 + (\mathscr{E}/\mathscr{E}_c)^\beta} \right]^{1/\beta} \tag{1.2.12}$$

The parameters v_l, $\mathscr{E}_c$, and β in Equation 1.2.12 are given (as a function of the absolute temperature T) in Table 1.2.

EXAMPLE *Velocity Limitations*

Use Equation 1.2.12 to find the field (at 300 K) at which the effective electron mobility (defined as the ratio of the drift velocity to the field) is reduced to half its low-field value.

Solution At low fields, the drift velocity v_d is proportional to the field (Equation 1.2.3) and $|v_d| = \mu_n \mathscr{E}$. Applying Equation 1.2.12 at low fields $(\mathscr{E}/\mathscr{E}_c \ll 1)$, we find that the low-field mobility can be expressed in terms of $\mathscr{E}_c$ and v_l; $\mu_n = |v_d/\mathscr{E}| = |v_l/\mathscr{E}_c|$. Using the values from Table 1.2 in this expression, we calculate $\mu_n = 1548$ cm^2 V^{-1} s^{-1} for the low-field mobility, which is about 10% higher than the value shown in Figure 1.16. This lack of correspondence is not unusual when parameters are obtained by curve-fitting.

To find the field at which the effective mobility $\mu_n(\mathscr{E})$ is reduced by 50% from the low-field value, we note that $\mu_n(\mathscr{E}) = |v_d/\mathscr{E}| = (1/2) \times |v_l/\mathscr{E}_c|$. Hence, from Equation 1.2.12

$$\frac{1}{2} = \left[\frac{1}{1 + (\mathscr{E}/\mathscr{E}_c)^\beta} \right]^{1/\beta}$$

where $\beta = 1.11$ and $\mathscr{E}_c = 6.91 \times 10^3$ V cm^{-1}. Solving this equation, we find $\mathscr{E}/\mathscr{E}_c = 1.142$ or $\mathscr{E} = 7.89 \times 10^3$ V cm^{-1}. ■

Before leaving the topic of carrier transport at high fields, we make a final point about hot electrons. The hot-electron temperature T_e, introduced earlier in our discussion, describes an ensemble of electrons that is interchanging energy by colliding with the lattice. Some electrons may, by chance, avoid collisions for relatively long times, and thus achieve velocities exceeding v_l and kinetic energies corresponding to temperatures greater than T_e.

Although these electrons are few in number, they can have important physical effects because of their very high energies and the long distances that they have traveled without scattering. When the distance that a significant number of "lucky" electrons travel before scattering is comparable to a device dimension, the charge transport by these carriers is said to be *ballistic* (unscattered).

Diffusion Current

In the previous section we discussed drift current, which flows when an electric field is applied and which follows Ohm's law. Ohmic behavior is observed in metals and semiconductors and is probably familiar from direct experience. In semiconductors, an additional important component of current can flow if a spatial variation of carrier energies or densities exists within the material. This component of current is called *diffusion current*. Diffusion current is generally not an important consideration in metals because metals have very high conductivities. The lower conductivity and the possibility of nonuniform densities of carriers and of carrier energies, however, often makes diffusion an important process affecting current flow in semiconductors.

To understand the origin of diffusion current, we consider the hypothetical case of an n-type semiconductor with an electron density that varies only in one dimension (Figure 1.19). We assume that the semiconductor is at a uniform temperature so that the average energy of electrons does not vary with x; only the density $n(x)$ varies. We consider the number of electrons crossing the plane at $x = 0$ per unit time per unit area. Because they are at finite temperature, the electrons have random thermal motion along the single dimension, but we assume that no electric fields are applied. On the average, the electrons crossing the plane $x = 0$ from the left in Figure 1.19 start at approximately $x = -\lambda$ after a collision where λ is the *mean-free path* of an electron, given by $\lambda = v_{th}\tau_{cn}$. The average rate (per unit area) of electrons crossing the plane $x = 0$ from the left, therefore, depends on the density of electrons that started at $x = -\lambda$ and is

$$\tfrac{1}{2}n(-\lambda)v_{th} \qquad (1.2.13)$$

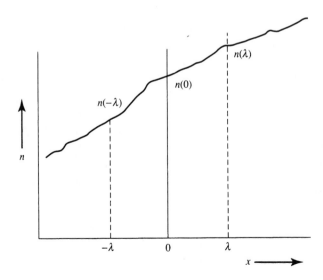

FIGURE 1.19 Electron concentration n versus distance x in a hypothetical one-dimensional solid. Boundaries are demarcated λ units on either side of the origin where λ is a collision mean-free path.

The factor ($\frac{1}{2}$) appears because half of the electrons travel to the left and half travel to the right after a collision at $x = -\lambda$. Similarly, the rate at which electrons cross the plane $x = 0$ from the right is given by

$$\tfrac{1}{2}n(\lambda)v_{th} \tag{1.2.14}$$

so that the net rate or *flux* of particle flow per unit area from the left (denoted F) is

$$F = \tfrac{1}{2}v_{th}[n(-\lambda) - n(\lambda)] \tag{1.2.15}$$

Approximating the densities at $x = \pm\lambda$ by the first two terms of a Taylor-series expansion, we find

$$F = \tfrac{1}{2}v_{th}\left\{\left[n(0) - \frac{dn}{dx}\lambda\right] - \left[n(0) + \frac{dn}{dx}\lambda\right]\right\} = -v_{th}\lambda\frac{dn}{dx} \tag{1.2.16}$$

Because each electron carries a charge $-q$, the particle flow corresponds to a current

$$J_n = -qF = q\lambda v_{th}\frac{dn}{dx} \tag{1.2.17}$$

We see that diffusion current is proportional to the spatial derivative of the electron density and arises because of the random thermal motion of charged particles in a concentration gradient. For an electron density that increases with increasing x, the gradient is positive, as is the current. Because we expect electrons to flow from the higher density region at the right to the lower density region at the left and current flows in the direction opposite to that of the electrons, the direction of current indicated by Equation 1.2.17 is physically reasonable.

We can write Equation 1.2.17 in a more useful form by applying the theorem for the equipartition of energy to this one-dimensional case. This allows us to write

$$\tfrac{1}{2}m_n^* v_{th}^2 = \tfrac{1}{2}kT \tag{1.2.18}$$

We now use the relationship $\lambda = v_{th}\tau_{cn}$ together with Equation 1.2.4 to write Equation 1.2.17 in the form

$$J_n = q\left(\frac{kT}{q}\mu_n\right)\frac{dn}{dx} \tag{1.2.19}$$

The quantity in parentheses on the right side of Equation 1.2.19 is defined as the *diffusion coefficient* D_n and our short derivation has established that

$$D_n = \left(\frac{kT}{q}\right)\mu_n \tag{1.2.20}$$

Equation (1.2.20) is known as the *Einstein relation*. It relates the two important coefficients that characterize free-carrier transport by drift and by diffusion in a solid. Its validity can be established rigorously by considering the statistical mechanics of solids in detail. Our derivation is aimed at intuition, not physical exactness.

If a field is present, drift, as well as diffusion, occurs. The total current is then the sum of the drift and diffusion currents.

$$J_{nx} = q\mu_n n\mathscr{E}_x + qD_n\frac{dn}{dx} \tag{1.2.21}$$

where $\mathscr{E}_x$ is the component of the electric field in the x-direction.

A similar expression can be found for the hole diffusion current so that the total hole current is written as

$$J_{px} = q\mu_p p \mathscr{E}_x - qD_p \frac{dp}{dx} \tag{1.2.22}$$

where the negative sign arises because of the positive charge of a hole. The Einstein relation (Equation 1.2.20) also relates D_p to μ_p.

Our discussion of diffusion current has been framed in terms of nonuniform carrier concentrations. Although this is the most frequent situation encountered in device analysis and Equations 1.2.21 and 1.2.22 are usually adequate, diffusion also occurs if carriers are equal in density in a semiconductor but are more energetic in one region than in another. In this case, other formulations must be used. Problem 1.16 considers a practical situation where carriers have unequal energies.

EXAMPLE   *Diffusion Current*

An electric field has a non-zero value at plane x_1 (perpendicular to the x-axis) inside a silicon crystal. At x_1, the electron density is 10^6 cm^{-3} and the electron density is nonuniform in the x direction. We observe that no electron current flows across the plane.

(a) Explain why no current is flowing.
(b) If the electric field is -10^3 V cm^{-1} (i.e., 10^3 V cm^{-1} in the negative x-direction), what is the electron gradient perpendicular to the plane?

Solution

(a) An electric field $\mathscr{E}$ would result in a drift current of magnitude $J_n = q\mu_n n\mathscr{E}$ according to Equation 1.2.5. Because no current flows, there must be a diffusion-current component that is equal in magnitude but opposite in direction. The balance of these two components leads to zero electron current.

(b) Using Equation 1.2.21, we have

$$J_n = 0 = q\mu_n n\mathscr{E}_x + qD_n \frac{dn}{dx}$$

$$\frac{dn}{dx} = -\frac{\mu_n}{D_n} n\mathscr{E}$$

$$= -\frac{q}{kT} n\mathscr{E}$$

$$= -\frac{10^6 \times (-10^3)}{0.0258}$$

$$= 3.88 \times 10^{10} \text{ cm}^{-4}$$

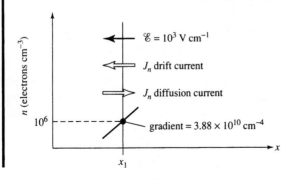

$\mathscr{E} = 10^3$ V cm^{-1}

J_n drift current

J_n diffusion current

gradient $= 3.88 \times 10^{10}$ cm^{-4}

Total Currents and Quasi-Fermi Levels.[†] Quasi-Fermi levels (defined in Equations 1.1.28 and 1.1.29) are useful in the analysis of semiconductors that are not at thermal equilibrium. A semiconductor in which a current is flowing is such a nonequilibrium case, and the quasi-Fermi levels can be used to define both drift and diffusion currents in a compact form.

To illustrate this, we first develop an expression for the electric field in a semiconductor in terms of the electron energy. The presence of an electric field causes charged particles to have position-dependent energies. Consequently, when a field is present, the energy bands for electrons slope away from the horizontal (constant energy), as seen in Figure 1.14. It is convenient to express the field in terms of the intrinsic Fermi level E_i by observing that the electron energy is obtained by multiplying the potential by the charge $-q$. The field $\mathscr{E}$, which is the negative derivative of the potential, can therefore be written (in one dimension) as

$$\mathscr{E} = \frac{1}{q}\frac{dE_i}{dx} = -\frac{d\phi_{fi}}{dx} \tag{1.2.23}$$

where the quasi-Fermi potential $\phi_{fi} = -E_i/q$ was introduced in Equation 1.1.28. The electron and hole concentrations are expressed in terms of the quasi-Fermi levels by

$$n = n_i \exp(E_{fn} - E_i)/kT = n_i \exp\, q(\phi_{fi} - \phi_{fn})/kT$$
$$p = n_i \exp(E_i - E_{fp})/kT = n_i \exp\, q(\phi_{fp} - \phi_{fi})/kT \tag{1.2.24}$$

Using Equations 1.2.23 and 1.2.24 in Equation 1.2.21 together with the Einstein relation (Equation 1.2.20), we obtain

$$J_n = \mu_n n \frac{dE_{fn}}{dx} = -q\mu_n n \frac{d\phi_{fn}}{dx} \tag{1.2.25}$$

A similar derivation for holes leads to

$$J_p = \mu_p p \frac{dE_{fp}}{dx} = -q\mu_p p \frac{d\phi_{fp}}{dx} \tag{1.2.26}$$

Equations 1.2.25 and 1.2.26 show that the total current (the sum of both drift and diffusion components) for each carrier is proportional to the gradient of the quasi-Fermi level of that carrier type. This compact representation can be very helpful in using energy-band diagrams to visualize the total current in a device. The equations are also useful for mathematical analysis, and their form simplifies a number of complex problems.

1.3 DEVICE: HALL-EFFECT MAGNETIC SENSOR

As is our pattern throughout the book, we conclude this chapter with a discussion of an integrated-circuit device, in this case a Hall-effect sensor for magnetic fields. Although Hall-effect sensors are unconventional integrated-circuit devices, they are commercially important. Hundreds of millions of integrated Hall circuits are in use, mainly as contactless switches (e.g. in computer-terminal keyboards) and as mechanical proximity detectors. Hall-effect, magnetic sensing, integrated circuits are highly successful examples of *integrated sensors*, that is, integrated circuits having intentional sensitivity to nonelectrical signals. This sensitivity is achieved by incorporating sensing elements on a silicon chip together with bias, amplifying, and signal-processing circuitry. The field of integrated sensors is developing rapidly because it capitalizes on the extraordinary refinements already achieved by purely electrical integrated circuits and economically applies these circuits when the input signals are not electrical quantities. For example, microprocessors integrated with nonelectrical sensors are powerful components in control systems. In addition to magnetic fields, other nonelectrical inputs for which integrated sensors show appreciable promise include visible and infrared radiation, temperature, pressure, force, acceleration, chemical vapors, and humidity.

Physics of the Hall Effect

The *Hall effect*, named for American physicist E. H. Hall who discovered it in 1879, is a direct consequence of the force exerted on charged carriers moving in a magnetic field. The force on a particle having charge q and moving in a magnetic field $\vec{B}$ with velocity $\vec{v}$ (both variables being vector quantities) is written

$$\vec{F} = q\vec{v} \times \vec{B} \qquad (1.3.1)$$

where the vector cross product ($\times$) signifies the product of the vector magnitudes times the sine of the angle between them.

The resulting Hall effect is illustrated schematically in Figure 1.20. The Hall effect is usually used with an extrinsic semiconductor so that one carrier dominates and the other has a negligible density. To aid our discussion, however, both electrons and holes are shown in Figure 1.20.

As expressed in Equation 1.3.1, the current carriers (electrons or holes) in a conductor experience a force in a direction perpendicular to both the magnetic field and the carrier velocity. In the steady state, this force is balanced by an induced electric field that results from a slight charge redistribution. These forces must balance because there can be no net steady-state motion of the carriers in the transverse direction. The induced electric field is called the *Hall field* $\mathscr{E}_H$. Integrating the Hall field with respect to position across the width of the conductor produces the *Hall voltage* V_H, which can be detected by

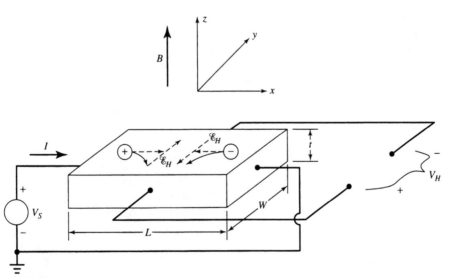

FIGURE 1.20 Schematic of the Hall effect. A current flows along the positive *x*-direction. A magnetic field *B* along the positive *z*-direction deflects holes and electrons along the negative *y*-direction. This leads to a Hall field $\mathscr{E}_H$ along the positive *y*-direction for holes and along the negative *y*-direction for electrons.

contacts placed at opposite sides of the conductor. For a uniform conductor and a uniform magnetic field, the magnitude of the Hall voltage is just the product of the Hall field and the conductor width W: $V_H = \mathscr{E}_H W$.

In Figure 1.20, current in the positive x-direction can be carried either by holes flowing toward positive x or else by electrons moving in the opposite direction. Because both the charge and the velocity are of opposite signs for the two carrier types, the magnetic force on both holes and electrons has the same sign for a given current direction. The charge separation caused by this force and the resulting Hall voltage is, therefore, of opposite polarity for holes and electrons. The sign dependence of the Hall voltage can thus be used to determine whether a semiconductor is p- or n-type.

We can derive the basic theory for the Hall effect using the quantities shown in Figure 1.20. In the figure, current flow is in the positive x-direction, the magnetic field is in the positive z-direction, and the Hall field is therefore along the y-direction. The magnetic force deflects both holes and electrons in the negative y-direction and induces a Hall field toward positive y for holes, and in the opposite direction for electrons. Consider the drift velocity v_d of the current carriers. Equating the magnitudes of the magnetic and Hall-field forces, we have $q\mathscr{E}_H = qv_dB$. The velocity v_d is related in magnitude to the current by $v_d = J_x/qp$ for holes or by $v_d = -J_x/qn$ for electrons. Thus, the Hall field can be written in terms of the current and the applied magnetic field as

$$\mathscr{E}_H = \frac{J_x B}{qp} \tag{1.3.2}$$

for holes, and

$$\mathscr{E}_H = -\frac{J_x B}{qn} \tag{1.3.3}$$

for electrons. Both Equations 1.3.2 and 1.3.3 can be expressed as

$$\mathscr{E}_H = R_H J_x B \tag{1.3.4}$$

where R_H, the *Hall coefficient,* is equal to $1/qp$ for holes and to $-1/qn$ for electrons in this simplified derivation. In practice, Equation 1.3.4 predicts the Hall field accurately if the Hall coefficient is modified to account properly for statistical variations in the velocities of free carriers. This modification introduces a new factor r into the expression for the Hall coefficient which now becomes

$$R_H = r/qp \tag{1.3.5}$$

for holes, and

$$R_H = -r/qn \tag{1.3.6}$$

for electrons. The factor r is typically between 1 and 2 (theoretically 1.18 for lattice scattering and 1.93 for ionized impurity scattering).

The Hall voltage V_H is given by the product of $\mathscr{E}_H$ and W, which can be written in terms of the total current I as

$$V_H = \frac{R_H I B}{10^8 t} \tag{1.3.7}$$

with R_H measured in $\text{cm}^3 \, \text{C}^{-1}$, I in amperes, B in Gauss, t in cm, and V_H in volts. (The factor 10^8 is needed to convert the MKS units meter and Tesla (or Webers m^{-2}) to more conventional semiconductor units centimeter and Gauss.)

From Equation 1.3.7 we see that in an unknown semiconductor, the Hall coefficient can be determined by measuring the Hall voltage for a given magnetic field and current. Equations 1.3.5 and 1.3.6 then permit the calculation of the unknown carrier types and densities. From the carrier densities and known currents, the material conductivity and *Hall mobility* ($\mu_H = \sigma|R_H|$) can then be found. The Hall effect is thus a powerful experimental technique for the study of semiconductors, and it is frequently used for this purpose [7].

Integrated Hall-Effect Magnetic Sensor

To use the Hall effect in an integrated circuit, it is necessary to isolate a conducting pattern similar to the region sketched in Figure 1.20. This is typically accomplished by using epitaxy and oxide or junction isolation (procedures described in Chapter 2) as shown in Figure 1.21. For the simplest Hall-effect theory to apply, the width W should be much greater than the length L so that the current density J is uniform over the sample cross section. As a practical matter, however, shorting of the Hall voltage by the ohmic end contacts is reduced if $L \gg W$. In production-integrated circuits, W is usually made comparable to L. For a rectangular geometry the theory that we have derived can be corrected by multiplying the expression for V_H (Equation 1.3.7) by a factor K that is typically approximately unity and is a function only of the ratio W/L [8]. This refinement is not included in our treatment of the Hall effect.

One important consideration when designing a Hall element for an integrated circuit is the power dissipated in the device. To consider power consumption, we express the resistance of the Hall element in terms of the Hall coefficient. For a p-type element, for example, we can write an expression for the resistance R

$$R = \frac{\rho L}{A} = \frac{L}{q\mu_p p W t} = \frac{L R_H}{r \mu_p W t} \tag{1.3.8}$$

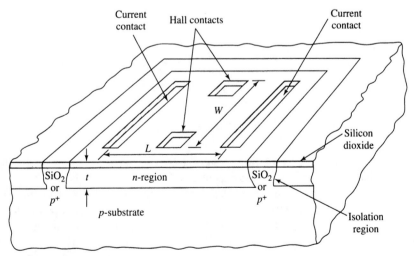

FIGURE 1.21 Hall-effect element for an integrated-sensor circuit. The element is fabricated in high-resistivity n-type silicon which is isolated by oxide regions or by p-type regions as described in Chapter 2. The sketch shows the element prior to the application of contact metal.

For a supply voltage V_S, we can write $I = V_S/R$ or

$$I = \frac{rV_S\mu_p Wt}{R_H L} \tag{1.3.9}$$

which allows us to write Equation 1.3.7 as

$$V_H = r\mu_p V_S \frac{W}{L} B \times 10^{-8} \tag{1.3.10}$$

where, as in Equation 1.3.7, B is measured in Gauss and length is measured in cm.

EXAMPLE *Hall Element Figure of Merit*

(a) Derive a figure of merit M_H for a Hall element that expresses the Hall voltage per unit of magnetic field per unit of power dissipation. Consider a p-type element having $R_H = 8 \times 10^3$ cm^3 C^{-1}, $W/L = 1$, $r = 1.2$, and $t = 8$ μm.

(b) Calculate the resistance of the element and the value of M_H if $B = 500$ Gauss and the power dissipated in the element is 1.43 mW.

Solution

(a) From Equation 1.3.10, we have V_H in terms of the supply voltage V_S. The dissipated power P_H is equal to $V_S \times I$.
Therefore,

$$P_H = \frac{rV_S^2 \mu_p Wt}{R_H L}$$

and

$$M_H = \frac{V_H}{P_H B} = \frac{R_H \times 10^{-8}}{V_S t} = \frac{r \times 10^{-8}}{qpV_S t}$$

in units of volts per Gauss-watt. From this derived result, we see that M_H is improved by decreasing the supply voltage and reducing both the dopant density and the thickness of the Hall element.

(b) Calculating the parameters, we have
1. $p = r/qR_H = 9.38 \times 10^{14}$
2. From Figure 1.16, $\mu_p = 475$ cm^2 V^{-1} s^{-1}
3. From Equation 1.3.8, $R = 17.5$ kΩ
4. Because $P_H = V_S^2/R$, $V_S = \sqrt{P_H R}$ or $V_S = 5$ V
5. Hence, $M_H = \dfrac{8 \times 10^3 \times 10^{-8}}{5 \times 8 \times 10^{-4}}$ or $M_H = 0.02$ ∎

As seen in this example, the performance of the Hall sensor is improved by decreasing the dopant density. As the dopant density is reduced, the mobilities of both holes and electrons increase (Figure 1.16), leading to an increased Hall voltage for a given bias (Equation 1.3.10). Most IC Hall sensors are fabricated in n-type Si doped with about 10^{15} donors cm^{-3}. From Figure 1.16, we see that at this concentration we are below the region in which the mobility varies strongly with the impurity density. Hence, for this dopant concentration, unintentional processing variations in the dopant density lead to only small variations in mobility, and thus in the Hall coefficient. Typical sensitivities obtained for Hall elements are of the order of 30 μV per Gauss with excellent linearity up to tens of kiloGauss.

FIGURE 1.22   A precision, linear-output, Hall-sensor chip. The integrated circuit contains bias elements, temperature-compensation circuitry, and on-chip amplification. The chip area is 1.12 by 1.98 mm^2 and the Hall element (large pattern on the lower right-hand side) measures 230 by 335 μm^2. (Courtesy: G. B. Hocker, Honeywell Corporation)

The major use of integrated Hall circuits is for sensing the position of a device or element. Most of the circuits produced are used in contactless keyboard switches. In a typical keyboard switching application, a permanent magnet on a plunger is activated by depressing a key. This action introduces a magnetic field of the order of 500 Gauss near the sensing Hall element, producing a Hall voltage of roughly 15 mV which is easily detected by an on-chip amplifier. Integrated Hall sensors are also produced for magnetometer applications, in which a signal proportional to the magnetic field is desired. Figure 1.22 shows a commercial Hall-effect integrated sensor which has been designed for magnetometer use.

SUMMARY

One of the cornerstones of solid-state electronics is the band structure of solids. This key concept is related to the quantized energy levels of isolated atoms. Huge differences in the electrical conductivities of metals, semiconductors, and insulators result from basic differences in the band structures of these three classes of materials.

By considering the band structure of semiconductors, we can deduce the existence of two types of current carriers, *holes* and *electrons*. In most cases of practical interest, holes and electrons can be considered as classical free particles inside the semiconductor crystal. The motion of the holes and electrons and the statistical distributions that characterize their energies can be calculated if their effective masses are

modified from those of truly free particles. The populations of holes and electrons can be controlled by adding impurities to otherwise pure semiconductor crystals. In this process, known as *doping*, extra holes can be added to single-crystal silicon, for example, by introducing ions with valence three, which are incorporated in the lattice in place of silicon, which has valence four. The resultant crystal is known as *p*-type silicon. To add electrons and thereby obtain *n*-type silicon by doping, impurities with valence five can be substituted for silicon atoms in the lattice.

An understanding of many electronic properties can be gained by applying the laws of statistical physics to the populations of electrons and holes in semiconductors. The *Fermi level* is an important parameter used to describe the concentrations of free carriers in a crystal at thermal equilibrium. Under usual conditions, the Fermi level in a semiconductor exists within a *forbidden gap* of energies that is ideally free of any allowed energy states for electrons. Semiconductors in this condition are described as *nondegenerate*. If the Fermi level lies within a band of allowed electronic states, the semiconductor is called *degenerate* because its conduction properties degenerate towards those of a metal. When thermal equilibrium does not apply, it is useful to define two *quasi-Fermi levels*, which are different for holes and electrons. The separation of the two quasi-Fermi levels is a measure of the degree to which thermal equilibrium in the semiconductor is disturbed. When the semiconductor returns to thermal equilibrium, the hole and electron quasi-Fermi levels merge into a single Fermi level that characterizes both carrier densities. Quasi-Fermi levels are useful in considering photogeneration of electrons and holes in semiconductors, and in simplifying the representation of free-carrier diffusion and drift.

In the basic theory of electrons and holes in semiconductors, it is assumed that only very low concentrations of dopants are present and that they do not influence the band structure of the host crystal. At *heavy-doping* concentrations, the impurity atoms are no longer situated far apart from one another, and their presence modifies the band structure. One of the most important effects of heavy doping is that n_{ie}, the effective value of the *intrinsic-carrier density*, increases with increasing doping density.

When free carriers in a semiconductor are subjected to a low or moderate electric field, they move with a constant *drift* velocity. The ease of carrier motion in response to the field expressed as the velocity per unit field is known as the *mobility* and denoted by μ. The magnitude of the mobility is determined by the nature of the scattering events experienced by free carriers as they interact with the lattice. The thermal energy of free carriers in solids causes them to move randomly and results in a net flux of carriers if either the carrier densities or their energies are not homogeneous in space. The net motion resulting from this process is called *diffusion*. When the crystal is at a uniform temperature, diffusion occurs from regions where the carriers are more dense to regions where their density is lower. In nondegenerate semiconductors, the *diffusion coefficient D* for a given type of carrier is related to the mobility μ by the *Einstein relation*, $D/\mu = kT/q$.

At high electric fields, scattering of electrons and holes is no longer determined mainly by the thermal velocity, and the drift motion is no longer proportional to the applied field. Mobility, instead of being constant, begins to decrease as the field increases. At fields in the range of 10^4 to 10^5 V cm^{-1}, additional modes of free-carrier scattering must be considered, leading to upper limits for hole and electron drift velocities.

The force on a moving charged particle in a magnetic field gives rise to the *Hall effect* when a semiconductor carries a current perpendicular to a magnetic field. Hall-effect experiments provide a powerful means of studying free-carrier properties in semiconductors. *Integrated-sensor* circuits in which Hall elements, together with bias circuits, temperature-compensating circuits, and amplifiers are all fabricated on a single silicon chip, are important commercial products.

REFERENCES

1. (a) F. MORIN and J. P. MAITA, *Phys. Rev.* **96**, 28 (1954).
   (b) A. S. GROVE, *Physics and Technology of Semiconductor Devices*, Wiley, New York (1967).
2. P. P. DEBYE and E. M. CONWELL, *Phys. Rev.* **93**, 693 (1954).
3. (a) D. M. CAUGHEY and R. F. THOMAS, *Proc. IEEE* **55**, 2192 (December 1967).
   (b) G. MASETTI, M. SEVERI, and S. SOLMI, *IEEE Trans. Electr. Devices* **ED-30**, 764 (July 1983).
4. N. D. ARORA, J. R. HAUSER, and D. J. ROULSTON, *IEEE Trans. Electr. Devices* **ED-29**, 292 (February 1982).
5. C. JACOBONI, C. CANALI, G. OTTAVIANI, and A. A. QUARANTA, *Solid-State Electronics* **20**, 77 (February 1977).
6. J. L. MOLL, *Physics of Semiconductors*, McGraw-Hill, New York (1964), p.198.
7. E. H. PUTLEY, *The Hall Effect and Related Phenomena*, Butterworths, London (1960).
8. J. T. MAUPIN and M. L. GESKE, *The Hall Effect in Silicon Circuits*, Symposium on the Hall Effect, Johns Hopkins University, Baltimore, Md. (1981), p. 421.
9. S. WANG, *Solid-State Electronics*, McGraw-Hill, New York (1966), p. 263.
10. S. M. SZE and J. C. IRVIN, *Solid-State Electronics* **11**, 599 (1968).
11. J. L. MOLL, *Physics of Semiconductors*, McGraw-Hill, New York (1964), p. 99.
12. R. B. ADLER, A. C. SMITH, and R. L. LONGINI, *SEEC*, Vol. 1, *Introduction to Semiconductor Physics*, Wiley, New York (1964).

BOOKS

B. G. STREETMAN and S. BANERJEE *Solid-State Electronic Devices*, 5th ed. Prentice-Hall, Simon and Schuster Co., N.J. (2000).

R. F. PIERRET, *Semiconductor Fundamentals*, Vol. I, *Modular Series on Solid-State Devices*, Addison-Wesley, Reading, Mass. (1983).

J. P. MCKELVEY, *Solid-State and Semiconductor Physics*. Original publisher Harper & Row, New York (1966), reprinted by Dover Publications, New York (1984).

W. E. BEADLE, J. C. C. TSAI, and R. D. PLUMMER, Editors, *Quick Reference Manual for Silicon Integrated-Circuit Technology*, Wiley-Interscience, New York (1985) (for silicon properties).

PROBLEMS

1.1 Phosphorus donor atoms with a concentration of 10^{16} cm^{-3} are added to a pure sample of silicon. Assume that the phosphorus atoms are distributed homogeneously throughout the silicon. The atomic weight of phosphorus is 31.

(a) What is the sample resistivity at 300 K?

(b) What proportion by weight does the donor impurity comprise?

(c) If 10^{17} atoms cm^{-3} of boron are included in addition to the phosphorus, and distributed uniformly, what is the resultant resistivity and conductivity type (i.e., *p*- or *n*-type material)?

(d) Sketch the energy-band diagram under the condition of (c) and show the position of the Fermi level.

1.2* Find the equilibrium electron and hole concentrations and the location of the Fermi level for silicon at 27°C if the silicon contains the following concentrations of shallow dopant atoms:

(a) 1×10^{16} cm^{-3} boron atoms

(b) 3×10^{16} cm^{-3} arsenic atoms and 2.9×10^{16} cm^{-3} boron atoms

1.3 An *n*-type sample of silicon has a uniform density $N_d = 10^{16}$ atoms cm^{-3} of arsenic, and a *p*-type silicon sample has $N_a = 10^{15}$ atoms cm^{-3} of boron. For each semiconductor material determine the following:

(a) The temperature at which half the impurity atoms are ionized. Assume that all mobile electrons and holes come from dopant impurities. [*Hint.* Use Equations 1.1.21 and 1.1.22 together with the fact that $f_D(E_f) = \frac{1}{2}$.]

(b) The temperature at which the intrinsic concentration n_i exceeds the impurity density by a factor of 10. See Table 1.4 for $n_i(T)$.

(c) The equilibrium minority-carrier concentrations at 300 K. Assume full ionization of impurities.

(d) The Fermi level referred to the valence-band edge E_v in each material at 300 K. The Fermi level if both types of impurities are present in a single sample.

1.4* A piece of *n*-type silicon has a resistivity of 5 Ω-cm at 27°C. Find the thermal-equilibrium hole concentrations at 27, 100, and 500°C. (Refer to the tables and figures for needed quantities.)

1.5† Grain boundaries and other structural defects introduce allowed energy states deep within the forbidden gap of polycrystalline silicon. Assume that each defect introduces two discrete levels: an acceptor level 0.51 eV above the top of the valence band and a donor level 0.27 eV above the top of the valence band. (Note that $E_a > E_d$ and that these are *not* shallow levels.) The ratio of the number of defects with each charge state ($+$, $-$, or neutral) at thermal equilibrium is given by the relation (see Ref. 11)

$$N_d^+ : N_0 : N_a^- = \exp\frac{E_d - E_f}{kT} : 1 : \exp\frac{E_f - E_a}{kT}$$

(a) Sketch the densities of these three species as the Fermi level moves from E_v to E_c. Which species dominates in heavily doped *p*-type material? In *n*-type material?

(b) What is the effect of the defects on the majority-carrier concentrations?

(c) Using the above information, determine the charge state of the defect levels and the position of the Fermi level in a silicon crystal containing no shallow dopant atoms. Is the sample *p*- or *n*-type?

(d) What are the electron and hole concentrations and the location of the Fermi level in a sample with 2×10^{17} cm^{-3} phosphorus atoms and 5×10^{16} cm^{-3} defects?

1.6* Two scattering mechanisms are important in a piece of semiconductor. If only the first scattering process were present, the mobility would be 800 cm^2 (Vs)$^{-1}$. If only the second were present, the mobility would be 200 cm^2 (Vs)$^{-1}$. What is the mobility considering both scattering processes?

* Answers to problems marked by an asterisk appear at the end of the book.

1.7 Find the mobility of electrons in aluminum with resistivity 2.8×10^{-6} Ω-cm and density 2.7 g cm^{-3}. The atomic weight of aluminum is 27. Of the three valence electrons in aluminum, on the average 0.9 electrons are free to participate in conduction at room temperature. If $m^* = m_0$, find the mean time between collisions and compare this value to the corresponding value in lightly doped silicon.

1.8* An electron is moving in a piece of lightly doped silicon under an applied field at 27°C so that its drift velocity is one-tenth of its thermal velocity. Calculate the average number of collisions it will experience in traversing by drift a region 1 μm wide. What is the voltage applied across this region?

1.9 In an experiment the voltage across a uniform, 2 μm-long region of 1 Ω-cm, n-type silicon is doubled, but the current only increases by 50%. Explain. (The silicon remains neutral at both current levels.)

1.10* The electron concentration in a piece of uniform, lightly-doped, n-type silicon at room temperature varies linearly from 10^{17} cm^{-3} at $x = 0$ to 6×10^{16} cm^{-3} at $x = 2$ μm. Electrons are supplied to keep this concentration constant with time. Calculate the electron current density in the silicon if no electric field is present. Assume $\mu_n = 1000$ cm^2 V^{-1} s^{-1} and T = 300 K.

1.11† Silicon atoms are added to a piece of gallium arsenide. The silicon can replace either trivalent gallium or pentavalent arsenic atoms. Assume that silicon atoms act as fully ionized dopant atoms and that 5% of the 10^{10} cm^{-3} silicon atoms added replace gallium atoms and 95% replace arsenic atoms. The sample temperature is 300 K.

(a) Calculate the donor and acceptor concentrations.

(b) Find the electron and hole concentrations and the location of the Fermi level.

(c) Find the conductivity of the gallium arsenide assuming that lattice scattering is dominant.

See Table 1.3 for properties of GaAs.

1.12† (Dielectric relaxation in solids.) Consider a homogeneous one-carrier conductor of conductivity σ and permittivity ϵ. Imagine a given distribution of the mobile charge density $\rho(x, y, z; t = 0)$ in space at $t = 0$. We know the following facts from electromagnetism, provided we neglect diffusion current:

$$\nabla \cdot D = \rho; \quad D = \epsilon \mathscr{E}; \quad J = \sigma \mathscr{E}; \quad \nabla \cdot J = \frac{-d\rho}{dt}$$

(a) Show from these facts that $\rho(x, y, z; t) = \rho(x, y, z; t = 0) \exp[-t/(\epsilon/\sigma)]$. This result shows that uncompensated charge cannot remain in a uniform conducting material, but must accumulate at discontinuous interfaces, surfaces, or other places of nonuniformity.

(b) Compute the value of the *dielectric relaxation time* ϵ/σ for intrinsic silicon; for silicon doped with 10^{16} donors cm^{-3}; and for thermal SiO$_2$ with $\sigma = 10^{-16}$ $(\Omega$-cm$)^{-1}$ [12].

1.13† Because of their thermal energies, free carriers are continually moving throughout a crystal lattice. While the net flow of all carriers across any plane is zero at thermal equilibrium, it is useful to consider the directed components that balance to zero. The component values are physically significant in that they measure the quantity of current that can be delivered by diffusion alone. This is relevant if, for example, one is able to unbalance the thermal equilibrium condition by intercepting all carriers flowing in a given direction. By considering that $J_x = -qn_0v_x$, show that the current in a solid in any random direction resulting from thermal processes is

$$J = \frac{-qn_0v_{th}}{4}$$

where v_{th} is the mean thermal velocity and n_0 is the free-electron density. (*Hint.* Consider the flux through a solid angle of 2π steradians.) [12]

1.14* Calculate the wavelengths of radiation needed to create hole-electron pairs in intrinsic germanium, silicon, gallium arsenide, and SiO$_2$. Identify the spectrum range (e.g., infrared, visible, UV, and X ray) for each case.

1.15† The relation between D and μ is given by

$$\frac{D}{\mu} = \frac{1}{q} \frac{dE_f}{d(\ln n)}$$

for a material that may be *degenerate*. (That is, the Fermi-Dirac distribution function must be used because the Fermi level may enter an allowed energy band.) Show that this relation reduces to the simpler Einstein relation $D/\mu = kT/q$ if the material is *nondegenerate* so that Boltzmann statistics can be used.

1.16 A *hot-probe* setup is a useful laboratory apparatus. It is used to determine the conductivity type of a semiconductor sample. It consists of two probes and an ammeter that indicates the direction of current flow. One of the probes is heated (most simply, a soldering iron tip is used), and the other is at room temperature. No voltage is applied, but a current flows when the probes touch the semiconductor. Considering the role of diffusion currents, explain the operation of the apparatus and draw diagrams to indicate the current directions for p- and n-type semiconductor samples.

1.17 Consider a simple model for diffusion in one dimension. Assume that all particles must move only at discrete, regularly spaced times, one time unit apart. When they move, each particle can only jump one space unit either to the left or to the right, with *equal* probability. Start with 1024 particles at $x = 0$, $t = 0$, and build up, step by step, the pattern of particles in space after 10 time units. Plot the distribution in space after each jump and measure the corresponding width W between points of one-half maximum concentration. Then plot W^2 versus time. What does your result suggest about the rate of "spread" of particles by the diffusion process (in *one* dimension at least)? [12]

1.18 The values in Tables 1.3 and 1.4 are not entirely consistent. To see that this is the case,

(a) Calculate E_g ($T = 300$ K) using the equation in Table 1.4 and compare the result with E_g in Table 1.3.

(b) Use both values of E_g to calculate n_i from Equation 1.1.25 with N_c and N_v as given in Table 1.3.

(c) Calculate n_i at 300 K using the temperature-variation formula in Table 1.4.

[The calculated inconsistencies reflect the fact that values in the tables are determined from measurements using a variety of experimental methods. Based on many experiments, a value of n_i equal to 1.45×10^{10} cm^{-3} for silicon at 300 K is widely used. The degree of precision of many parameters is unknown.]

1.19[†] In a nearly intrinsic material, the motions of both electrons and holes are significant for the Hall effect. Show that the Hall coefficient in the case that both holes and electrons are present becomes

$$R_H = \frac{\mathcal{E}_H}{J_x B} = \frac{\mu_p^2 p - \mu_n^2 n}{q(\mu_p p + \mu_n n)^2}$$

Prove that this result is consistent with the simpler theory derived in Section 1.3.

APPENDIX

Electric Fields, Charge Configuration, and Gauss' Law

As discussed in Sec. 1.2, an internal electric field in a solid moves electrons and holes and thereby causes a drift current. If we know the size and direction of the field, we can predict the drift current in terms of material properties. This is but one of many circumstances in which knowledge of the electric field in a solid has crucial importance. Although electric-field analysis is typically taught in basic physics courses, the analysis techniques have so many important applications to device electronics that we review them briefly here. Readers who require amplification of these ideas can consult any of the many excellent physics texts that discuss electricity and magnetism in greater detail.

For device analysis, the most useful fundamental relationship is that between electric charge and the electric field. Expressed in one dimension (which very often suffices for device analysis), the differential equation relating a charge density $\rho(x)$ to the field $\mathcal{E}$ is

$$d(\epsilon\mathcal{E})/dx = \rho \qquad (1A.1)$$

In Equation 1A.1, $\epsilon = \epsilon_r\epsilon_0$ is the permittivity of the material which is, in turn, equal to the product of the relative permittivity ϵ_r and the permittivity of free space ϵ_0.

Michael Faraday (1791–1867), the famous English engineer-scientist, provided an extremely useful interpretation of Equation 1A.1 by introducing the concept of electrical *lines of force* that act on charges and that emanate from positive charges and terminate on negative charges. The lines of force are always parallel to the electric field and the densities of the lines of force throughout a region are proportional to the local magnitude of the electric field in the same region. Faraday's concept provides a straightforward means to visualize the correctness of Gauss' Law which states that the integrated electric field over a closed surface is equal to the net electric charge contained within that surface divided by the permittivity within the closed surface.

To develop these ideas in a useful manner for device electronics, we can integrate Equation 1A.1 along x to write a one-dimensional form of *Gauss' Law* (assuming the usual case that ϵ is not a function of x).

$$\epsilon(\mathcal{E}_2 - \mathcal{E}_1) = \int_{x_1}^{x_2} \rho(x)\,dx \qquad (1A.2)$$

Equation 1A.2 is readily interpreted using Faraday's concept of lines of force because it states mathematically that the product of the permittivity and the electric field at the boundaries of a region (between x_2 and x_1) changes by an amount equal to the total charge contained in that region. Field lines are added by positive charges and subtracted by negative charges; hence, the change in field between two boundaries (x_1 and x_2) multiplied by the permittivity equals the sum of the charges contained in the region between these two boundaries. The special case of charge σ distributed uniformly in density on a planar surface within a bounded volume (therefore at a single value of $x = x_0$) leads, from Equation 1A.2, to a stepwise change in the field $\mathscr{E}$ at $x = x_0$ such that $\Delta\mathscr{E}$ at $x = x_0$ is σ/ϵ. Not evident in Equation 1A.2 is the effect of a change in the permittivity, for example at a boundary between two different materials. Because a material's permittivity ϵ represents the polarizing effect of an electric field on its molecules, a change in ϵ at a boundary results in an effective surface charge at that boundary. As a result, if the permittivity changes from ϵ_a to ϵ_b at $x = x_1$, the field must also change from its value $\mathscr{E}_a$ in region a to $\mathscr{E}_b$ in region b at $x = x_1$ such that $\epsilon_a\mathscr{E}_a = \epsilon_b\mathscr{E}_b$.

The force F on an electron in a field $\mathscr{E}$ is the product of its charge $-q$ and the field.

$$F = -q\mathscr{E} \tag{1A.3}$$

The work dW needed to move the charge an incremental distance dx is

$$dW = -q\mathscr{E}dx = q\,d\phi \tag{1A.4}$$

where the field $\mathscr{E}$ is related to the electric potential ϕ by

$$\mathscr{E} = -d\phi/dx \tag{1A.5}$$

Equation 1A.5 can be used together with Equation 1A.1 to write *Poisson's Equation.*

$$d^2\phi/dx^2 = -\rho/\epsilon \tag{1A.6}$$

In Equation 1A.6, we have assumed that the permittivity ϵ does not depend on x.

Equation 1A.6 is a second-order linear differential equation for the potential ϕ, and hence its solutions have two constants that need to be evaluated using boundary conditions. Because Equation 1A.6 is linear, its solutions can be added; i.e., a potential ϕ_1 that is a solution corresponding to a charge density ρ_1 can be added to a potential ϕ_2 that is a solution corresponding to a charge density ρ_2 so that $\phi = \phi_1 + \phi_2$ represents the potential for a charge density $\rho = \rho_1 + \rho_2$. This additive property of solutions for potential, known as the *superposition principle*, is often a valuable problem-solving aid.

EXAMPLE   *Capacitor Electronics Using Gauss' Law*

(a) Consider two parallel perfectly conducting plates of area A that are separated a distance W in air and connected to electrodes to form a capacitor. A voltage V_0 is applied between the two plates. Sketch and label plots showing (*i*) the field, (*ii*) charges, and (*iii*) the potential as functions of the distance x between the two conducting plates ($0 \leq x \leq W$).

(b) A perfectly conducting plate of area A and thickness $W/2$, having no electrical connections, is inserted midway between the plates. Repeat the calculations requested in part (a).

(c) Instead of the conducting plate of part (b), a dielectric layer with relative permittivity $\epsilon_r = 2$ and thickness $W/2$ is inserted midway between the plates. Repeat the sketches requested in part (a).

(d) What capacitance would be measured between the two conducting plates in parts (a), (b), and (c) of this example?

Solution

(a)

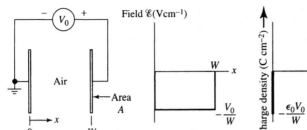

(b)

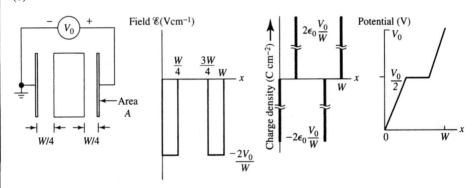

(c) Call the field in air $\mathcal{E}_a$ and the field in the dielectric $\mathcal{E}_d$.
At the boundary ($x = W/4$), we have $\epsilon_0 \mathcal{E}_a = \epsilon_r \epsilon_0 \mathcal{E}_d$, hence

$$\mathcal{E}_d = \mathcal{E}_a/\epsilon_r = \mathcal{E}_a/2$$

The voltage between the plates $= V_0$

Therefore, $\mathcal{E}_a\left(\dfrac{W}{4}\right) + \dfrac{\mathcal{E}_a}{2}\left(\dfrac{W}{2}\right) + \mathcal{E}_a\left(\dfrac{W}{4}\right) = V_0$ and $\mathcal{E}_a = \dfrac{4V_0}{3W}$

Charge on the plates $= \epsilon_0 \mathcal{E}_a = \epsilon_0 \dfrac{4V_0}{3W}$

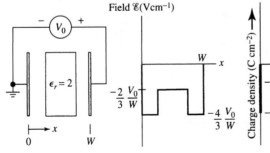

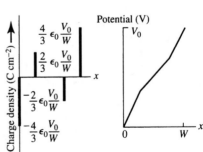

(d) Capacitance $C = Q/V$.   For **(a)** $Q = \epsilon_0 V_0/W$, $C = \epsilon_0/W$;

  (b) $Q = 2\epsilon_0 V_0/W$, $C = 2\epsilon_0/W$;   **(c)** $Q = \dfrac{4}{3}\epsilon_0 V_0/W$, $C = \dfrac{4}{3}\epsilon_0/W$.

APPENDIX PROBLEMS

A1.1 The charge distribution shown below is that of an idealized metal-semiconductor contact. A thin sheet (delta function) of positive charge with charge/(unit area) $Q' = \rho_1 \times x_d$ is located at $x = 0$ to balance the uniformly distributed negative charge in the semiconductor. The permittivity of semiconductor is ϵ_s.

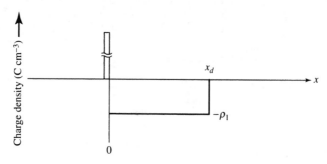

(a) Derive an expression for $\mathscr{E}(x)$ from $-\infty < x < +\infty$ and sketch $\mathscr{E}(x)$.

(b) Derive the expression for $\phi(x)$ from $-\infty < x < +\infty$ and sketch $\phi(x)$.

(c) Find the potential difference $\Delta\phi$ (i.e., $\phi(x = 0) - \phi(x_d)$) between $x = 0$ and $x = x_d$.

(d) Sketch $\mathscr{E}(x)$ and $\phi(x)$ for the charge distribution shown below. Your intuition should tell you that you don't have to repeat the previous calculations.

ALL of your sketches must be to the proper proportions.

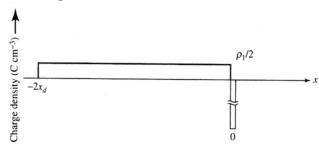

A1.2 The following metal-semiconductor-contact charge distribution has two regions of different charge concentrations in the semiconductor. The thin sheet of negative charge at $x = 0$ balances the total positive charge in the semiconductor. The permittivity inside the semiconductor is ϵ_s.

(a) Derive an expression for $\mathscr{E}(x)$ from $-\infty < x < +\infty$ and sketch $\mathscr{E}(x)$.

(b) Derive the expression for $\phi(x)$ from $-\infty < x < +\infty$ and sketch $\phi(x)$.

(c) Find the potential differences $\Delta\phi_1$ (i.e., $\phi(0) - \phi(x_{d1})$) between $x = 0$ and $x = x_{d1}$, and $\Delta\phi_2$ (i.e., $\phi(0) - \phi(x_{d2})$) between $x = 0$ and $x = x_{d2}$.

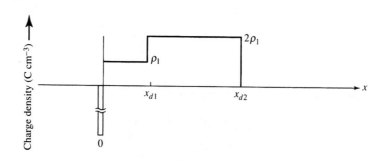

A1.3 The charge distribution shown is related to semiconductor *pn* junctions. The total positive charge in the *n*–type semiconductor balances the total negative charge in the *p*–type semiconductor (i.e., $\rho_2 \times x_{d2} = \rho_1 \times x_{d1}$). The permittivity in both regions of the semiconductor is ϵ_s.

(a) Derive an expression for $\mathscr{E}(x)$ from $-\infty < x < +\infty$ and sketch $\mathscr{E}(x)$.

(b) Derive an expression for $\phi(x)$ from $-\infty < x < +\infty$ and sketch $\phi(x)$.

(c) Find the potential difference $\Delta\phi$ (i.e., $\phi(x_{d2}) - \phi(-x_{d1})$) between $x = x_{d2}$ and $x = -x_{d1}$.

(d) Use the *superposition principle* and the results of Problem A1.1 to solve for $\mathscr{E}(x)$ and $\phi(x)$.

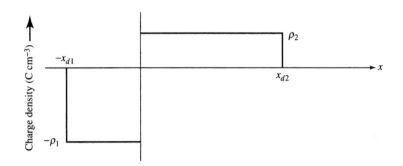

A1.4 The figure represents the charge distribution of a metal-oxide-semiconductor (MOS) structure. The oxide has a permittivity ϵ_{ox} and the semiconductor has a permittivity ϵ_s. There is a thin sheet of charge at $x = 0$ in the metal, balancing the total semiconductor charge of concentration ρ and thickness x_d. Note that $\rho(x) = 0$ inside the oxide.

(a) Derive an expression for $\mathscr{E}(x)$ from $-\infty < x < +\infty$ and sketch $\mathscr{E}(x)$.

(b) Derive an expression for $\phi(x)$ from $-\infty < x < +\infty$ and sketch $\phi(x)$.

(c) Find the potential difference $\Delta\phi_0$ (i.e., $\phi(0) - \phi(x_0)$). This is the voltage drop across the oxide.

(d) Find the potential difference $\Delta\phi_d$ (i.e., $\phi(x_0) - \phi(x_0 + x_d)$). This is the voltage drop across the semiconductor.

(e) Use the *superposition principle* to solve for $\mathscr{E}(x)$ and $\phi(x)$ by decomposing the charge in a convenient way to make use of the previous results.

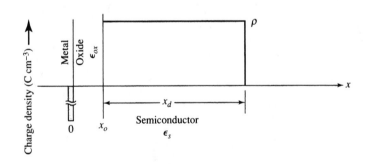

TABLE 1.3 Properties of Semiconductors and Insulators (at 300 K Unless Otherwise Noted)

Property	Symbol	Units	Si	Ge	GaAs	GaP	SiO$_2$	Si$_3$N$_4$
Crystal structure			Diamond	Diamond	Zincblende	Zincblende	Amorphous for most IC applications	
Atoms per unit cell			8	8	8	8		
Atomic number	Z		14	32	31/33	31/15	14/8	14/7
Atomic or molecular weight	MW	g/g-mole	28.09	72.59	144.64	100.70	60.08	140.28
Lattice constant	a_0	nm	0.54307	0.56575	0.56532	0.54505		0.775
Atomic or molecular density	N_0	cm^{-3}	5.00×10^{22}	4.42×10^{22}	2.21×10^{22}	2.47×10^{22}	2.20×10^{22}	1.48×10^{22}
Density		g cm^{-3}	2.328	5.323	5.316	4.13	2.19	3.44
Energy gap 300K	E_g	eV	1.124	0.67	1.42	2.24	~8 to 9	4.7
0K	E_g	eV	1.170	0.744	1.52	2.40		
Temperature dependence	$\Delta E_g / \Delta T$	eV K^{-1}	-2.7×10^{-4}	-3.7×10^{-4}	-5.0×10^{-4}	-5.4×10^{-4}		
Relative permittivity	ϵ_r		11.7	16.0	13.1	10.2	3.9	7.5
Index of refraction	n		3.44	3.97	3.3	3.3	1.46	2.0
Melting point	T_m	°C	1412	937	1237	1467	~1700	~1900
Vapor pressure		Torr (mm Hg) (at °C)	10^{-7} (1050) 10^{-5} (1250)	10^{-9} (750) 10^{-7} (880)	1 (1050) 100 (1220)	10^{-6} (770) 10^{-4} (920)		
Specific heat	C_p	J (g K)$^{-1}$	0.70	0.32	0.35		1.4	0.17
Thermal conductivity	κ	W(cm K)$^{-1}$	1.412	0.606	0.455	0.97	0.014	0.185(?)
Thermal diffusivity	D_{th}	cm^2 s^{-1}	0.87	0.36	0.44		0.004	0.32(?)
Coefficient of linear thermal expansion	α'	K^{-1}	2.5×10^{-6}	5.7×10^{-6}	5.9×10^{-6}	5.3×10^{-6}	5×10^{-7}	2.8×10^{-6}
Intrinsic carrier concentration*	n_i	cm^{-3}	1.45×10^{10}	2.4×10^{13}	9.0×10^{6}			
Lattice mobility Electron	μ_n	cm^2(V s)$^{-1}$	1417	3900	8800	300	20	
Hole	μ_p	cm^2(V s)$^{-1}$	471	1900	400	100	~10^{-8}	

	Symbol	Units					
Effective density of states							
Conduction band	N_c	cm^{-3}	2.8×10^{19}	1.04×10^{19}	4.7×10^{17}		
Valence band	N_v	cm^{-3}	1.04×10^{19}	6.0×10^{18}	7.0×10^{18}		
Electric field at breakdown	$\mathscr{E}_1$	V cm^{-1}	3×10^5	8×10^4	3.5×10^5		$6 - 9 \times 10^6$
Effective mass							
Electron	m_n^*/m_0		1.08^a 0.26^b	0.55^a 0.12^b	0.068	0.5	
Hole	m_p^*/m_0		0.81^a 0.386^b	0.3	0.5	0.5	
Electron affinity	qX	eV	4.05	4.00	4.07	~ 4.3	1.0
Average energy loss per phonon scattering		eV	0.063	0.037	0.035		
Optical phonon mean-free path							
Electron	λ_{ph}	nm	6.2	6.5	3.5		
Hole	λ_{ph}	nm	4.5	6.5	3.5		

Sources: A. S. Grove, *Physics and Technology of Semiconductors*, Wiley, New York (1967); S. M. Sze, *Physics of Semiconductor Devices*, Wiley, New York (1981); D. E. Hill, *Some Properties of Semiconductors* (table), Monsanto Co., St. Peters, Mo. (1971); H. Wolf, *Semiconductors*, Wiley, New York (1971); W. E. Beade, J. C. C. Tsai, R. D. Plummer, *Quick Reference Manual for Silicon Integrated Circuit Technology*, Wiley-Interscience, New York, 1985; F. Shimura and H. R. Huff, "VLSI Silicon Material Criteria," Chapter 15 in *VLSI Handbook* (ed. N. G. Einspruch, Academic Press, 1985).

^{a} Used in density-of-states calculations.

^{b} Used in conductivity calculations. [9]

* The intrinsic free-carrier density n_i in silicon is reported as $n_i = 1 \times 10^{10}$ cm^{-3} in several fairly recent papers [M. A. Green, J. Appl. Phys., vol. 67, p. 2944 (1990); A. B. Sproul and M. A. Green, J. Appl. Phys., vol. 70, p. 846 (1991); and S. M. Sze, *Semiconductor Devices: Physics and Technology* (2nd ed. Wiley, New York 2002), pp. 36–37.] The intrinsic density that we give in the table, $n_i = 1.45 \times 10^{10}$ cm^{-3} is, however, in wide use and is the value taken in problems and further discussion in this book.

TABLE 1.4 Additional Properties of Silicon (Lightly Doped, at 300 K Unless Otherwise Noted)

Property	Symbol	Units	Values
Tetrahedral radius		nm	0.117
Pressure coefficient of energy gap	$\Delta E_g/\Delta p$	eV (atm)$^{-1}$	-1.5×10^{-6}
Intrinsic carrier density[a]	n_i	(T in K)	$3.87 \times 10^{16}\, T^{3/2} \exp\left[\dfrac{-7014}{T}\right]$
Bandgap	E_g	eV	$1.17 - \dfrac{4.73 \times 10^{-4}\, T^2}{T + 651}$
Temperature coefficient of lattice mobility			
Electron	$\Delta\mu_n/\Delta T$	cm^2(V s K)$^{-1}$	-11.6
Hole	$\Delta\mu_p/\Delta T$	cm^2(V s K)$^{-1}$	-4.3
Diffusion coefficient			
Electron	D_n	cm^2 s^{-1}	34.6
Hole	D_p	cm^2 s^{-1}	12.3
Hardness	H	Mohs	7.0
Elastic constants	c_{11}	dyne cm^{-2}	1.656×10^{12}
	c_{12}		0.639×10^{12}
	c_{44}		0.796×10^{12}
Young's modulus (<111> direction)	Y	dyne cm^{-2}	1.9×10^{12}
Surface tension (at 1412°C)	σ_0	dyne cm^{-2}	720
Latent heat of fusion	H_F	eV	0.41
Expansion on freezing		%	9.0
Cut-off frequency of lattice vibrations	ν_0	Hz	1.39×10^{13}

Main Source: H. F. Wolf, *Semiconductors,* Wiley, New York (1971), p. 45.

Energy Levels of Elemental Impurities in Si[b]

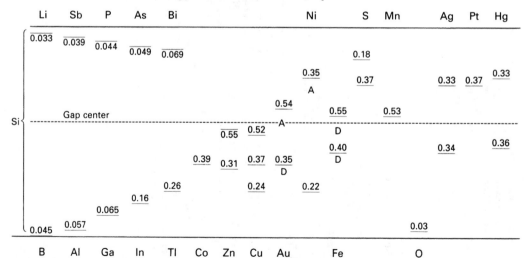

[a] a plot of n_i versus T at higher temperatures is given in Figure 2.10.

[b] The levels below the gap center are measured from the top of the valence band and are acceptor levels unless indicated by *D* for donor level. The levels above the gap center are measured from the bottom of the conduction band and are donor levels unless indicated by *A* for acceptor level [10].

SILICON TECHNOLOGY

Successful engineering rests on two foundations. One is a mastery of underlying physical concepts; a second foundation, at least of equal importance, is a perfected technology—a means to translate engineering concepts into useful structures. In Chapter 1 we reviewed the physical principles needed for integrated-circuit electronics. In this chapter we discuss the technology to produce devices in silicon, a technology now so powerful that the entire modern world has felt its impact. In addition to describing IC technology as it is now practiced, we attempt to provide a perspective on continuing developments in this fast-moving area.

Technological evolution toward integrated circuits began with development of an understanding of diode action and the invention of the transistor in the late 1940s. At that time the semiconductor of greatest interest was germanium. Experiments with germanium produced important knowledge about the growth of large single crystals having chemical purity and crystalline perfection that were previously unachievable.

Germanium is an element that crystallizes in a diamond-like lattice structure, in which each atom forms covalent bonds with its four nearest neighbors. The crystal structure is shown in Figure 1.8. Germanium has a band gap of 0.67 eV and an intrinsic carrier density equal to 2.5×10^{13} cm^{-3} at 300 K. Because of the relatively small band gap in germanium, its intrinsic-carrier density increases rapidly with increasing temperature, growing roughly to 10^{15} cm^{-3} at 400 K (see Figure 1.9).

Because most devices are no longer useful when the intrinsic-carrier concentration becomes comparable to the dopant density, germanium devices are limited to operating temperatures below about 70°C (343 K). As early as 1950, the temperature limitations of germanium devices motivated research on several other semiconductors that crystallize with similar lattice structures but can be used at higher temperatures. In the intervening decades technological development for integrated circuits has focused on the elemental semiconductor silicon ($E_g = 1.12$ eV) and the compound semiconductor gallium arsenide ($E_g = 1.42$ eV). Silicon is used in the overwhelming majority of integrated circuits, while compound semiconductors, such as gallium arsenide, find application in specialized, high-performance circuits. Compound semiconductors are especially useful in optical devices that rely on the efficient light emission from direct-bandgap compound semiconductors.

Most of this chapter deals with silicon technology, but the technology associated with compound semiconductors will be briefly discussed later in this chapter. Properties of germanium, silicon, gallium arsenide, and several other useful electronic materials are summarized in Table 1.3 (page 52–53). Table 1.4 (page 54) lists some additional properties of silicon.

2.1 THE SILICON PLANAR PROCESS

In addition to the good semiconducting properties of silicon, the major reason for its widespread use is the ability to form on it a stable, controllable oxide film (silicon dioxide SiO$_2$) that has excellent insulating properties. This capability, which is not matched by any other semiconductor-insulator combination, makes it possible to introduce controlled amounts of dopant impurities into small, selected areas of a silicon sample while the oxide blocks the impurities from the remainder of the silicon. The ability to dope small regions of the silicon is the key to producing dense arrays of devices in integrated circuits.

Two chemical properties of the Si–SiO$_2$ system are of basic importance to silicon technology. First, *selective etching* is possible using liquid or gaseous etchants that attack only one of the two materials. For example, hydrofluoric acid dissolves silicon dioxide but not silicon. Second, silicon dioxide can be used to shield an underlying silicon crystal from dopant impurity atoms brought to the surface either by high-energy ion beams or from a high-temperature gaseous diffusion source.

Using these features, dopant atoms can be introduced into areas on the silicon that are not shielded by thick silicon dioxide. The shielded regions can be accurately defined by using photosensitive polymer films exposed with photographic masks. The polymer pattern protects selected oxide regions on the silicon surface when it is immersed in a hydrofluoric-acid bath, or exposed to a gas-phase etchant, forming a surface consisting of bare silicon *windows* in a silicon-dioxide layer. This selective etching process, originally developed for lithographic printing applications, permits delineation of very small patterns. When the silicon sample is placed in an ambient that deposits dopant atoms on the surface, these atoms enter the silicon only at the exposed silicon windows.

Proper sequencing and repetition of the oxidation, patterning, and dopant-addition operations just described can be used to introduce *p*- and *n*-type dopant atoms selectively into regions having dimensions ranging down to the few hundred nanometer range. These steps are the basic elements of the *silicon planar process*, so called because it is a process that produces device structures through a sequence of steps carried out near the surface plane of the silicon crystal.

In addition to providing a means of limiting the area of dopant introduction, a well-formed oxide on silicon improves the electrical properties at the surface of the silicon substrate. Because of the termination of the crystal lattice of the silicon substrate, uncompleted or *dangling* bonds exist at an ideal free surface. As we will see in Chapter 4, these broken bonds can introduce allowed states into the energy gap of the silicon substrate at its surface and degrade the electrical behavior of device regions near the surface. However, a well-formed silicon-dioxide layer on the silicon surface electrically passivates almost all of these surface states, allowing nearly ideal behavior of the surface region of the silicon. Although the area density of bonds at the silicon surface is about 10^{15} cm^{-2}, the number of electrically active bonds can be reduced to less than 10^{11} cm^{-2} by properly growing a silicon-dioxide layer on the surface. The ability to remove virtually all the electrically active states at the silicon surface allows the successful operation of the ubiquitous silicon metal-oxide-semiconductor (MOS) transistor, which is the basis of most large-scale integrated circuits today.

The silicon crystals used for the planar process consist of slices (called *wafers*) that are made from a large single crystal of silicon. Dopant atoms are typically added to the silicon by *depositing* them in selected regions on or near the wafer surface, and then *diffusing* them into the silicon. Because dopant atoms are introduced from the surface and typical diffusion dimensions are very small, the active regions of the planar-processed devices are themselves within a few micrometers* of the wafer surface. The remaining thickness (generally several hundred micrometers) serves simply as a mechanical support for the important surface region.

A major advantage of the planar process is that each fabrication step (prior to packaging) is typically applied to the entire wafer. Hence, it is possible to make and interconnect many devices with high precision to build an *integrated circuit* (IC). At present, individual ICs typically measure up to 20 mm on a side, so that a wafer (most often 20–30 cm in diameter) can contain many ICs. There is clear economic advantage to increasing the area of each wafer and, at the same time, reducing the area of each integrated circuit built on the wafer.

The most important steps in the planar process, shown in Figure 2.1, include (*a*) formation of a masking oxide layer, (*b*) its selective removal, (*c*) deposition of dopant atoms on or near the wafer surface, and (*d*) their diffusion into the exposed silicon regions. These processes determine the location of the dopants which, in turn, determines the electrical characteristics of the devices and ICs.

---

* 1 micrometer = 1 μm, sometimes called a *micron*, equals 10^{-4} cm = 10^{3} nm = 10^{4} Å (angstrom).

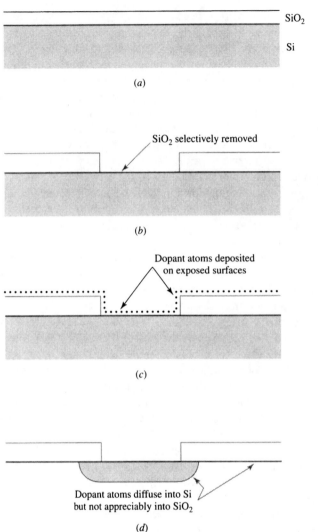

FIGURE 2.1   Basic fabrication steps in the silicon planar process: (*a*) oxide formation, (*b*) selective oxide removal, (*c*) deposition of dopant atoms on wafer, (*d*) diffusion of dopant atoms into exposed regions of silicon.

Refinements of the planar process and, along with it, the growth of silicon-based electronics has been amazingly rapid and continuous. Figure 2.2 gives some perspective to the development of silicon integrated-circuit technology over the past 40 years.

The minimum feature size on an integrated circuit has continued to evolve at a rapid rate, decreasing from 8 μm in 1969 to 130 nm* today. The rate of evolution can be appreciated by plotting the minimum feature size (on a logarithmic scale) versus year of first commercial production, as shown in Figure 2.2*f*. The straight line on this plot indicates the exponential decrease in the minimum feature size. The exponential change was first quantified by Gordon Moore of Intel Corporation and is known as "Moore's Law." (Moore's law is so well recognized and frequently cited that it has even been mentioned in the cartoon "Dilbert" [1]. Along with the rapid decrease in minimum feature size, the size of the IC chip continues to increase, although less rapidly than the minimum feature size decreases. Because of the combination of smaller features and larger chips, the number

---

* 1 nanometer (nm) $= 10^{-9}$ m $= 10^{-7}$ cm $= 10^{-3}$ μm.

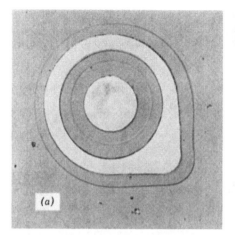
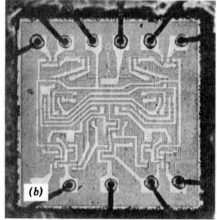

FIGURE 2.2 Evolution of IC technology. (a) First commercial silicon planar transistor (1959) (outer diameter 0.87 mm). (b) Diode-transistor logic (DTL) circuit (1964) (chip size 1.9 mm square). (c) 256-bit bipolar random access memory (RAM) circuit (1970) (chip size 2.8 × 3.6 mm). (d) VLSI central-processor computer chip containing 450,000 transistors (1981). The different functions carried out by the IC are labeled on the figure (chip size 6.3 mm square). [(a), (b), (c) courtesy of B.E. Deal—Fairchild Semiconductor, (d) Courtesy of Hewlett-Packard Co.] (e) Block diagram of Pentium 4 processor with 42 million transistors (2000); the corresponding chips photo is shown on the book cover. (Courtesy of Intel Corporation.) (f) Minimum feature size versus year of first commercial production. (g) Another embodiment of Moore's law shows that the number of transistors per chip has doubled every 18–24 months for approximately 30 years. (h) Along with decreasing feature size, the number of electrons in each device decreases. [(f)–(h) adapted from Mark Bohr, Intel; Howard Huff, Sematech; Joel Birnbaum, Hewlett-Packard; Motorola.]

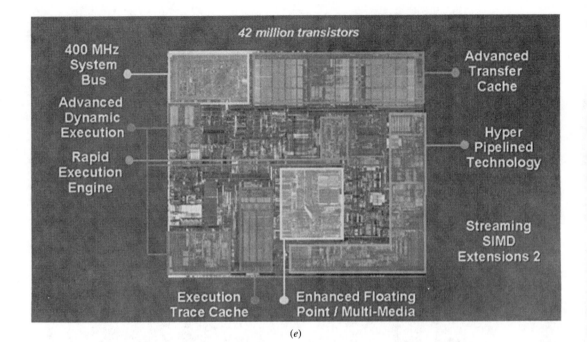

(e)

(f)

FIGURE 2.2 (continued)

of transistors on an IC chip increases even more rapidly—doubling every 18–24 months (Figure 2.2g). This rapid increase is made possible by continuing technological development, while the basic physics of the transistors remains relatively constant. (However, secondary effects that were less important in large transistors can dominate the behavior of smaller transistors.)

Because different manufacturing equipment is often needed to produce circuits with smaller features, the decrease of feature size is not continuous. Rather, the area of a transistor for each device "generation" decreases by a ratio that provides enough benefit to justify the cost of new equipment. Typically the area decreases by a factor of two, so the linear dimension decreases by a factor of 1.4 (i.e., the dimension "scales" by 0.7). Device features of 60 nm are produced today, and features of 20–30 nm have been demonstrated [2].

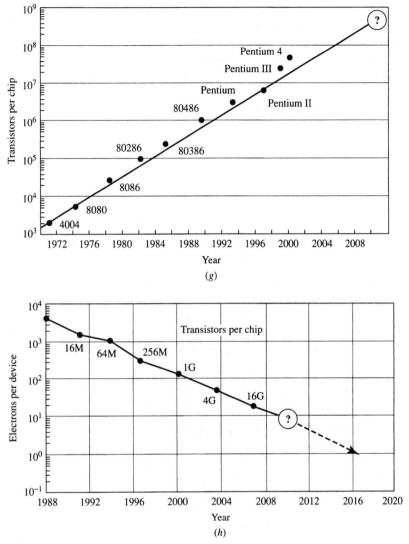

FIGURE 2.2 (continued)

Once Moore's law was well accepted, it became almost a self-fulfilling prophecy. By extrapolating from the past evolution of feature size shown in Figure 2.2*f*, a target value for minimum feature size is determined for each future year. Projections for feature size and other physical and electrical characteristics have been quantified in an "International Technology Roadmap for Semiconductors" (ITRS) [3], which is updated frequently. Semiconductor manufacturers then devote the resources needed to develop the technology required to produce features of the predetermined size. In fact, the changes can sometimes exceed the predicted rate of improvement. If the predetermined size is typical for the industry, then each company tries to develop the technology more rapidly than predicted to give it a competitive advantage in the market.

Device scaling cannot continue indefinitely, however. It will be limited by two factors. First, as the device features scale to smaller dimensions, the number of electrons within each transistor decreases, as shown in Figure 2.2*h*. As the number of electrons n

decreases, the statistical fluctuations in the number ($\sim \sqrt{n}$) becomes an increasing fraction of the total, limiting circuit performance and making circuit design more difficult. Several years ago, these statistical fluctuations began to impact the performance of analog circuits. Within a few years, similar fluctuations will influence digital circuit design. Considerably after statistical fluctuations become important, we will reach the time at which each transistor contains only one electron (perhaps about 2015). At that point, the entire concept of electronic devices must change. A number of different alternatives are being investigated in advanced research laboratories, but no favored approach has yet emerged.

Second, even in the shorter term with conventional devices, each generation of technology becomes more difficult and more expensive. The lithography needed to define increasingly small dimensions is often limiting. Eventually, the cost of technology development and manufacturing equipment is likely to limit the further evolution of IC technology. A modern IC manufacturing facility can cost several billion U.S. dollars today, and the cost is continuing to increase. The high cost of the manufacturing facilities limits the number of companies that can afford to manufacture ICs. In fact, many companies have the circuits they design manufactured by "foundries" that specialize in high-volume manufacturing of ICs for other companies.

However, even with these future limitations, the planar process will continue to dominate electronics for a number of years. The benefits it has provided for computers, communications, and consumer products have led to the continuing huge investment in research and development and manufacturing facilities. The planar process is the foundation for the production of silicon integrated circuits and continually evolves to allow production of increasingly complex circuits. Making the best use of its many degrees of freedom in understanding and designing devices requires a fairly thorough understanding of the basic elements of silicon technology. Much of the remainder of this chapter is directed toward providing such an understanding and a basis for the discussion of devices in the following chapters.

2.2 CRYSTAL GROWTH

Silicon used for the production of integrated circuits consists of large, high-quality, single crystals [4]. What is meant by "high quality" can be better understood from a brief consideration of the ultimate requirements placed on the silicon.

Typical IC applications require dopant concentrations ranging from roughly 10^{15} to 10^{20} atoms cm^{-3}. To control device properties, any unintentional or background concentrations of electrically active impurity atoms should be at least two orders of magnitude lower than the minimum intentional dopant concentration—that is, about 10^{13} cm^{-3} or lower. Because silicon contains roughly 5×10^{22} atoms cm^{-3}, the presence of only about one unintentional, electrically active, impurity atom per billion silicon atoms can be tolerated. This high purity is well beyond that required for the raw material in virtually any other industry; the routine production of material of this quality for the manufacture of integrated circuits demonstrates the extraordinary refinement of silicon technology.

High-purity silicon is obtained from two common materials: silicon dioxide (found in common sand) and elemental carbon. In a high-temperature ($\sim 2000°C$) electric-arc furnace, the carbon reduces the silicon dioxide to elemental silicon, which condenses as about 90% pure, metallurgical-grade silicon—still not pure enough for use in semiconductor devices. The metallurgical-grade silicon is purified by converting it into liquid trichlorosilane ($SiHCl_3$), which can be purified. Selective distillation (fractionation) separates the trichlorosilane from any other chloride complexes. The purified trichlorosilane is then re-duced by hydrogen to form high-purity, semiconductor-grade, solid silicon. At this point, the silicon is *polycrystalline*, composed of many small crystals with random orientations. This elemental polycrystalline silicon, also called *polysilicon*, is usually deposited on a high-purity rod of semiconductor-grade silicon to avoid contamination.

The silicon is then formed into a large (about 20–30 cm diameter), nearly perfect, single crystal because grain boundaries and other crystalline defects degrade device performance. Sophisticated techniques are needed to obtain single crystals of such high quality. These crystals can be formed-by either the *Czochralski* (*CZ*) technique or the *float-zone* (*FZ*) method.

Czochralski Silicon.   In the Czochralski technique, which is most widely used to form the starting material for integrated circuits, pieces of the polysilicon rod are first melted in a fused-silica crucible in an inert atmosphere (typically argon) and held at a temperature just above 1412°C, the melting temperature of silicon (Figure 2.3).

A high-quality seed crystal with the desired crystalline orientation is then lowered into the melt while being rotated, as indicated in Figure 2.3*a*. The crucible is simultane-ously turned in the opposite direction to induce mixing in the melt and to minimize temperature nonuniformities. A portion of the seed crystal is dissolved in the molten sil-icon to remove the strained outer portions and to expose fresh crystal surfaces. The seed is then slowly raised (or *pulled*) from the melt. As it is raised, it cools, and material from the melt solidifies on the seed, forming a larger crystal, as shown in Figure 2.3*b*. Under the carefully controlled conditions maintained during growth, the new silicon atoms con-tinue the crystal structure of the already solidified material. The desired crystal diameter is obtained by controlling the pull rate and temperature with automatic feedback mecha-nisms. In this manner cylindrical, single-crystal *ingots* of silicon can be fabricated. As crystal-growing technology has developed, the diameter of the cylindrical crystals has increased progressively from a few mm to the 20 or 30 cm common today.

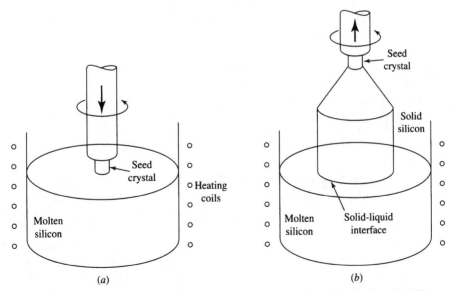

FIGURE 2.3 Formation of a single-crystal semiconductor ingot by the Czochralski process: (*a*) initiation of the crystal by a seed held at the melt surface, (*b*) withdrawal of the seed "pulls" a single crystal.

For many integrated-circuit processes, an initial dopant density of about 10^{15} cm^{-3} is desired in the silicon. This dopant concentration is obtained by incorporating a small, carefully controlled quantity of the desired dopant element, such as boron or phosphorus, into the melt. Typically, dopant impurities weighing about one-tenth milligram must be added to each kilogram of silicon. For accurate control, small quantities of heavily doped silicon, rather than the elemental dopant, are usually added to the undoped melt. The dopant concentration in the pulled crystal of silicon is always less than that in the melt because dopant is rejected from the growing crystal into the melt as the silicon solidifies. This *segregation* causes the dopant concentration in the melt to increase as the crystal grows, and the *seed* end of the crystal is less heavily doped than is the *tail* end. Slight dopant-concentration gradients can also exist along the crystal radius in Czochralski silicon.

Czochralski silicon contains a substantial quantity of oxygen, resulting from the slow dissolution of the fused-silica (silicon dioxide) crucible that holds the molten silicon. Oxygen does not contribute significantly to the net dopant density in the moderately doped wafers used for silicon integrated circuits. However, the typical inclusion of about 10^{18} cm^{-3} oxygen atoms (the solid solubility of oxygen in silicon at the solidification temperature) in silicon crystals produced by the Czochralski process can be used to control the movement of unintentional impurities (typically metals) in IC wafers. At the temperatures used in integrated-circuit processing, oxygen can precipitate, forming sites on which other impurities tend to accumulate. If the oxygen precipitates are located within an active device, they degrade its performance. However, if they are remote from active devices, precipitates can function as *gettering* sites for unwanted impurities, attracting them away from electrically active regions and improving device properties. Control of the location and size of the oxygen precipitates is important in determining the uniformity of device properties in high-density integrated circuits.

Convective flow of the molten silicon accelerates erosion of the crucible, increasing the oxygen content in the melt and in the solidifying crystal. The convective flow can also contribute to instabilities in the crystal growth process, degrading the structure of the growing crystal. The convective flow can be suppressed by superposing magnetic fields on the melt. This *magnetically confined* Czochralski growth technique offers the possibility of improved control over the crystal-growth process and increased purity of the resulting crystal.

A small fraction of the included oxygen, about 0.01% of the total, can act as donors after moderate-temperature heat treatments. This small concentration (about 10^{14} cm^{-3}) does not appreciably alter the resistivity of most integrated-circuit silicon. However, for the high-resistivity (20–100 Ω-cm) silicon used to produce power devices and for other specialized applications, oxygen can become a problem. For this reason, high-resistivity silicon is usually formed by the *float-zone process*.

Float-Zone Silicon. In the float-zone process, a rod of cast polycrystalline silicon is held in a vertical position and rotated while a melted zone (between 1 and 2 cm long) is slowly passed from the bottom of the rod to the top, as shown in Figure 2.4. The melted region is heated [usually by a *radio-frequency* (*RF*) induction heater] and moved through the rod starting from a *seed crystal* that initiates crystallization. The impurities tend to segregate in the molten portion so that the solidifying silicon is purified. In contrast to Czochralski-processed silicon, oxygen is not introduced into float-zone silicon because the silicon is not held in an oxygen-containing crucible. Although it is a more costly process and limited to smaller-diameter ingots, the silicon produced by the float-zone process typically contains about 1% as much oxygen as that produced by the Czochralski

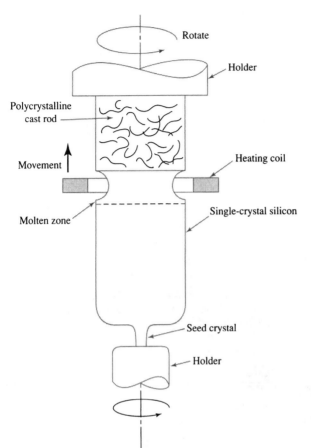

FIGURE 2.4 The float-zone process. A molten zone passes through a polycrystalline-silicon rod, and a single crystal grows from a seed at the bottom end.

method, and other impurity concentrations are also reduced. Multiple passes of the molten zone can be used to produce silicon with resistivities above 10^4 Ω-cm.

Wafer Production.   Once the single-crystal ingot is grown and ground to a precise diameter, it is sliced with a diamond saw into thin, circular *wafers*. The wafers are chemically etched to remove sawing damage and then polished with successively finer polishing grits and chemical etchants until a defect-free, mirror-like surface is obtained. The wafers are then ready for device fabrication.

Before slicing, index marks are placed on the wafer to facilitate the orientation of processed circuits along specific crystal directions. In particular, edges of an area on the wafer called a *die* or a *chip* are typically aligned in directions along which the wafer readily breaks so that the *dice* can be separated from one another after planar processing is completed. This separation is often accomplished by *scribing* between them with a sharp stylus and *breaking* them apart, so the orientation of the easy cleavage planes is important. On smaller wafers the crystal directions are indicated by grinding a *flat* on the wafer, usually perpendicular to an easy cleavage direction. In most cases, the *primary flat* is formed along a $\langle 110 \rangle$ direction. A smaller *secondary flat* is sometimes added to identify the orientation and conductivity type of the wafer, as shown in Table 2.1. On larger wafers, using wafer flats would appreciably decrease the wafer area and the number of chips on the wafer. Therefore, the wafer flats are omitted on larger wafers, and the crystal axes are indicated by a small notch placed at the edge of the wafer.

TABLE 2.1 Location of Secondary Wafer Flat

Crystal Orientation	Conductivity Type	Secondary Flat (Relative to Primary Flat)
(100)	n	180°
(100)	p	90°
(111)	n	45°
(111)	p	No secondary flat

After slicing, a unique wafer number is often marked on the wafer surface by vaporizing small spots of silicon with a laser. The marks identify the manufacturer, ingot, dopant species, and crystal orientation. Other digits are unique to a wafer, allowing each wafer to be identified at any point in the wafer-fabrication process. The laser marks can be read both manually by equipment operators and also by automated process equipment. The laser marks are added before the wafer is etched and polished so that strain from the marking process can be removed by chemical etching and stray spattered material does not contaminate the fine surface finish after polishing.

2.3 THERMAL OXIDATION

An oxide layer about 2 nm thick quickly forms on the surface of a bare silicon wafer in room-temperature air. The thicker (typically 8 nm to 1 μm) silicon dioxide layers used to protect the silicon surface during dopant incorporation can be formed either by *thermal oxidation* or by *deposition*. When silicon dioxide is formed by deposition, both silicon and oxygen are conveyed to the wafer surface and reacted there (Sec. 2.6). In thermal oxidation, however, there is a direct reaction between atoms near the surface of the wafer and oxygen supplied in a high-temperature furnace. Thermally grown oxides are generally of a higher quality than deposited oxides. Although their structure is amorphous, they typically have an exact stoichiometric ratio (SiO_2), and they are strongly bonded to the silicon surface. The interface between silicon and thermally grown SiO_2 has stable and controllable electrical properties. As we will see in Chapter 8, the quality of this excellent semiconductor-insulator interface is fundamental to the successful production of metal-oxide-semiconductor (MOS) transistors.

To form a thermal oxide, the wafer is placed inside a quartz tube that is set within the cylindrical opening of a resistance-heated furnace. This furnace can be oriented horizontally, as shown in Figure 2.5, or vertically. The wafer surface is usually perpendicular to the main gas flow. Temperatures in the range of 850 to 1100°C are typical, the reaction proceeding more rapidly at higher temperatures. Silicon itself does not melt until the temperature reaches 1412°C, but oxidation temperatures are kept considerably lower to reduce the generation of crystalline defects and the movement of previously introduced dopant atoms. In addition, the quartz furnace tube and other fixtures start to soften and degrade above 1150°C.

The oxidizing ambient can be dry oxygen, or it can contain water vapor, which is generally produced by reaction of oxygen and hydrogen in the high-temperature furnace. Use of such *pyrogenic steam* requires careful safety procedures to handle explosive hydrogen gas, and a slight excess of oxygen is generally introduced to avoid having unreacted

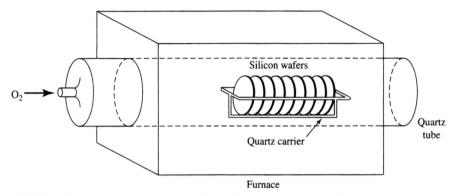

FIGURE 2.5 An insulating layer of silicon dioxide is grown on silicon wafers by exposing them to oxidizing gases in a high-temperature furnace.

hydrogen in the furnace. A steam environment can also be formed by passing high-purity, dry oxygen or nitrogen through water heated almost to its boiling point. The overall oxidation reactions are

$$Si(s) + O_2(g) \rightarrow SiO_2(s) \tag{2.3.1a}$$

and

$$Si(s) + 2H_2O(g) \rightarrow SiO_2(s) + 2H_2(g) \tag{2.3.1b}$$

Oxidation proceeds much more rapidly in a steam ambient, which is consequently used for the formation of thicker protective layers of silicon dioxide. Growth of a thick oxide in the slower, dry-oxygen environment can lead to undesirable movement (*redistribution*) of impurities introduced into the wafer during previous processing.

Oxidation takes place at the Si–SiO$_2$ interface so that oxidizing species must diffuse through any previously formed oxide and then react with silicon at this interface (Figure 2.6). At lower temperatures and for thinner oxides, the surface reaction rate at the Si–SiO$_2$ interface limits the growth rate, and the thickness of the oxide layer increases linearly with increasing oxidation time.

At higher temperatures and for thicker oxides, the oxidation process is limited by diffusion of the oxidizing species through the previously formed oxide. In this case the grown oxide thickness is approximately proportional to the square root of the oxidation time. This square-root dependence is characteristic of diffusion processes and sets a practical upper limit on the thickness that can be conveniently obtained.

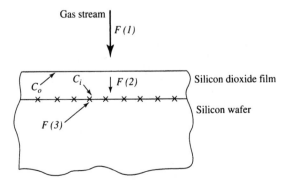

FIGURE 2.6 Three fluxes that characterize the oxidation rate: $F(1)$ the flow from the gas stream to the surface, $F(2)$ the diffusion of oxidizing species through the already formed oxide, and $F(3)$ the reaction at the Si–SiO$_2$ interface. The concentration of the oxidizing species varies in the film from C_o near the gas interface to C_i near the silicon interface.

Oxidation Kinetics

For most of the oxidation process (after formation of a thin layer which obeys different oxidation kinetics), the relation between the grown oxide thickness x_{ox}, oxidation time t, and temperature T can be found by equating the rates at which oxygen atoms (1) transfer from the gas phase to the growing oxide, (2) move through the already formed oxide, and (3) react according to Equations 2.3.1 at the Si–SiO$_2$ interface (Figure 2.6). These considerations form the basis of the classic Deal-Grove model [5] which is outlined below.

First, consider transfer of the oxidizing species (either oxygen or water vapor) from the gas phase to the outer layer of the already formed oxide. This transfer rate is proportional to the difference between the actual concentration C_o of the oxidizing species in the solid at its surface and C^*, the concentration that would be in equilibrium with the gas-phase oxidizing species:

$$F(1) = h(C^* - C_o) \qquad (2.3.2)$$

where h is the gas-phase, mass-transfer coefficient, and the concentrations C^* and C_o are related to corresponding gas-phase partial pressures p by the ideal gas law $C = p/kT$.

Transport of the oxidizing species across the growing oxide to the Si–SiO$_2$ interface occurs by diffusion, a process analogous to the hole and electron diffusion discussed in Sec. 1.2 (Equation 1.2.15). The flux of the diffusing species can be written as the product of the concentration gradient across the oxide $(C_o - C_i)/x_{ox}$ (Figure 2.6) and the diffusivity D, which describes the ease of diffusion of the oxidizing species through the already formed oxide. C_i is the concentration of the oxidizing species in the oxide near the Si–SiO$_2$ interface. The diffusing flux is therefore

$$F(2) = D\frac{(C_o - C_i)}{x_{ox}} \qquad (2.3.3)$$

Reaction of the oxidizing species at the Si–SiO$_2$ interface is characterized by a rate constant k_s so that the reaction rate $F(3)$ of the oxidant is

$$F(3) = k_s C_i \qquad (2.3.4)$$

In steady state, $F(1) = F(2) = F(3) = F$. The oxidation rate can be found from the flux and expressed in terms of N_{ox}, the density of oxidant molecules per unit volume of oxide. Eliminating C_o and C_i from Equations 2.3.2–2.3.4, we find the oxide growth rate R to be

$$R = \frac{dx_{ox}}{dt} = \frac{F}{N_{ox}} = \frac{k_s C^*/N_{ox}}{(1 + k_s/h + k_s x_{ox}/D)} \qquad (2.3.5)$$

Equation 2.3.5 can be solved to find the oxide thickness grown in a time t (Problem 2.6) [5].

$$x_{ox} = \frac{A}{2}\left[\sqrt{1 + \frac{(t + \tau)}{A^2/4B}} - 1\right] \qquad (2.3.6)$$

where

$$A = 2D\left[\frac{1}{k_s} + \frac{1}{h}\right] \qquad (2.3.7)$$

and

$$B = \frac{2DC^*}{N_{ox}} \qquad (2.3.8)$$

The parameter τ depends mainly on the thickness of the oxide initially present on the surface. For short oxidation times the surface reaction rate $F(3)$ limits oxide growth, and Equation 2.3.6 can be approximated by a linear relationship between x_{ox} and t:

$$x_{ox} = \frac{B}{A}(t + \tau) \qquad (2.3.9)$$

The proportionality factor B/A in Equation 2.3.9, called the *linear rate coefficient*, is related to breaking bonds at the Si–SiO$_2$ interface [$F(3)$], and therefore depends on crystal orientation. The most commonly used crystals for ICs are (100)- or (111)-oriented. The linear rate coefficient is larger for the (111) orientation, which has fewer bonds between adjacent planes than does (100)-oriented silicon.

For long oxidation times a square-root relationship is obtained from Equation 2.3.6:

$$x_{ox} = \sqrt{B(t + \tau)} \approx \sqrt{Bt} \qquad (2.3.10)$$

The coefficient B in Equation 2.3.10, called the *parabolic rate coefficient*, depends on diffusion across the already formed oxide [flux $F(2)$], and is independent of the orientation of the silicon crystal. Experimental values for the linear rate coefficient B/A and the parabolic rate coefficient B are shown in Figures 2.7a and 2.7b.

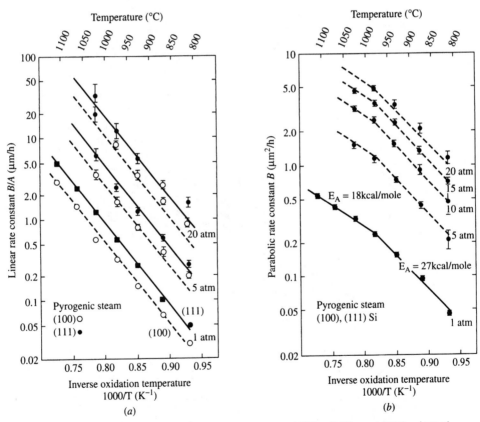

FIGURE 2.7 (a) Linear rate coefficient B/A versus $1000/T$ for (100)- and (111)-oriented silicon oxidized in pyrogenic steam at 1, 5, and 20 atm. (b) Parabolic rate coefficient B versus $1000/T$ for (100)- and (111)-oriented silicon oxidized in pyrogenic steam at 1, 5, 10, 15, and 20 atm [6].

In practice, values of A, B, and τ are determined experimentally from measurements of oxide thickness versus time at various temperatures. Oxide thicknesses as a function of oxidation time are shown in Figures 2.8a (for dry oxidation) and 2.8b (for steam oxidation).

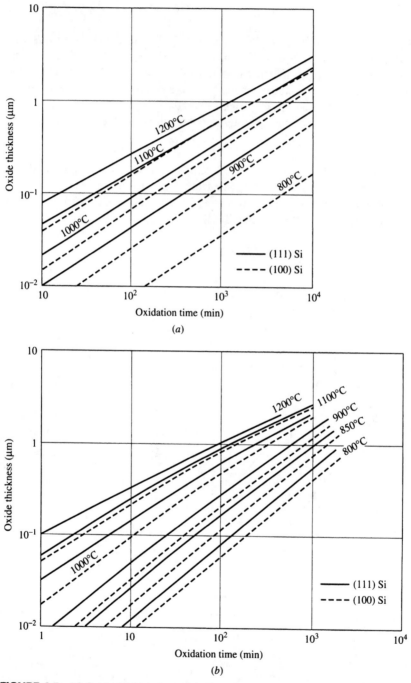

FIGURE 2.8 (a) Oxide thickness as a function of oxidation time in dry oxygen for several commonly used oxidation temperatures for (111)- and (100)-oriented silicon. (b) Oxide thickness x_{ox} as a function of time in a steam ambient for (111)- and (100)-oriented silicon (typically obtained by reacting hydrogen and oxygen near the entrance to the oxidation furnace). Adapted from [7].

The data in both figures apply to oxides grown on (111)- and (100)-oriented Si. The oxide thicknesses is somewhat less for (100)-oriented material, especially for thin oxides and at lower oxidation temperatures at which growth is limited for a longer time by the surface oxidation rate [F(3)]. As an example of the use of Figures 2.8a and b, we can compare the time it takes to grow 300 nm (0.3 μm) of oxide in dry oxygen at 1100°C on Si(100) (4.4 h from Figure 2.8a) to the growth time in steam (17 min from Figure 2.8b). Oxide thicknesses of a few hundred nanometers are often used, with 1 to 2 μm being the upper practical limit when conventional oxidation techniques are employed.

In IC fabrication we must frequently determine the thickness added to an already formed oxide layer during additional oxidation. To do this, we can use the parameter τ in Equation 2.3.6 as the time needed to grow an oxide of the thickness already present *under the conditions of the additional oxidation*. The final thickness is then determined by adding the oxidation time t and calculating the oxide thickness corresponding to the total time $(t + \tau)$.

EXAMPLE *Calculation of Oxide Thickness*

Find the final oxide thickness on Si(111) after an additional oxidation in dry oxygen for 2 hours at 1000°C of a region that is covered initially by 100 nm of SiO_2.

Solution From Figure 2.8a,

$$x_i = 100 \text{ nm} \qquad \tau = 120 \text{ min}$$
$$(t + \tau) = 240 \text{ min} \qquad x_{ox} = 153 \text{ nm}$$

Note that the oxide thickness increases less than linearly with time. The initial 100 nm of SiO_2 would have grown in 2.0 h at 1000°C, whereas another 2 h of oxidation only forms an additional 53 nm of oxide. ∎

Although Equation 2.3.5 describes the oxidation rate over a wide range of oxide thicknesses, the initial stage of oxidation is more rapid than described by Equation 2.3.5. This accelerated growth of thin oxides can be incorporated by adding a term to Equation 2.3.5 of the form [8]

$$\frac{dx_{ox}}{dt} = K \exp\left(-\frac{x_{ox}}{L}\right) \tag{2.3.11}$$

where L is a characteristic length of the order of 7 nm and K depends on temperature [8]. The rapid initial oxidation is possibly related to an altered layer near the silicon surface containing additional sites for oxidation. For thicker oxides, the accelerated initial oxide growth can be approximately incorporated into the experimentally determined parameter τ shown in Equation 2.3.6.

During thermal oxidation, silicon is consumed from the wafer and incorporated in the growing oxide. Because SiO_2 contains 2.2×10^{22} molecules cm^{-3} (and an equal number of silicon atoms), while pure silicon contains 5.0×10^{22} atoms cm^{-3}, the thickness of silicon consumed is 0.44 times the thickness of SiO_2 formed. This relation holds for all orientations of silicon and also for polycrystalline silicon because it depends only on volume densities.

High-Pressure Oxidation. Growth of thick oxides is time consuming because the thickness of thick oxides increases only as the square root of time (Equation 2.3.10). Excessive movement of dopant atoms during extended oxidation often makes more rapid oxide formation desirable. This rapid formation can be accomplished by using high pressures of the oxidizing species.

Although atmospheric-pressure oxidation processes are described by Equation 2.3.6 with well-known values of A and B, the same equation is also valid at other pressures (with different values of A and B). The oxidation rate for thick oxides (Equation 2.3.10) is determined by the parameter B. As seen from Equation 2.3.8, B increases when C^*, the concentration of oxidizing species in equilibrium with the gas phase increases. C^* can be increased by supplying the oxidant at a high pressure. Typical elevated pressures are 10 to 20 atmospheres, causing a 10- to 20-fold increase in the parameter B. Although the diffusivity D of the oxidant in oxide depends somewhat on pressure, the major effect on oxidation rate is through the increased concentration C^*.

High-pressure oxidation allows formation of a desired oxide thickness at the same temperature in a shorter time than at one atmosphere; alternatively, a desired thickness can be formed in the same time at a lower temperature. Using a lower oxidation temperature can reduce the number of crystal defects introduced during thermal oxidation.

Concentration-Enhanced Oxidation.[†]

Heavily doped n-type silicon oxidizes more rapidly than does lightly doped silicon (Figure 2.9). The different oxidation rates of heavily and lightly doped regions of an IC can be used for selective definition of desired areas; conversely, the different oxide thicknesses can make uniform etching of oxides grown on different parts of the circuit difficult.

High concentrations of some dopants cause isolated *point defects* in the silicon lattice consisting of missing silicon atoms (*vacancies*) or extra silicon atoms (silicon *interstitials*). These point defects affect the surface reaction rate (and therefore the linear rate coefficient B/A) if their concentration is greater than the intrinsic carrier concentration n_i *at the oxidation temperature*. As shown in Figure 2.10, the value of n_i increases rapidly with increasing temperature and is 10^{19} cm^{-3} at 1000°C.

The parabolic rate coefficient B depends on properties of the SiO_2, so it is not directly affected by point defects in the silicon. However, B does depend on the diffusivity of the oxidizing species in the oxide. Because some dopant from the silicon can enter the

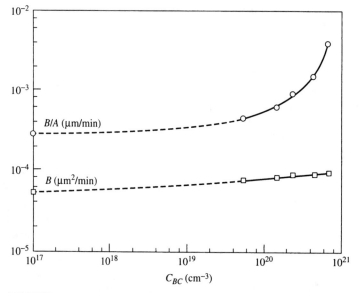

FIGURE 2.9   Linear (B/A) and parabolic (B) rate coefficients as functions of the initial phosphorus concentration in the substrate for oxidation at 900°C [9].

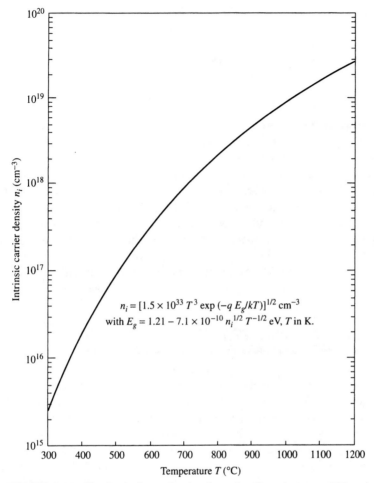

FIGURE 2.10 The intrinsic carrier density n_i in silicon between 300 and 1200°C [10].

oxide during oxidation and weaken its structure, the diffusivity (and hence the parabolic rate coefficient B) can increase. This increase, however, is small compared to the increase in the linear rate coefficient B/A. Thus, for thicker oxides, oxidation-rate enhancement resulting from high concentrations of dopant in the silicon is small.

Chlorine Oxidation.[†]   The oxidation rate also increases (typically by 10–20%) when a chlorine-containing species is added to the oxidant. The chlorine, usually obtained from HCl, Cl_2, or organic compounds, is incorporated near the Si–SiO_2 interface where it can improve the electrical properties of devices, especially the MOS devices discussed in Chapter 8. The addition of chlorine to the oxidant to improve the properties of the Si–SiO_2 system must be done very carefully. Only a small improvement occurs when the concentration is too low, but it is easy to provide an excess concentration that can etch the silicon surface or form gas bubbles at the Si–SiO_2 interface and rupture the oxide.

Nitrided oxides.   Although oxides of a moderate thickness effectively block diffusion of dopant species, dopant atoms—especially boron—can penetrate very thin oxides (<5 nm-thick).

As the gate oxides of MOS transistors become thinner, this unwanted dopant penetration can change the electrical behavior of the transistor. The dopant diffusion can be reduced by adding nitrogen to a portion of the oxide by a thermal treatment in a nitrogen-containing gas such as ammonia (NH_3) after the oxide is grown. Nitrogen-rich layers form at the top and perhaps at the bottom of the oxide layer, increasing its resistance to dopant penetration.

High-permittivity oxides. As the requirement for better capacitive coupling of a gate electrode of an MOS transistor or memory cell to the channel or storage region becomes more stringent, silicon dioxide and silicon nitride cannot provide the required capacitance, and insulators with a higher permittivity are needed. Materials such as tantalum oxide (Ta_2O_5) can be used, and oxides of zirconium and hafnium may be useful. Materials such as barium-strontium titanate (BST), with its even higher permittivity have also been suggested. To take advantage of the higher permittivity of these advanced oxides, they must be in intimate contact with the conducting electrode. Because of the stability of silicon dioxide, a thin layer of SiO_2 often forms between a silicon electrode and the high-permittivity oxide. The associated parasitic series capacitance decreases the total capacitance below that expected from the advanced oxide alone.

2.4 LITHOGRAPHY AND PATTERN TRANSFER

Photolithography. Once the protective layer of SiO_2 has been formed on the silicon wafer, it must be selectively removed from those areas in which dopant atoms are to be introduced. Selective removal is usually accomplished by using a light-sensitive polymer material called a *resist*. The oxidized wafer is first coated with the liquid resist by placing a few drops on a rapidly spinning wafer. After drying the resist, a glass plate with transparent and opaque features (called a *mask* or *photomask*) is placed over the wafer as shown in Figure 2.11a, and aligned using a microscope. The resist is then exposed to ultraviolet light that changes its structure. For a *positive resist*, molecular bonds are broken where the resist is illuminated, while the molecules of a *negative resist* are cross linked (polymerized) in areas that are exposed to light. The weakly bonded or unpolymerized areas of the resist are then selectively dissolved using a solvent, so that the remaining, acid-resistant, hardened coating reproduces the mask pattern on the SiO_2 (Figure 2.11b). Similarly, resist patterns can be formed on top of other layers used in the IC process.

In the most straightforward implementation of planar processing, the patterns on all the dice on a wafer are exposed simultaneously. However, as device dimensions become smaller, not only does the minimum feature size that must be resolved decrease, but the registration of one patterned layer to another must become more exact. During heat cycles, the wafer can be slightly distorted by thermal stress or by stress from incorporated dopant atoms or added layers of material. Thus, a subsequent photomask may not exactly match a pattern previously formed on the wafer. This distortion can limit the accuracy of registration from one masking level to another. One tool used to improve the registration accuracy is the *optical wafer stepper* (Figure 2.11c), which exposes one die on a wafer at a time. After one die is exposed, the wafer is moved or *stepped* to the next die, which is then exposed. Although this technique is mechanically more complex and slower than full-wafer exposure, it allows improved registration. In addition, by demagnifying the pattern (about five or ten times) from a large mask or *reticle* containing the pattern for a single die, smaller device features can be defined.

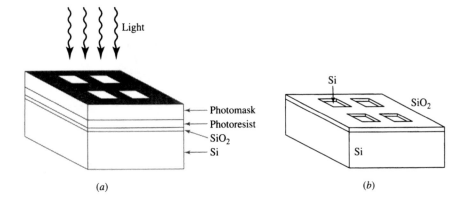

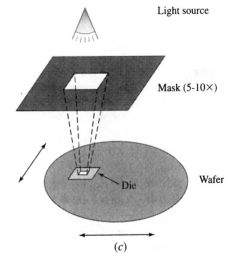

FIGURE 2.11 (*a*) The areas from which the oxide is to be etched are defined by exposing a light-sensitive resist through a photographic negative (*mask*). (*b*) The hardened resist protects the oxide in the masked areas from chemical removal. (*c*) In a *stepper*-type lithography system, exposure light passes through features on a mask. The image of each feature is reduced and focused on one die on the wafer, and all features on the die are exposed simultaneously. The wafer is then moved (*stepped*) to the next die position where the exposure process is repeated.

Advanced Lithography. To place more transistors on an integrated circuit, the minimum size of the features to be defined by lithography must continuously be reduced. Diffraction limits the size of features defined by straightforward exposure techniques to approximately the wavelength of the exposing illumination. Because of diffraction, the electric field and intensity of the illumination reaching the wafer surface vary gradually over a distance related to the wavelength of the exposing illumination (Figure 2.12*a*), making definition of sharp edges difficult. Consequently, finer features can be defined if shorter-wavelength light is used. To obtain intense illumination, light from mercury arc lamps is often used. Three strong emission wavelengths in the UV wavelength range occur at 436 nm (G-line), 405 nm (H-line), and 365 nm (I-line). Finer features can be defined by using shorter-wavelength I-line illumination, instead of G-line illumination. Even shorter wavelengths can be obtained by using light emitted by laser sources, such as a KrF or ArF laser; commonly used wavelengths with these laser sources are 248 and 193 nm, and using an F_2 source at 157 nm is being explored. Reducing the wavelength further is more difficult because most materials used for lenses and masks become opaque at shorter wavelengths. Consequently, reflective optics must be used, complicating the design and use of the tools. However, considerable effort is being devoted to developing very short wavelength, *extreme ultraviolet (EUV)* exposure systems operating at a wavelength of 13 nm (in the soft x-ray regime) while still simultaneously exposing all features within a sizable exposure

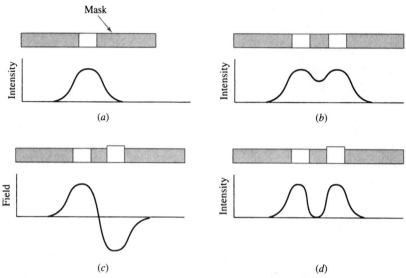

FIGURE 2.12 (*a*) The illumination intensity varies gradually near the edge of a fine feature because of diffraction. (*b*) The intensity between two closely spaced features does not reach zero. Shifting the phase of the electric field by 180° by locally changing the path length through the mask (*c*) allows the intensity to become zero between the features (*d*).

field (usually the size of one or more chips). In addition to developing the complex optics, obtaining a high-intensity source at a short wavelength is a major challenge.

In addition to using shorter wavelengths, smaller features can be defined by using more complex processing of the resist system on the wafer. Multiple reflections can be reduced by *anti-reflection coatings* below the resist layer. Sharper edges of a defined resist feature can be obtained by using *phase-shift* masks. As shown in Figure 2.12*a*, the illumination varies gradually near the edges of fine features and does not reach zero if the spacing between two features is small (Figure 2.12*b*). However, the illumination intensity is proportional to the square of the electric field associated with the light used. If the sign of the electric field can be changed from one feature to the next, then the field must be zero at some point between the features (Figure 2.12*c*), and its square (the illumination intensity) must also be zero (Figure 2.12*d*). The sign of the electric field can be changed by changing (*shifting*) its phase by 180° by adding a coating to parts of the mask so that the light passing through the clear regions of the mask must travel an extra half wavelength in some features, but not in adjacent ones. The electric field associated with adjacent features then differs in sign, as desired. Nonlinear contrast-enhancement techniques can also be used so that a small change in the light intensity near the edge of a feature produces a large chemical change in the resist layer.

The uniformity of a fine line can also be limited by the nonplanar surface of the partially fabricated IC chip. A line crossing a high step formed by previous processing may be poorly defined because of the limited depth over which the exposing beam can be focused. Better line definition can be obtained with a *multilayer resist* structure consisting of two or three layers. The thick first layer fills in the recessed portions of the chip, producing a more uniform level surface. Fine features can then be exposed in a thin resist layer on top, and the features can be transferred into the thicker underlying layer by directional etching.

By a combination of shorter wavelength illumination and more complex resist systems, "optical" lithography can form features as small as 130 nm. The major advantage

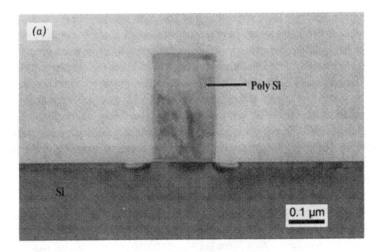

FIGURE 2.13 (a) Cross-sectional transmission electron micrograph of a polysilicon gate approximately 180 nm across, the gate oxide under the polysilicon, and the surrounding shallow junctions in the silicon substrate. (*Courtesy of Accurel Systems International Corp.*) (b) Anisotropically etched lines 500 nm wide spaced 1.5 μm apart. Resist covers the double-layer structure consisting of 180 nm TaSi$_2$ over 260 nm of polycrystalline silicon. Note the uniformity of the vertical surface through the various layers. (*Courtesy of G. Dorda, Siemens Corporation*).

of optical lithography remains its ability to simultaneously expose all the features within a sizable area. Figure 2.13a shows the cross section of a polysilicon gate approximately 180 nm across formed by optical lithography and anisotropic etching.

Alternatives to optical lithography are also being considered to form features with even smaller sizes. Electron beams, x-rays, or ion beams can be used to expose the resist.

In electron-beam lithography a focused stream of electrons delivers energy to the resist and exposes it. Rather than exposing all features of a complex pattern simultaneously, as is done in optical lithography, the electron beam is deflected to expose the elements of the desired pattern sequentially. The information necessary to guide the electron beam is stored in a computer and no mask is needed. Electron-beam lithography is most frequently used to fabricate fine-geometry masks, which are then used with more-conventional photolithographic exposure techniques. The electron beam can be finely focused to a size much smaller than the minimum feature size and moved across the die, with each feature composed by multiple, slightly offset scans of the electron beam. Alternatively, the beam can be formed into a rectangular shape and the pattern built up by repeated block-like exposures. In either case the sequential nature of electron-beam exposure limits the speed with which a wafer can be processed. The increased time needed to expose a wafer is, however, at least partially compensated by the higher circuit density that can be obtained.

With *x-ray lithography*, a beam of x rays passes through a mask to expose a resist layer. X-ray lithography, like optical lithography, exposes many features simultaneously, but the short x-ray wavelengths allow finer features to be formed. X-ray lithography is less well developed than either optical lithography or electron-beam lithography and is limited by the demanding requirements of x-ray sources and masks. Most common x-ray sources are point sources which cause the beam to diverge as it travels to the mask and wafer, limiting the possible mask-to-wafer spacing. Synchrotrons, which produce partially collimated beams of x-rays, are possible, but expensive, alternative high-intensity x-ray sources.

X-ray masks are also difficult to fabricate. Because the beam must readily penetrate the "transparent" sections of a mask, a thin membrane is generally used, making handling difficult. A more fundamental limitation to the use of x rays is that the area exposed at one time may be limited by wafer distortion during processing. Any large-area exposure technique suffers from this limitation. As with optical exposure, single-die exposure may allow adequate registration of fine features between layers, even when a small amount of wafer distortion is present. Finally, possible damage to the active device regions in the silicon by high-energy x-rays must be considered.

Pattern Transfer. After the pattern is formed in the resist, the unprotected regions of the SiO_2 or other material are etched to transfer the pattern to the chip. If SiO_2 is being defined, the exposed layer can be dissolved in a hydrofluoric-acid (HF) containing liquid etchant to expose the bare silicon surface. The resist is then removed from the remaining areas where it protected the oxide from etching. At this point, portions of the wafer are protected by SiO_2, and bare silicon is exposed at windows in the oxide (Figure 2.11) through which the dopant impurities are subsequently introduced). There are a large number of different liquid chemical solutions that etch materials selectively with very little attack of underlying materials. This high *selectivity* is an advantage of liquid-based or *wet etching*.

Dry Etching. As the dimensions to be defined on an IC die decrease, several limitations of the wet-etching process become apparent. One major problem is that wet etching is usually *isotropic*, proceeding laterally under the masking layer as well as vertically toward the silicon surface (Figure 2.14a). Thus, the etched features are generally larger than the dimensions on the mask. Dry (*plasma* or *reactive-ion*) etching techniques can be *anisotropic* and minimize this difficulty (Figure 2.14b). (Reactive-ion etching is a special case of plasma etching, in which the conditions are optimized to produce a highly anisotropic etch profile.)

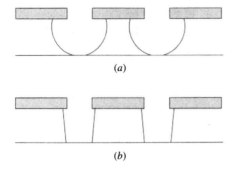

(a)

(b)

FIGURE 2.14 (*a*) *Isotropic* wet etching or dry etching that is dominated by chemical reactions cause significant undercut of the masking layer. (*b*) *Anisotropic*, ion-assisted, dry etching creates a near-vertical profile, retaining the dimensions of the masking layer.

A plasma is a nearly neutral mixture composed of excited neutral species (*radicals*,) ions, and electrons created in a high-frequency electric field. Because less energy is required to create excited neutral species than to ionize molecules, radicals are more numerous than ions. Although the plasma itself is essentially neutral, it is often physically separated from the wafer surface, and a substantial electric field between the plasma region and the wafer can accelerate ions toward the surface.

For dry etching an etching gas is chosen that reacts with the material to be removed. One dominant consideration in chosing the etching gas is that the reaction products must be volatile. Silicon and its compounds are effectively etched by gases containing fluorine, while aluminum is removed with chlorine-containing species. The etch products of copper are generally not volatile at room temperature, making copper difficult to etch; the alternative *Damascene* process used to define copper patterns will be discussed in Sec. 2.7. Organic resists are dry etched in oxygen plasmas to produce water vapor and carbon dioxide. Non-reactive diluent gases are often added to aid the gas flow.

In dry etching, the masked wafer is exposed to the plasma. The excited neutral species interact chemically with exposed regions of the material to be etched, while the ions in the plasma bombard the surface and physically remove exposed material. However, the vertical etch rate is more than just the sum of individual chemical and physical material removal rates. The ions bombarding horizontal surfaces can assist reaction products to desorb so that fresh reactant can reach the surface more readily, increasing the etch rate. Alternatively, excited species striking horizontal surfaces can create lattice damage, weakening bonds and allowing chemical removal to proceed more easily. When the ratio of ions to excited neutral species reaching the surface increases, as in a "high-density plasma," physical sputtering of material from the surface also becomes important.

By proper choice of the reactant gases, electric field, and reactor geometry, the etching reaction can be made anisotropic so that nearly vertical sidewalls are formed in the etched material at the edges of the mask, as shown in Figure 2.14*b*. During directional dry etching, reactant species, partially decomposed species, and reaction products can deposit on the sidewalls of features being etched. Few ions strike the nearly vertical surfaces, so these deposits remain on the surface to inhibit lateral etching and allow highly anisotropic etching. Directional electric fields and low pressures (less ion scattering) tend to enhance vertical ion bombardment and anisotropic etching. By varying the importance of chemical and ion-assisted removal, the amount of anisotropy be varied.

Although highly anisotropic etching is possible, the chemical selectivity of dry-etching processes is not as great as that of wet etching and degrades as the process becomes more anisotropic (i.e., ion bombardment becomes more important). The masking material can be significantly attacked during the etching process. The limited selectivity also causes attack of the material under the layer being etched; therefore, dry etching processes must be terminated soon after the desired layer has been removed. This can be done effectively by monitoring the characteristic light emitted by the reaction. Alternatively, we can observe changes in the reflection of a laser beam incident on the wafer resulting from changes in the optical interference as the thickness of the material being etched decreases. When the material is completely removed, the characteristics of the reflected signal change, allowing the etch process to be terminated after the desired amount of overetching.

Etching multiple layers of materials is especially challenging. The differing anisotropy of the etching of different materials can create notches and undercut, complicating subsequent processing. Figure 2.13*b* shows a uniform profile etched through layers of different materials.

The effect of the energetic environment in which dry etching occurs must also be considered; the excited ions and high-energy photons in the reactor bombard the silicon surface and can damage the devices being fabricated. Plasma-enhanced processes will be discussed further in Sec. 2.6, where we consider plasma-enhanced chemical vapor deposition.

2.5 DOPANT ADDITION AND DIFFUSION

The distribution of the dopant atoms added to the silicon (*dopant profile*) is generally determined in two steps. First, the dopant atoms are placed on or near the surface of the wafer by *ion implantation*, *gaseous deposition*, or possibly by coating the wafer with a layer containing the desired dopant impurity. This step is followed by a *drive-in diffusion* that rearranges the dopant atoms within the wafer. The shape of the resulting dopant distribution is determined primarily by the manner in which the dopant is placed near the surface, while the diffusion depth depends chiefly on the temperature and time of the drive-in diffusion. Because the characteristics of semiconductor devices depend strongly on the dopant profiles, we discuss the processes used for dopant addition in some detail.

Ion Implantation

Ion implantation is a highly controlled method of introducing dopant atoms into semiconductors. The desired dopant atoms are first ionized and then accelerated by an electric field to a high energy (typically from 25 to 200 keV). A beam of these high-energy ions strikes the semiconductor surface (Figure 2.15) and penetrates into exposed regions of the wafer. The masking material can be an oxide or other layer used in the IC

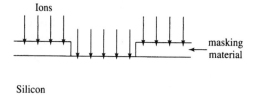

FIGURE 2.15 In ion implantation, a beam of high-energy ions strikes selected regions of the semiconductor surface, penetrating into these exposed regions.

structure itself. Because ion implantation only heats the wafer moderately, photoresist can also be used to block the implant from selected regions of the chip. Higher density materials block the implant more effectively, allowing a thinner layer to be used.

The ions typically penetrate less than 1 μm below the surface, and considerable damage is done to the crystal during implantation. Consequently, an annealing step is necessary to restore the lattice quality and to ensure that the implanted dopant atoms are located on substitutional sites, where they act as donors or acceptors. After the ions are implanted they may be redistributed by a subsequent diffusion if desired.

Ion implantation makes possible precise control of the area density (atoms cm^{-2}) of dopants entering the wafer. Because the dopants are transported as electrically charged ions, they can be counted during the implantation by a relatively simple charge-sensing apparatus placed adjacent to the beam path. The beam can then be turned off when the desired number of ions has been introduced.

Dopant *doses* (number of dopant atoms cm^{-2}) ranging from mid–10^{11} to more than 10^{16} cm^{-2} can be introduced into a wafer with fine control by ion implantation. The lowest doses permit precise tailoring of device properties, while very high doses can provide the dopant concentrations needed to make low-resistance ohmic contact to the silicon. The upper limit is comparable to the quantity of dopant that can be introduced by a gaseous deposition. Typical ion-beam currents in an implanter are of the order of 1 mA, which corresponds to a flux of 6.25×10^{15} singly charged ions s^{-1}. For high doses, implanters capable of much higher currents are used to reduce the time needed for an implantation.

Besides controlling the total dose very exactly, ion implantation also allows introduction of extremely pure dopant species. This purity is achieved by using a mass spectrometer near the dopant source to sort ionic species and allow only the desired dopant species to reach the wafer.

Because the implanted ions penetrate beneath the exposed surface, ion implantation can be used to introduce dopant atoms into silicon through an overlying layer of another material such as SiO_2. This capability is useful for adding the implanted atoms *after* the high temperature heat cycles needed to form a thermal oxide. The location of the dopant atoms is then determined by the energy of the implant, rather than by diffusion during an extended thermal oxidation. Only a mild heat cycle is needed to remove the implantation damage and to activate the dopant. The lateral spread of the dopant atoms is, therefore, also reduced. However, as the ion beam travels through the oxide layer, it transfers energy and momentum to the oxygen atoms. Some oxygen atoms can be pushed ("knocked") into the underlying silicon, where they can interfere with device operation.

The distribution of implanted ions in an amorphous layer is nearly *Gaussian* in shape [i.e., having the form A exp $-(x/\lambda)^2$]. The Gaussian distribution has a maximum value beneath the surface at a mean penetration depth called the *projected range* R_p. The width of the distribution is described by its projected standard deviation or *straggle* ΔR_p. The total distribution as a function of the depth x below the surface is

$$C(x) = C_p \exp\left[-\frac{(x - R_p)^2}{2 \Delta R_p^2} \right]$$ (2.5.1)

where C_p the peak dopant concentration (atoms cm^{-3}) is related to the implanted dose N' (atoms cm^{-2}) by

$$C_p = \frac{N'}{\sqrt{\pi}(\sqrt{2} \Delta R_p)}$$ (2.5.2)

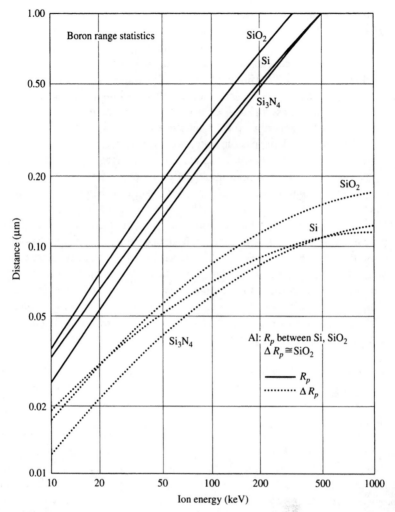

FIGURE 2.16 Projected range R_p and its standard deviation ΔR_p for implantation of boron into Si, SiO_2, Si_3N_4, and Al [11].

As seen from Equation 2.5.1, $\sqrt{2}\,\Delta R_p$ is a characteristic length describing the spatial broadening of the implanted distribution. Extensive tables have been published giving values of R_p and ΔR_p [11]. Values of these parameters for the three most common dopants in silicon are shown in Figures 2.16 through 2.18. The penetration of an ion through several layers can be approximated by considering the energy lost in each successive layer.

Equations 2.5.1 and 2.5.2 apply strictly only to implantation in amorphous material, in which the ion scattering can be assumed to be isotropic. In crystalline material the ion penetration can be significantly larger than R_p if the ions enter the lattice in a direction that allows them to *channel* along directions with widely spaced columns of atoms. Channeling is undesirable for most IC processing because of its sensitivity to the angle of incidence of the ion beam on the wafer surface. It can be reduced if the wafer is tilted until the beam and the crystal axis make an angle of approximately 7° during the implantation or if the wafer surface is coated with a thin amorphous

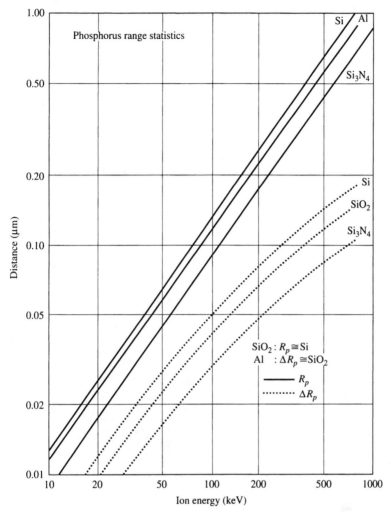

FIGURE 2.17 Projected range R_p and its standard deviation ΔR_p for implantation of phosphorus into Si, SiO$_2$, Si$_3$N$_4$ and Al [11].

layer such as SiO$_2$. When either of these measures is taken, the range parameters for amorphous implantation can be used. The many different orientations in a polycrystalline material make it impossible to avoid channeling in all crystallites by tilting the wafer.

Although masks for ion implantation can be made of thick layers of SiO$_2$, it is also possible to use metals or organic resists because the increase of the wafer temperature can be limited during the implantation. The use of resist is convenient, but care must be taken to insure that the implant itself does not heat the organic materials sufficiently to make them flow and lose definition. This is a special hazard for high-current implants, and the equipment must be designed to remove heat from the wafer during the implantation. The energy in the implant beam may also change the resist structure so that it cannot subsequently be removed easily with wet chemicals. In this case it necessary to etch the resist in an oxygen plasma.

The excellent control over the quantity, purity, and position of implanted dopant atoms has made ion implantation the dominant method of introducing dopant atoms into

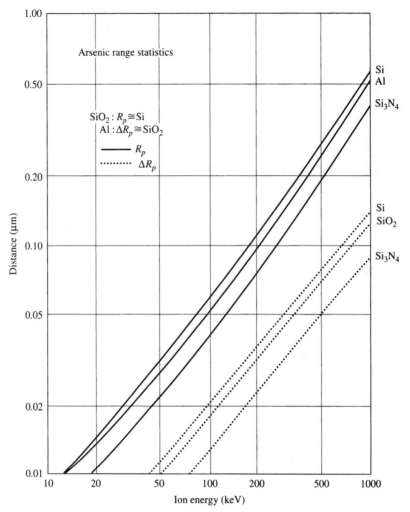

FIGURE 2.18 Projected range R_p and its standard deviation ΔR_p for implantation of arsenic into Si, SiO$_2$, Si$_3$N$_4$ and Al [11].

selected regions of an integrated circuit. We discuss implantation further in Chapter 9 and show that it is an essential part of advanced MOS processing.

Diffusion

Dopant atoms in a silicon wafer can migrate through the crystal if they have enough thermal energy. They can move from the region of high concentration where they are first deposited toward regions of lower concentration, usually deeper in the wafer. The diffusion of dopant atoms is similar to the diffusion flow of free carriers discussed in Sec. 1.2. The primary difference between the two cases is the temperature necessary to cause appreciable motion to occur. Dopant atoms must typically make their way through the lattice by meeting point defects (generally silicon vacancies or interstitial silicon atoms) while free carriers in the valence or conduction band can move without interacting with point defects. Temperatures of the order of 800–1000°C are required before appreciable diffusion of typical dopant atoms occurs.

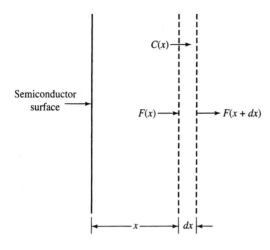

FIGURE 2.19 The increase in dopant concentration in a region dx is related to the net flux of atoms into the region: $F(x) - F(x + dx)$.

The change in the concentration C of the dopant atoms with time in a narrow region of width dx at a depth x from the surface (Figure 2.19) can be written as the difference between the flux of dopant atoms per unit area entering the region from the left and that leaving at the right

$$\frac{\partial C}{\partial t} dx = F(x) - F(x + dx) \tag{2.5.3}$$

We can approximate the last term by the first two terms of a Taylor-series expansion.

$$F(x + dx) \approx F(x) + \left(\frac{\partial F}{\partial x}\right) dx \tag{2.5.4}$$

to obtain

$$\frac{\partial C(x)}{\partial t} = -\frac{\partial F}{\partial x} \tag{2.5.5}$$

As we saw in Sec. 1.2, the first-order expression for the flux density is proportional to the concentration gradient

$$F = -D\frac{\partial C}{\partial x} \tag{2.5.6}$$

The parameter D, called the *diffusivity*, describes the ease with which the dopant atoms move in the lattice and is a strong function of temperature. The diffusivities of several common dopant impurities are shown in Figure 2.20. The simple model considered here neglects many aspects of diffusion that become important at higher dopant concentrations. Consequently, the values given in Figure 2.20 are only valid at low and moderate dopant concentrations. They must be used with caution when the dopant densities become high ($\gtrsim n_i$) because the diffusivity itself is then a function of the dopant concentration. Phosphorus is especially troublesome in this respect. More details of these second-order considerations are discussed later. Combining Equations 2.5.5 and 2.5.6, we find that

$$\frac{\partial C}{\partial t} = D\frac{\partial^2 C}{\partial x^2} \tag{2.5.7}$$

Equation 2.5.7 (sometimes called *Fick's second law*) can be solved explicitly for $C(x,t)$.

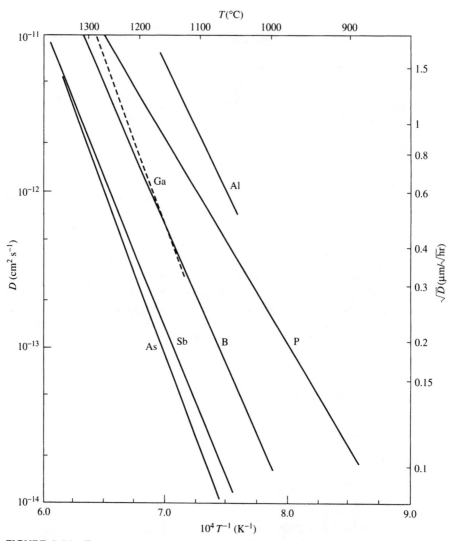

FIGURE 2.20 Temperature dependence of the diffusivities (at low concentrations) of commonly used dopant impurities in silicon [12].

Solutions of Equation 2.5.7 are obtained in specific cases by considering relevant boundary conditions. For the diffusion conditions most common in semiconductor processing, two different boundary conditions for Equation 2.5.7 are widely used, providing two solutions to this equation. A complementary error function is obtained for diffusion with a fixed surface dopant concentration (atoms cm^{-3}), and a Gaussian distribution describes the redistribution of a constant total number of diffusing atoms (atoms cm^{-2}); these two distributions are shown in Figure 2.21.

Gaseous Deposition. When a gaseous deposition source is used to introduce dopant atoms into a semiconductor, the patterned wafer is placed in a diffusion furnace similar to the furnace used for oxidation, and a gas containing the desired dopant impurity—typically phosphorus or boron—is passed over it. The quantity of dopant that enters the wafer is limited to values less than or equal to the solid solubility of the dopant in silicon at the furnace temperature.

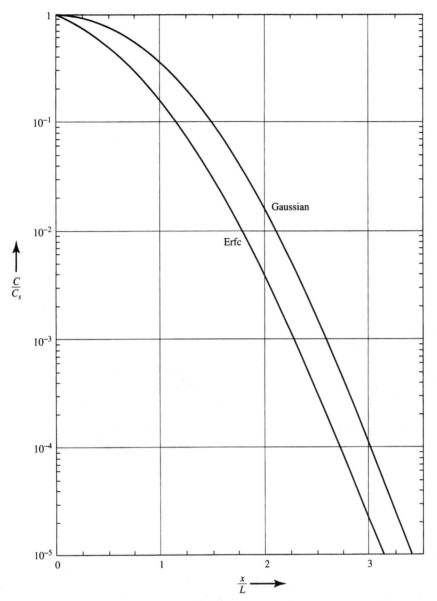

FIGURE 2.21 Complementary-error-function and Gaussian distributions; the vertical axis is normalized to the peak concentration C_s, while the horizontal axis is normalized to the characteristic length $L = 2\sqrt{Dt}$.

The best way to control the number of dopant atoms entering the silicon is to adjust the gas flows so that the dopant concentration at the silicon surface reaches its solid solubility. The solid solubilities in silicon for several common dopant species are shown in Figure 2.22.

Relatively low temperatures and short times are usually used for gaseous deposition to limit the number of dopant atoms introduced (atoms cm^{-2}) to the values needed for proper device operation. The penetration of the dopant atoms during this deposition step is generally small, and a *drive-in diffusion* is subsequently used to distribute the deposited atoms over the desired depth.

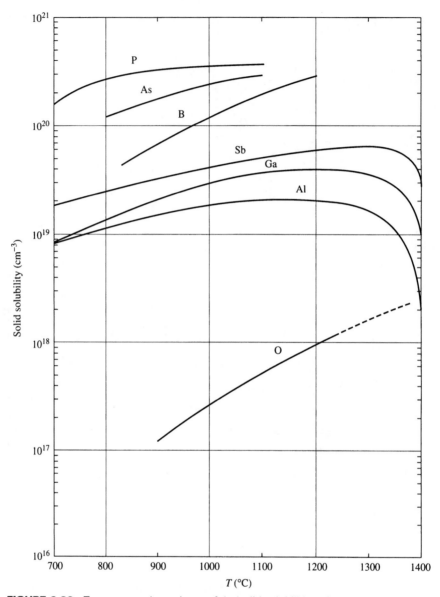

FIGURE 2.22 Temperature dependence of the solid solubilities of several elements in silicon [13].

During gaseous deposition, the silicon surface is exposed to a constant concentration of dopant atoms. The relevant solution of Equation 2.5.7 shows that the dopant atoms have a complementary-error-function distribution along x (the dimension measured away from the surface) after deposition.

$$C(x,t) = C_s \operatorname{erfc}\left(\frac{x}{2\sqrt{Dt}}\right) = \frac{2C_s}{\sqrt{\pi}} \int_{x/2\sqrt{Dt}}^{\infty} \exp(-v^2)\, dv \tag{2.5.8}$$

where C_s is the surface concentration of dopant atoms. Note that

$$\operatorname{erfc}(\eta) = 1 - \operatorname{erf}(\eta) = 1 - \frac{2}{\sqrt{\pi}} \int_0^{\eta} \exp(-v^2)\, dv \tag{2.5.9}$$

Thus, erfc (0) = 1, and the complementary error function decreases rapidly with increasing values of its argument η (Figure 2.21).

The complementary error function is a solution to Equation 2.5.7 for any method of dopant introduction that provides a constant value of dopant at the surface throughout the process [$C(0,t) = C_s$]. The combination of parameters $2\sqrt{Dt}$ used to normalize the x-axis of Figure 2.21 represents the characteristic diffusion length L associated with a particular diffusion cycle and describes the depth of penetration of the dopant. Note that the depth increases only as the square root of the diffusion time t.

If we know the surface concentration, the solid solubility, the diffusivity, and the diffusion time, we can calculate the impurity distribution from Equation 2.5.8. More important for the deposition cycle, we can find the total density N' of dopant atoms per unit surface area introduced by the diffusion. This quantity is calculated by integrating $C(x,t)$ over x to find

$$N' = \int_0^\infty C(x,t)\,dx = 2\sqrt{Dt/\pi}\ C_s \tag{2.5.10}$$

for the complementary-error-function distribution. Equation 2.5.10 shows that the area density of impurities, like the diffusion depth, increases as the square root of the diffusion time t.

Dopant Redistribution.   Dopant atoms are _redistributed_ by subsequent heat treatments after deposition.

After ion implantation, dopants have a Gaussian distribution (Equation 2.5.1). The subsequent _drive-in_ diffusion broadens this initial distribution, reducing the peak concentration because the total number of dopant atoms is fixed. The new distribution is also Gaussian, but it is described by a new characteristic length that is a function of both the initial characteristic length $\sqrt{2}\,\Delta R_p$ and the further dopant spread described by the diffusion length $2\sqrt{Dt}$. These quantities are combined in a root-mean-square (rms) fashion to obtain a new characteristic length L'

$$L' = \sqrt{2\,\Delta R_p^2 + 4Dt} \tag{2.5.11}$$

The new peak concentration is found by replacing the initial characteristic length $\sqrt{2}\,\Delta R_p$ by L', so that the final dopant distribution is

$$C(x) = \frac{N'}{L'\sqrt{\pi}}\exp\left[-\left(\frac{x - R_p}{L'}\right)^2\right] \tag{2.5.12}$$

Redistributing dopant atoms added by gaseous deposition also leads to a Gaussian distribution. Because a typical drive-in diffusion often produces an impurity distribution much deeper than that resulting from a gaseous deposition step, we often approximate the distribution at the beginning of the drive-in diffusion by a sheet of dopant at the semiconductor surface with a total concentration N' per unit area determined by the deposition cycle. We then assume that the drive-in diffusion simply redistributes this fixed amount of dopant impurity. With this boundary condition, the solution of the diffusion equation is again a Gaussian distribution

$$C(x,t) = \frac{N'}{\sqrt{\pi Dt}}\exp\left(-\frac{x^2}{4Dt}\right) \tag{2.5.13}$$

where the characteristic diffusion length $2\sqrt{Dt}$ is now determined by the temperature and the time of the drive-in diffusion.

When the dopant is successively redistributed by two or more diffusion steps, the over-all characteristic length is determined by an rms combination of the characteristic lengths associated with each process. For example, for m steps

$$L' = \left(\sum_{i=1}^{m} L_i^2 \right)^{1/2} = \left(\sum_{i=1}^{m} 4D_i t_i \right)^{1/2} \tag{2.5.14}$$

where the diffusivities are evaluated at the temperature corresponding to each heat cycle. (Note from Equation 2.5.14 that the squares of the diffusion lengths, *not the lengths themselves*, are summed, but that the individual $D_i t_i$ products are added.) In this manner, an approximation to the dopant distribution can be found after a number of different heat cycles, such as those needed to carry out a complete IC fabrication process.

For a given area density of dopant atoms added to a wafer, the peak concentration is twice as large when a gaseous predeposition is used as when the atoms are added by ion implantation. This is the case because the dopant concentration is a maximum at the silicon surface for gaseous predeposition, permitting diffusion only into the wafer (assuming we prevent evaporation of dopant from the surface). The ion-implanted dopant, on the other hand, has a maximum concentration beneath the silicon surface so that diffusion proceeds both toward the surface and into the bulk, spreading the fixed number of dopant atoms over a larger volume.

As device performance improves, shallower junctions are needed. In some cases the depth associated with ion implantation and the subsequent anneal needed to remove the lattice damage is too large. To obtain very shallow junctions, a material with a very high diffusivity is placed over the single-crystal silicon. Dopant is implanted into the material with high diffusivity, and then diffused from this material into the underlying single-crystal silicon. Because of the high diffusivity in the overlayer, the dopant readily spreads through it, achieving a nearly constant concentration in it. In addition, as dopant diffuses from the overlayer into the single-crystal silicon, it is readily transported within the overlayer to the interface to replenish dopant diffusing into the single-crystal silicon. The overlying material thus acts as a dopant source providing a constant surface concentration in the single-crystal silicon, and the dopant profile in the single-crystal silicon is described by a complementary-error-function distribution. Although the implantation process damages the overlayer, little of the damage propagates into the underlying single-crystal silicon.

Polycrystalline-Silicon Doping Sources. One convenient material for this application is the *polycrystalline* silicon that we will discuss in Sec. 2.6. The grain boundaries in polycrystalline silicon allow dopant atoms to diffuse much more readily than in single-crystal silicon. Its rapid dopant diffusivity, coupled with its compatibility with silicon processing, makes polycrystalline silicon nearly ideal as a dopant diffusion source.

Oxide Doping Sources. In an alternate method of dopant addition, a layer of oxide containing the dopant impurity can be deposited on the wafer surface. The dopant atoms are then diffused into the silicon from this glassy layer. One convenient method of forming the doped oxide layer is *chemical vapor deposition* (CVD) of SiO_2 with a dopant species added to the oxide during its deposition. We discuss the CVD process in greater detail in Sec. 2.6. It is also possible to incorporate the dopant within particles of glass dispersed in an organic solvent. This material can be "spun-on" the wafer, dried, and the organic residue then driven off by heating the film to about 200°C. After the doped oxide has been deposited on the wafer by either of these two methods, the impurity must be diffused into the silicon with a drive-in diffusion step.

When a doped oxide is used, the concentration of dopant atoms at the silicon surface generally remains constant during the entire drive-in diffusion at a fixed fraction of the concentration of dopant in the oxide. This condition leads once again to a complementary-error-function distribution after the drive-in step (Equation 2.5.8 and Figure 2.21).

Unlike the case of the polycrystalline silicon diffusion source, the diffusivity of dopant in the glasses is low. The constant concentration source approximation is valid because the dopant concentration in the glass is so high (up to tens of percent) that only a very small fraction of the dopant is transported into the silicon, and the concentration in the oxide remains approximately constant.

Diffusivity Variations. Because dopant diffusion occurs through the interaction of the diffusing species and point defects (chiefly silicon vacancies or interstitial silicon atoms), any factor that changes either the density of point defects or the charges associated with them can modify the diffusion process. We briefly consider two important cases in which this occurs: (1) *oxidation-enhanced diffusion* and (2) *concentration-dependent diffusion* of a dopant species.

If the point-defect concentration varies with position in the crystal, the diffusivity also depends on position. In this case Equations 2.5.5 and 2.5.6 can be combined to express the diffusion equation as

$$\frac{\partial C}{\partial t} = \frac{\partial}{\partial x}\left(D\frac{\partial C}{\partial x}\right) \tag{2.5.15}$$

instead of the simpler form of Fick's Law (Equation 2.5.7).

The solutions to Equation 2.5.15 are more complicated than the simple Gaussian or complementary-error-function distributions discussed previously. The exact profiles cannot usually be written in closed mathematical form and are found from Equation 2.5.15 using numerical techniques. A few qualitative comments can, however, provide useful insight into this more complex diffusion process.

Localized charge is present in the crystal near a silicon vacancy or interstitial. The charge state of these point defects depends on the position of the Fermi level in the crystal and thus depends on the dopant concentration and the temperature. Because the diffusing dopant atoms move by interacting with charged, as well as neutral, point defects, this dependence on the Fermi level affects the dopant diffusivity. The diffusivity in this case can be written as the sum of components that account separately for interactions of the dopant atoms with different charge states of the point defects [14]. For example, for diffusion dominated by interaction with vacancies, the effective diffusivity D_{eff} can be written

$$D_{eff} = h\left[D_i^0 + D_i^-\left(\frac{n}{n_i}\right) + D_i^=\left(\frac{n}{n_i}\right)^2\right] \tag{2.5.16}$$

where h is a parameter that accounts for the effect of electric field on the diffusivities, and each D term is associated with a different charge state of a point-defect site.

From Equation 2.5.16 we see that the diffusivity depends on the dopant concentration when the terms involving the charged vacancies are comparable to the neutral vacancy term D_i^0. Because the various D_i values are of comparable magnitudes, this occurs when n/n_i becomes of order unity at the diffusion temperature. For example, consider diffusion at 1000°C; Figure 2.10 shows that $n_i \approx 9 \times 10^{18}$ cm^{-3}. Hence, for dopant densities above this value, we expect to observe an increase in D_{eff}. Concentration-dependent diffusion can multiply the diffusivity by a factor of 10 or 20, greatly enhancing the diffusion rate in regions of high-dopant concentration. The resultant dopant profile then differs from the Gaussian or complementary error functions which apply to low-concentration doping profiles.

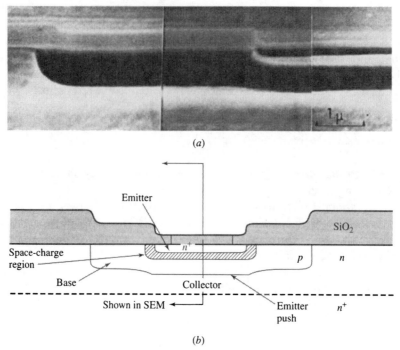

(a)

(b)

FIGURE 2.23 (a) Scanning electron micrograph showing cross section through a bipolar transistor, (b) sketch identifying the regions shown. The boron-doped base region has been pushed ahead (emitter push) by the concentration-dependent diffusion effects associated with heavy phosphorus doping in the emitter [15].

An example of concentration-dependent diffusion, often observed in bipolar integrated circuits, is called the *emitter-push* effect. It occurs when the point defects associated with the heavy phosphorus or arsenic doping used to form the emitter region increases the diffusivity of the boron base dopant. This emitter push of the base dopant occurs primarily directly under the emitter region, but it also enhances diffusion a short distance to the sides. The details of bipolar transistor design and operation are described in Chapter 6. Figure 2.23 is a scanning electron microscope (SEM) picture showing the cross section of a transistor in which emitter push of the base is seen under the emitter region. Bulging dopant profiles caused by effects similar to emitter push can appear in other regions of some ICs.

In addition to concentration-dependent diffusion, interaction with point defects created by other mechanisms can also increase the diffusion rate. When silicon is oxidized, many bonds are broken at the surface, and point defects are generated. Some of these point defects migrate into the underlying silicon until they meet a different type of point defect that is capable of annihilating them. For example, a vacancy can recombine with (be annihilated by) an interstitial in similar fashion to the recombination of an electron and a hole. Before recombining, however, the point defects can migrate significant distances into the silicon, modifying the diffusion rate of any dopants in this region. This effect causes *oxidation-enhanced diffusion*, which is readily observed when portions of a silicon wafer are covered with a nonoxidizing layer while other regions are oxidized (Figure 2.24). Diffusion under the oxidizing regions of a wafer can be considerably greater than that under portions that are not oxidized (and consequently do not contain large densities of point defects).

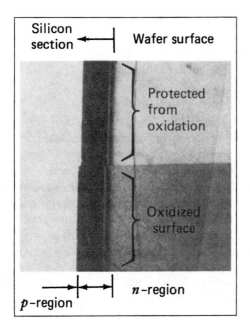

FIGURE 2.24 Section of a silicon wafer showing deeper diffused n–type region (dark area) under oxidized silicon surface (bottom) than underneath a surface protected from oxidation (top) [16].

In addition to enhancement by oxidation, diffusion can be greatly enhanced by lattice damage created during ion implantation. During a subsequent heat treatment the lattice damage is removed, but, at the same time, the dopant atoms diffuse. During the early stages of the annealing process, the lattice damage promotes rapid dopant migration (*transient-enhanced diffusion*), but the diffusion slows as the damage is removed. A quantitative description of the combined effects is extremely complex and must be treated by computer modeling. This interaction of lattice damage and dopant motion is especially severe when using rapid thermal processing (RTP) techniques, in which the wafer is heated to a high temperature by radiation for only a few seconds to remove lattice damage and activate the implanted dopant, while trying to minimize diffusion. The effect of lattice damage and point defects on dopant motion varies widely at different temperatures. By properly removing the lattice damage (and the point defects) before the major portion of the dopant activation process, transient-enhanced diffusion can be minimized.

Solid Solubility. The amount of dopant entering the silicon surface by gaseous deposition is limited by thermodynamics to the solid-solubility concentration (Figure 2.22). When dopant atoms are introduced by ion implantation, however, the impurity concentration can exceed the solid-solubility value because thermodynamic equilibrium is not involved. During the subsequent damage-removal and dopant-activation anneal, however, the amount of dopant entering substitutional sites is limited to the solid solubility at the annealing temperature. Excess dopant can form clusters or precipitates. As seen in Figure 2.22, the solid solubility increases with increasing temperature for typical dopants over temperature ranges of interest. Hence, a dopant concentration that is at its solid-solubility limit at high temperatures may exceed the limit during a subsequent, lower-temperature anneal. When this occurs, the excess dopant tends to move out of electrically active, substitutional sites. If the temperatures are sufficiently low (e.g., room temperature), however, expulsion of excessive dopant from the lattice can be very slow. In practice, doping concentrations can remain higher than the room-temperature solubility limits indefinitely if the dopants have been introduced at elevated temperatures and the wafer is cooled rapidly.

Segregation. During thermal oxidation, silicon is consumed, and any dopant in the wafer must redistribute between the silicon and the growing oxide. At equilibrium, a constant ratio exists between the dopant concentrations on the two sides of the $Si-SiO_2$ interface. The *segregation coefficient* (*m*) describes this ratio:

$$m = \frac{C_{Si}}{C_{SiO_2}} \tag{2.5.17}$$

The *n*-type dopants phosphorus and arsenic tend to segregate into the silicon ($m > 1$) and to be pushed ahead of a growing layer of SiO_2. Boron tends to deplete from the surface regions of silicon into the growing oxide layer ($m < 1$). However, because the boundary between silicon and SiO_2 is moving during thermal oxidation, equilibrium at the interface is only approached at very low oxidation rates. For typical processing conditions, the amount of segregation differs significantly from that predicted by thermal-equilibrium considerations. Dopant segregation also depends on the diffusion rate of the dopant species in the oxide. Equilibrium is approached only if the dopant diffuses slowly in the oxide so that any loss at the SiO_2–gas interface does not affect the $Si-SiO_2$ interface. Dopant transport from the silicon bulk to the interface by diffusion must also be considered. The ratio of the dopant concentration at the interface to its value in the bulk is primarily determined by the relative rates of oxidation and diffusion; the less able is the dopant to transfer between the bulk and the interface, the greater is the accumulation or depletion of dopant near the silicon surface.

Thus, the three important parameters that must be considered when determining the amount of surface segregation are (1) the segregation coefficient *m*, (2) the ratio of the oxidation rate to the square root of the dopant diffusivity in the silicon (which measures diffusion in the silicon), and (3) the ratio of the dopant diffusivities in silicon and in SiO_2.

Rapid Thermal Processing and Single-Wafer Processing.

Although IC technology is based on "batch processing" of a substantial number of wafers (often 25) at one time, two conflicting trends are encouraging the development and implementation of *single-wafer processing* (i.e., processing one wafer at a time). As wafer diameters increase, less thermal stresses can be tolerated by the wafer before permanent crystal damage occurs. In conventional furnaces this requires that the wafers be heated more slowly; in addition, the thermal heat capacity associated with the wafers and fixtures delay achieving thermal equilibrium at the process temperature. These considerations place a lower limit on the duration of a heat cycle because the wafers must reach their steady-state temperature in a time short compared with the total heat cycle. On the other hand, the longest time allowed at a given temperature is limited by diffusion of previously introduced dopant atoms. The allowable time can be increased by using lower temperatures, but some processes produce better results at higher temperatures; for example, the quality of oxides grown at higher temperatures is better than that of oxides grown at lower temperatures.

To achieve the quality associated with high temperatures and the corresponding short times allowed at high temperatures, *rapid thermal processing* is becoming more popular. In this technique a wafer is inserted into a processing system at a low temperature and held on a support with very low thermal mass (e.g., 3 quartz posts). Lamps are then used to increase the wafer temperature rapidly, with typical heating rates of 50–100°C/sec. The heating must, of course, be very uniform to avoid crystal damage from thermal stress. Heat loss from the edges of the wafer make uniform heating a challenging task. If the heating is uniform, however, the wafer can be heated to a reasonably high temperature (perhaps 1000°C) in about 10–20 sec. Processing times of a few tens of seconds become practical, allowing processing to occur at higher temperatures without excessive dopant diffusion.

Although using single-wafer processing seems to sacrifice the advantages gained by batch processing, a short cycle time per wafer can be used (*and is required*) to offset the disadvantages of processing one wafer at a time. For example, if a single-wafer process can be completed in one minute, the throughput can be similar to a batch process in which 100 wafers are processed in 100 minutes—a time typical for a batch furnace process. In addition, when small quantities of specialized products are processed, single-wafer processing is especially attractive because enough wafers to fill a large furnace are not available, limiting the economic benefits of batch processing.

Single-wafer processing offers advantages even when rapid heating is not critical. As layers become thinner in more advanced circuits, control of interfaces between layers becomes more critical, and keeping the wafer in a well controlled, inert ambient between process steps is advantageous. With single-wafer processing, various process chambers can readily be *clustered* around a central automated wafer handler (a non-anthropomorphic "robot"). The wafer is transported between process chambers in a vacuum or inert gas ambient. Because related process steps can be completed without exposing the wafer surface to air, better control of interfaces can be achieved. For example, the wafer surface may be cleaned in one chamber before a chemical vapor deposition or physical vapor deposition process in another chamber. Although reliability of complex equipment remains a concern, cluster tools are now common in IC processing.

2.6 CHEMICAL VAPOR DEPOSITION

Although the basic elements of an integrated circuit can be formed by oxidation, lithography, and diffusion, more advanced structures require the flexibility of adding a conducting, semiconducting or insulating layer on top of a partially formed integrated circuit. Deposited insulators can be used to avoid high-temperature oxidation after dopant atoms have been introduced, while lightly doped, single-crystal silicon layers or polycrystalline-silicon films may be useful in other locations. These added layers can be formed by *chemical vapor deposition* (CVD) or by physical vapor deposition (PVD). In CVD all constituents forming the deposited layer are introduced into the reactor in the vapor phase; none come from the silicon wafer itself. The structure of a CVD film depends on the substrate on which it is deposited (amorphous or crystalline) and on the deposition conditions (mainly temperature, deposition rate, and gas pressure). CVD layers can be formed over a wide temperature range. The deposition reactions are usually promoted by heating the substrate, but energy can also be introduced into the system electrically by generating a plasma within the deposition chamber.

Epitaxy

We have described how impurities can be added to a wafer by ion implantation or by gaseous deposition and diffusion. These processes can be used to increase the dopant concentration in a layer near the surface. They do not, however, permit us to produce a layer that is less heavily doped near the surface than it is underneath. In theory this could be done by adding an approximately equal concentration of impurities of the opposite conductivity type, using compensation as described in Sec. 1.1. However, limited control of the accuracy of the diffusion process generally makes it impractical to achieve nearly balanced compensation of dopants. An additional drawback to using a balanced compensation is that the carrier mobility in such a layer is degraded because the mobility is limited by scattering from the total number of ionized impurities $N_d + N_a$, rather than by the net number of impurities $|N_d - N_a|$, which determines the carrier concentration.

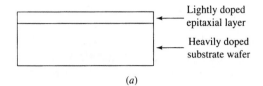

(a)

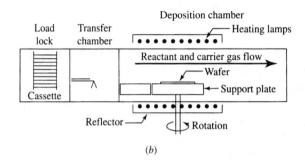

(b)

FIGURE 2.25 (a) Lightly doped epitaxial layer grown on heavily doped silicon substrate. (b) Single-wafer, epitaxial deposition system showing silicon wafer on support plate that is heated by infrared lamps located outside the quartz deposition chamber.

We can fabricate the desired lightly doped layer above a heavily doped region by the process of *epitaxy* (Figure 2.25), which is the controlled growth of single-crystal silicon on a single-crystal wafer or *substrate*. To grow an epitaxial layer, the wafer is placed in a heated chamber where a gas such as silane (SiH_4) or dichlorosilane (SiH_2Cl_2) passes over its surface. The gas decomposes on the surface of the wafer, and a layer of silicon is deposited there. Silane decomposes *pyrolytically*, that is, by the addition of heat alone

$$SiH_4(g) \rightarrow Si(s) + 2H_2(g) \qquad (2.6.1)$$

with the byproduct H_2, while the decomposition of dichlorosilane creates the corrosive byproduct HCl:

$$SiH_2Cl_2(g) \rightarrow Si(s) + 2HCl(g) \qquad (2.6.2)$$

Note that all the incoming species in reactions 2.6.1 and 2.6.2 are gases, which accounts for the name chemical *vapor* deposition (CVD).

To make the CVD of silicon an *epitaxial* deposition on single-crystal silicon, it is necessary to heat the wafer sufficiently so that the depositing silicon atoms gain enough thermal energy to move into low-energy positions where they form covalent bonds to the substrate and extend the single-crystal lattice before they become buried and immobilized by subsequently arriving atoms. Single-crystal growth or epitaxy is usually carried out at temperatures between 850 and 1200°C. The epitaxial film can be more lightly doped than the substrate; therefore, epitaxy provides a means for obtaining a low concentration of dopant above a high-concentration region. This capability is especially important in optimizing the structure of bipolar transistors. Dopant atoms can be aded to the growing film during its deposition by introducing dopant-containing gases such as arsine (AsH_3), phosphine (PH_3), or diborane (B_2H_6) into the reactor along with the silicon-containing gas.

Nonepitaxial Films

In addition to epitaxial silicon, a number of other CVD films are useful for IC applications. For example, in any IC, conducting layers are needed to interconnect devices. These layers must, of course, be isolated from the substrate. Aluminum is very often used for these conducting layers, but its low melting point (660°C) and its reactivity with other elements generally precludes heating it higher than 500°C after it is deposited. If, as an alternative, a thin layer of silicon is used as an interconnecting path, subsequent heat treatments can be carried out at 1000°C or higher. This high-temperature capability is extremely important in a number of IC applications.

Polycrystalline Silicon.   To deposit silicon for interconnections, methods similar to those used for epitaxy are often employed. However, because these layers are not deposited directly on the single-crystal silicon wafer, but usually over an amorphous SiO_2 layer, they cannot grow epitaxially. These CVD films are typically composed of many small crystallites (often with ~50–100 nm dimensions) and are therefore called *polycrystalline silicon*, or simply *polysilicon*.

Polysilicon has special importance in MOS processing where it is used as a transistor electrode in *silicon-gate* MOSFETs, which are described in Chapter 9. Polysilicon is also found in bipolar ICs where it is used as a diffusion source, as discussed in Sec. 6.5. For this application, polysilicon is formed over the single-crystal substrate by lowering the CVD temperature to about 600–700°C. The limited thermal energy of the silicon atoms available for surface migration at these temperatures and the residual oxide on the silicon surface favor forming a polycrystalline, rather than a single-crystal, layer. The behavior of the layer during subsequent processing depends critically on the thin oxide layer (~1 atom-layer thick) between the polysilicon and the underlying single-crystal silicon.

Amorphous Silicon.   When CVD silicon is deposited at still lower temperatures (below about 600°C), an amorphous film forms regardless of the nature of the substrate. Amorphous silicon has only very short-range order (typically over only a few atomic spacings), and no crystalline regions can be observed. During heat treatments at moderate temperatures ($\geq$600°C), silicon deposited in an amorphous form crystallizes to become polycrystalline silicon. Depositing silicon in an amorphous form and subsequently crystallizing it is becoming increasingly important in a number of applications, such as the active transistor matrix used in liquid-crystal displays.

Somewhat below 600°C amorphous silicon can be deposited by thermal decomposition of silane. However, at markedly lower temperatures in the 300°C range, additional energy in the form of a plasma is used to decompose the silane and produce an amorphous silicon layer. At this lower temperature, significant amounts of hydrogen from the decomposition of the silane remain in the deposited layer and modify its electrical properties.

Insulating Films.   Insulating, as well as conducting, films can be formed by CVD. Especially important to IC processing are CVD films of silicon dioxide (SiO_2) and silicon nitride (Si_3N_4) formed by reacting a gas such as silane (SiH_4) or dichlorosilane (SiH_2Cl_2) with oxygen or nitrous oxide (N_2O) (for SiO_2) or with ammonia (NH_3) (for Si_3N_4). Although CVD oxides are not usually as pure nor of equivalent electrical quality as thermally grown oxides, they do not require the high-temperature processing needed for thermal oxidation. For example, CVD oxide can be used above a polysilicon interconnection layer before aluminum or copper is deposited to avoid the dopant diffusion that would occur during a thermal oxidation. CVD oxide deposited at even lower temperatures can be used above aluminum or copper layers to separate different conducting layers that interconnect active devices and also to protect the finished integrated circuit from contamination. In the latter case, the oxide is usually doped with phosphorus to impede the migration of any contaminant through the oxide to the circuit. A deposited oxide commonly used between conducting layers is formed by the reaction of silane and oxygen at about 400°C. The reaction

$$SiH_4(g) + O_2(g) \rightarrow SiO_2(s) + 2H_2(g) \tag{2.6.3}$$

produces an oxide that is less dense and less chemically resistant than thermally grown oxide, but is very useful when high temperatures must be avoided.

Because silicon-nitride (Si_3N_4) layers do not oxidize as readily as does silicon, silicon-nitride layers are useful for limiting the regions where thermal oxide is grown in a *local oxidation (LOCOS)* process used for device isolation in integrated circuits [17]. In the LOCOS process, a layer of silicon nitride is deposited on the silicon substrate and lithographically defined to retain the nitride in the device regions (Figure 2.26a). The nitride is removed from the area between the devices where a thick, isolating oxide layer is to be grown. A thin, *stress-relief* SiO_2 layer is usually inserted between the nitride and the silicon wafer to prevent stress from the nitride from creating defects in the silicon wafer. After the nitride is defined, the wafer is inserted into an oxidation furnace, and a thick oxide is grown in the exposed silicon regions, usually in a steam or pyrogenic ($H_2:O_2$) ambient (Figure 2.26b). The nitride prevents oxidation of the device regions. A layer of SiO_2 more than one μm thick can be grown on exposed silicon while only a few tens of nm of Si_3N_4 are converted to SiO_2. After the oxidation, the thin oxide over the nitride is removed, and the nitride and thin stress-relief oxide

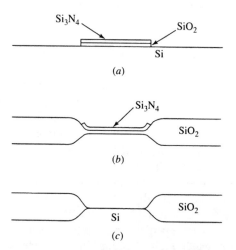

(a)

(b)

(c)

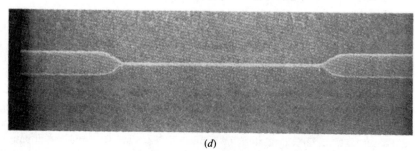

(d)

FIGURE 2.26 *LOCal Oxidation of Silicon (LOCOS).* (a) Defined pattern consisting of stress-relief oxide and Si_3N_4 covering the area over which further oxidation is not desired, (b) thick oxide layer grown over the bare silicon region, (c) stress-relief oxide and Si_3N_4 removed by etching to permit device fabrication, (d) scanning electron micrograph (5000 ×) showing LOCOS-processed wafer at step (b).

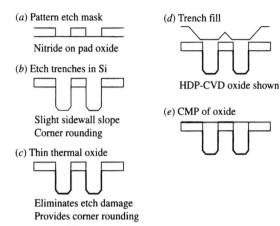

(a) Pattern etch mask

Nitride on pad oxide

(b) Etch trenches in Si

Slight sidewall slope
Corner rounding

(c) Thin thermal oxide

Eliminates etch damage
Provides corner rounding

(d) Trench fill

HDP-CVD oxide shown

(e) CMP of oxide

FIGURE 2.27 Trench isolation is used to form very narrow isolation regions between adjacent devices. After the trench pattern is etched in a masking material (a), the trench is etched directionally using a reactive ion etch process (b). A thin, high-quality oxide is formed (c), and the trench is filled with polysilicon or with oxide [shown for oxide in (d)]. The excess material is removed by *chemical-mechanical polishing* (e).

are etched away to expose the bare silicon in the device regions (Figure 2.26c). Figure 2.26d shows a cross section scanning-electron micrograph of a device region formed by a LOCOS process. The oxide thickness tapers gradually from the isolation region to the device region. This taper is an advantage in avoiding sharp edges that make continuous film coverage difficult. For devices made with very small dimensions, however, the taper is a disadvantage and limits formation of device regions with very small (submicrometer) dimensions.

To form isolation regions suitable for very fine features, *trench isolation* can be used, as shown in Figure 2.27. As the name implies, a very narrow and relatively deep trench is etched around the device to be isolated. Directional, *reactive ion etching* is used to form the nearly vertical sidewalls of the trench needed to allow the top of the trench to have the minimum feature size that can be defined and the trench to be adequately deep (Figure 2.27b). After a thin, high-quality, insulating oxide is formed on the walls and bottom of the trench (Figure 2.27c), the trench is filled with either an oxide or possibly with polysilicon. For deep trenches, polysilicon is useful because its thermal coefficient of expansion is virtually identical to that of single-crystal silicon, reducing thermal stress and bending of the wafer. For shallow trenches, oxide can be used with minimum effect on the stress. When oxide is used, it is often deposited by a CVD process using a *high-density plasma* (Figure 2.27d). The directionality of the high-density plasma allows filling of very deep and narrow trenches. The ion bombardment occurring during the deposition process improves the quality and density of the oxide by knocking weakly bound atoms from the surface of the depositing layer.

Reaction Kinetics. Layers deposited by CVD are usually formed in open-flow reactors, as illustrated in (Figure 2.25b). The gases flow continuously through the reaction chamber where the deposition takes place; gaseous byproducts are exhausted along with unused reactant gases. A *carrier gas* is often used to push the reactants through the chamber. The gases are usually mixed before entering the reaction chamber unless they react at low temperatures. At the surface of the wafers, there is a gas-phase boundary layer through which the reactants must diffuse (Figure 2.28a). This boundary layer is a transition region between the unrestricted flow region and the walls and fixtures in the chamber where the gas velocity is reduced by viscous forces that tend to pull against the moving gas and retard its flow (Figure 2.28b).

The reactants must pass through the boundary layer to reach the surface, where the reaction is promoted by heat. The deposition takes place on all heated surfaces reached by the gases. For example, if the walls of the chamber are hot, a generally undesired and troublesome film forms on the walls, as well as on the wafers. Either the rate of diffusion

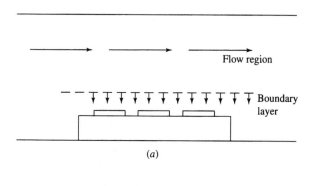

(a)

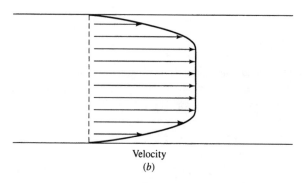

Velocity

(b)

FIGURE 2.28 (a) Section along horizontal, open-flow reactor showing gas flow parallel to the wafer surface and indicating the location of the boundary layer in which the gas flow is nearly perpendicular to the wafer surface, (b) representation of gas velocity distribution across the reaction chamber.

through the boundary layer or the rate of reaction at the surface can limit the overall deposition rate. We discussed similar transfer-rate limitations in the case of thermal oxidation in Sec. 2.3. Unlike the case of thermal oxidation, however, the diffusion process for CVD occurs in the gas phase rather than in the solid SiO_2 layer. As in the case of oxidation, we write expressions for the fluxes of molecules diffusing through the boundary layer [$F(1)$] and reacting at the surface [$F(2)$] (Figure 2.29):

$$F(1) = D\frac{C_g - C_s}{\delta} \tag{2.6.4}$$

and

$$F(2) = k_s C_s \tag{2.6.5}$$

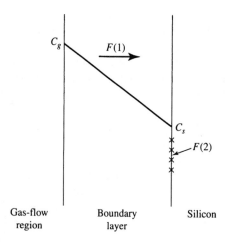

Gas-flow region   Boundary layer   Silicon

FIGURE 2.29 Reactant gases diffuse through the boundary layer to the wafer surface at a rate $F(1)$ and react there at a rate $F(2)$.

In Equation 2.6.4, D is the gas-phase diffusivity of the reactant (only weakly temperature dependent), δ is the boundary-layer thickness, and the concentrations C_g and C_s occur at the outer edge of the gaseous boundary layer and in the gas phase near the solid surface, respectively. In Equation 2.6.5. k_s is the surface reaction-rate coefficient, which depends exponentially on the reciprocal of temperature, with an activation energy E_a [$k_s = k_{so} \exp(-E_a/kT)$].

In steady state, $F(1) = F(2) = F$, and the overall deposition rate R_d can be written

$$R_d = \frac{F}{N} = \frac{C_g/N}{\delta/D + 1/k_s} \tag{2.6.6}$$

where N is the number of atoms per unit volume in the deposited film. The first term in the denominator of Equation 2.6.6 represents the impedance to gas-phase diffusion, while the second term is the impedance to the surface reaction. Reactions limited by surface processes depend strongly on k_s and typically dominate CVD processes at low temperatures and low pressures ($p \ll 1$ atm). However, k_s increases rapidly with increasing temperature so that the surface reaction no longer limits the overall process at higher temperatures; gas-phase diffusion through the boundary layer then becomes limiting. Typical temperature behavior is shown in Figure 2.30. Gas-phase diffusion becomes more difficult as pressure increases (toward one atmosphere) so that higher-pressure operation also tends to cause gas-phase diffusion to limit the overall deposition process.

Choice of the appropriate limiting process is influenced by the geometry of the CVD reactor. In reactors operating near atmospheric pressure with the gas flow nearly parallel to the wafer surface, such as that shown in Figure 2.25b, the gas-flow is readily controlled, while the temperature is difficult to control. Consequently, the temperature-insensitive, gas-phase diffusion process is generally designed to be the limitation of the overall CVD process. In the horizontal reactor, throughput is very limited, especially in *single-wafer reactors*, where only one wafer is processed at a time.

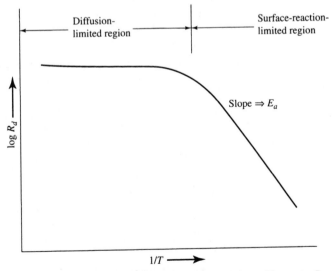

FIGURE 2.30 Typical dependence of overall deposition rate R_d as a function of reciprocal temperature $1/T$; the surface reaction rate limits the deposition at low temperatures, and gaseous diffusion (mass transport) limits at high temperatures.

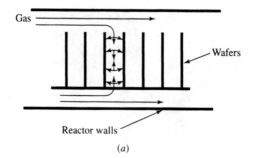

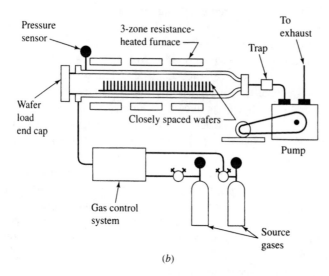

FIGURE 2.31 (*a*) Gases in a high-capacity reactor flow through the annular space between the wafers and reactor wall and then diffuse between the closely spaced wafers. (*b*) The basic elements of a *LPCVD* reactor.

For high wafer capacity, wafers are closely spaced, as in a diffusion or oxidation furnace (Figure 2.5). With this geometry, however, the gas flow is difficult to control because the gases first flow through the annular space surrounding the wafers and then into the narrow spaces between wafers (Figure 2.31a). Consequently, diffusion-limited-operation leads to very nonuniform film thicknesses. On the other hand, the temperature in this type of furnace can be well controlled so that operation in the surface-reaction-limited region is feasible. Hence, high-capacity reactors with closely spaced wafers are generally operated at lower temperatures where the surface reaction limits the deposition rate. The diffusion limitation is further eased by operating at a reduced pressure (~1 Torr). Because the gas-phase diffusivity is inversely proportional to the pressure, the first term in the denominator of Equation 2.6.6 then becomes even less of a limitation.

High-capacity, low-pressure CVD (LPCVD) reactors are routinely used to deposit polysilicon and silicon nitride. The basic elements of an LPCVD system are shown in Figure 2.31b. The chamber can be oriented as shown in Figure 2.31b, with the main gas flow in the horizontal direction and the wafers held vertically. Alternatively, the reactor can be rotated by 90° so that the chamber and main gas flow are vertical and the wafers are held horizontally. The latter arrangement is more amenable to automated wafer handling.

This type of reactor has not been widely used for silicon epitaxy because of the higher temperatures generally needed to assure single-crystal growth. Using higher temperatures conflicts with avoiding a diffusion limitation, as mentioned above. However, ultra-high purity reactors of this geometry can be used for specialized epitaxial deposition at lower temperatures.

Plasma-Enhanced CVD.   For some applications, a layer must be deposited at a low temperature. For example, a passivating layer over aluminum or copper interconnections already on the chip must be deposited below ~400°C. At this low temperature, the thermal deposition rate can be unacceptably low or the layer may not cover irregular features adequately. In these cases, thermally activated CVD is inadequate, and an additional source of energy must be supplied to the wafer surface to allow the necessary chemical reactions to proceed or to modify the characteristics of the material being deposited. A high-frequency electric field can supply electrical energy to the gas mixture to create a plasma, which enhances the deposition process in *plasma-enhanced* CVD (PECVD), as it enhanced the etching processes discussed in Sec. 2.4. In both cases, ion bombardment can be important. During deposition, the ion bombardment can modify the characteristics of the depositing layer and even physically sputter material from the depositing layer.

Plasma-enhanced CVD is increasingly important in IC fabrication, being used for the deposition of silicon oxide, silicon nitride, and silicon. Because of the lower temperatures used in PECVD, the layers produced are not of equivalent quality to those deposited by thermal oxidation at higher temperatures. Oxide and nitride layers deposited by PECVD are often not stoichiometric (i.e., not SiO_2 or Si_3N_4, respectively), and the structure of silicon layers is poorly defined.

The nature of the plasma depends on many independent variables such as electron concentration, electron-energy distribution, gas density, and residence time of the excited species within the plasma. These microscopic variables are controlled by macroscopic parameters such as (1) reactor geometry, (2) intensity and frequency of the high-frequency power used to excite the plasma, (3) pump speed, (4) electrode temperature, and (5) flow rates of the reactant and diluent gases. Unfortunately, the quantitative relation between the macroscopic and microscopic parameters is not straightforward.

In the simplest realization of plasma reactors, the plasma is created by exciting a mixture of all the reactant and diluent gases. The excited neutral species diffuse to the wafer surface, and the ions are accelerated toward the surface by the electric field between the plasma region and the wafer. In more advanced reactors, an attempt is made to separately control the plasma generation, the chemical reaction, and the ion bombardment. In these *remote* plasma reactors (Figure 2.32a), the plasma is generated in one region, usually

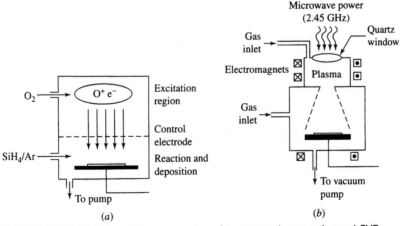

FIGURE 2.32 (*a*) Schematic cross section of a remote plasma-enhanced CVD reactor, in which the plasma generation, the chemical reaction, and the ion bombardment are partially decoupled. (*b*) Schematic cross section of an electron cyclotron resonance, high-density plasma reactor.

by exciting a gas which does not form a deposit by itself (e.g., oxygen). The plasma is then transported toward the wafer surface. Near the wafer surface, it is mixed with the more reactive gas (e.g., silane), so that the desired layer (e.g., SiO_2) forms on the wafer surface. This remote generation of the plasma allows more flexibility and control of the process. It reduces premature reaction (and particles), and very importantly it allows controlled ion bombardment of the surface.

When the mean free path of the ions approaches a characteristic dimension of the reactor, few ions are scattered, and most approach the surface in the direction of the applied electric field. They can dislodge weakly bound atoms from the depositing layer, increasing its density and quality, and they can physically sputter overhanging deposits that would prevent good filling between closely spaced features. A plasma that contains enough ions to significantly modify the characteristics of the depositing layer is called a *high-density plasma* (*HDP*). Figure 2.32*b* shows a cross section of one type of HDP reactor. In this *electron cyclotron resonance* (*ECR*) reactor, an interaction between the electric and magnetic fields confines much of the plasma, allowing efficient power absorption and plasma generation.

2.7 INTERCONNECTION AND PACKAGING

Interconnections

To build an integrated circuit, the individual devices formed by the planar process must be interconnected by a conducting path, as shown in Figure 2.33. This procedure is usually called *interconnection* or *metallization*. As the performance of individual transistors improves, the overall circuit performance can be limited by the interconnections between the transistors, rather than by the transistors themselves.

The simplest and most widely employed interconnection method is the *subtractive* process. First, the SiO_2 is removed from areas where a contact is to be made to the silicon. Then a layer of metal is deposited over the surface, usually by physical vapor deposition (PVD). A solid source of the material to be deposited is vaporized by *electron-beam* (*EB*) bombardment in an evacuated chamber or by ion bombardment (sputtering) in a low-pressure ambient. The vaporized metal atoms travel to the wafer where they condense to form a uniform film. The metal, often aluminum or an aluminum alloy (such as aluminum with a few percent of silicon or copper), is then removed from areas where it is not desired by lithography and etching operations similar to those already discussed. The aluminum is usually etched by anisotropic, dry etching techniques, although it can be etched in aqueous phosphoric-acid solutions when the feature sizes are large.

With shallow junctions, penetration of the metal even slightly into the silicon cannot be tolerated. Silicon is slightly soluble in aluminum so a small amount of silicon from the substrate dissolves into pure aluminum to satisfy the solid-solubility requirement. The resulting voids in the substrate are then filled by aluminum. When this "spiking" occurs,

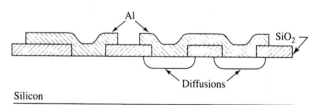

FIGURE 2.33 A thin layer of aluminum can be used to connect various doped regions of a semiconductor device.

the aluminum can penetrate through a *pn* junction, creating a high leakage-current path. In the past, silicon dissolution into the aluminum was prevented by purposely including a few percent of silicon in the aluminum film during deposition. With very fine features, however, the use of silicon-doped aluminum is less attractive. Any silicon in excess of the solid-solubility limit can precipitate in the very small contacts, occupying an appreciable fraction of the area and significantly increasing the contact resistance. For this reason, an additional layer is often inserted between the main aluminum metallization and the silicon. As discussed below, contacts are often made to silicide layers, which are needed to lower series resistance of diffused regions. In other cases, thin conducting *barrier layers* are included between the silicon and the aluminum. Alloys of titanium and tungsten or titanium and nitrogen are widely used.

As junctions become shallower, the resistivity of diffused regions increases, and the current through a device can be limited by the lateral resistance of a diffused region in series with the active region of the device, rather than by the active region itself. To reduce this series resistance, lower-resistance materials can be placed on top of (in parallel with) diffused regions. In one method a *self-aligned* metal silicide (a *salicide*) is formed on top of exposed silicon, as shown in Figure 2.34a. A metal layer, such as titanium, is deposited over the entire wafer. The wafer is then heated so that the titanium reacts with silicon in the exposed silicon regions to form $TiSi_2$, while no reaction occurs where the titanium is over oxide. The unreacted metal is then removed from the oxide by wet etching in a solution that removes the metal, but not its silicide. A further heat treatment converts the silicide to its low resistivity form, so that a low-resistance silicide ($\rho \sim 15\ \mu\Omega$-cm) remains over the exposed silicon while no conducting material remains over the insulating oxide regions. By using this self-aligned silicide formation process, no additional area is needed for aligning photomasks, providing a very compact arrangement. Other metals, such as cobalt, can also be used to form self-aligned silicide regions. The salicide process is especially valuable for forming low-resistance layers over the electrodes of MOS transistors. The silicide forms over the single-crystal source and drain regions and also over the polycrystalline-silicon gate electrode, as shown in Figure 2.34b.

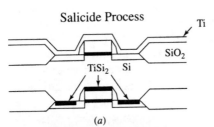

(a)

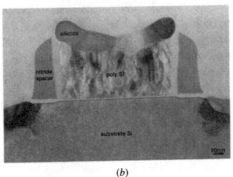

(b)

FIGURE 2.34 (a) In the salicide process Ti is deposited over the entire wafer and annealed to form $TiSi_2$ over the exposed silicon. The unreacted Ti over the oxide is then removed by wet chemical etching. (b) Cross-sectional transmission electron micrograph of silicide formed by the salicide process over the gate, source, and drain regions of an MOS transistor. (*Courtesy of Accurel Systems International Corp.*)

In virtually all MOS circuits and frequently in bipolar circuits, one interconnection layer is composed of polycrystalline silicon, while another is made of aluminum or copper. With these two interconnection layers plus a possible diffused interconnection line in the surface of the silicon wafer, current can be carried at three different vertical levels—an important degree of freedom in circuits having many thousands of devices. As we will see below, additional layers of metal are used in complex circuits.

However, the resistivity of polycrystalline silicon is limited to about 500 $\mu\Omega$-cm so that significant voltage drops can occur across long polysilicon conductors. Perhaps more significant, however, the RC time constants associated with the resistance of a long polysilicon interconnection and its capacitance to the substrate can slow signal propagation through the IC. Therefore, the polysilicon is only used for very short *local* interconnections. Alternative materials that are more conductive than polysilicon are used for longer interconnections. Silicides of the refractory metals, such as tungsten silicide (WSi_2), tantalum silicide ($TaSi_2$), and titanium silicide ($TiSi_2$), as well as the refractory metals themselves, are employed for intermediate-length interconnections, especially when they are to be subjected to higher-temperature processing.

Multilevel Interconnections.

Multilevel Interconnections. With the increasing complexity of modern ICs, interconnecting transistors and other electronic devices becomes more difficult. In logic circuits, especially, signals must often be moved from one part of the chip to another part. When the signals must be moved over a long distance, metal interconnections must be used because of their low resistance. However, a single layer of metal is no longer adequate because of the large number of intersecting interconnections, and more than one layer of metal must be used. Typically, a first layer of metal makes contact to the devices themselves (usually to a barrier layer over the silicon or polysilicon). The first layer of metal is covered with an insulating layer—often silicon dioxide. The insulating layer is removed from regions where connections between the metal layers are to be formed, and a second layer of metal is deposited and patterned. Sequential deposition and patterning of insulating and metal layers can be repeated a number of times to form a complex multilevel metallization system. Circuits with five metal layers are common, and the use of eight or possibly more layers of metal is expected.

Figure 2.35 shows a cross-sectional transmission electron micrograph of three levels of a multilevel interconnect system. The overall impression is that the interconnect system dominates the volume of the chip. Polysilicon lines covered with silicide are visible at the bottom of the micrograph. The three levels of aluminum metallization (gray regions with grain structure) are separated from the transistors and from each other by layers of amorphous silicon dioxide (featureless, light-gray regions). The dark regions above and sometimes below the aluminum are barrier layers, which act as an etch-stop layer and also prevent interdiffusion of the materials in the interconnect system. These regions appear darker because they are composed of heavier metals, which allow fewer electrons to penetrate during transmission electron microscopy. The dark regions connecting the different conducting lines are tungsten *plugs* filling the *vias* between conductors. These plugs allow current flow between the different metal layers and to the devices at the bottom of the micrograph. The dark regions above the polysilicon lines are self-aligned silicide, which reduces the resistance of the polysilicon interconnections. Figure 2.36 shows a more-detailed transmission electron micrograph of tungsten plugs connecting an underlying $TiSi_2$/TiN contact layer and an overlying aluminum layer.

As the lateral dimensions of the features decrease, the vertical dimensions do not scale as rapidly (to minimize resistance in the metal lines and interlevel capacitance between metal lines on different layers of interconnections, both of which can limit the

FIGURE 2.35 Cross-sectional transmission electron micrograph of three levels of a multilevel interconnection system. Three levels of aluminum metallization and associated barrier layers are visible, along with the tungsten-filled (black) vias between metal layers. Polysilicon lines are visible just above the substrate. (*Courtesy of Rudolph Technologies, Inc.*)

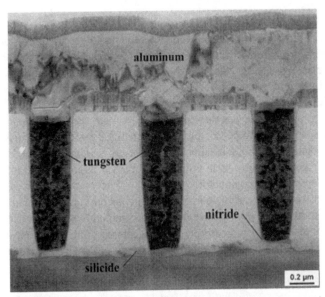

FIGURE 2.36 Cross-sectional transmission electron micrograph showing a more detailed view of tungsten plugs connecting the underlying silicide layer and the overlying aluminum first metallization layer. (*Courtesy of Accurel Systems International Corp.*)

frequency response of the circuit). Therefore, the "aspect ratio" (height/width) of the vias that must be filled with metal and the gaps between metal lines that must be filled with insulators increases, making fabrication more difficult. Advanced deposition techniques, such as the directional "high-density plasma" deposition discussed above have been developed to aid the filling of these small spaces.

With smaller features and more metal layers, the difference in the vertical elevation of the different regions of the IC becomes a concern. As the resolution of lithographic tools improves, their depth of focus decreases. On the other hand, with a complex metallization system, contacts must be patterned at several different levels. To reduce this incompatibility, several techniques are being used to achieve a more level (*planar*) surface.

One *planarization* technique polishes the deposited dielectric layers at each step of the interconnection process to provide a smooth and level surface for subsequent layers of metal. This "chemical-mechanical" polishing (CMP) acts by a combination of chemical and mechanical removal of material to provide a very smooth surface with minimum damage to the underlying structures. CMP is especially useful for providing the flat surface that aids formation of a subsequent layer of interconnections in a multilevel metal system. It is similarly valuable to remove the rough topography (Figure 2.27d) formed by the high-density plasma oxides used to fill narrow gaps between adjacent metal lines. The CMP technique produces a flat surface over arrays of fine features, but does not produce a level surface between widely spaced features because of the "dishing" of a softer material during the final stages of polishing when both softer and harder materials are being removed.

In a less frequently used planarization technique, a thick oxide is deposited by CVD and a photoresist layer is applied to a spinning wafer in the conventional manner. The photoresist tends to fill spaces between nearby features and creates a more level surface locally. The combination of photoresist and oxide is then etched by a dry etching technique that removes both resist and oxide at approximately the same rate. The smooth surface of the resist is replicated in the oxide after the etching removes all the resist and some of the oxide.

Copper. As device dimensions decrease, the requirements on the metallization system become more severe. For example, we show in Chapter 9 that the most usual technique for systematically reducing (scaling) the feature sizes in an IC results in an increased current density in the interconnections. If device dimensions on the surface are reduced by a factor K, the scaled current should also decrease by the same factor (Table 9.1). However, the interconnection cross section decreases by K^2 so that the current density flowing through the interconnection increases by K. This increase results in a larger voltage drop in the interconnections so that a smaller fraction of the externally applied voltage is effective in activating an IC device. To minimize this effect, it is important to reduce the resistivity of the interconnecting conductor.

During most of the evolution of integrated-circuit technology, aluminum was the dominant interconnecting metal. As we have discussed, small amounts of silicon were often added to prevent interaction with the underlying silicon contact region, and small amounts of copper were added to reduce electromigration, as will be discussed below. However, the basic material was aluminum with its low resistivity of 3 $\mu\Omega$-cm (in bulk form). As the interconnection system with its associated RC time constants becomes an increasing limitation on the overall integrated-circuit performance, metals with even lower resistivity are needed. The resistivity of copper is about 1.7 $\mu\Omega$-cm (in bulk form). This low resistivity makes copper attractive for use in the IC metallization system despite the difficulty of integrating it into an IC fabrication process.

A major concern is that copper introduces allowed deep levels in the bandgap of silicon, as we saw in Table 1.4. These deep levels increase the leakage current and degrade transistor gain. Therefore, care is needed to ensure that no copper can diffuse into the silicon itself and that it is all confined to the metallization system. The problem is compounded by the rapid movement of copper through silicon dioxide. Therefore, the copper must be surrounded by diffusion barriers, complicating the fabrication process. In addition, copper is difficult to etch using the anisotropic, reactive ion etching needed for very fine lines; the reaction products of the etching process are not volatile, so they do not leave the surface, and the etching effectively stops. Therefore an alternative to the conventional etched-metal interconnection system is used for copper.

After an insulating layer (usually silicon dioxide) is formed, grooves or trenches are etched into the insulator, and suitable barrier layers are formed on the bottom and sides of the grooves. The remainder of each groove is filled with copper. The barrier material and copper also deposit on the surface of the insulator outside the trench. This excess material is removed by chemical-mechanical polishing, which also produces a flat surface that aids forming the next metallization layer. This *Damascene* process (named after the inlaid jewelry-making technique associated with Damascus, Syria) can be refined to form vias between metal lines (the *dual Damascene* process), as well as the lines themselves.

When the overall circuit performance is limited by the *RC* time constant of the interconnection system, reducing the capacitance, as well as the resistance, is beneficial. Consequently materials with lower relative permittivities (ϵ_r) than the common silicon dioxide are increasingly attractive. The relatively permittivity can be reduced from the 3.9–4.1 of silicon oxide to about 3.5 by adding fluorine to the oxide. However, the decreasing stability of the oxide as more fluorine is added, limits the amount that can be included. Other materials with lower permittivities include mixtures of silicon, oxygen, and carbon, and also amorphous carbon or carbon-fluorine mixtures. Organic materials with reduced permittivities are also used. With any change of the materials, however, the overall process used to form the interconnection system must be modified to ensure that the interconnection system can be produced with high yield on chips with millions of lines and vias and also is reliable for many years of normal operation.

EXAMPLE   *Interconnection Delay*

Consider a device with a time constant τ_d of 10 ps driving a metal interconnection line with a capacitance C and a resistance R. The width W of the interconnection is 0.3 μm and its thickness t_M is 0.2 μm. The metal line is separated from an underlying conductor by an insulator with thickness $t_I = 0.3$ μm. Find the total time constant for the device and a line length 1 mm long and also the interconnection length at which the interconnection delay equals the device delay for the following cases.

(a) the metal is aluminum and the insulator is silicon dioxide

(b) the metal is aluminum and the insulator has a relative permittivity of 2.5

(c) the metal is copper and the insulator is silicon dioxide

(d) the metal is copper and the insulator has a relative permittivity of 2.5

Assume the following material properties:

$\rho_{Al} = 3.2$ $\mu\Omega$-cm

$\rho_{Cu} = 1.7$ $\mu\Omega$-cm

$\epsilon_{r\text{-}SiO_2} = 3.9$

and use the formula for a simple parallel plate capacitor.

Solution The RC time constant of the metal line is

$$\tau = \left(\frac{\rho L}{W t_M}\right)\left(\frac{\epsilon_0 \epsilon_r WL}{t_I}\right) + \tau_d$$

$$\tau = \frac{\rho \epsilon_0 \epsilon_r}{t_M t_I} L^2 + \tau_d$$

(a) $\tau = 18$ ps $+ 10$ ps $= 28$ ps

(b) $\tau = 12$ ps $+ 10$ ps $= 22$ ps

(c) $\tau = 9.8$ ps $+ 10$ ps $= 19.8$ ps

(d) $\tau = 6.3$ ps $+ 10$ ps $= 16.3$ ps

Length at which the interconnection delay equals the device delay:

(a) 0.74 mm

(b) 1.01 mm

(c) 0.92 mm

(d) 1.26 mm ∎

Electromigration. Electromigration is an IC interconnection reliability problem that can cause a circuit to fail when an interconnection become discontinuous after hundreds of hours of successful operation. Electromigration refers to the movement of atoms of the conducting material as a result of momentum exchange between the mobile carriers and the atomic lattice. The moving electrons in the metal collide with atoms and push them toward the positively biased electrode (Figure 2.37). As a result, the metal piles up near this electrode and is depleted from other parts of the conductor, especially from the regions near the intersection of grain boundaries in a polycrystalline metal film. This transfer of material eventually causes voids in the film and a discontinuous interconnection. Electromigration occurs more rapidly at higher current densities and in severe temperature gradients.

Electromigration is a major concern with aluminum, where current density is often restricted by design to be less than 10^5 A cm^{-2}. Electromigration can be reduced by adding small quantities of a second metal, such as copper, to the aluminum to inhibit the movement of aluminum atoms along the grain boundaries. The addition of 2–3% copper can increase the long-term, current-handling capability by two orders of magnitude without greatly increasing the resistivity of the film. Copper metallization is more resistant to electromigration than is aluminum metallization, and the higher current-handling capability of copper is another motivation to use copper instead of aluminum. Alternatively, higher-temperature metals, such as tungsten, which are also more electromigration resistant, can be used for metallization.

After the metal interconnection layer is deposited and defined, the wafer is placed in a low-temperature furnace (at about 450°C for aluminum) to improve the interface between metals in the interconnect system and ensure good ohmic contact. This heat treatment also improves the quality of the Si-SiO$_2$ interfaces. With the completion of the interconnection patterning, the processing of the IC wafer is complete.

In addition to resistivity and electromigration, other points to consider in choosing an interconnection material include the following: (1) ability to make ohmic contacts to both n- and p-type silicon; (2) stability in contact with silicon after the circuit is completed; (3) adhesion to both silicon and silicon dioxide; (4) ability to be patterned using available lithography and etching (especially dry etching) or polishing techniques; (5) resistance to corrosion by reaction with the environment; (6) ability to be bonded to make connection

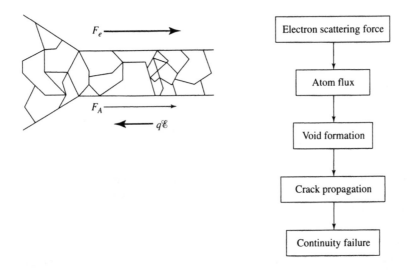

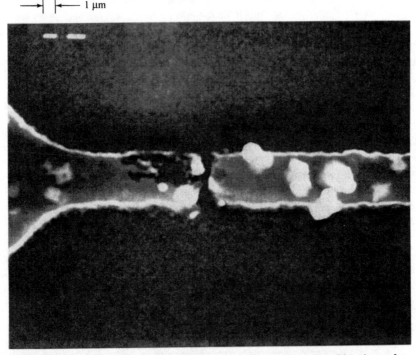

FIGURE 2.37 Electromigration mechanism in a conducting stripe. Directions of electron flux F_e, electrostatic force $q\mathcal{E}$ and resultant atomic flux F_A (upper left). Scanning electron micrograph showing void formation to the left of the break and accumulation of material in the form of hillocks to the right of the break (lower figure). The steps leading to electromigration failure are indicated at the upper right [18].

to a suitable package; (7) coverage of steps in the IC; and (8) ability to be deposited without degrading the characteristics of devices already present. While no single interconnection material is optimum for all these requirements, aluminum and its alloys satisfied enough criteria to become widely used for many years. However, as the requirements on the interconnections became more severe, the limitations of aluminum (especially electromigration) became more evident, leading to the search for alternate materials and the widespread adoption of copper.

Testing and Packaging

After the wafer fabrication process is complete, the ICs are electrically tested to determine which ones are working correctly so that only the functioning chips are packaged. As the complexity of ICs increases, testing, which is typically done under computer control, becomes much more difficult. Ease of electrical testing is an important consideration in the initial circuit design and layout, and a modest amount of extra circuitry can be included to allow efficient testing. The chip is often designed so that critical internal voltages can be accessed externally to determine whether the circuit is operating properly.

After this preliminary *functional testing*, the wafer is diced into individual circuits or chips, often by fracturing the silicon along weak crystallographic planes after scribing the surface with a sharp, diamond-tipped instrument (a diamond *scribe*). Other techniques for dicing include sawing the wafer apart or melting part way through the thickness of the wafer with a laser before breaking it. In the most straightforward packaging approach, the back of each functioning chip is then soldered to a package, and wires are connected or *bonded* from the leads on the package to the metal pads on the face of the semiconductor chip (Figure 2.38). Finally, the package is sealed with a protective ceramic or metal cover or with plastic, and the circuit undergoes further electrical testing to ensure that it is still functioning and often to determine its frequency response.

As larger systems containing more complex IC chips are designed, the number of interconnections required for communication between chips increases, and alternative packaging techniques are used. The requirement for more interconnections and greater reliability and packing density has led to the development of ceramic substrates containing several layers of metal interconnections. The IC chips can then be bonded face down on these substrates so that the metal pads on the IC chip are directly above corresponding pads on the ceramic. All leads are then simultaneously bonded by melting pre-formed solder bumps on the IC pads in what is called *flip-chip bonding*. Flip-chip bonding connects hundreds of leads simultaneously and allows connections near the center of the chip.

Power dissipation is an increasingly important factor in designing a high-performance IC and in choosing the proper packaging technique. Conventional packaging techniques limit the power dissipated in a chip to a few watts; more elaborate packaging and cooling techniques (e.g., heat sinks and even fans) can allow several tens of watts to be dissipated. As

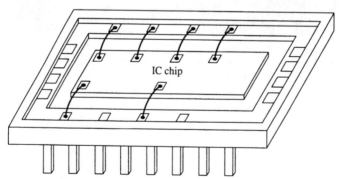

FIGURE 2.38 The IC chip is mounted in a package, and wires are connected to the external leads.

device dimensions decrease, the heat generated within a given size chip remains constant at best and often increases. Larger chip sizes increase the power dissipation per chip even more. Therefore, power dissipation (and supplying the required power) can limit the complexity of an IC chip (especially for handheld applications) and can lead to the choice of one type of circuit technology in preference to another.

Bipolar circuits generally dissipate more power than do MOS circuits, and single-channel MOS circuits dissipate more than complementary MOS (CMOS) circuits, in which little dc power flows. However, even in CMOS circuits, power dissipation is an increasing constraint. Even though little dc power flows in most CMOS ICs, the ac power to charge interconnecting lines and drive transistors can be large. The ac power is

$$P_{ac} = \tfrac{1}{2}CV^2 f \tag{2.7.1}$$

where C is the capacitance, V is the voltage change (typically almost the entire supply voltage for a CMOS circuit), and f is the frequency. Reducing the supply voltage decreases the power, but increasing the frequency (as required in high-performance circuits) increases the power greatly. The various IC technologies are discussed further in the following chapters.

Contamination

As the die size increases and the feature size decreases, the damage caused by stray particles or "dirt" becomes more serious. With larger chips fewer particle can be tolerated. With smaller feature sizes the size of particle that can destroy the chip decreases. Wafer fabrication facilities and equipment are designed to minimize the number of particles near the wafer. Class 10 (10 particles per cubic foot) *clean rooms* and class 1 (1 particle per cubic foot) areas are widely used, although at a high cost.

2.8 COMPOUND-SEMICONDUCTOR PROCESSING

Our discussion in this chapter focuses on the well-developed silicon technology, which is used for the vast majority of semiconductor devices. In specialized applications, however, other semiconductors must be used. Extremely high speed transistors benefit from the better transport properties available in selected compound semiconductors. Materials composed of column III and column V elements of the periodic table, such as gallium arsenide, can be used in specific IC applications where high carrier mobility is critical; II-VI materials are less frequently used. In addition, of course, direct-bandgap compound semiconductors, such as GaAs and GaN, are used for efficient light emission in optical emitters such as light-emitting diodes (LEDs) and semiconductor lasers. Silicon with its indirect bandgap is unlikely to become suitable for these applications.

Because of their importance for high-speed and optoelectronic devices, we briefly consider some of the processing associated with compound semiconductors. We focus our discussion on GaAs because it is the most widely used compound semiconductor material and illustrates some of the differences in processing between silicon and compound semiconductors.

The limited volume of the specialized applications for which GaAs excels has restricted development of GaAs technology, and less attention has been paid to achieving high levels of integration for GaAs integrated circuits than for silicon ICs. Consequently, the number of transistors in GaAs ICs is several orders of magnitude lower than that of silicon ICs. The focus of compound-semiconductor technology has been on achieving high performance with very fine feature sizes or accomplishing functions not realizable with silicon, rather than achieving high levels of integration.

Compound-semiconductor technology evolved along different lines than silicon technology, leading to somewhat different approaches to processing. However, the investment in silicon processing technology is so much greater than that possible for the lower-volume compound semiconductors that much of compound-semiconductor processing is evolving toward silicon technology. This convergence allows leveraging the extensive processing knowledge developed for silicon technology. Conversely, as thermal budgets for silicon processing become more limited, the use of thick thermally grown oxides is decreasing, and silicon technology is adopting some of the characteristics of the processing used for compound-semiconductor technology. Consequently, some of the previous differences between silicon and compound-semiconductor processing are becoming less important.

Because GaAs is the most common compound semiconductor, we focus on it in this discussion. Several fundamental characteristics of GaAs distinguish its processing from silicon processing. First, GaAs is a compound so that slight deviations from equal amounts of Ga and As can create deviations from an ideal *stochiometric* material, which lead to defects that degrade its electrical properties. Second, a stable oxide does not readily form on GaAs, making control of surface properties difficult.

Crystal Growth. Like silicon, GaAs can be grown by the Czochralski technique. However, the task is more complex when growing a compound semiconductor because the stoichiometry of the semiconductor, as well as its crystal quality, must be controlled. Slight deviations from equal amounts of Ga and As can lead to electrically active species. Because arsenic is much more volatile than is gallium, arsenic rapidly evaporates from the melt, leading to a gallium-rich crystal. To prevent the evaporation of arsenic from the molten material, the crystal growth apparatus can be contained in a sealed chamber containing a high vapor pressure of arsenic. The complexity and hazards of this approach led to the idea of covering the surface of the melt with another liquid which floats on the molten GaAs and blocks the evaporation of arsenic. This *liquid-encapsulated Czochralski (LEC)* technique has gained wide acceptance for forming GaAs crystals of larger diameter.

Single-crystal GaAs can also be grown by a float-zone technique similar to the one used for silicon. This *horizontal Bridgman technique* is sometimes used for smaller diameter wafers. Because the ingot is horizontal during crystal growth, it is not cylindrical, and slices from the ingot are more D-shaped than circular. As with silicon, float-zone growth became less popular as the control of the Czochralski process improved and wafer diameters increased.

Because controlling the growth of a multi-element material is more difficult than dealing with a single element, growth of GaAs and other compound semiconductors lags that of silicon. Wafers are generally markedly smaller and more expensive. The compound semiconductors also are more fragile than silicon, requiring more careful mechanical handling (especially important for automation). Thermal shock must also be minimized by controlling the heating and cooling rates more carefully than required for silicon. The compound semiconductors are less chemically robust than silicon and are attacked by many chemicals that do not damage silicon.

High Resistivity. GaAs has a wider bandgap than silicon, so its intrinsic resistivity is markedly higher. From Equation 1.1.25 we remember that to achieve the highest resistivity the Fermi level should be located near midgap to reduce the free-carrier concentrations.

While it is difficult to place the Fermi level near mid-gap in silicon, GaAs can be purposely grown with a high concentrations of specific impurities, such as Cr, which have deep energy levels near the middle of the band gap. The Fermi level is then located near mid-gap, far from either band edge to produce *semi-insulating* material.

The resulting high resistivity reduces the vertical capacitance between the substrate and devices built above it. This reduced capacitance is a significant advantage for high-performance GaAs devices compared to silicon devices. For silicon devices, the active device regions are separated from the substrate only by a *pn* junction of limited thickness, preventing the device-to-substrate capacitance from being significantly reduced.

Although insulating oxide can be grown on silicon to separate metal interconnections from the conducting substrate, the limited thickness of the oxide layer makes capacitance between the substrate and the interconnections significant. The thick, high-resistivity substrate possible with GaAs reduces this component of the capacitance also.

For silicon devices, an insulating oxide can be grown to laterally separate adjacent device regions. For GaAs the substrate can be virtually insulating, so that isolation between devices can be achieved by adding dopant by ion implantation only into the active device areas, and retaining the semi-insulating regions between devices. If semiconductor layers are grown over the entire semi-insulating wafer, good isolation can be obtained by etching these layers from the regions between devices. The resistivity of GaAs can also be increased in specific locations by disrupting the crystal lattice with a hydrogen ion implantation and not subsequently annealing to remove the purposely introduced damage. The high resistivity thus achieved can be useful for lateral isolation between devices. Although these techniques provide dc isolation between adjacent devices, the lateral capacitance between device islands can still be significant because of the high relative permittivity of GaAs.

Epitaxial Deposition. Because other compound semiconductors can be grown epitaxially on GaAs, well-controlled epitaxial growth techniques have been more extensively developed for the compound semiconductors than for silicon. Because of the large number of compound semiconductors, materials with different bandgaps can be combined (to improve electrical performance of devices) while still retaining the same lattice constant (to limit strain and the resulting crystal defects), as shown in Figure 2.39. For example, the lattice constant of $Al_{1-x}Ga_xAs$ is virtually the same as that of GaAs while its bandgap is considerably greater. Thus, unstrained epitaxial layers of $Al_{1-x}Ga_xAs$ can be grown on GaAs by epitaxial techniques similar to those discussed for silicon. (A fraction x of the Ga is replaced by Al, so the composition is indicated by the formula $Al_{1-x}Ga_xAs$. Pure

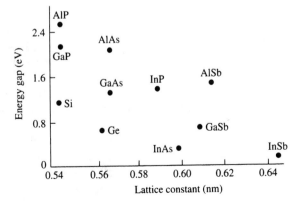

FIGURE 2.39 Electrical device performance can be improved by combining two semiconductors with different bandgaps. However, lattice mismatch and the associated strain limit the useful *heteroepitaxial* combinations of materials.

AlAs has an indirect bandgap, so the Al fraction is usually limited to a value that retains the direct bandgap). By using different materials for different parts of a device, we can take advantage of the difference in bandgaps between the two materials to build useful electronic and optoelectronic devices. These *heterojunction* devices, which use the purposely formed discontinuities in the bandgap at the interfaces, will be briefly discussed in Chapters 4 through 7. After general discussion of heterostructure devices, we will focus on the example of heterojunction devices composed of silicon and alloys of silicon and germanium both because of their importance and to stay close to our focus on silicon ICs. (Note however, that the $Si/Si_{1-x}Ge_x$ system is *not* lattice matched; the resulting strain limits the thickness of the layers than can be used.)

Because the interfaces between the epitaxial layers of heterojunction devices strongly influence the electrical behavior, good control of the transition region between the different materials is needed. Techniques have been developed to control the interfaces within a few nanometers, providing much finer control than typically used for conventional silicon devices. Conversely, processing of compound semiconductors after epitaxial growth is more difficult.

For device fabrication of structures with layers of different semiconducting materials, specific layers must often be removed from parts of the device by etching without attacking very thin underlying layers of other materials. To obtain high selectivity, wet chemical etching techniques have been well developed for GaAs and related materials.

Dielectrics. One significant difficulty with GaAs and other compound semiconductors is the lack of a stable thermally grown oxide to form isolation between the semiconductor and overlying metal layers and also to provide an electrically stable semiconductor interface. As we will see in Chapter 8, the SiO_2 that readily grows on silicon terminates most of the broken bonds at the silicon surface. This oxide reduces the number of allowed states within the forbidden gap at the silicon surface that would otherwise be formed by the disruption of the periodic crystal structure. Without an analogous oxide for GaAs, these *surface states* make controlling interface properties much more difficult, and they can dominate device operation. Thin epitaxial layers of wider bandgap materials, such as $Al_{1-x}Ga_xAs$, can be grown on the GaAs to provide more stable surfaces.

The lack of a stable thermally grown oxide for GaAs and other compound semiconductors also complicates controlled introduction of dopant into selected small regions of the surface, dictating extensive use of CVD dielectrics for compound semiconductor fabrication. Plasma-enhanced chemical vapor deposition is especially popular because the low temperatures suitable for plasma deposition minimize evaporation of the constituents of the compound semiconductors. Silicon oxides and nitrides are typically used, often taking advantage of the flexible composition of material formed by PECVD.

Dopant Addition. Within limits, dopant can be added to GaAs by gas-phase diffusion techniques. Alternatively, as for silicon, ion implantation is increasingly being used to add dopant atoms to selected regions of a device. As in silicon technology, photoresist can be used to define the areas into which the the dopant is implanted.

However, the low solid solubility of many dopants in GaAs often makes obtaining high concentrations of electrically active dopant species difficult, thus limiting device options to the few for which GaAs is suitable and enables devices with significantly better electrical performance than silicon devices. The limited solid solubility also makes difficult achieving the high dopant concentrations needed to form good ohmic contacts.

Annealing to remove crystal damage from ion implantation and activate dopant atoms is more difficult for GaAs than for silicon because of the limited thermal stability of GaAs. Arsenic tends to evaporate from GaAs at temperatures above about 600°C. If temperatures in this range are needed for implant annealing, the surface must be well protected by a deposited layer to prevent arsenic evaporation and the resulting loss of stochiometry near the surface. Nonstoichiometric silicon nitride layers deposited by plasma-enhanced CVD at a low temperature (~300°C) can block the evaporation, while oxides are less effective. Alternatively, an unprotected surface can be annealed in an overpressure of arsenic—a difficult processing requirement.

Interconnections.   Once dopant atoms are in place and activated, additional layers of deposited dielectrics can be used to isolate the overlying conductors from the active device regions. After the dielectrics are formed, contact windows are opened, as for silicon, and metal layers are deposited. To make good electrical contact to GaAs, metal systems containing several different layers are usually employed. In the past this multilayer metallization was a drawback compared to the simpler metallization systems used for silicon, but silicon metallization is now more complex also, with intermediate layers used between the main aluminum or copper metallization system and the silicon.

For high performance with compound semiconductors, as with silicon, considerable attention must be paid to the interconnection system. For highest performance, low resistivity metals and low permittivity dielectrics should be used in the interconnection system. Because of the lower level of integration of GaAs ICs, the metallization system is more flexible. In addition to aluminum, gold is often used as the main conductor. Because gold does not make an ohmic contact, intermediate layers of metal are used between the gold and the semiconductor.

To reduce the capacitance, low-permittivity dielectrics are used between the metal and the conducting region; organic materials are sometimes used, as they are for silicon ICs. In extreme cases, the dielectric between the metal and the device can be removed by wet chemical etching after the metal is defined. Although the relative permittivity of the air between the metal and the semiconductor is unity, such "air bridges" are fragile and only suitable for specialized and small ICs.

To take advantage of the high performance of compound-semiconductor devices, the packages in which they are placed must not seriously degrade the overall performance. Specialized packages with very low capacitance and inductance are available for high-frequency devices. For optoelectronic applications, packaging has different requirements. If light must be able to leave the chip, parts of the package must be transparent while still providing adequate protection to the electronics inside. The light must also be reflected and focused so that it leaves the chip and package, rather than being reabsorbed in various regions of the chip or in the packaging material.

2.9 NUMERICAL SIMULATION

Basic Concept of Simulation

As integrated-circuit processes and devices become more complex, solving the analytical equations associated with device physics or processing becomes more difficult. Devices with finer dimensions and shallower junctions require second-order physical and electrical effects to be considered. At the same time that predicting the device and process behavior is becoming more complex, the difficulty of conducting meaningful experiments is also increasing. The experimental control needed to clearly reveal the effects being studied can exceed that possible with even carefully designed experiments, and the cost of experiments is increasing as process equipment becomes more expensive. At the same time, the decreasing flexibility of increasingly automated equipment makes experiments more difficult to perform. The overwhelming number of possible variations preclude experimentally studying all the relevant combinations of variables.

In contrast, numerical simulation becomes increasingly attractive as readily available computers become more powerful, and simulation is increasingly being used in place of extensive experimentation. Both physical fabrication processes and electrical device behavior can be numerically simulated with increasing accuracy. The ability to rapidly simulate many process and device variations allows the most promising ones to be identified and experimental work to focus on these limited number of possibilities. In addition, modern devices require that the structural parameters and dopant profiles be known more accurately than previously. Device performance depends increasingly on these parameters and on more accurate understanding of the behavior of carriers under electric fields in these structures.

The simulation techniques can be strictly numerical or they can combine numerical methods with analytical expressions. When analytical expressions are available, their use aids the designer by providing better insight into the process chemistry and device physics, as well as decreasing the computer time required. However, using more accurate representations of the process or device (e.g., using the more complex diffusion equation of Equation 2.5.15 in place of the simpler Fick's-law expression of Equation 2.5.7) typically leads to equations that can only be solved by numerical techniques.

The numerical simulation tools available are increasing in capability. Process and device simulators are incorporating more realistic physical models and can consider two and even three dimensions. As device dimensions decrease, the influence of the device edges extends over a significant fraction of the device and noticeably modifies device fabrication and behavior. Lateral effects between adjacent device elements are also becoming increasing important. Consequently, including two-dimensional effects is critical for simulation of modern processes and devices. As the dimensions of the simulation increase, however, the numerical computing power needed increases rapidly, and multi-dimensional simulators typically cannot include as detailed models as can one-dimensional simulators. Considering effects at corners is also becoming more important, requiring three-dimensional simulation, with its increased computational complexity. Because the computer capability needed increases rapidly when two and three-dimensional effects are included, numerically evaluated analytical models are attractive when they can be used. When two-dimensional effects are considered, the shape of the features and dopant distributions near the edges should be considered. For example, the shape of chemically vapor deposited layers in deep features is important and should be modeled. Etching of fine features also requires considering edge profiles and, therefore, two-dimensional effects.

The computer simulation programs perform computation based on physical models and numerical parameters and are only as accurate as the values of the parameters used. Accurate simulation requires accurate knowledge of the large number of parameters that enter into the models. Knowing these parameters is a major challenge. Default values are usually built into the simulation program based on the model developers' "best guess" from the literature or experiments. Because the values of the parameters can be refined as more accurate experimental values become available, many of the programs are written so that the numerical parameters can be modified by the user. In addition to the limited knowledge of the parameters, some must be modified for particular fabrication facilities; for example, oxidation depends on the oxygen pressure, which is significantly lower at higher elevations (e.g., Colorado, compared to California).

Grids

Numerical simulation is based on calculating properties at a number of points or nodes in the device regions of interest. The properties are calculated based on time variations and values at neighboring nodes. The nodes are the intersection points of a *grid* superposed on the device. Grid generation is one of the most difficult tasks in computer simulation. The grid must be fine enough to capture details of interest, which may vary rapidly with position in the internal regions of active devices, without requiring excessive computation time in regions with less-rapid spatial variations. The grid spacing depends on the dimensions of the physical features and also on the characteristic lengths of moving species. Simultaneously simulating species moving fractions of a micrometer (e.g., dopant atoms) and species moving hundreds of micrometers (e.g., point defects) is difficult because of the different length scales involved. When boundaries between regions move during the simulation, as during oxidation or silicide formation, specifying the grid is especially troublesome. The boundaries separating regions of different materials no longer remain at the same distances from nodes, and a node may physically move into a different material. When species with widely varying time constants must be considered in the same simulation, specifying time increments is also difficult.

In one-dimensional simulation, the physical separation of nodes is chosen consistent with the size of the expected spatial variations. The distance between grid lines can vary so that the nodes can be closely spaced in regions of expected rapid variation of structural or electrical parameters; they can be more widely spaced in regions with slowly varying properties to reduce computational time. In two dimensions, specifying the grid is more complex. To avoid excessively long computation times, the number of nodes should be reduced as much as possible. A regular set of equally spaced grid lines provides more nodes than needed, and nonuniform grids are again used so that the nodes can be concentrated in regions of rapidly varying physical features or electrical properties. The capability of including arbitrary geometries becomes especially important as device dimensions decrease and the actual shapes of various regions of the device have more influence on the device behavior. However, the need to handle arbitrary shapes makes grid generation more difficult.

To use a typical two-dimensional simulator, the user initializes the grid by specifying a nonuniform rectangular grid. The lines of the rectangular grid can be nonuniformly spaced to place more nodes in regions of rapidly changing features, and portions of some grid lines can be removed. The modified coarse rectangular grid is then automatically converted into a triangular grid by adding diagonals. With a triangular grid, a node exists wherever three or more line segments meet. The area of each triangle is shared by the three nodes that form the triangle. The simulator can increase the grid density in regions with rapidly varying parameters. For example, the grid can be refined near the edges of the source and drain regions of an MOS transistor where the doping changes rapidly. In device simulation, after Poisson's equation is solved for a specific bias condition, the grid can be refined in regions where potential or charge density change rapidly.

As the structure changes during processing, the grid must also change. For example, during oxidation the Si/SiO_2 interface can move through existing nodes. These nodes must be removed from the silicon, and additional nodes must be added to the growing oxide. Similarly, deposition and etching processes require that nodes be added and removed. These modifications are done by the simulation program itself.

The grid should be fine enough to resolve rapid changes in structure or dopant profile (for process simulation) or potential and carrier concentrations (for device simulation). It should fit the device shape to resolve the geometric features of the structure being simulated. However, the computation time increases rapidly as the number of nodes increases, varying approximately as N_p^α, where N_p is the number of nodes and α is between 1.5 and 2. Irregularities in the grid array also cause convergence problems. Figure 2.40 [19] shows a triangular grid applied to non-rectangular geometries. It would be difficult to adapt a rectangular grid to these nonrectangular structures without greatly increasing the number of nodes.

Process Models

Process simulators generally compute changes caused by each individual process step in the same sequence as the operations would be physically performed on a silicon wafer.

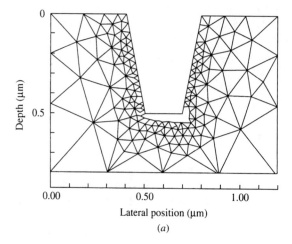

Lateral position (μm)

(a)

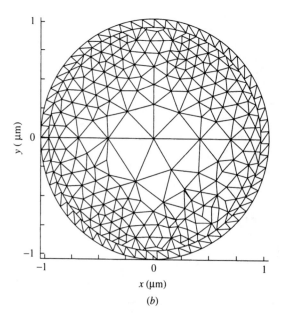

x (μm)

(b)

FIGURE 2.40 Triangular grid applied to nonrectangular geometries. (*a*) Trench with sloping sidewalls. (*b*) Circle [19].

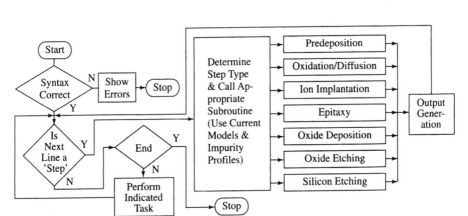

FIGURE 2.41   Block flow diagram for the SUPREM process-modeling computer program.

In addition to predicting the final dopant profiles, the process simulators calculate some simple electrical parameters. The calculated dopant profiles can also be used as the input to a device-modeling program for more extensive device simulation.

At the same time that more accurate models are needed for process and device design, the simulators are also moving from the realm of research and development into manufacturing-oriented engineering. In this environment, the complexity of simulation tasks grows, with the need to simulate complete device-fabrication process sequences and the resulting device characteristics. To aid simulation of long process sequences, shorter modules containing a limited number of related process steps are first developed. These are then tied together to allow simulation of the entire device-fabrication process. Modules describing a group of process steps are first defined. Then, the modules are linked together (often by moving icons representing the individual modules on a graphical interface) to simulate a complete process flow. Parameters can then be varied within each module without rewriting the entire specification.

When simulation moves into manufacturing environments, the effect of expected process variations on device characteristics must be considered. The effect of statistical distributions of the results of individual process steps on device performance is needed to design a process and device with economically acceptable yield. That is, the process and device parameters must be chosen so that the device characteristics are not overly sensitive to the normal fluctuations of parameters from any process step. Simulations provide a practical way of studying the effect of the process variations on device characteristics.

SUPREM.   One widely used process modeling program is called SUPREM (for Stanford University Process Engineering Model) [20].* The program input for SUPREM is a description of the individual processing steps. This input specifies a sequence of times, temperatures, ambients, and other parameters for diffusion, oxidation, implantation, deposition, and etching. The output is the impurity profile in the silicon substrate and the overlying layers, such as SiO_2 or polysilicon. The basic structure of the SUPREM program, illustrated in Figure 2.41, is designed so that process steps can be simulated either individually or sequentially, with the dopant profile predicted at the end of one process

---

* Several commercial derivatives and variations of SUPREM are available.

step used as the input for the next. This program includes detailed models for nonlinear diffusion, dopant segregation during oxidation, evaporation at the solid–gas interface, effects of the moving Si–SiO$_2$ boundary beneath a growing SiO$_2$ layer, impurity clustering during diffusion, concentration-enhanced oxidation, epitaxy, and ion implantation, as well as several other models that go beyond the first-order descriptions that can be treated analytically. Complex point-defect models are included because of the importance of point defects on dopant diffusion for shallow junctions.

To find the dopant concentrations at each point and each time, SUPREM considers the dopant redistribution and other processes occurring during each process step. Values at each node are calculated based on initial conditions at the boundaries of the simulated region or the values known at each node from the previous process step. The process step being simulated is then divided into small time intervals Δt, and the changes occurring in each time interval are calculated from the equations governing the behavior. For example, in its simplest form diffusion is governed by Fick's second law (Equation 2.5.7):

$$\frac{\partial C(x)}{\partial t} = D \frac{\partial^2 C}{\partial x^2} \tag{2.9.1}$$

in one dimension. The spatial derivatives can be replaced by differences between concentration values at nearby nodes; and the time derivative, by the time interval Δt:

$$\Delta C(x) = D \frac{\Delta^2 C}{\Delta x^2} \Delta t \tag{2.9.2}$$

The concentration at a node at time $t + \Delta t$ is then found from the values at the same node and nearby nodes at time t:

$$C(x, t + \Delta t) = C(x, t) + \Delta C(x) \tag{2.9.3}$$

In the example above, finding the value of the second derivative requires using concentrations at more than the nearest nodes [$C(x - \Delta x, t)$ and $C(x + \Delta x, t)$] so that the curvature of the concentration profile can be determined. In numerical simulators, more complex models of the diffusion process can be included, and approximation by a constant diffusivity is not required; the diffusion coefficient D can be a function of the concentrations C, as in Equation 2.5.15.

Deposition, etching, impurity deposition, oxidation, and out-diffusion occur at the exposed surfaces; ion implantation occurs near the top surface; and diffusion occurs throughout the structure. The structure is composed of regions of single-crystal silicon, polycrystalline silicon, silicon dioxide, silicon nitride, silicon oxynitride, aluminum, and photoresist. These layers can be deposited or removed and oxide can flow under stress at high temperatures. Therefore, during each process step, several different physical process must be considered. For example, during oxidation we must consider diffusion of dopant in the substrate, transfer of dopant from the substrate into the growing oxide, and possibly viscous flow of the growing oxide, as well as the growth of the oxide itself. At points where two or more materials meet, there are multiple nodes, one for each material at that point, so that multiple values of the parameters can be represented. There is also an extra node at each point on an exposed boundary to represent concentrations in the ambient gas. The coordinate system is fixed to the substrate, rather than the surface of the wafer, to avoid problems when the location of the surface changes during oxidation or etching.

Diffusion. Typically considered dopant species in silicon include boron, phosphorus, arsenic, and antimony. Diffusion of dopant atoms occurs during heat cycles in either an inert or an oxidizing ambient. In the latter case, both the dopant atoms and the surface of the silicon move with respect to the fixed substrate. As we saw in Sec. 2.5 the diffusion coefficient can depend on the dopant concentration and the point defects (silicon interstitials and vacancies) caused by oxidation or implantation. A concentration-dependent diffusion coefficient precludes general analytical solutions, but can readily be treated by numerical simulation techniques. Oxidation injects excess silicon atoms (self-interstitials) from the oxidizing interface into the underlying silicon. These point defects can diffuse long distances and affect the diffusion of nearby dopant atoms. Similarly, the damage caused by ion implantation can create excess interstitials and vacancies and greatly change the diffusion of nearby dopant atoms. The effect of point defects on the diffusion must be included in the models to obtain reasonable accuracy. The effect of built-in electric fields on dopant diffusion can also be included; this electric field couples the diffusion equations for the different dopant species present.

To include the effect of point defects, the diffusion coefficient is modeled by

$$D = \left[D^0 + D^+\left(\frac{n_i}{n}\right) + D^-\left(\frac{n}{n_i}\right) + D^=\left(\frac{n}{n_i}\right)^2 \right] F_{IV} \tag{2.9.4}$$

The components of D were discussed in Sec. 2.5; not all are used for each element. F_{IV} models the amount of enhancement or retardation by nonequilibrium concentrations of point defects and is related to the fraction of the diffusion that occurs by vacancy or interstitial mechanisms. At high dopant concentrations, the electrically active dopant concentration can be less than the total concentration physically present, and this effect can also be included, as can the segregation of dopant atoms across a Si/SiO_2 boundary.

Oxidation. Oxidation can be simulated using the Deal-Grove model with the increment of oxide grown in a time Δt given by the rate (Equations 2.3.5–2.3.8) times the time increment. Because numerical technique are being used and closed-form analytical solutions are not required, the rate expression of the basic Deal-Grove model can be modified to include a term expressing the initial oxidation rate for thin oxides, which decays exponentially with increasing oxide thickness (Equation 2.3.11) [8]:

$$\Delta x_{ox} = \left[\frac{B}{A + 2x_{ox}} + K \exp\left(-\frac{x_{ox}}{L_o}\right) \right] \times \Delta t \tag{2.9.5}$$

Different values of the parameters can be used for differently oriented single-crystal silicon and also for polysilicon. In two-dimensional models the gradual variation between an oxidizing region and a region protected from oxidation (e.g., by a silicon nitride layer) can be specified by considering lateral transport of the oxidizing species. The shape of the resulting oxide region formed is shown in Figure 2.42.

Ion Implantation. Ion implantation can be modeled either by using the analytical methods described in Sec. 2.5 or by Monte Carlo techniques. In the former case, the Gaussian distributions can be modified by using different values for the projected standard deviation or straggle ΔR_p in the region between the surface and the peak concentration than in the region beneath the peak concentration and also by including higher-order moments. Some limited capability to model channeling of ions along open crystal directions in the lattice can also be included. Two-dimensional effects are included by considering lateral straggle (y-direction), as well as vertical straggle (x-direction), so that the implanted profile can be written as

$$C(x, y) = C_p \exp\left[-\frac{x - R_p}{2\Delta R_p} \right] \left[\frac{1}{2} \text{erfc}\left(\frac{y - a}{\sqrt{2}\Delta R_p} \right) \right] \tag{2.9.6}$$

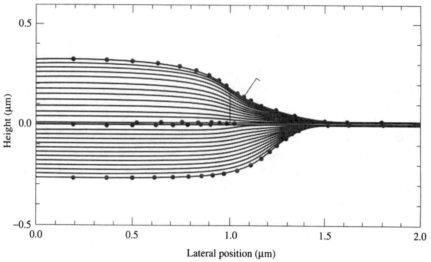

FIGURE 2.42 Two-dimensional shape of the oxide formed by lateral transport of the oxidizing species near the boundary between an oxidizing region and a region protected from oxidation by silicon nitride. [K. M. Cham, et al., *Computer-Aided Design and VLSI Device Development* (Kluwer Academic Publishers, Boston 1986), [21], used by permission.]

for $y > |a|$, where C_p is the peak concentration in the unmasked region, and a is the half width of the mask opening.

Alternatively, Monte Carlo techniques can be used. The paths of many individual particles are considered until a general picture of the behavior caused by the sum of the individual particles emerges. Monte Carlo techniques are especially valuable if the implanted species must travel through a number of layer of different materials. For each implanted ion the energy loss and direction change by each scattering event is found so that the energy and direction of the ion is known as it enters the next layer. The ion reaches its final stable position when its energy approaches zero, and the distribution is found by counting the number of ions stopping within each depth interval after the trajectories of a large number of ions are calculated. Both nuclear and electronic energy-loss mechanisms are generally included. Because the calculations are repeated for a large number of ions to determine the expected distribution, Monte Carlo techniques are time consuming and are used only when the computationally more efficient analytical models do not give adequate accuracy. To reduce the computational time, the combination of parameters can sometimes be reduced to a single variable, and the solution computed for values of this variable and stored in a table so that the solution for a given value of the variable can be found from the table, rather than computed each time it is needed.

By properly combining models, the dopant concentration near complex transition regions can be modeled in two dimensions after implantation and annealing. Figure 2.43 shows lines of constant dopant concentration near the transition between a device region and an isolation region.

Deposition and Etching. During deposition processes at elevated temperatures (e.g., epitaxy), dopant already present in the structure diffuses. Consequently, the diffusion of all dopant species must be calculated, in addition to adding the desired layer to the structure being simulated. For lower temperature depositions, diffusion can be neglected to save computational time. Etching can also be specified. In basic models, the material

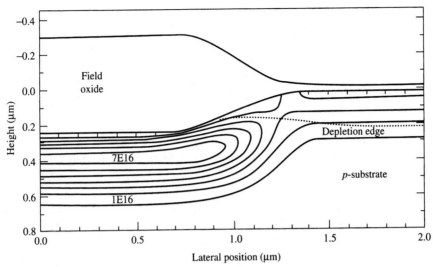

FIGURE 2.43 Lines of constant dopant concentration near the edge of a LOCOS oxide isolation region, such as that shown in Figure 2.42. [Cham, et al., [22], used by permission.]

is simply removed to a given depth with simple assumptions about the profile of the edges of the remaining material. Separate, detailed models have been developed to describe the shape of etched edges (and also the shape of layers deposited on irregular surfaces and into trenches).

Stress. Calculating stress can be important. If the stresses become too great during a processing step (e.g., thermal oxidation), crystal defects can be generated. For example, during LOCOS oxidation (Sec. 2.6), large amounts of stress can be generated in the silicon if the stress-relief oxide separating the silicon and the silicon nitride is not sufficiently thick. These stresses can create linear crystal defects (dislocations) that degrade the electrical properties of junctions through which they pass. [The defects caused by these stresses are the reason the stress-relief oxide (also called the *pad oxide*) is used even though it allows lateral expansion of the oxidizing region and makes control of fine dimensions difficult.] Stresses are calculated using Young's modulus, Poisson's ratio, and the linear coefficients of thermal expansion for each material. Intrinsic stresses resulting from deposition processes can also be included.

The calculated stress during oxidation is used to modify the oxidation reaction-rate coefficient and the diffusion coefficient of the oxidizing species through the already formed oxide. At high temperatures, SiO_2 can flow viscously, and this flow can be included in models. The viscous flow is especially important when considering oxidation of a surface partially covered with a masking layer because viscous flow can change the shape of the transition region. The models can also predict the residual stress remaining in the film after the sample cools. Because of the different thermal coefficients of expansion of silicon, SiO_2, and other materials present, the stress can increase as the sample cools even if the sample was stress free at the oxidation temperature.

Surfaces. Purely numerical simulations are useful for studying the concentrations of dopant atoms within silicon and their changes during processing. Changes at a point are usually dominated by the conditions at neighboring points at a previous time step. However, near a surface, changes at a point are dominated by species arriving from distant locations. For example, both deposition and vapor etching are influenced by species within a mean free path in the gas phase, a distance often large compared to the device features. In some such cases, Monte Carlo techniques are useful. The trajectory and effect of individual particles arriving at the wafer with random velocity and direction is simulated by considering their behavior under the known laws of physics and chemistry that affect them. After a large number of particles is considered, the average behavior is determined; however, the computational resources needed for extensive Monte Carlo simulations, limits their use to specific detailed studies.

For more general use, quasi-analytical solutions are often attractive. For example, the deposition or etching rate at a given point can be determined by integrating the effect of particles (ions, radicals, etc.) arriving at the point from the distant gas-phase source (e.g., a physical vapor deposition source or a chemical species for CVD) and from all other nearby surfaces from which species are reflected after arriving from the distant gas-phase source. The changes at other points on the surface are then similarly calculated to determine the time evolution of the surface.

Postprocessing. Once the simulation is complete, the data are often sent to a separate postprocessing program which allows plotting and manipulation of the data to aid in understanding the results of the simulation.

EXAMPLE *Process Modeling: One-Dimensional Boron Diffusion* [23]

Use the modeling program SUPREM to investigate a silicon wafer that is (1) ion implanted with boron (B) and (2) subjected to several subsequent high-temperature process steps. Find the distribution of the boron after processing is completed.

Solution The input to SUPREM includes several lines of computer code. First, a title such as the following is assigned to identify the analysis.

```
TITLE: BORON IMPLANT AND REDISTRIBUTION
```

Next, the silicon substrate is described. The dopant species and concentration are specified together with the crystal orientation.

```
SUBS ELEM = P, CONC = 2E15, ORNT = 100
```

In this example the substrate is doped with phosphorus (P) to a concentration $N_d = 2 \times 10^{15}$ cm^{-3} and has a (100) crystal orientation.

Parameters relating to the grid spacing in the vertical direction are then described. The vertical dimension in the silicon is divided into two portions, a high-resolution region just beneath the surface, and a lower-resolution region farther from the surface. The grid spacing, depth of the high-resolution region, and total depth to be simulated are specified by the user. There are typically 350 to 400 grid points along the vertical dimension. Those points not specified for the high-resolution region are automatically distributed throughout the lower-resolution region.

```
GRID DYSI = 0.005, DPTH = 1.5, YMAX = 2.5
```

This line specifies the grid spacing ΔY to be 0.005 μm in the high-resolution region, which is 1.5 μm deep (thus using 300 points in this region). The total region to be simulated extends to 2.5 μm beneath the silicon surface, so the lower resolution region can contain 50 to 100 grid points.

At this point, details of the output format for printing and plotting results can be entered. In addition, process models that are not included in the basic SUPREM program can be specified. If they are not, built-in models are used in the calculations.

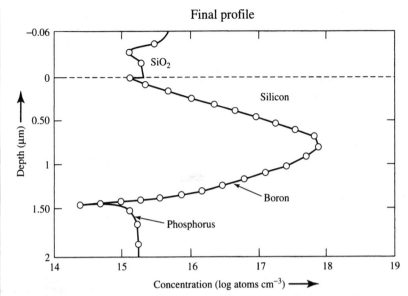

FIGURE 2.44 Boron and phosphorus profiles at the end of processing as calculated by SUPREM.

After these initial parameters are entered, the actual implantation, diffusion, and oxidation operations are specified. A line is included for each operation or *step*.

```
STEP TYPE = IMPL, ELEM = B, DOSE = 3.2E13, AKEV = 380
```

In this example, boron is to be implanted with a dose of 3.2×10^{13} cm^{-2} at an implant energy of 380 keV. Subsequent heat-treatment steps are then specified. One step might be the following:

```
STEP TYPE = OXID, TEMP = 1000, TIME = 30, MODL = DRY1
```

Here an oxidation at 1000°C for 30 min is specified. The oxidation kinetics are described by the model called DRY1, which has been previously entered. It describes the dilute-oxygen ambient used during insertion of the wafers into the furnace. Other oxidation and inert-ambient heat treatments are then specified.

At the end of the sequence of steps, parameters calculated by the program are printed and the final dopant profile is plotted. In this case the relevant parameters are the fraction of the boron dose remaining in the silicon, the portion in the oxide, and the dopant concentration at the surface. Figure 2.44 shows the SUPREM prediction for the impurity profile representing boron subjected to the two process steps described above followed by four subsequent heat treatments similar to those used during fabrication of an MOS transistor.   ■

Device Simulation

Process simulation provides the essential first step in the overall simulation of an electronic device. It can tell us the detailed physical structure and the dopant profiles in one, two, or three dimensions. These attributes are then used as the starting point for detailed simulation of the electrical behavior of the device. The output of the process-simulation program is usually saved in a computer file that can be directly used as the input to the device-simulation program. This coupling of process and device simulation greatly enhances our ability to predict the behavior of a device from the processing sequence. Automatic optimization of the process sequence to obtain desired device characteristics is the next logical step, but much harder.

As for process simulation, device simulation can also be one-dimensional or multi-dimensional. One-dimensional models can simulate more sophisticated physical effects; inclusion of such effects is limited by the available computer capability in two and three dimensions. Device-simulation programs can provide information about behavior in both steady state and during transient conditions. The more general models consider both electrons and holes and are based on solution of Poisson's equation and the continuity equations. The physical structure to be simulated can be specified directly as the input to the device simulator. This structure can be estimated from nominal dimensions and analytical models or approximations for dopant profiles, or the structure predicted by the output of a process simulator can be used as the input of a device simulator. Device simulation will be discussed in more detail in Chapter 5 after additional device concepts are introduced.

Simulation Challenges

Process and device simulation is becoming ever more essential as devices dimensions decrease and higher-order physical effects must be considered in designing advanced processes and devices. As devices become smaller, however, the detailed physical models for process and device behavior become increasingly complex and, in some cases, are still not completely understood. Although models are continually being improved, having the appropriate physical models available when needed for advanced process and device simulation is a major challenge. Simulating tomorrow's processes, devices, and circuits with yesterday's, or even today's, physical models does not provide adequate insight. Accurate computation based on inadequate physical models can be misleading, especially as the user is shielded more and more from the details included in the models.

Because of these practical limitations, simulation cannot be used to replace understanding of the physical mechanisms involved. Instead, it is useful for gaining better understanding than can be obtained by experiments alone. It can also provide more information than analytical computation by reducing the need for approximations imposed by the limited number of equations that can be solved analytically. Even when some parameters are not known accurately, carefully used simulation can provide insight by revealing trends in the behavior. In addition, process models must often be simplified so that the calculations can be completed in a reasonable amount of time. When many sequential process steps are simulated, the small errors introduced in approximating each step typically compound so that details of the final predicted behavior can be considerably in error. Experimental measurements are essential to confirm computer predictions at critical points, to determine model parameters, and to point the way to improved models. Ideally, simulation provides the tools to couple our physical understanding of the mechanisms with experimental results.

2.10 DEVICE: INTEGRATED-CIRCUIT RESISTOR

Resistors are simple electronic elements that are important in many integrated circuits. There are several different ways to fabricate them using the processing steps discussed in this chapter. We describe resistor fabrication techniques after a general discussion of the electrical behavior of diffused IC resistors.

In Chapter 1 we noted that the resistance of a bar of uniform conducting solid material is given by the equation

$$R = \frac{\rho L}{A} \tag{2.10.1}$$

where the resistivity ρ is the reciprocal of the conductivity given by Equation 1.2.7, which is repeated here:

$$\sigma = \frac{1}{\rho} = (q\mu_n n + q\mu_p p) \tag{2.10.2}$$

A frequently used method for forming an integrated-circuit resistor is to define an opening in a protective SiO_2 layer above a uniformly doped silicon wafer and to introduce dopant impurities of the opposite conductivity type into the wafer, as shown in Figure 2.45. In Chapter 4 we will see that the junction between two regions of opposite conductivity type presents a barrier to current flow. Therefore, if contacts are made near the two ends of the p-type region and a voltage is applied, a current flows parallel to the surface in this region. It is not possible to use Equation 2.10.1 to calculate the resistance of this resistor because it is not a uniform bar. As shown in Figure 2.21, the dopant concentration resulting from the processing described above decreases from a maximum near the surface as one moves into the silicon. To calculate the resistance in this case, it is useful to consider the conductance parallel to the surface.

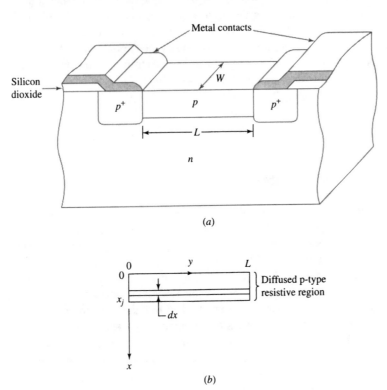

FIGURE 2.45 (a) An IC resistor defined by diffusing acceptors into selected regions of an n–type wafer. The p^+-regions are highly doped to assure good contact between the metal electrodes and the p–type resistor region. (b) The dimensions of a thin region in the resistor having conductance dG given by Equation 2.10.3.

Conductance.   Consider a p-type resistor made by introducing a p-type dopant into an n-type wafer as shown in Figure 2.45. The differential conductance dG of a thin layer of the p-type region of thickness dx parallel to the surface and at a depth x (shown in Figure 2.45b) is

$$dG(x) = q\mu_p p(x) \frac{W}{L} dx \qquad (2.10.3)$$

We can find the conductance G of the entire p-type region by summing the conductance of each thin slab from the surface down to the bottom of the layer. This sum becomes an integral in the limit of many thin slabs.

$$G = \frac{W}{L} \int_0^{x_j} q\mu_p p(x)\, dx \qquad (2.10.4)$$

where x_j is the depth at which the hole concentration becomes negligible (close to the point where $N_a = N_d$).

If the p-region was formed by a gaseous deposition cycle followed by a drive-in diffusion, we can approximate the dopant profile $N_a(x)$ by a Gaussian distribution (Equation 2.5.13). As shown in Sec. 2.5, the total density of dopant N' per unit area diffused into the wafer is determined by the nature of the deposition process, while the characteristic length of the diffused region $2\sqrt{Dt}$ is associated with the drive-in diffusion. The hole concentration at any depth into the wafer is approximately given by the net concentration of p-type dopant atoms in excess of the original n-type dopant concentration of the starting wafer. In practice, most of the current in the diffused area is carried in the regions with the highest dopant concentration, which is usually of the order of 10^{18} cm^{-3} or greater. Because the starting wafer often has a dopant concentration of about 10^{15} cm^{-3}, we can neglect the background concentration and assume that $p(x) = N_a(x)$.* Substituting Equation 2.5.13 into Equation 2.10.4 we obtain

$$G = \frac{qN_p'}{\sqrt{\pi Dt}} \frac{W}{L} \int_0^{x_j} \mu_p \left[\exp\left(\frac{-x^2}{4Dt}\right) \right] dx \qquad (2.10.5)$$

The mobility μ_p in Equation 2.10.5 is a function of the dopant concentration (Figure 1.16) and, therefore, a function of the distance from the surface. Consequently, we cannot move it outside the integral.

The most straightforward and accurate means of evaluating Equation 2.10.5 is by numerical integration using a computer. An alternative technique that is often valuable is to approximate various regions of the curve of Figure 1.16 by analytical expressions that can then be used in Equation 2.10.5. Both of these techniques produce results of high accuracy within the limits of the original assumptions. However, as described in Sec. 2.5, real diffused profiles can deviate from a simple Gaussian distribution. For this reason, detailed numerical analysis is only warranted when the nature of the diffusion profile has been investigated experimentally for the particular dopant species under consideration or more accurate process models, such as those discussed in Sec. 2.9, are available.

We can obtain an approximate value of the conductance by using an average value $\overline{\mu_p}$ for the mobility. Because much of the current is carried in a region with a dopant concentration close to the maximum, selecting a value of mobility corresponding to perhaps half the maximum dopant concentration is reasonable and consistent with the use of simplified diffused

---

* We have assumed here that the holes and the acceptor atoms have the same depth distribution. In Sec. 4.1 we show that this assumption is slightly inaccurate. In the present case, however, the small difference between N_a and p can be ignored.

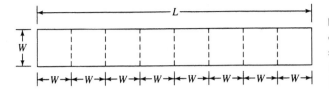

FIGURE 2.46 The number of squares describing the surface dimensions of a resistor is given by the ratio L/W.

dopant profiles. The expression for the conductance (Equation 2.10.4) then reduces to

$$G = N'q\overline{\mu_p}\frac{W}{L} = g\frac{W}{L} \tag{2.10.6}$$

where $g \equiv N'q\overline{\mu_p}$ is the conductance of a square resistor pattern ($L = W$). The conductance, in turn, is determined by the product of the average mobility $\overline{\mu_p}$ and the total dopant density per unit surface area N' (Equation 2.5.10). The resistance R is therefore

$$R = \frac{1}{G} = \frac{L}{W}\frac{1}{g} \tag{2.10.7}$$

Generally many resistors in an integrated circuit are fabricated simultaneously by defining different geometric patterns on the same mask. Because the same diffusion cycle is used for all of these resistors, it is convenient to separate the magnitude of the resistance into two parts; the ratio L/W determined by the mask dimensions, and $1/g$ determined by the diffusion process.

Sheet Resistance.   Any resistor pattern on the mask can be divided into squares of dimension W on each side (Figure 2.46). The number of squares in any pattern is just equal to the ratio L/W. The value of a resistor is thus equal to the product of the number of *squares* into which it can be divided and the parameter $1/g$, which is usually denoted by the symbol $R_\square$ (or sometimes R_s), called the *sheet resistance*. The sheet resistance has units of ohms, but it is conventionally specified in units of ohms per square ($\Omega/\square$) to emphasize that the value of a resistor is given by the product of the number of squares times the sheet resistance. For example, a resistor 100 μm long and 5 μm wide contains 20 squares (20 $\square$). If the diffusion process used produces a diffused layer with a sheet resistance of 200 $\Omega/\square$, the value of the resistor is 20 $\square$ × 200 $\Omega/\square$ = 4.0 kΩ.

Controllable values of sheet resistance are such that resistors in the kilohm range and higher require patterns containing many squares. Because the width of the pattern is determined by the ability to mask and etch very narrow lines, the length of the resistor can become large to obtain the required number of squares for a high-value resistor. The large area needed for high-value resistors is a practical limitation in an integrated circuit, and circuits are usually designed to avoid large-value resistors. Transistors are usually used in place of high-value resistors whenever the circuit allows because transistors are less "expensive" (i.e., occupy less surface area). When a resistor containing a large number of squares must be used, it is generally designed to minimize area by using a serpentine pattern, as shown in Figure 2.47. The current flow across the corner square of

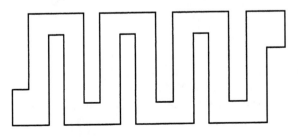

FIGURE 2.47 A serpentine pattern can be used when a long, high-value resistor must be designed.

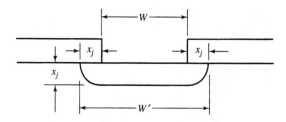

FIGURE 2.48 Lateral diffusion changes the dimensions of the resistor from the nominal mask dimensions.

such a pattern is not uniform. The resistance contributed by a corner square is estimated by taking it to be approximately 65% of the straight-path value.

For large-area resistors, the dimensions L and W are simply determined by the mask dimensions. However, for very narrow resistors (small W) the effective width W_{eff} can differ substantially from the mask dimension because impurities diffuse laterally under the oxide, as well as vertically (Figure 2.48). If W is much greater than the diffusion depth x_j, we can neglect this effect. However, W is often made as small as masking tolerances allow so that a given number of squares fit into the smallest area. In this case x_j can be an appreciable fraction of W, and an effective value W_{eff} must be used for the width of the resistor to account for the lateral diffusion.

EXAMPLE *Diffused Resistor*

A p-type resistor pattern is laid out for an IC with two highly conducting p-type regions contacting a resistive stripe that extends 4 μm between contacts and is 1 μm wide. The stripe has a junction depth x_j of 1 μm. The desired value of the resistance is 1kΩ.

Determine the sheet resistance and the average resistivity required to meet the specifications given. (Neglect lateral diffusion in this example.)

Solution The number of squares in the resistor pattern is

$$L/W = 4/1 = 4$$

Hence, the sheet resistance R is

$$R = 1000/4 = 250 \ \Omega/\square$$

From Equation 2.10.6 and the relationship $R_\square = (1/g)$, we calculate the required dopant density per unit area N_a':

$$N_a' = (q\overline{\mu_p}R_\square)^{-1} \text{ dopant atoms cm}^{-2}$$

The average volume density of impurity atoms $\overline{N_a}$ is related to the area density N_a' through the equation

$$\overline{N_a} = \frac{N_a'}{x_j}$$

and the average resistivity in the p-doped diffused resistor is (from Equation 2.10.2):

$$\overline{\rho} = (q\overline{\mu_p}\overline{N_a})^{-1} = \left[\frac{q\overline{\mu_p}N_a'}{x_j}\right]^{-1} = R_\square x_j = 250 \times 1 \times 10^{-4} = 0.025 \ \Omega\text{-cm} \qquad \blacksquare$$

Precision of Resistance Values. An important point should be made before leaving the subject of diffused resistors. As seen in Equation 2.10.7 the resistance is a function of two factors L/W, which is controlled by the lithography used, and the sheet resistance $1/g$ or $R_\square$, which depends on the dopant deposition and redistribution. Control of the sheet resistance is usually the most limiting factor in the design of a precision resistor. Although the value of $R_\square$ obtained in a process may vary somewhat, it will be nearly constant over a typical die area. Thus, two nearby resistors in an IC can be expected to have the same value of sheet resistance, and the resistance ratio between them is determined by their relative dimensions. With suitable processing care these dimensions can be controlled accurately, and significantly greater precision can be maintained in the ratios of paired resistors than in the resistance values themselves. For this reason, ICs are often designed so that their critical behavior depends on the ratio of two resistor values rather than on the absolute value of a specific resistor in the circuit.

Diffused resistors are commonly used in silicon ICs because of their compatibility with the remainder of the planar process. They are usually formed at the same time as other circuit elements and, therefore, do not add to the fabrication cost. Instead of diffusion, however, it is also possible to make resistors in ICs by patterning epitaxial material that forms an isolating *pn*-junction with the underlying substrate. Bipolar transistor ICs typically are built with this type of epitaxial layer. Because the epitaxial material has the highest resistivity of the silicon used to form the circuit, we can obtain sheet resistances by this method that are several times larger ($\sim 1000\ \Omega/\square$) than those normally attainable in *p*-type diffused resistors ($\sim 200\ \Omega/\square$). For even higher values of sheet resistance, a double-diffused structure is sometimes used in which an *n*-region is formed over the *p*-region to reduce the vertical dimension of the diffused resistor. Resistors of this type, known as *pinch resistors*, can increase the sheet resistance by a factor of roughly 40 or 50. Making pinch resistors is clearly more complicated, and reproducibility of the sheet resistance is typically poor.

If a particular circuit requires large or accurate values of resistance, alternative techniques can be used. Instead of forming a resistor by adding dopant atoms to the wafer, a resistive film can be deposited on top of the insulating silicon dioxide that covers most of the circuit. This film is then defined by masking and connected to the rest of the circuit by metal interconnections. The use of deposited resistors provides flexibility in the design of circuits. Sometimes these resistors can be formed in a layer of material that is already included in an IC process. An example is the use of polysilicon that is deposited as part of the production process for silicon-gate MOS ICs. If a usable resistive layer is not a part of a given IC process, the added cost of depositing and patterning another layer of resistive material must be carefully considered.

In this discussion of IC resistors we have seen that particular constraints imposed by technology influence circuit design. We will see this pattern repeated when we consider other IC devices in subsequent chapters.

SUMMARY

The overwhelming importance of silicon to electronics is a result of both its advantageous material properties and the superb control that has been achieved over its technology. The ability to produce on silicon a highly insulating oxide with excellent, repeatable properties and with a well-controlled interface between the element and its oxide is unique. The *planar process* used to fabricate silicon integrated circuits is the basis for the accurate delineation of small-geometry devices. It makes possible the simultaneous fabrication of many devices, and thus is the key to uniformity, reliability, and economy in IC production.

Large, single crystals of silicon are usually grown by the *Czochralski* method to provide the starting material for IC production. *Float-zone* refining is used when the oxygen content must be low. The single crystals are sliced into *wafers* before beginning the planar process. The oxidation of silicon to produce SiO_2 can be carried out in a dry ambient at temperatures near 1000°C or (at a significantly faster rate) in a steam ambient at similar temperatures. Production of the intricate patterns for an IC requires the definition of patterns in polymer *resist* films by lithographic techniques. *Selective* removal of oxide layers and other materials in ICs is accomplished using the patterned resist as a mask. Dopant impurities that can alter the conductivity (changing its magnitude as well as conductivity type) can then be added in patterns defined in the resist on through patterns defined by the resist in other materials. Both *ion implantation* and *gaseous deposition* are used to deposit the dopant atoms in the desired areas of the IC. Subsequent *diffusion* of the dopant atoms into the wafer can be modeled most simply by *Fick's Law*, which is a partial differential equation having two analytic solutions (*Gaussian* and *complementary error function*) that apply to IC diffusion processes. Fick's Law, however, does not account for several more complex aspects of diffusion, and computer modeling is necessary to predict the diffusion process for more accurate IC design. *Chemical vapor deposition* is an important IC technology for producing single-crystal silicon on a single-crystal substrate (*epitaxy*) as well as for depositing *polycrystalline silicon* and insulating films.

Interconnection of the devices in an IC is an important step, which places severe requirements on the conducting material used. Aluminum is susceptible to *electromigration*, which becomes more important as device and conductor dimensions decrease. Copper, with its lower electrical resistivity and greater resistance to electromigration, is used in high-performance circuits.

Compound semiconductors, often composed of elements from Columns III and V of the periodic table, are used in specialized devices. Selected compound semiconductors are especially useful for optical devices because their *direct bandgap* allows efficient light emission when electrons and holes recombine. They are also useful for high-frequency devices because of their high mobility. Combinations of different semiconductors in a device are most readily achieved and useful when they have the same lattice constant, but different bandgaps.

Process and device simulation using sophisticated computer programs provides an essential tool for the engineer to investigate and design physical structures and impurity profiles and to make predictions about the device behavior that can be expected without carrying through all the increasingly expensive processing steps. Powerful numerical techniques allow an increasing number of physical effects to be included in the models. As devices become smaller, the importance of these effects increases.

Diffused resistors are widely used in ICs. They are typically fabricated by employing the planar process to delineate patterns that define the current path for drifting carriers. Resistance values can be calculated from the *sheet resistance*, which is measured in ohms per square ($\Omega/\square$), and given the symbol $R_\square$. The sheet resistance represents the resistance across the opposite edges of a square pattern; therefore, the resistance of a diffused resistor is obtained by multiplying $R_\square$ by the number of squares making up the resistor pattern.

REFERENCES

1. *Dilbert* by Scott Adams, United Features Syndicate, 7/15/1997.
2. R. Chau, et al., Tech. Digest Int'l. Electron Devices Mtg. (San Francisco, December 10–13, 2000), pp. 45–48; 2001 Silicon Nanoelectronics Workshop, Kyoto, June 10, 2001.
3. International Technology Roadmap for Semiconductors: http://public.itrs.net; 2001 update: http://public.itrs.net/Files/2001ITRS/Home.htm.
4. W. C. O'Mara, R. B. Herring, and L. P. Hunt, eds., *Handbook of Semiconductor Silicon Technology*, (Noyes Publications, Park Ridge NJ, 1990).
5. B. E. Deal and A. S. Grove, J. Appl. Phys. **36**, 3770 (1965).
6. R. R. Razouk, L. N. Lie, and B. E. Deal, *J. Electrochem. Soc.* **128**, 2214 (October 1981).
7. D. Hess and B. E. Deal, *J. Electrochem. Soc.* **124**, 735 (1977); J. J. Barnes, J. M. DeBlasi, and B. E. Deal, *J. Electrochem. Soc.* **126**, 1779 (1979); B. E. Deal, J. Electrochem. Soc. **125**, 576 (1978); F. Shimura and

H. R. Huff in *VLSI Handbook* (ed. N. G. Einspruch), Academic Press, Orlando, FL (1985), Chapter 15.
8. H. Z. Massoud, J. D. Plummer, and E. A. Irene, *J. Electrochem. Soc.* **132**, 2685 (1985).
9. C. Ho, J. D. Plummer, and J. D. Meindl, *J. Electrochem. Soc.* **125**, 665 (April 1978).
10. M. C. H. M. Wouters, H. M. Eijkman, and L. J. van Ruyven, Philips Research Rep. **31**, 278 (1976).
11. J. F. Gibbons, W. S. Johnson, and S. W. Mylroie, *Projected Range Statistics*, 2nd Ed. Dowden, Hutchinson, and Ross, New York (1975).
12. Research Triangle Institute, *Integrated Silicon Device Technology, Vol. IV, Diffusion* (ASD-TDR-63-316). Research Triangle Institute, Durham N. C. (1964). Reprinted by permission of the publisher.
13. F. A. Trumbore, *Bell System Tech. J.* **39**, 205 (1960); G. Masetti, D. Nobili, and S. Solmi, *Semiconductor Silicon 1977*, Electrochemical Society (1977), p. 648; A. Armigliato, D. Nobili, P. Ostoja, M. Servidori, and

S. Solmi, *Semiconductor Silicon 1977*, Electrochemical Society (1977), p. 638; D. Nobili, A. Carabelas, G. Celotti, and S. Solmi, *J. Electrochem. Soc.* **130**, 922 (April 1983); R. A. Craven, *Semiconductor Silicon 1981*, Electrochemical Society (1981), p. 254.

14. R. B. Fair, *Semiconductor Silicon 1977*. Electrochemical Society (1977), p. 968.

15. E. S. Meieran and T. I. Kamins, Solid-State Electr. **16**, 545 (1973).

16. B. Swaminathan, *Doctoral dissertation*, Department of Elect. Eng., Stanford University (April 1983).

17. J. A. Appels, E. Kooi, M. M. Paffen, J. J. H. Schatorje, and W. H. C. G. Verkuylen, Philips Research Rep. **25**, 118 (1970).

18. P. P. Merchant, *Hewlett-Packard Journal*, **33**, 28 (August 1982).

19. P. Vande Voorde, S.-Y. Oh, and R. W. Dutton, private communication.

20. D. A. Antoniadis and R. W. Dutton, *IEEE J. Solid-State Circuits* **SC-14**, 412 (April 1979); R. W. Dutton and Z. Yu, *Technology CAD: Computer Simulation of IC Processes and Devices*, Kluwer Academic Publishers, Boston (1993).

21. K. M. Cham, S. Y. Oh, D. Chin, and J. L. Moll, *Computer-Aided Design and VLSI Device Development*, Kluwer Academic Publishers, Boston, 1986, p. 60.

22. Cham, et al., p. 221.

23. R. D. Rung, "Silicon IC Technology Series: Computer Simulation of Silicon Processing," Videotape No. 90862, Hewlett-Packard Co. (1979).

24. F. M. Smits, *Bell System Tech. J.* **37**, 711 (1958).

BOOKS

S. M. Sze, *Semiconductor Devices: Physics and Technology*, second edition, Wiley, New York (2002).

J. D. Plummer, *Silicon VLSI Technology: Fundamentals, Practice, and Modeling*, Prentice Hall, 1999.

S. Wolf and R. N. Tauber, *Silicon Processing for the VLSI Era: Vols. 1–3*, Lattice Press, Sunset Beach CA (1990–2000).

S. K. Ghandhi, *VLSI Fabrication Principles: Silicon and Gallium Arsenide*, second edition, Wiley-Interscience, New York (1994).

W. E. Beadle, J. C. C. Tsai, and R. D. Plummer (Editors), *Quick Reference Manual for Silicon Integrated-Circuit Technology*, Wiley-Interscience, New York (1985).

S. M. Sze *Physics of Semiconductor Devices*, second edition, Wiley, New York (1981).

S. M. Sze (Editor), *VLSI Technology*, McGraw-Hill, New York (1983).

R. A. Colclasser, *Microelectronics, Processing, and Device Design*, Wiley, New York (1980).

PROBLEMS

2.1* A crystal of silicon is to be grown using the Czochralski technique. Prior to initiating the crystal growth, 1 mg of phosphorus is added to 10 kg of melted silicon in the crucible.

(a) What is the initial dopant concentration in the solid at the beginning of the crystal growth?

(b) What is the dopant concentration at the surface of the silicon crystal after 5 kg of the melt has solidified?

$$\left[\text{The segregation coefficient } \frac{C_{solid}}{C_{liquid}} \right.$$

$$\left. \text{for phosphorus in silicon is } 0.3. \right]$$

2.2 A Czochralski-grown silicon wafer is heated in a nitrogen ambient at a high temperature to evaporate oxygen from the regions of the wafer near the surface. It is then heated at a low temperature to cause the remaining oxygen to precipitate in "clumps." Explain how and why this process improves the electrical properties of devices subsequently fabricated in the wafer.

2.3 A (111)-oriented silicon wafer is oxidized several times during an *IC* process. Find the total thickness of silicon dioxide after each of the following steps, which are carried out in sequence.

(a) 60 min at 1100°C in dry O_2 and HCl (enough HCl is added to enhance the oxidation rate by 10% over the rate in pure O_2).

(b) 2 h at 1000°C in pyrogenic steam (at 1 atm).

(c) 6 h at 1000°C in dry O_2.

2.4* A silicon wafer is covered with a 200-nm-thick layer of silicon dioxide. What is the added time required to grow an additional 100 nm of silicon dioxide in dry O_2 at 1200°C?

2.5 (a) How long does it take to grow 1 μm of silicon dioxide in steam at 1000°C and one atm? Consider (111) orientation for parts (a) through (d)].

(b) How long does it take to grow 1 μm of silicon dioxide in steam at 1000°C and 10 atm?

(c) How long does it take to grow 1 μm of silicon dioxide in steam at 800°C and 1 atm?

(d) How long does it take to grow 1 μm of silicon dioxide in steam at 800°C and 10 atm?

[This problem shows that a thick oxide (1 μm) can be grown at a reduced temperature (800°C) by using elevated pressures.]

2.6 Carry out a derivation of Equation 2.3.6 using Equation 2.3.5.

2.7* In a *LOCOS* process (as described in Sec. 2.6), an 8-h oxidation in a steam ambient (at 1 atm and 1000°C) is carried out after a 50-nm layer of silicon nitride has been deposited and patterned. After the oxidation, the nitride layer is removed, exposing the original silicon surface. How far above the silicon surface is the top of the grown oxide layer? [About 24 nm of the silicon nitride is converted to silicon dioxide during the 8-h steam oxidation.] Calculate for a (100)-oriented wafer.

2.8 Calculate the thickness of silicon dioxide produced on the surface of a silicon-nitride layer for every nm of silicon nitride that is oxidized in the *LOCOS* process.

2.9* A deep vertical groove 1 μm wide and several micrometers deep is etched in a silicon substrate. The grooved surface is bare silicon, but the plane silicon surface is covered with a thin layer of silicon nitride which serves as an oxidation mask (Figure P2.9). The wafer is then oxidized in steam at 1 atm and 1100°C to fill the groove with oxide.

(a) What is the width of the stripe of silicon dioxide that results when the groove is completely filled?

(b) How long does it take to fill the groove with silicon dioxide?

[*Hint*: Note, for part a that an oxide stripe x units wide is formed from $(x - 1)$ units of silicon. Do part b by applying Equation 2.3.6 with $\tau = 0$ and the data for (100) silicon in Figures 2.7a and b.]

2.10† A dose of phosphorus ions equal to 3 $\times$ 10^{16} cm^{-2} is implanted into a silicon wafer at an energy of 50 keV (R_p = 63 nm, ΔR_p = 27 nm) to form contacts to a transistor.

(a) If the wafer is now oxidized, is it important to consider the effects of concentration-enhanced oxidation?

(b) The phosphorus is now diffused for 60 min at 1000°C prior to an oxidation step. Is concentration-enhanced oxidation important in this case?

(c) Reconsider parts a and b if phosphorus is implanted at 150 keV (R_p = 180 nm, ΔR_p = 64 nm).

2.11 After a high concentration of boron is added near the surface of an *n*–type silicon wafer, a portion of the wafer is covered with a layer of polycrystalline silicon containing a high concentration of crystalline defects. The wafer is then oxidized, and the junction depth is found to be much greater in the regions which were *not* covered with polysilicon during the oxidation. Explain this result.

2.12 The linear coefficient of expansion of glass is 9 $\times$ 10^{-6} per°C. Assume that there is a 1°C increase in the mask temperature between two photolithography steps but that the wafer temperature stays the same. A wafer 20 cm in diameter is processed under these conditions with its center prefectly aligned. What is the minimum misalignment at the edges of the wafer?

2.13 Phosphorus is added to a silicon wafer from a gaseous source at 975°C for 30 min. Determine the junction depth for

(a) a 0.3 Ω-cm *p*–type substrate,

(b) a 20 Ω-cm *p*–type substrate.

Assume that the diffusion coefficient of phosphorus is 10^{-13} cm^2 s^{-1} and that its solid solubility is 10^{21} cm^{-3} at 975°C.

2.14† Boron atoms with an area concentration of 10^{15} cm^{-2} are introduced into an *n*-type wafer from a BCl$_3$ source in a carrier gas. The starting wafer has a uniform donor concentration of 5 $\times$ 10^{15} cm^{-3}. The subsequent drive-in diffusion is carried out at 1100°C in a nitrogen ambient. The desired junction depth is 2 μm.

(a) How long should the drive-in diffusion be continued?

(b) Sketch the resulting boron distribution on log N_a and linear N_a versus x plots.

(c) It is required to lay out a 500 Ω resistor. The minimum mask dimension is 4 μm. What is the length of the resistor as indicated on the mask? What is the silicon surface area used? An exact, closed form solution to this part of the problem is not possible, but the sketches of part b should indicate the method of solution. State your approximations.

(d) Qualitatively describe the effect of doubling the time of the drive-in diffusion. Note that the net num-

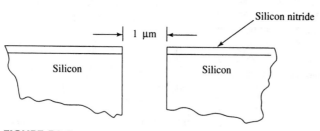

FIGURE P2.9

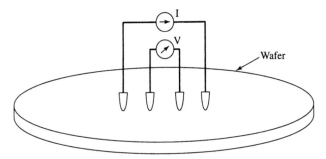

FIGURE P2.15

ber of acceptor impurities $N_a - N_d$ can contribute to the conduction. Also note that the mobility decreases as the impurity concentration increases above about 10^{16} cm^{-3} for Si.

2.15 A *four-point probe* can be used to measure the sheet resistance without making ohmic contact to the semiconductor. The four probes are spaced along a line (as shown in Figure P2.15) and a current is passed through the two end probes. The resulting voltage drop between the inner two probes is measured using a high-impedance meter which draws negligible current. The sheet resistance is given by

$$R_\Box = \frac{\pi}{\ln 2} \times \frac{V}{I} = 4.53 \times \frac{V}{I}$$

if the probe spacing is large compared to the thickness of the sample, but small compared to its surface dimensions. (A correction factor can be used for small or thick samples. [24])

If the four-point probe current $I = 1$ mA, what probe voltage V would be measured on a region in which a density $N' = 10^{12}$ phosphorus atoms cm^{-2} has been diffused into a very high resistivity p-type wafer? Assume that the junction depth $x_j = 1$ μm and that $\bar{\mu}_n$ is the mobility associated with the average phosphorus density.

2.16 A dose of boron ions equal to 10^{12} cm^{-2} is implanted into a 5 Ω-cm n-type silicon wafer at 100 keV ($R_p = 290$ nm, $\Delta R_p = 70$ nm) and then diffused for 2 h at 1000°C ($D = 2 \times 10^{-14}$ cm^2 s^{-1}).

(a) What is the peak boron concentration and how wide is the p-type region immediately after the implant?

(b) What is the peak boron concentration after the subsequent diffusion?

2.17[†] A dose N' cm^{-2} of a dopant is implanted into silicon with a background dopant concentration C_B to create a pn junction.

(a) Show that the vertical junction depth is

$$x_j = R_p + \Delta R_p \left[2 \ln\left(\frac{N'}{\sqrt{2\pi} \, \Delta R_p C_B} \right) \right]^{1/2}$$

(b) Calculate x_j for a dose of 10^{15} As atoms cm^{-2} implanted at 60 keV into a wafer doped with $N_a = 10^{16}$ cm^{-3} boron atoms.

2.18* A polycrystalline silicon interconnection line having a resistivity of 500 μΩ-cm is 5 μm wide and 0.5 μm thick. Current is supplied through a 1 mm length of this line to charge a capacitor that measures 0.1×0.5 mm^2 on its surface and which has plates spaced on either side of a silicon–dioxide layer that is 100 nm-thick. What is the RC time constant for the resulting resistor-capacitor series connection? [The resistivity for polysilicon given in this problem is about the minimum attainable; hence, this problem indicates a limitation to the use of polysilicon for interconnections in VLSI.]

2.19 Assume that uniform concentrations of 6.55 μm thickness can be imbedded in silicon of the opposite conductivity type to form an integrated-circuit resistor (Figure P2.19).

Make calculations for two cases: (i) for $N_d = 10^{16}$ cm^{-3} and (ii) for $N_a = 10^{16}$ cm^{-3} to obtain:

(a) The relationships between L and W for resistances of 100 Ω, 1 kΩ, and 10 kΩ between contacts at 25°C.

(b) The actual dimensions if the resistors should dissipate 10 mW of power each and the maximum power dissipation is 1 μW/μm^3.

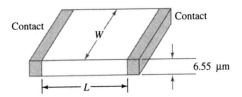

FIGURE P2.19

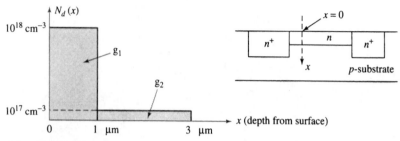

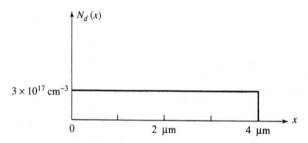

FIGURE P2.20A

FIGURE P2.20B

(c) The temperature coefficients of the resistors (TCR) around 25°C. The TCR for a resistor R is defined by $(1/R)(\partial R/\partial T) \times 100$ (in % per degree).

2.20* Assume that the stepped doping profile shown in Figure P2.20a is attained at the surface of a silicon wafer.

(a) Calculate the sheet resistance without using an average mobility.

After some further processing steps, assume that the dopant profile along x changes to the shape shown in Figure P2.20b.

(b) Assume that it is somehow possible to introduce extra dopant atoms uniformly between $x = 0$ and $x = 4$ μm. What type of dopant (donor or acceptor) and what concentration should be added to make the sheet resistance for a constant dopant profile (Figure P2.20b) equal to the sheet resistance obtained in part a?

2.21† An acceptor diffusion is carried out, introducing an acceptor profile $N_a = N_s$ erfc (x/λ) with $N_s = 10^{18}$ cm^{-3} and $\lambda = 0.05$ μm in to a lightly doped n-type silicon crystal. This diffusion will form a resistor in an integrated circuit.

(a) Show that the resistance across any square pattern on the surface of the resistor is approximately given by

$$R_\square \approx \left(q\mu \int_0^\infty N_s \mathrm{erfc}\left(\frac{x}{\lambda}\right) dx \right)^{-1}$$

State clearly where the approximation is made. This expression can be integrated by parts and the value for $R_\square$ written as

$$R_\square = \frac{\sqrt{\pi}}{q\mu N_s \lambda}$$

(b) Derive this form.

(c) What is the approximate resistance if the resistor is made 4 μm wide and 200 μm long?

(d) What is the largest resistance that can be fabricated in a surface area that measures 20 by 7 μm if 1 μm line widths and 1 μm spacings are the minimum dimensions? Sketch the resistor pattern. (Consider each corner square to be 65% effective.) Neglect lateral diffusion in this problem.

METAL-SEMICONDUCTOR CONTACTS

Most of the electronic devices that make up an integrated circuit are connected by means of metal-semiconductor contacts. Moreover, all integrated circuits communicate with the rest of an electrical system via metal-semiconductor contacts. As we will see, the properties of these contacts can vary considerably, and it is necessary to consider several factors in order to understand them. We will narrow our discussion whenever necessary to consider only metal contacts to silicon, but first let us consider in general the nature of the thermal equilibrium that is established when a metal and a semiconductor are in intimate contact. The concepts that are developed to understand this equilibrium are very important. They will prove useful many times in our discussion of devices because they underlie the basic properties of semiconductor *pn* junctions as well as the properties of interfaces between semiconductors and insulators and between metals and insulators.

Application of the equilibrium principles to metals and semiconductors provides a simple theory (Schottky theory) of ohmic and rectifying behavior in various metal-semiconductor systems. The Schottky theory is, however, not adequate in many cases: most notably not for metal-silicon systems without further consideration of the real nature of solid interfaces. The important influence of surface states and the origins of these states are therefore discussed. Several applications are then described, but special emphasis is given to the Schottky-diode clamp that has been so widely used in fast logic circuits.

3.1   EQUILIBRIUM IN ELECTRONIC SYSTEMS

Metal-Semiconductor System

We will find it very profitable to construct an energy-band representation for a metal-semiconductor contact. To do so we employ at first a general viewpoint and consider the metal and the semiconductor as representing two systems of allowed electronic energy states. Using the concepts developed in Chapter 1, we recognize that these systems of allowed states can be regarded as being almost entirely populated at energies less than the Fermi energy and nearly vacant at higher energies. When the metal and the semiconductor are remote from one another and therefore do not interact, both the systems of electronic states and their Fermi levels are independent. Let us designate the metal as state-system number 1 and the semiconductor as state-system number 2. Each system has a density of allowed states $g(E)$ per unit energy. Of these $g(E)$ states, $n(E)$ are full and $v(E)$ are empty. (The variables are all functions of energy; hence, the use of E in parentheses.) Our terminology is indicated in Figure 3.1. The Fermi-Dirac distribution functions

$$f_{D1.2} = \frac{1}{1 + \exp[(E - E_{f1.2})/kT]} \tag{3.1.1}$$

allow us to relate $v_{1,2}(E)$, $n_{1,2}(E)$, and $g_{1,2}(E)$.* The filled state density is

$$n_{1,2} = g_{1,2} \times f_{D1,2} \tag{3.1.2}$$

and the vacant state density is

$$v_{1,2} = g_{1,2} \times (1 - f_{D1,2}) \tag{3.1.3}$$

In Equations 3.1.1 through 3.1.3 we have omitted the unnecessary parentheses that have served only to emphasize the dependence on energy.

We now consider that the two remote systems are brought into intimate contact. The systems then begin interacting, and electrons transfer between them. Equilibrium is reached when there is no net transfer of electrons at any energy. As we noted in Sec. 1.1, this does not mean a cessation of all processes; rather it means that every process and its inverse are occurring at the same rate. To make this point clear, we consider transfer in space of electrons that are at a given energy E_x. Equilibrium can be expressed mathematically by noting that the transfer probability is proportional to the population of electrons $n(E_x)$ available for transfer at a given energy and also proportional to the density of available states $v(E_x)$ to which the electrons can transfer. Because of the Pauli exclusion principle, the available state density is not $g(E_x)$, the state density, but $v(E_x)$, the vacant-state density.

---

* A more complete discussion of Fermi–Dirac statistics would show that f_D may have slightly altered forms depending upon specific properties of the allowed electronic states. The alternate forms for f_D typically have factors of 2 or $\frac{1}{2}$ preceding the exponential in Equation 3.1.1. We will not be concerned with this refinement.

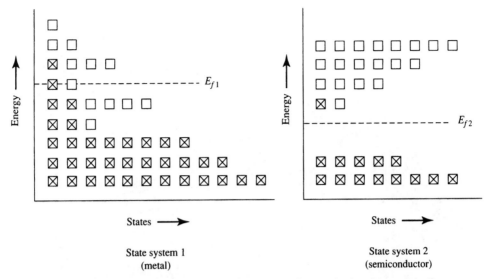

FIGURE 3.1 Allowed electronic-energy-state systems for two isolated materials. States marked with an × are filled; those unmarked are empty. System 1 is a qualitative representation of a metal; system 2 qualitatively represents a semiconductor.

The proportionality factor relating the transfer probability to these two densities is related to the detailed quantum nature of the states and is the same for transfer from system 1 to system 2 as it is for transfer from system 2 to system 1. Therefore, at thermal equilibrium,

$$n_1 \times v_2 = n_2 \times v_1 \tag{3.1.4}$$

at any given energy.

Using Equations 3.1.2 and 3.1.3 in Equation 3.1.4, we have

$$f_{D1}g_1(1 - f_{D2})g_2 = f_{D2}g_2(1 - f_{D1})g_1$$

or

$$f_{D1}g_1g_2 = f_{D2}g_2g_1 \tag{3.1.5}$$

Equation 3.1.5 can only be true if $f_{D1} = f_{D2}$ or, by Equation 3.1.1, if $E_{f1} = E_{f2}$. We therefore have established the important property of any two systems in thermal equilibrium: they have the same Fermi energy. Note, from our derivation, that no constraints exist on the density-of-states functions g_1 or g_2. Regardless of the detailed nature of these functions, a statement that their Fermi levels are equal is equivalent to stating that the two systems are in thermal equilibrium. Generalization of this statement to more than two systems or its reduction to apply to one system of states is straightforward: *At thermal equilibrium the Fermi level is constant throughout a system.*

3.2 **IDEALIZED METAL-SEMICONDUCTOR JUNCTIONS**

Band Diagram

We now turn to the particular features of metal-semiconductor energy-band diagrams. For the purposes of our present discussion, the most distinctive characteristic of the electronic-energy states of the metal and the semiconductor is the relative positions of the Fermi

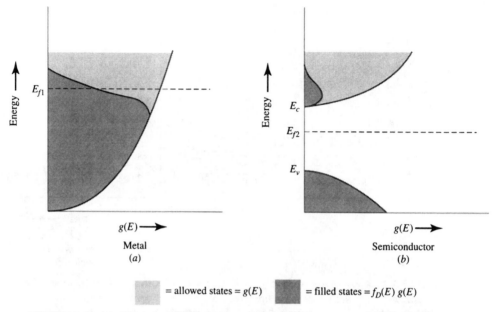

FIGURE 3.2 (a) Allowed electronic-energy states $g(E)$ for an ideal metal. The states indicated by cross-hatching are occupied. Note the Fermi level E_{f1} immersed in the continuum of allowed states. (b) Allowed electronic-energy states $g(E)$ for a semiconductor. The Fermi level E_{f2} is at an intermediate energy between that of the conduction-band edge and that of the valence-band edge.

levels within the densities of allowed states $g(E)$. In the metal, the Fermi level is immersed within a continuum of allowed states, while in a semiconductor, under usual circumstances, the density of states is negligible at the Fermi level. Plots of $g(E)$ versus energy for idealized metals and semiconductors are shown in Figures 3.2a and 3.2b.

These points, which were made in general in our discussion of Figure 1.3, are apparent in the specific cases of the allowed energy states for gold and for silicon shown in Figures 3.3a and 3.3b, respectively. Note that these sketches differ from those in

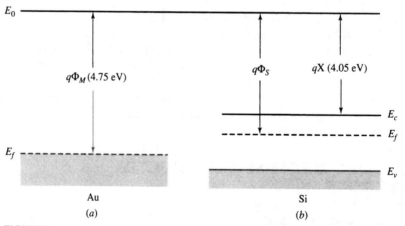

FIGURE 3.3 Pertinent energy levels for the metal gold and the semiconductor silicon. Only the work function is given for the metal, whereas the semiconductor is described by the work function $q\Phi_S$, the electron affinity qX_S, and the band gap $(E_c - E_v)$.

Figure 3.2 in that they represent allowed energies in the bulk of the material versus position and not the density of allowed energy states versus energy. These two types of sketches are both frequently used and are sometimes confused.

New quantities also appear in these figures. First, a convenient reference level for energy is taken to be the vacuum or free-electron energy E_0. E_0 represents the energy that an electron would have if it were just free of the influence of the given material. The difference between E_0 and E_f is called the *work function*, usually given the symbol $q\Phi$ in energy units and often listed as Φ in volts for particular materials. In the case of the semiconductor, the difference between E_0 and E_f is a function of the dopant concentration of the semiconductor, because E_f changes position within the gap separating E_v and E_c as the doping is varied. The difference between the vacuum level and the conduction-band edge is, however, a constant of the material. This quantity is called the *electron affinity*, and is conventionally denoted by qX in energy units. Tables of X in volts exist for many materials. (The symbol X is a Greek capital letter *chi*.)

The choice of E_0 as a common energy reference makes clear that if Φ_M is less than Φ_S and the materials do not interact, an electron in the metal has, on average, a total energy that is higher than the average total energy of an electron in the semiconductor. On the contrary, if Φ_M is greater than Φ_S, the average total energy of an electron in the semiconductor is higher than it is in the metal. For the sake of discussion, we consider the latter case where $\Phi_M > \Phi_S$. When an intimate contact is established, the disparity in the average energies can be expected to cause the transfer of electrons from the semiconductor into the metal.

Another way of looking at the establishment of equilibrium for this case where $\Phi_M > \Phi_S$ is to make use of the concepts developed in Section 3.1. The left-hand side of Equation 3.1.5 in that section is proportional to the flow of electrons from state-system 1 to state-system 2 while the right-hand side of the same equation is proportional to the inverse flow. The net flow is easily seen to be in the direction 2 to 1 if $f_{D2} > f_{D1}$ or, therefore, if $E_{f2} > E_{f1}$. The charge transfer continues until equilibrium is obtained, and a single Fermi level characterizes both the metal and the semiconductor. At equilibrium, the semiconductor, having lost electrons, is charged positively with respect to the metal.

To construct a proper band diagram for the metal and the semiconductor in thermal equilibrium, we need to note two additional facts. The first is that the vacuum level E_0 must be drawn as a continuous curve. This is because E_0 represents the energy of a "just-free" electron and thus must be a continuous, single-valued function in space. If it were not, one could conceive of means to extract work from an equilibrium situation by emitting electrons and then reabsorbing them an infinitesimal distance away where E_0 had changed value. Second, we note that electron affinity is a property associated with the crystal lattice like the forbidden-gap energy. Hence, it is a constant in a given material. Considering these three factors: constancy of E_f, continuity of E_0, and constancy of X in the semiconductor, we can sketch the general shape of the band diagram for the metal-semiconductor system. The sketch is given in Figure 3.4a for an *n*-type semiconductor for which $\Phi_M > \Phi_S$.

We see from Figure 3.4a that there is an abrupt discontinuity of allowed energy states at the interface. The magnitude of this step is $q\phi_B$ electron volts with

$$q\phi_B = q(\Phi_M - X) \tag{3.2.1}$$

Figure 3.4a indicates that electrons at the band edges (E_c and E_v) in the vicinity of the junction in the semiconductor are at higher energies than are those in more remote regions. This is a consequence of the transfer of negative charge from the semiconductor

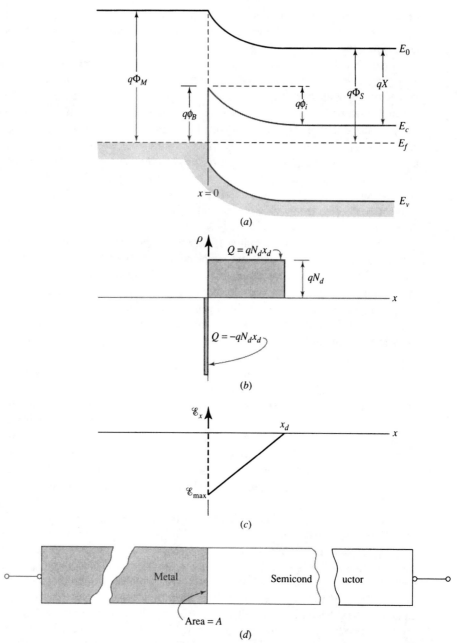

FIGURE 3.4 (a) Idealized equilibrium band diagram (energy versus distance) for a metal-semiconductor rectifying contact (Schottky barrier). The physical junction is at $x = 0$. (b) Charge at an idealized metal-semiconductor junction. The negative charge is approximately a delta function at the metal surface. The positive charge consists entirely of ionized donors (here assumed constant in space) in the depletion approximation. (c) Field at an idealized metal-semiconductor junction.

into the metal. Because of the charge exchange, there is a field at the junction and a net increase in the potential energies of electrons within the band structure of the semiconductor. The free-electron population is thus depleted near the junction, as indicated by the increased separation between E_c and E_f at the surface compared to that in the bulk.

Before considering the electrical properties of the metal-semiconductor junction we note that our development thus far has relied on the important idealization that the basic band structures of the two materials are unchanged near their surfaces. We shall gain some useful results by working with this idealized model, but it will be necessary to consider conditions at the surfaces more carefully later, and then to construct a nonidealized band picture.

Charge, Depletion Region, and Capacitance

The charge and field diagrams for an ideal metal-semiconductor junction are sketched in Figures 3.4b and 3.4c. To the extent that the metal is a perfect conductor, the charge transferred to it from the semiconductor exists in a plane at the metal surface. In the idealized n-type semiconductor, positive charge can consist either of ionized donors or of free holes while electrons make up the negative charge. We have made several assumptions about the semiconductor charge in drawing Figures 3.4b and 3.4c. First, the free-hole population is assumed to be everywhere so small that it need not be considered; second, the electron density is much less than the donor density from the interface to a plane at $x = x_d$. Beyond x_d, the donor density N_d is taken to be equal to n. These assumptions make up what is usually called the *depletion approximation*. Although they are not precisely true, they are generally sufficiently valid to permit the development of very useful relationships. We will reconsider the space-charge distributions at a junction in equilibrium in Chapter 4 and consider the depletion approximation in more detail there.

Under the depletion approximation, the extent of the space-charge region is exactly x_d units, and the magnitude of the field (for this case of a constant doping in the semiconductor) is a decreasing linear function of position (Figure 3.4c). The maximum field $\mathscr{E}_{max}$ is located at the interface and, by Gauss' law, its value is given by

$$\mathscr{E}_{max} = \frac{-qN_dx_d}{\epsilon_s} \tag{3.2.2}$$

where ϵ_s is the permittivity of the semiconductor. The voltage across the space-charge region, which is represented by the negative of the area under the field curve in Figure 3.4c, is given by

$$\phi_i = -\frac{1}{2}\mathscr{E}_{max}x_d = \frac{1}{2}\frac{qN_dx_d^2}{\epsilon_s} \tag{3.2.3}$$

It is also frequently useful to express x_d in terms of ϕ_i. From Equation 3.2.3, we can write $x_d = \sqrt{2\phi_i\epsilon_s/qN_d}$. From Figure 3.4a, the built-in voltage ϕ_i is also seen to be $\Phi_M - \Phi_S = \Phi_M - X - (E_c - E_f)/q$. The space charge Q_s (per unit area) in the semiconductor is

$$Q_s = qN_dx_d = \sqrt{2q\epsilon_sN_d\phi_i} \tag{3.2.4}$$

Applied Bias. Up to this point, we have been considering thermal equilibrium conditions at the metal-semiconductor junction. Now we add an applied voltage and consider the resulting nonequilibrium condition. We saw in Figure 3.4a that there is an abrupt step in allowed electron energies at the metal-semiconductor interface. This step makes it more difficult to cause a net transfer of free electrons from the metal into the semiconductor than it is to obtain a net flow of electrons in the opposite direction. There is a barrier of $q\phi_B$ electron volts between electrons at the Fermi level in the metal and the conduction-band states in the semiconductor near its surface (Figure 3.4a). To first order this barrier height is independent of bias. Referring to Figure 3.4c, we see that the voltage drop across the near delta function of space charge in the metal (equivalent to the area between the $\mathscr{E}$-field curve and the axis) is effectively zero in equilibrium; that is, no voltage drop can be sustained across the metal. The total voltage drop across the space-charge region (ϕ_i) occurs within the semiconductor, as can be seen in Figure 3.4a. An applied voltage is similarly dropped entirely within the semiconductor and alters the equilibrium-band diagram (Figure 3.4a) by changing the total curvature of the bands, modifying the potential drop from ϕ_i. Thus, electrons in the bulk of the semiconductor at the conduction-band edge are impeded from transferring to the metal by a barrier that can be changed readily from its equilibrium value $q\phi_i$ by an applied bias. The barrier is reduced when the metal is biased positively with respect to the semiconductor, and it is increased when the metal is more negative. Energy-band diagrams for two cases of bias are shown in Figures 3.5a and 3.5b. Because these diagrams correspond to nonequilibrium conditions, they are not drawn with a single Fermi level. The Fermi energy in the region from which electrons flow is higher than is the Fermi energy in the region into which electrons flow. Currents, of course, move in the direction opposite to the electron flow.

To investigate bias effects on the barrier, we consider the semiconductor to be grounded and take forward bias to correspond to the metal electrode being made positive. The applied voltage is called V_a, and the bias polarity is indicated in Figure 3.5c. Under reverse bias the metal is negatively biased ($V_a < 0$). If the metal-semiconductor junction is placed under reverse bias so that the voltage drop across the space-charge region

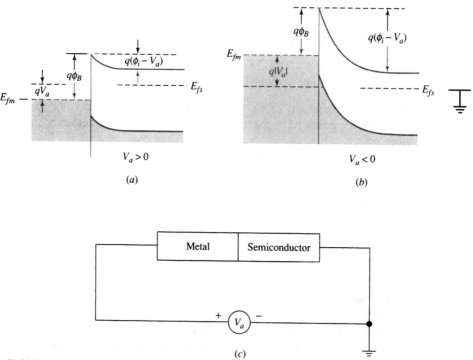

FIGURE 3.5   Idealized band diagrams (energy versus distance) at a metal-semiconductor junction (a) under applied forward bias ($V_a > 0$) and (b) under applied reverse bias ($V_a < 0$). The semiconductor is taken as the reference (voltage ground) as shown in (c). The vacuum levels for the two cases are not shown.

increases to $(\phi_i - V_a)$, then the space-charge density in the semiconductor increases from its equilibrium value [Equation (3.2.4)] to

$$Q_s = \sqrt{2q\epsilon_s N_d(\phi_i - V_a)} \tag{3.2.5}$$

With a small-signal ac bias added to the fixed dc bias V_a, the junction shows a capacitive behavior that can be calculated by using Equation 3.2.5:

$$C = \left|\frac{\partial Q_s}{\partial V_a}\right| = \sqrt{\frac{q\epsilon_s N_d}{2(\phi_i - V_a)}} = \frac{\epsilon_s}{x_d} \tag{3.2.6}$$

where C in Equation 3.2.6 represents capacitance per unit area. The last form of Equation 3.2.6 is a general result for small-signal capacitance C. Because C represents the ratio of a *differential* charge (hence a charge with limited linear extent) to differential voltage, it can always be expressed by the ratio of the permittivity to the total space-charge width. This result is discussed further in Chapter 4.

Solving Equation 3.2.6 for the total voltage across the junction, we obtain

$$(\phi_i - V_a) = \frac{q\epsilon_s N_d}{2C^2} \tag{3.2.7}$$

The form of Equation 3.2.7 indicates that a plot of the square of the reciprocal of the small-signal capacitance versus the reverse bias voltage should be a straight line, as sketched in Figure 3.6. The slope of the straight line can be used to obtain the doping in the semiconductor, and the intercept of the straight line with the voltage axis should equal ϕ_i. Measurements of small-signal capacitance as a function of dc bias and the subsequent construction of plots similar to Figure 3.6 from the data are often used to study semiconductors. In practice, the most serious inaccuracy comes from obtaining ϕ_i from the intercept with the voltage axis. The slope of the curve usually gives an accurate indication of the semiconductor doping.

Measurements of the small-signal capacitance are useful even in cases where the semiconductor doping varies with distance. In that circumstance, the space-charge picture is altered from the one shown in Figure 3.4b to that sketched in Figure 3.7. For a given dc (reverse) bias $(V_a < 0)$, the space-charge layer is of width x_d. A small increase in the magnitude of V_a causes a small increase in Q_s where

$$Q_s = q \int_0^{x_d} N(x)\,dx \tag{3.2.8}$$

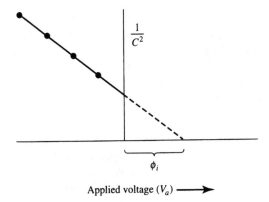

ϕ_i

Applied voltage (V_a) $\longrightarrow$

FIGURE 3.6 Plot of $1/C^2$ versus applied voltage for an ideal metal-semiconductor junction.

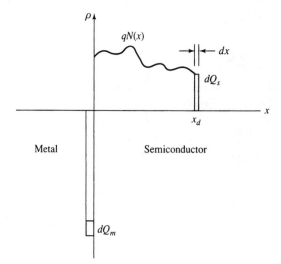

FIGURE 3.7 Schematic representation of space charge at a metal-semiconductor junction with nonuniform doping in the semiconductor.

Consequently, an increment in voltage dV_a results in an increment in Q_s of value

$$dQ_s = qN(x_d)\,dx = -C\,dV_a$$

or

$$N(x_d) = -\frac{C}{q\,(dx/dV_a)} \tag{3.2.9}$$

where x_d is the depletion-region width that corresponds to the applied dc voltage V_a at which the small-signal capacitance C is to be measured. We can rewrite Equation 3.2.9 by noting that because the small-signal capacitance C is given by $C = \epsilon_s/x_d$, the derivative (dx/dV_a) can be written $dx/dV_a = (dx/dC)(dC/dV_a) = -(\epsilon_s/C^2)\,dC/dV_a$. Thus,

$$N(x_d) = \frac{C^3}{q\epsilon_s(dC/dV_a)} \tag{3.2.10}$$

Equation 3.2.10 can be placed in a more useful form by recognizing that

$$\frac{d(1/C^2)}{dV_a} = -\left(\frac{2}{C^3}\right)\frac{dC}{dV_a}$$

so that

$$N(x_d) = \frac{-2}{q\epsilon_s[d(1/C^2)/dV_a]} \tag{3.2.11}$$

The result that we have derived in Equation 3.2.11 shows that the slope of a plot of $1/C^2$ versus reverse-bias voltage directly indicates the doping concentration at the edge of the space-charge layer. This slope, when divided into $(2/q\epsilon_s)$, gives $N(x_d)$ directly. Several commercial "semiconductor profilers" use this technique to determine the dopant concentration as a function of position. Some have direct-reading outputs that convert the slope to its equivalent doping concentration.

EXAMPLE *Schottky Barrier Diode*

A gold Schottky blocking contact is made to *n*-type silicon at the plane $x = 0$. The silicon has a surface layer extending from $x = 0$ to $x = x_1$ ($x_1 = 20$ nm) with doping $N_{d0} = 5 \times 10^{14}$ cm^{-3}. Below this layer ($x > x_1$) the silicon is *n*-type with a higher doping density (N_{d1} cm^{-3}) as shown in the accompanying sketch. The built-in potential across the metal-silicon junction ϕ_i is 0.50 V. Assume that the work function for gold $q\Phi_M$ is 4.75 eV.

(a) What is the value of N_{d1}, the dopant density in the interior of the silicon?

(b) Sketch the charge density and the electric field in the diode at thermal equilibrium ($V_a = 0$).

(c) Calculate the maximum field assuming that the voltage ΔV_s across the thin surface layer is negligible. Using this calculated maximum field, estimate the voltage between $x = 0$ and $x = x_1$ to determine if ΔV_s is negligible.

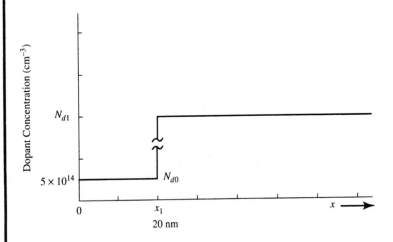

Solution   The surface layer is thin and very lightly doped. We assume that it is fully depleted and that the space-charge region extends into the bulk of the silicon crystal. If this assumption is incorrect our calculations will reveal it. From Figure 3.4,

$$q(\Phi_M - \Phi_S) = q\phi_i = 0.50 \text{ eV}$$

Hence,

$$q\Phi_S = 4.75 - 0.50 = 4.25 \text{ eV}$$

Again, from Figure 3.4,

$$q\Phi_S = qX + (E_c - E_f)$$

or

$$(E_c - E_f) = 4.25 - 4.05 = 0.20 \text{ eV}$$

and

$$(E_f - E_i) = (E_c - E_i) - (E_c - E_f)$$

so that

$$(E_f - E_i) = 0.562 - 0.20 = 0.362 \text{ eV}$$

(a) Using Equation 1.1.26, we calculate

$$n = N_{d1} = n_i \exp\left(\frac{0.362}{0.0258}\right) = 1.8 \times 10^{16} \text{ cm}^{-3}$$

The calculated density N_{d1} is much higher than N_{d0}, supporting our assumption that the surface layer is fully depleted.

(b) The charge and field diagrams should appear as sketched below.

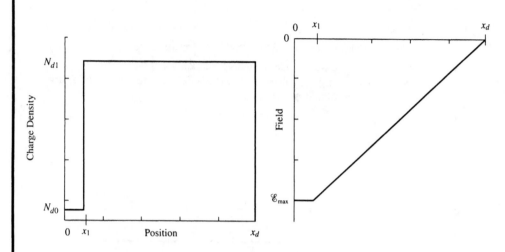

(c) If we neglect the voltage drop across the surface layer, we have, from Equation 3.2.3, $\mathcal{E}_{max} = -2\phi_i/x_d$ where $x_d = \sqrt{2\phi_i\epsilon_s/qN_d} = 190$ nm. Hence, $\mathcal{E}(x = 0) = \mathcal{E}_{max} = -5.27 \times 10^4$ V cm^{-1}. To estimate the voltage across the thin surface layer, we note that the change in the field in this region can be found from Gauss' Law.

$$\Delta\mathcal{E} = \frac{qN_{d0}}{\epsilon_s}\Delta x = \mathcal{E}(x = 0) - \mathcal{E}(x_1) = 154 \text{ V cm}^{-1}$$

Thus the field can be considered essentially constant across the surface layer, and we can calculate $\Delta V_s = -\mathcal{E}_{max} \times x_1 = 0.105$ V. Thus, ΔV_s is about 20% of ϕ_i, sufficiently high to make it reasonable to consider redoing the problem without assuming that the voltage drop is negligible across the surface layer. ∎

Schottky Barrier Lowering.[†]

We now reconsider our earlier statement that the barrier to electron flow from the metal into the semiconductor is "to first order" unchanged by an applied bias. The small dependence on applied voltage of this barrier height can be observed particularly under reverse bias. The dependence corresponds to an effect that was explained many years ago by Walter Schottky when electron emission into a vacuum was studied.

In our derivation we consider that within the semiconductor close to the interface it should be possible to approximate the electron energy by using the free-electron theory and by treating the metal as a plane conducting sheet. In this model we take account of the semiconductor in two ways: (1) by assigning the electron an effective mass m_n^*, and (2) by using a relative permittivity that differs from unity ($\epsilon_r = 11.7$ for Si). A sketch of the appropriate energy picture is given in Figure 3.8 under equilibrium conditions and also for the case of an applied field that tends to move electrons away from the metal surface. The function $E_1(x)$ that represents electron energy in the figure is derived by the classical technique; the conducting metal plane has the same effect on the electron as an image charge of opposite sign placed equidistant behind the plane $x = 0$. In the presence of a field $-\mathcal{E}$ tending to move electrons away from the metal surface into the semiconductor, the electron energy $E_2(x)$ is

$$E_2 = \frac{-q^2}{16\pi\epsilon_s x} - q\mathcal{E}x \qquad (3.2.12)$$

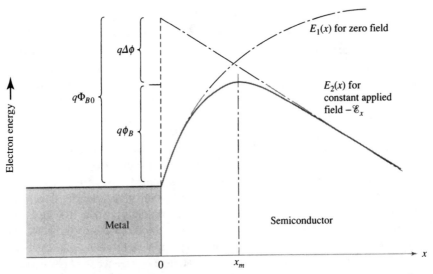

FIGURE 3.8 Classical energy diagram for a free electron near a plane metal surface at thermal equilibrium [$E_1(x)$], and with an applied field $-\mathscr{E}_x[E_2(x)]$.

From studies of metal-vacuum systems, Equation 3.2.12 has been shown to be accurate for distances greater than a few nanometers. The plane at which E_2 is a maximum is easily found, as is the energy $q\Delta\phi$ in Figure 3.8:

$$q\Delta\phi = \sqrt{\frac{q^3\mathscr{E}}{4\pi\epsilon_s}} \tag{3.2.13}$$

According to this model the height of the barrier $q\phi_B$ to electron flow out of the metal decreases as $q\Delta\phi$ increases. The current flow depends exponentially on this height because only that fraction of the Boltzmann-distributed electrons in the metal with energies above the barrier maximum can pass across it. We therefore expect the current emitted from the metal under reverse bias to vary as

$$J = J_0 \exp\frac{\sqrt{q^3\mathscr{E}/4\pi\epsilon_s}}{kT} \tag{3.2.14}$$

Using the depletion approximation, we can relate the field $\mathscr{E}$ to the bias V_a and built-in potential ϕ_i by

$$\mathscr{E} = \sqrt{\frac{2qN_d}{\epsilon_s}(\phi_i - V_a)} \tag{3.2.15}$$

Equations 3.2.14 and 3.2.15 show that, because of Schottky barrier lowering, the current emitted over a reverse-biased blocking contact depends exponentially on the fourth root of voltage at higher biases. Although this type of an exponential dependence is sometimes observed in practice, reverse currents that result from the generation of free carriers in the space-charge region may be larger than the component of current that we have considered here. In that case, the dependence on bias is more gradual. We will consider generation as a source of reverse current when we discuss current flow in *pn* junctions in Chapter 5.

3.3 CURRENT-VOLTAGE CHARACTERISTICS

The basic dependence of current on voltage in a Schottky-barrier diode can be deduced from qualitative arguments. These arguments provide fundamental insight into the nature of the equilibrium behavior of the metal-semiconductor system, and we therefore consider them before calculating the *I–V* characteristic more rigorously.

The band diagram at thermal equilibrium shown in Figure 3.4 is the starting point for the derivation. At equilibrium, the rate at which electrons cross over the barrier into the semiconductor from the metal is balanced by the rate at which electrons cross the barrier into the metal from the semiconductor. From the discussion of diffusion in Chapter 1, we know that free carriers in crystals are constantly in motion because of their thermal energies. In Problem 1.13, for example, this fact was used to show that a density n_0 of free carriers in thermal motion can be considered to cause a current density equal to $-qn_0v_{th}/4$ in an arbitrary direction. At thermal equilibrium, of course, this current density is balanced by an equal and opposite flow, and there is zero net current. Applying this concept to the boundary plane of the band diagram in Figure 3.4, we see that there is a tendency of electrons to flow from the semiconductor into the metal and an opposing balanced flux of electrons from the metal into the semiconductor. These currents are proportional to the density of electrons at the boundary. In the semiconductor, this density n_s (from Equation 1.1.21) is

$$n_s = N_c \exp\left(-\frac{q\phi_B}{kT}\right) \tag{3.3.1}$$

which can be written in terms of the bulk density $n = N_d$ by applying Equation 1.1.21 in the bulk of the semiconductor and noting from Figure 3.4 that $q\phi_B = q\phi_i + E_c - E_f$ in the semiconductor.

$$n_s = N_d \exp\left(-\frac{q\phi_i}{kT}\right) \tag{3.3.2}$$

Thus, equilibrium at the junction corresponds to

$$|J_{MS}| = |J_{SM}| = KN_d \exp\left(-\frac{q\phi_i}{kT}\right) \tag{3.3.3}$$

where J_{MS} and J_{SM} are the thermally induced current densities (current per area) directed from the metal toward the semiconductor and vice versa, and K is a proportionality factor.

When a bias V_a is applied to the junction as in Figure 3.5, the potential drop within the semiconductor is changed, and we can expect the flux of electrons from the semiconductor toward the metal to be modified. If we assume that the surface and the bulk of the semiconductor remain nearly at thermal equilibrium under bias, then the equation for n_s is modified from Equation 3.3.2 to

$$n_s = N_d \exp\left[-\frac{q(\phi_i - V_a)}{kT}\right] \tag{3.3.4}$$

The current arising from the electron flow out of the semiconductor is therefore altered by the same factor. The flux of electrons from the metal to the semiconductor, however, is not affected by the applied bias because the barrier ($q\phi_B$) remains at its equilibrium value.

Subtracting these two components, we obtain an expression for the net current from the metal into the semiconductor under applied bias:

$$J = J_{MS} - J_{SM}$$
$$= KN_d \exp\left[-\frac{q(\phi_i - V_a)}{kT}\right] - KN_d \exp\left(\frac{-q\phi_i}{kT}\right) \tag{3.3.5}$$

which can be written

$$J = J_0[\exp(qV_a/kT) - 1] \qquad (3.3.6)$$

where $J_0 = KN_d \exp(-q\phi_i/kT)$ is a new constant.

Equation 3.3.6 is often called the ideal diode equation. As this derivation makes clear, the ideal diode equation applies when a barrier to electron flow affects the thermal flux of carriers asymmetrically. Although more detailed analysis modifies the current equation slightly, the essential dependence of current on voltage for metal-semiconductor barrier junctions is contained in Equation 3.3.6. The ideal diode equation predicts a saturation current $-J_0$ for V_a negative and a very steeply rising current when V_a is positive (Problem 3.10).

We will again meet the ideal-diode behavior (Equation 3.3.6) when we consider currents across junctions made between p- and n-type semiconductor regions in Chapter 5. To build the arguments for the pn junction, we will find it useful to consider the interrelationships between energy bands, voltages, and doping levels in semiconductors as is done in Chapter 4. Readers may wish to go from this point directly to Chapter 4 and to complete Chapters 4 and 5 before returning to finish our discussion of metal-semiconductor junctions in the remainder of Chapter 3. Neither logic nor text continuity will be interrupted by this seemingly circuitous route; it merely provides an alternative path that may better accomodate specific reader interests.

More detailed consideration of the $J-V$ characteristics at metal-semiconductor barrier junctions shows that the saturation current J_0 is not completely independent of applied voltage. Analyses that lead to this result are carried out in the remainder of this section.

Schottky Barrier[†]

The dependence of current on applied voltage in a metal-semiconductor junction can be obtained by integrating the equations for carrier diffusion and drift across the depletion region near the contact. This approach, which was first used by Schottky [1], assumes that the dimensions of this space-charge region are sufficiently large so that the use of a diffusion constant and a mobility are meaningful—basically, that the width of the region is at least a few electronic mean-free paths and that the field strength is less than that at which the drift velocity saturates. An alternative physical approach, adopted first by Bethe [2a,b] and based on carrier emission from the metal, is valid even if these constraints are not met, and it leads to the same $J-V$ dependence [3].

If we refer to the metal-semiconductor contact under bias as shown in Figure 3.9 and consider one-dimensional electron flow through the barrier region, then (as shown by Equation 1.2.21) we can write

$$J_x = q\left[n\mu_n \mathscr{E}_x + D_n \frac{dn}{dx}\right] \qquad (3.3.7)$$

If we denote potential in the region as ϕ, then by using $\mathscr{E}_x = -d\phi/dx$ and the Einstein relation (Equation 1.2.20), we can write

$$J_x = qD_n\left[\frac{-qn}{kT}\frac{d\phi}{dx} + \frac{dn}{dx}\right] \qquad (3.3.8)$$

Figure 3.9*b* is a sketch of ϕ with the metal taken as ground.

We next rewrite Equation 3.3.8 in an integrated form that can be evaluated at the two ends of the depletion region ($x = 0$ and $x = x_d$ in Figure 3.9). Multiplying both sides

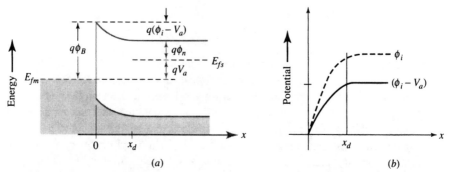

FIGURE 3.9 (a) Band diagram of a rectifying metal-semiconductor junction under forward bias. The applied voltage V_a displaces the Fermi levels: $qV_a = E_{fs} - E_{fm}$. (b) The potential across the surface depletion layer is decreased to $\phi_i - V_a$.

of Equation 3.3.8 by an integrating factor $\exp(-q\phi/kT)$ permits direct integration of the right-hand side. Using the limits of the depletion region, we obtain

$$J_x \int_0^{x_d} \exp\left(\frac{-q\phi}{kT}\right) dx = qD_n \left[n \exp\left(\frac{-q\phi}{kT}\right) \right]_0^{x_d} \tag{3.3.9}$$

In writing Equation 3.3.9 we have assumed that the particle current represented by J_x is not a function of position and can be placed outside the integral. This assumption has wide validity. Since the reference for potential is adjacent to the metal, the boundary conditions on voltage for use in Equation 3.3.9 are

$$\phi(0) = 0 \quad \text{and} \quad \phi(x_d) = (\phi_i - V_a) \tag{3.3.10}$$

As we see from Figure 3.9a, $\phi(x_d)$ can also be written as

$$\phi(x_d) = (\phi_B - \phi_n - V_a) \tag{3.3.11}$$

where $q\phi_n$ is $(E_c - E_f)$ in the semiconductor bulk. Boundary conditions for n are also necessary in order to evaluate Equation 3.3.9. Using Equation 1.1.21, we can express these boundary conditions as

$$n(0) = N_c \exp\left(\frac{-q\phi_B}{kT}\right)$$

and

$$n(x_d) = N_d = N_c \exp\left(\frac{-q\phi_n}{kT}\right) \tag{3.3.12}$$

Substituting the boundary values into Equation 3.3.9, we find

$$J_x = qD_n N_c \exp\left(\frac{-q\phi_B}{kT}\right) \left[\exp\left(\frac{qV_a}{kT}\right) - 1 \right] \Big/ \int_0^{x_d} \exp\left(\frac{-q\phi(x)}{kT}\right) dx \tag{3.3.13}$$

To obtain current as a function of voltage, the functional dependence of ϕ on x must be inserted into the integrand in the denominator of Equation 3.3.13 and the integration must be carried out. This functional dependence of ϕ on x is determined by the doping profile in the semiconductor near the contact.

The contact that we have been considering, a barrier to electron flow from a metal into a semiconductor of constant doping, is called a Schottky barrier in recognition of its initial analysis by Walter Schottky [1]. For the Schottky barrier we can use the depletion approximation, as in Sec. 3.2, to derive an expression for the potential in the depletion region.

$$\phi(x) = \frac{qN_d}{\epsilon_s} x \left(x_d - \frac{x}{2} \right) \qquad (0 < x < x_d) \qquad (3.3.14)$$

When Equation 3.3.14 is inserted into Equation 3.3.13 and the integration is performed, an explicit solution for J_x as a function of V_a is obtained.

$$J_x = J_S \left[\exp\left(\frac{qV_a}{kT} \right) - 1 \right] \qquad (3.3.15)$$

where

$$J_S = \frac{q^2 D_n N_c}{kT} \left[\frac{2q(\phi_i - V_a)N_d}{\epsilon_s} \right]^{1/2} \exp\left(\frac{-q\phi_B}{kT} \right) \qquad (3.3.16)$$

Equation 3.3.16 shows that J_S is not independent of voltage; hence, some of the voltage dependence in Equation 3.3.15 is implicit in J_S. The square-root dependence of J_S on voltage is, however, weak compared to the term that depends exponentially on voltage in Equation 3.3.15. We can, therefore, approximate the current-voltage characteristic by writing, instead of Equation 3.3.15:

$$J_x = J_S' \left[\exp\left(\frac{qV_a}{nkT} \right) - 1 \right] \qquad (3.3.17)$$

where J_S' is independent of voltage and n is taken to be a parameter having a value that is usually found experimentally to be between 1.02 and 1.15. (See Problem 3.9). Experimental measurements for a forward-biased, aluminum-silicon Schottky barrier are shown in Figure 3.10. The good fit between the measured data in Figure 3.10 and Equation 3.3.17 with n = 1.07 is typical.

The analyses that we have just carried out have made use of a number of assumptions, several of which have been explicitly noted. One important implicit assumption, however, should be mentioned before proceeding because it is very frequently used in first-order device analysis. This is the assumption that the system is in *quasi-equilibrium* (that is, almost at thermal equilibrium) even though currents are flowing. Quasi-equilibrium was implicitly invoked at a number of points in our analysis; for example, in writing Equation 3.3.4 and in using the Einstein relationship to write Equation 3.3.8. Logically we expect quasi-equilibrium to be more valid under low-bias conditions when currents are small—and, in fact, this is the case. Often, a quasi-equilibrium analysis suffices or else the analysis can be extended to cover all currents within the range of interest by making slight modifications to the theory. The ultimate test of any assumption is, of course, good agreement between measurements and predictions, as seen in Figure 3.10.

Mott Barrier[†]

Our derivation thus far has been for a metal-semiconductor junction in which the semiconductor doping is constant throughout the space-charge region; that is, for a Schottky barrier. To derive the current-voltage characteristic for other dopant profiles, we can use the equations through 3.3.13, but we must modify the expression for voltage in the depletion region (Equation 3.3.14). One case of doping that is useful for some metal-semiconductor junctions is known as the Mott barrier after N.F. Mott, who analyzed it in connection with studies of diodes composed of oxide compounds [4].

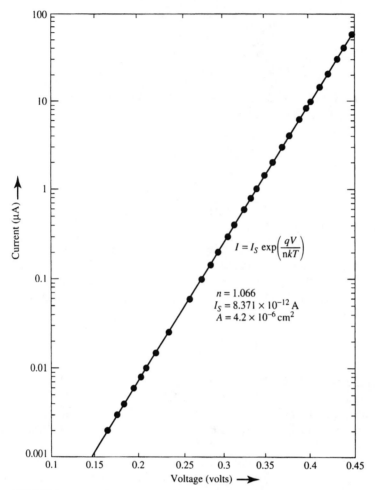

FIGURE 3.10 Measured values of current (plotted on a logarithmic scale) versus voltage for an aluminum-silicon Schottky barrier. Values for $I_S = J_S' A$ and n are obtained from an empirical fit of the data to Equation 3.3.17.

In the Mott-barrier approximation, the semiconductor is characterized by an abrupt change in doping from a low value near the metal interface to a high value a very short distance from the surface. The distance is short in the sense that essentially no field lines terminate in the lightly doped region. (This condition can be discussed by saying that the distance is much shorter than a Debye length L_D, where L_D is discussed in Sec. 3.4.) An electron-energy diagram for this situation is sketched in Figure 3.11. Because the length of the lightly doped region is very short, the electric field can be assumed to be constant throughout it. The field lines are assumed not to penetrate into the highly doped region because the donor density there is so high. This condition might model a case in which the doping near the contact is altered during the fabrication of the junction. It might also represent the situation in which the metal contact is made to a thin, lightly doped, epitaxial film above a highly doped region in the crystal; this latter case is encountered in the design of bipolar integrated circuits.

To obtain the dependence of current on voltage for the Mott barrier, we proceed as for the Schottky barrier, first expressing the potential ϕ as a function of x and then performing the integration indicated in Equation 3.3.13. Because the field in the lightly doped

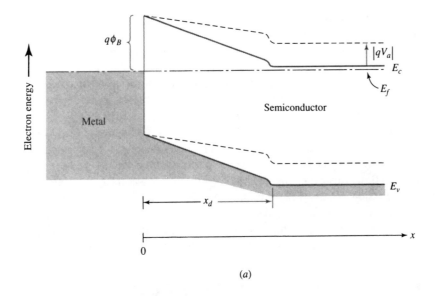

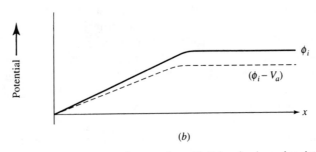

FIGURE 3.11 Electron-energy diagram for a Mott barrier (near-insulating region at the surface with a sharp transition at $x = x_d$ to a highly conducting region). The solid line indicates thermal equilibrium; the dotted line, a forward bias of V_a volts with the metal held at ground potential. (b) Potential diagram for the Mott barrier.

region is constant, ϕ depends linearly on x:

$$\phi(x) = (\phi_i - V_a)\frac{x}{x_d} \qquad (0 < x < x_d) \qquad (3.3.18)$$

Using Equation 3.3.18 in 3.3.13, we obtain a result that can be written in the form of Equation 3.3.15

$$J_x = J_M\left[\exp\left(\frac{qV_a}{kT}\right) - 1\right] \qquad (3.3.19)$$

where J_M depends more strongly on V_a than does the parameter J_S, derived for the Schottky barrier (Equation 3.3.16)

$$J_M = \frac{q^2 D_n N_c(\phi_i - V_a)\exp\left(\dfrac{-q\phi_B}{kT}\right)}{x_d kT\left\{1 - \exp\left[\dfrac{-q(\phi_i - V_a)}{kT}\right]\right\}} \qquad (3.3.20)$$

Mott and Schottky barriers represent idealized metal-semiconductor rectifiers. In many cases, these idealizations suffice. To derive the J–V characteristic in cases for which they are not adequate, an accurate representation for potential throughout the barrier region must be found and used in place of Equation 3.3.14 or 3.3.18. In general, the major dependence on voltage is contained in an exponential form, as we derived in both Equations 3.3.15 and 3.3.19. An empirical fit to measured data is generally possible using a form similar to Equation 3.3.17 with a value of n that is nearly unity.

Both equations that have been derived for saturation current (J_S in Equation 3.3.16 and J_M in Equation 3.3.20) become physically unreasonable as V_a approaches ϕ_i. In practice, if V_a were to equal ϕ_i, there would be no barrier at the junction and very large currents could then flow. However, when a large forward bias is applied to a practical diode, a significant fraction of the voltage is dropped across the resistance of the semiconductor regions in series with the junction so that the actual forward voltage applied to the Schottky barrier is never as large as the built-in voltage. The effect of series resistance will be discussed in more detail when currents in *pn* junctions are described in Chapter 5.

3.4 NONRECTIFYING (OHMIC) CONTACTS

In our discussion of metal-semiconductor contacts, we have thus far considered cases in which the semiconductor near the metal has a lower majority-carrier density than the bulk and in which there is a barrier to electron transfers from the metal. In such cases any applied voltage is dropped mainly across the junction region, and currents are contact limited. *The inverse case, in which the contact itself offers negligible resistance to current flow when compared to the bulk, defines an ohmic contact.* Although this definition of an ohmic contact may sound awkward, it emphasizes one essential aspect: when voltage is applied across a device with ohmic contacts, the voltage dropped across the ohmic contacts is negligible compared to voltage drops elsewhere in the device. Thus, no power is dissipated in the contacts, and the ohmic contact can be described as being at thermal equilibrium even when currents are flowing. An important and useful consequence of this property is that all free-carrier densities at an ohmic contact are unchanged by current flow; the densities remain at their thermal-equilibrium values.

Tunnel Contacts

The metal-semiconductor contacts that we have considered in the previous section can, for example, be made ohmic if the effect of the barrier on carrier flow can be made negligible. In practice this is accomplished by heavily doping the semiconductor so that the barrier width x_d is very small. To see this, we refer to Equation 3.2.4 and solve for x_d:

$$x_d = \sqrt{\frac{2\epsilon_s \phi_i}{qN_d}} \tag{3.4.1}$$

The space-charge region, therefore, narrows as N_d increases. When the barrier width approaches a few nanometers, a new transport phenomenon, *tunneling* through the barrier, can take place.

Figure 3.12a is a schematic illustration of the tunneling process through a very thin Schottky barrier. When the barrier is of the order of nanometers and the metal is biased negatively with respect to the semiconductor, electrons in the metal need not be energetic enough to surmount the barrier ($q\phi_B$ units above the Fermi energy) to enter the semiconductor. Instead, they can tunnel through the barrier into the conduction-band states in the

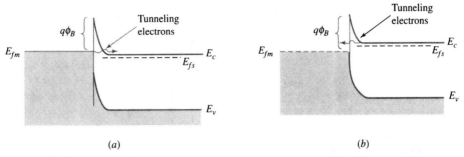

FIGURE 3.12 Metal-semiconductor barrier with a thin space-charge region through which electrons can tunnel. (*a*) Tunneling from metal to semiconductor. (*b*) Tunneling from semiconductor to metal.

semiconductor. Likewise, when the semiconductor is biased negatively with respect to the metal, electrons from the semiconductor can tunnel into electronic states in the metal (Figure 3.12*b*). Many electrons are available to take part in these processes and currents rise very rapidly as bias is applied. Hence, a metal-semiconductor contact at which tunneling is possible has a very small resistance. It is virtually always an ohmic contact. Ohmic contacts are frequently made in this way in practice. To assure a very thin barrier, the semiconductor is often doped until it is degenerate (i.e., until the Fermi level enters either the valence or the conduction band).

In modern devices the conductivity of the semiconducting regions is higher than in previous device generations. Consequently, ohmic contacts must have even lower resistance so that no appreciable voltage drop occurs across them.

Schottky Ohmic Contacts[†]

Another method of obtaining an ohmic contact is to cause the majority carriers to be more numerous near the contact than they are in the bulk of the semiconductor. An ohmic contact of this type results if the semiconductor surface is not depleted when it comes into equilibrium with the metal, but rather has an enhanced majority-carrier concentration. Using the ideal Schottky theory that we described in Sec. 3.2, we see that this condition occurs in a metal-semiconductor junction between a metal and an *n*-type semiconductor with a larger work function than that of the metal. In this case, electrons are transferred to the surface of the semiconductor and the metal is left with a skin of positive charge.[*] The relevant energy diagram is sketched in Figure 3.13*a*, and the corresponding charge and field diagrams are given in Figures 3.13*b* and 3.13*c*. There is a qualitative similarity between these figures and Figures 3.4*a* to 3.4*c* that referred to a rectifying contact; the important distinction between the two situations is that the semiconductor charge consists of free electrons in the case of the ohmic contact, but it is fixed (on positive donor sites) in the barrier case. The charge distribution, field, and potential for such a contact can be calculated using techniques similar to those employed for the rectifying contact [5]. Details of this procedure are considered in Problem 3.11. The results are summarized as follows.

To find a solution for the distribution of the space charge in the semiconductor, we take the reference for potential ϕ in the metal. We assume that the excess electrons in the

---

[*] For an ohmic contact to a *p*-type semiconductor, the relative sizes of the work functions in the two regions need to be reversed to achieve a net positive charge in the semiconductor and, thereby, an enhanced hole density near the contact.

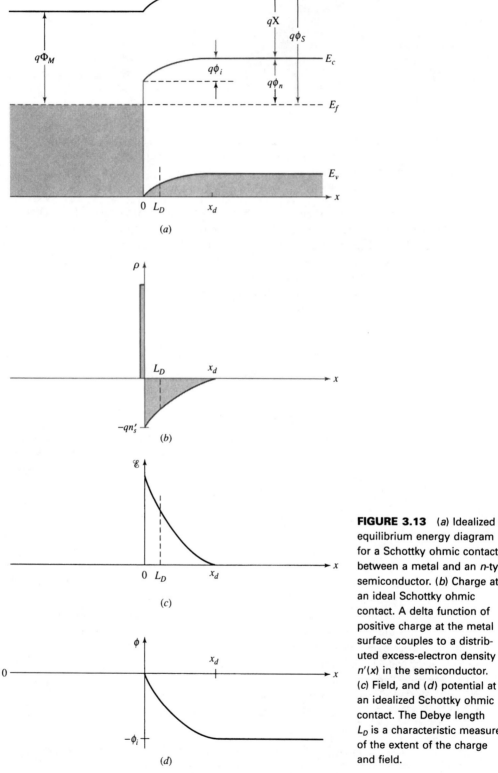

FIGURE 3.13 (a) Idealized equilibrium energy diagram for a Schottky ohmic contact between a metal and an n-type semiconductor. (b) Charge at an ideal Schottky ohmic contact. A delta function of positive charge at the metal surface couples to a distributed excess-electron density $n'(x)$ in the semiconductor. (c) Field, and (d) potential at an idealized Schottky ohmic contact. The Debye length L_D is a characteristic measure of the extent of the charge and field.

semiconductor surface region (denoted as n') are Boltzmann-distributed in energy so that $n' = n_s \exp(-q|\phi|/kT)$ where n_s represents the excess concentration in the semiconductor at the metal-semiconductor interface. We can write an integrable form of Poisson's equation if we represent the space-charge density in the semiconductor by qn' (which neglects the space charge contributed by donor ions). This approximation is reasonable for the major extent of the space-charge region (until the potential is within a few kT/q of the built-in value ϕ_i). Solving Poisson's equation, we find $\rho(x)$ the space-charge density.

$$\rho(x) = -qn_s \bigg/ \left(1 + \frac{x}{\sqrt{2}L_D}\right)^2 \tag{3.4.2}$$

where

$$L_D = \left(\frac{\epsilon_s kT}{q^2 n_s}\right)^{1/2} \tag{3.4.3}$$

is known as the *Debye length* at the surface. We will discuss the Debye length in more detail after completing our analysis of the contact.

The field varies with distance according to the equation

$$\mathscr{E}_x = \frac{\sqrt{2}kT}{L_D q}\left(1 + \frac{x}{\sqrt{2}L_D}\right)^{-1} \tag{3.4.4}$$

and the space-charge layer extends into the semiconductor a distance

$$x_d = \sqrt{2}L_D\left[\exp\left(\frac{q|\phi_i|}{2kT}\right) - 1\right] \tag{3.4.5}$$

The built-in voltage ϕ_i is seen from Figure 3.13a to be

$$|\phi_i| = \phi_n - (\Phi_M - X) \tag{3.4.6}$$

where $q\phi_n = (E_c - E_f)$ in the bulk. The condition $\Phi_M - X = \phi_n$ defines what can be called a neutral contact: that is, a contact with no built-in voltage and a surface that has the same density of free electrons as the bulk. In Problem 3.12 we show that a neutral contact can be considered to be "ohmic" for currents less than $qn_0 v_{th}/4$ where n_0 is the electron density and v_{th} is the thermal electron velocity. For $\Phi_M - X \ne \phi_n$, the current limit for ohmic behavior at the contact differs from $qn_0 v_{th}/4$ by a factor $\exp\{q[(\Phi_M - X) - \phi_n]/kT\}$. Thus, contacts to n-type material are properly divided into *ohmic* behavior for cases in which the surface bands bend down and into *blocking* behavior for cases in which the bands bend up. The inverse conditions apply for contacts to p-material.

In summary, we repeat the essential condition for ohmic behavior: an unimpeded transfer of majority carriers between the two materials forming the contact. At ohmic contacts there are generally built-in potentials. Unless penetrable barriers exist, majority carriers must be more numerous at an ohmic contact than they are in the bulk.

Debye Length.   The expressions that we derived for charge density (Equation 3.4.2), field (Equation 3.4.4), and space-charge-layer width (Equation 3.4.5) all contain the characteristic length L_D. Figures 3.13a through d show that L_D is an appropriate qualitative measure of the spatial extent of electrical effects at the boundary. The result of Problem 3.13 confirms this quantitatively by showing that 50% of the space charge in the semiconductor lies within $\sqrt{2}L_D$ units of the surface.

Considering a wider scope than this particular problem, we find that the solution of Poisson's equation in the presence of free charges always leads to a characteristic Debye length. If the charge configuration differs from the example given, the Debye length is still defined by Equation 3.4.3 except that n_s is replaced by the free-charge density appropriate to that configuration. We will see an example of this in Chapter 4. The general result that L_D qualitatively measures the spatial extent of space charge is always true. Problem 3.14 shows that a relationship exists between L_D in a given region and the dielectric relaxation time in the same region. This relationship can be interpreted physically in terms of a balance of two transport mechanisms for free carriers: diffusion (thermally induced motion), and drift (field-induced motion).

3.5 SURFACE EFFECTS

When we began our consideration of metal-semiconductor contacts in Sec. 3.2 and made use of Fermi-level equality to infer an equilibrium energy-band picture, we made an important idealization. We considered that both the semiconductor and the metal had allowed energy states (energy bands) at the surface that were no different from those in the bulk. The real situation is more complicated, and we will need to develop the theory further to derive a more practical physical model. The most important correction needed is to account for the effects of surface states. Although these effects modify some conclusions, we will be able to retain most of the ideas that were developed in previous sections.

Surface States

To begin, we should clarify the term *surface states*. Surface states are extra allowed states for electrons that are present at the semiconductor surface, but not in the bulk. These extra states arise from a number of sources. First, consider an absolutely clean surface composed purely of atoms of the host lattice. There are extra states on this surface because the energy field of the crystal is one-sided; that is, electrons in the surface region are bonded only from the side directed toward the bulk (Figure 3.14). We expect the characteristic energies

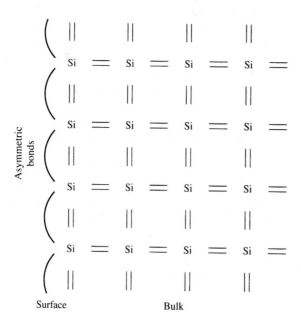

FIGURE 3.14 Bonding diagram for a silicon crystal near its surface (straight lines indicate coupled pairs of bonding electrons). The bonds at a clean semiconductor surface are anisotropic and, consequently, the allowed energy levels differ from those in the bulk.

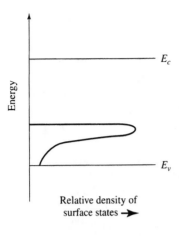

FIGURE 3.15 Approximate distribution of Tamm-Shockley states in the diamond lattice [8]. The distribution appears to peak sharply at an energy roughly one-third of the bandgap above E_v.

for electrons at such sites to differ from the characteristic energies in the bulk.* Surface states of this type are called Tamm or Shockley states after original investigations by these scientists [6,7]. The density of the Shockley-Tamm states in a particular semiconductor is of the order of the density of atoms at the surface, or roughly $N_0^{2/3}$ (cm^{-2}) where N_0 is the bulk atomic density (atoms cm^{-3}). For silicon, N_0 is 5×10^{22} cm^{-3} (see Table 1.3) and the density of the Tamm-Shockley states is about 10^{15} cm^{-2}. The energetic distribution of these states is not well established although studies on the diamond lattice [8] indicate their density to be peaked about one-third of the forbidden-gap energy above the valence band, as sketched in Figure 3.15.

Bonded foreign atoms at the surface and crystal defects are sources of other types of surface states. An example is oxygen, which is virtually always found at a silicon surface. Oxygen can produce surface states spread over a range of energies depending upon the nature of the specific bonds it shares with silicon atoms. Other surface complexes of metals, hydroxyl ions, etc., can also be expected on any processed silicon surface. The net effect of all of these sources is a density of available electronic surface states that is not zero at any energy, although there may be pronounced peaks at specific energies.

Besides varying in energy, surface states also vary in type and may be classified according to the charge they carry at equilibrium. For example, states that are neutral when occupied by electrons and positively charged when unoccupied are classified as donor states. States that are negative when occupied but neutral when empty are classified as acceptor states in a manner analogous to the bulk dopant atoms discussed in Sec. 1.1.

One other classification of surface states (sometimes called *interface states*) arises from the properties of real interfaces between solids. On an atomic scale (a fraction of a nanometer), such interfaces are not abrupt but consist rather of zones of intermediate materials and impurities that are several to tens of atomic layers in thickness. Within these intermediate zones some surface states are physically close to the bulk semiconductor, and they remain in thermal equilibrium with bulk states even when the potential is changed fairly rapidly. These are called *fast surface states* because the electrons occupying them come into equilibrium quickly. In contrast to these are so-called *slow states*, which are states situated more remote from the bulk semiconductor within the intermediate layer. These states take relatively longer times to reach thermal equilibrium with the bulk states. Although the demarcation between these categories of states is vague, general usage sets the boundary at a response time corresponding to about 1 kHz.

---

* In terms of quantum mechanics, the wave functions for the electrons are perturbed by the termination of the crystal potential; hence, the allowed energy states are different at the surface than in the bulk.

Surface Effects on Metal-Semiconductor Contacts[†]

The presence of surface states can modify the contact theory that we have presented. If the semiconductor surface states are not neutral or if they change their charge state when the contact is formed, then the charge configurations that we obtain by applying Schottky theory are not valid. A theory for metal-semiconductor systems with interface states has been developed [9]. We will not discuss this theory in detail, but rather give a summary and show the agreement of theory with experimental results.

To account for interface effects, the metal-semiconductor contact is treated as if it contained an intermediate region sandwiched between the two crystals. The intermediate region ranges from several to about ten atomic dimensions in thickness. This layer contains the impurities and added interface states. It is too thin to be an effective barrier to electron transfer (which can occur by tunneling), but it can sustain a voltage drop. Extra allowed electronic states are postulated to be distributed in energy at the assumed planar boundary between the interfacial layer and the semiconductor. The band structure and the surface states are sketched in Figure 3.16. In the figure, the states are assumed to be acceptor type and to have a density D_s states $cm^{-2} eV^{-1}$. Note in Figure 3.16 the thin layer of width δ, which sustains a voltage drop of Δ volts. Because this layer is thin enough for electron tunneling, the metal-semiconductor barrier height is measured between the Fermi level and the conduction band at the semiconductor surface. A surface layer should be associated with the semiconductor even when the metal is not present (Figure 3.17). When acceptor surface states are present as in Figure 3.17, the n-type semiconductor is depleted of electrons near its surface, and negative charge exists in the acceptor surface states.

Consider the formation of a blocking contact to a semiconductor with surface states such as that shown in Figure 3.17. Such a contact has an energy diagram similar to that in Figure 3.16. To draw this diagram we note that from Schottky theory a blocking contact is formed if contact is made to a metal having Φ_M greater than Φ_S, because electrons are transferred from the semiconductor into the metal. Transfer of electrons from the semiconductor bends the conduction band away from the Fermi level.

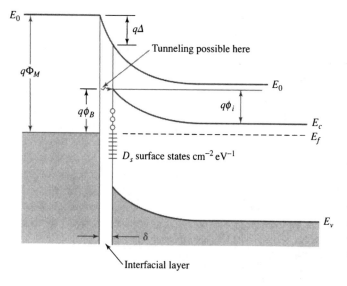

FIGURE 3.16 Band structure near a metal-semiconductor contact according to the model of Cowley and Sze [9]. The model considers a thin interfacial layer of thickness δ that sustains a voltage Δ at equilibrium. Acceptor-type surface states distributed in energy are assumed to be described by a distribution function D_s states $cm^{-2} eV^{-1}$.

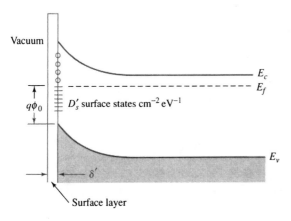

FIGURE 3.17 Band diagram for a semiconductor surface showing a thin surface layer containing acceptor-type surface states distributed in energy. A surface-depletion region is present because of charge in the surface states.

For the semiconductor in Figure 3.17 this removes charge from some of the surface states by lifting them above E_f. The larger is D_s, the greater is the charge removed for each incremental energy increase in E_c near the contact. If D_s is large, therefore, a negligible movement of the Fermi level at the semiconductor surface transfers sufficient charge to equalize the Fermi levels. In this case, the Fermi level is said to be *pinned* by the high density of states. Note that pinning of the Fermi level is not exclusively associated with the surface states being acceptor type. Any electronic states clustered near the Fermi energy cause the Fermi level to be pinned when the state density becomes very large because slight changes in the Fermi energy result in very sizable charge transfer.

When the Fermi level is pinned, the barrier height $q\phi_B$ becomes [9]

$$q\phi_B = (E_g - q\phi_0) \tag{3.5.1}$$

where $q\phi_0$ is $(E_f - E_v)$ at the surface when the semiconductor is not covered by metal, as shown in Figure 3.17. As D_s approaches zero, the barrier height $q\phi_B$ approaches the height predicted by the basic Schottky theory given in Equation 3.2.1 and repeated here for reference.

$$q\phi_B = q(\Phi_M - \mathrm{X}) \tag{3.5.2}$$

Whether a metal-semiconductor barrier height is predicted by Equation 3.5.1 or by Equation 3.5.2 or has an intermediate value depends on the magnitude of D_s at energies near the Fermi level and also on specific properties of the boundary layer, such as its precise thickness and permittivity [9].

In practice, most Schottky barrier heights for important semiconductors are predicted more accurately by Equation 3.5.1 than by Equation 3.5.2; there is only a small dependence on the metal work function. This is true for silicon, germanium, and especially for gallium arsenide and other III–V semiconductors. For silicon, as for germanium, gallium arsenide, and gallium phosphide, the quantity $q\phi_0$ is found experimentally to be about equal to $\frac{1}{3}E_g$ so that the barrier height $q\phi_B$ from Equation 3.5.1 is typically close to $\frac{2}{3}$ of the band gap or roughly 0.75 eV for silicon. The reason for this consistency may be the characteristic very high density of states that appears to be common to the diamond-type lattice and which pins the Fermi level at this energy (Figure 3.15). Some measured barriers for various metals contacting n- and p-type silicon crystals are shown in Table 3.1.

TABLE 3.1   Schottky Barriers to Silicon [10]
(qX for Silicon = 4.05 eV)

Silicon Type	Metal	$q\Phi_M$(eV)	$q\phi_B$(eV)
n	Al	4.1	0.69
p	Al	—	0.38
n	Pt	5.3	0.85
p	Pt	—	0.25
n	W	4.5	0.65
n	Au	4.75	0.79
p	Au	—	0.25

3.6   METAL-SEMICONDUCTOR DEVICES: SCHOTTKY DIODES

The industrial use of metal-semiconductor barrier devices, generally designated as Schottky diodes, is widespread. The area of greatest application is that of digital logic circuits, that is, circuits that perform binary arithmetic. Schottky diodes are often used in these circuits as fast switches that can be made on integrated-circuit chips within very small dimensions on the surface. There is also an increasing interest in Schottky-diode power rectifiers because large-area devices with an excellent thermal path through the contact metal allow operation at high currents. These high currents flow with lower voltage drops across the junction than in the diffused *pn* junction diodes to be discussed in Chapters 4 and 5.

Schottky diodes are also used as variable capacitors that can be operated efficiently in the microwave region of the spectrum. For variable capacitance applications, the diode is continuously kept under reverse bias. Voltage changes across it are then able to modulate the depletion-region width and capacitance in the manner discussed in Sec. 3.2.

Another application that makes use of modulation of the depletion width is the Schottky-barrier, field-effect transistor* also called the *MEtal-Semiconductor Field-Effect Transistor* (MESFET). This device, pictured in Figure 3.18, consists of a barrier junction at the input that acts as a control electrode (or *gate*), and two ohmic contacts through which output current flows (described as the *source* and *drain* electrodes). The output current varies when the cross section of the conducting path beneath the gate electrode is changed. This device is a special form of a junction field-effect transistor (JFET), an amplifying device in which the control electrode is usually made from a reverse-biased *pn* junction. Substituting a Schottky barrier for the *pn* junction is especially useful with semiconductor materials in which the fabrication of *pn* junctions is not practical. The major application of Schottky barriers to MESFETs is to make high-frequency, gallium-arsenide devices. We will discuss the operation of JFETs and MESFETs in more detail after we have considered *pn*-junction operation in Chapter 4.

---

* Since this is the first point at which we meet the term *transistor*, it is appropriate to point out that the word is an amalgamation of the descriptive terms *transfer resistor*. It was coined by W. Shockley and co-workers at the Bell Telephone Laboratories where the junction field-effect transistor and the bipolar transistor, to be discussed in Chapters 6 and 7, were both invented.

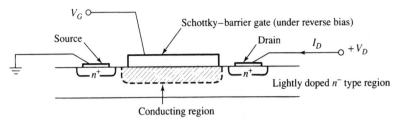

FIGURE 3.18   Schottky-barrier-gate, field-effect transistor. Current I_D flowing from drain to source is modulated by gate voltage V_G that controls the dimensions of the depletion region. This, in turn, modulates the cross-sectional conducting area for I_D. The source and drain contacts are ohmic because they are made to highly doped material.

Schottky Diodes in Integrated Circuits

Two fortuitous designers' choices, made for reasons having nothing to do with Schottky barriers, combine to allow very simple fabrication of Schottky diodes in silicon digital integrated circuits. These choices are the use of high resistivity n-type silicon in which to build npn bipolar transistors and the use of evaporated aluminum metal to form the "wire" interconnections in integrated circuits. The aluminum forms a blocking contact to the lightly doped, n-type silicon if the silicon surface has been thoroughly cleaned. It is only necessary that the doping of the silicon be sufficiently low so that the barrier cannot be penetrated by tunneling electrons, as was described in Sec. 3.4. Practically, this limits the doping to less than about 10^{17} cm^{-3} (Problem 3.7).

The barrier height between n-type silicon and aluminum is about 0.70 eV, and diodes made by depositing aluminum onto the silicon surface in a vacuum have characteristics that approximate theoretical predictions quite well (Figure 3.10). Because of the concentration of electric-field lines near the corners, however, reverse breakdown is not an abrupt function of voltage and occurs at a relatively low bias ($\sim$15 V). Several techniques, such as the use of diffused guard rings (Figure 3.19a) or field plates (Figure 3.19b), have been developed to improve the reverse characteristics. Because they complicate circuit processing, however, these techniques are avoided unless especially needed. Excellent Schottky diodes with high barriers can be made using refractory metals; platinum, in particular, is frequently employed. Because of their high vaporization temperatures, refractory metals are difficult to deposit by evaporation, and they are often deposited by

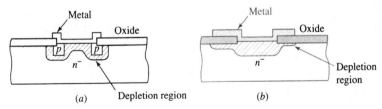

FIGURE 3.19   Special processing techniques improve the performance of Schottky diodes shown here in cross section. (*a*) The diffused *p*-type guard ring leads to a uniform electric field and eliminates breakdown at the junction edges and corners. (*b*) The metal field plate is an alternative means for achieving the same effect.

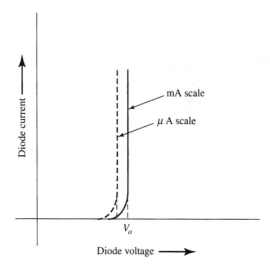

FIGURE 3.20 Linear plot of current versus voltage for a Schottky diode illustrating the concept of a diode "turn-on voltage."

sputtering.* More easily deposited metals are also frequently deposited by sputtering because of the high deposition rates possible. If sputtering is included in the manufacturing process, the silicon surface can be cleaned thoroughly by bombarding it with high-energy ions (*sputter-etching*). After sputter-etching, the subsequent sputter-deposition of platinum forms an excellent Schottky diode. Although Schottky diodes can be made to *p*-type silicon, the barrier heights are inherently smaller (roughly $\frac{1}{3}$ of the band gap or 0.36 V), and the yield and electrical performance are correspondingly poorer.

Designers who use Schottky diodes for digital logic circuits often call them *clamps* because they fix or clamp the voltage across one junction of a transistor and improve circuit performance. We will see the reasons in Chapter 6, but at this point we merely assert that the speed of a digital logic circuit can be substantially improved if clamps are put between the collector and base of a switching transistor to keep its collector-base junction from being forward biased. Schottky diodes are nearly ideal for this purpose.

To place this topic in context, we first discuss the concept of a *turn-on voltage*. This term refers to the forward voltage drop that must be placed across a diode to "turn-it-on"; that is, to cause it to pass substantial current. The current-voltage equations for metal-semiconductor barriers (Equations 3.3.17 and 3.3.19) show, however, a continuous dependence of current on voltage with no abrupt change of characteristics to identify as a turn-on voltage. From an engineering point of view, however, there is a threshold for conduction that becomes apparent when we plot the current-voltage characteristics of a Schottky diode using linear scales (Figure 3.20). The linear current scale permits only a small range of current to be plotted meaningfully. The strong dependence of current on voltage allows the data to be fit fairly well by two straight lines, one nearly horizontal at $J = 0$ and one nearly vertical. The intersection of the nearly vertical line with the voltage axis defines a turn-on voltage V_o. When digital circuits are designed, V_o is a good approximation to the voltage drop across any diode that is conducting. From Equation 3.3.17, at a given forward current density J_F, V_o is given by

$$V_o = \frac{nkT}{q} \ln\left(\frac{J_F}{J_S'} + 1\right) \tag{3.6.1}$$

---

* Sputtering is a technique for laying down thin films of materials in which a source, or target, is bombarded with high-energy gaseous ions (often argon).

Thus, from the designer's point of view, V_o depends most strongly on J'_S. For diodes designed to pass currents in the milliamp range, aluminum Schottky diodes fabricated on *n*-type silicon typically have a V_o of about 450 mV. Because this number is about 200 mV smaller than V_o for a comparable *pn* junction (Chapter 5), a Schottky diode placed in parallel with a *pn*-junction diode will not permit the forward bias to rise sufficiently for the junction diode to conduct. In the case of a Schottky-clamped collector junction, the bipolar transistor is kept out of a condition called *saturation*, in which switching transients are inherently slowed. Thus, digital circuits using Schottky clamps are faster by several nanoseconds than are unclamped circuits.

The only penalty paid in electrical performance for using the Schottky diode is the small amount of extra capacitance with which the reverse-biased Schottky diode loads the circuit. The clamped transistor takes up only slightly more surface area on the silicon chip than does the unclamped transistor, a big advantage for an integrated-circuit device. There is, however, generally some loss in fabrication yield when Schottky processing is used because metallization and surface preparation are more critical than in circuits that do not use Schottky barriers.

SUMMARY

An ensemble of electrons at thermal equilibrium is characterized by a single *Fermi energy*. The Fermi energy is therefore the appropriate reference for a diagram of allowed energy states versus position. It is particularly useful in drawing the appropriate thermal-equilibrium diagram for an inhomogeneous material and for systems of materials in intimate contact. If two materials, characterized by different Fermi energies (and therefore not in thermal equilibrium with each other) are brought sufficiently close to interact with one another, electrons will be transferred from the material with the higher Fermi energy to the material with the lower Fermi energy. Application of these principles to metals and semiconductors, under the idealized conditions that the bulk energy-state configuration continues to the surface and that the semiconductor is homogeneous, is the basis for Schottky, metal-semiconductor contact theory. This theory predicts blocking contacts and rectifying behavior for *n*-type semiconductors if the metal work function Φ_M exceeds the semiconductor work function Φ_S and ohmic behavior if Φ_S is greater than Φ_M. The inverse is true for metal contacts to *p*-type semiconductors. Diagrams for ideal Schottky contacts are shown in Figure 3.21. If the space-charge regions at a Schottky barrier become thin enough for substantial electron tunneling (resulting from highly doping the semiconductor), the contacts are also ohmic.

To develop a theory of metal-semiconductor contacts, it is necessary to make use of Poisson's equation in conjunction with the thermal-equilibrium band diagram. Simplifying assumptions such as the *depletion approximation* and *quasi-equilibrium* make the theory tractable. Basic Schottky theory predicts a number of observed properties successfully. Among these are the major dependences of the current-voltage characteristics and the reverse-bias capacitance behavior of the Schottky barrier. Contacts to a semiconductor having a high-resistivity region on top of a low-resistivity bulk are analyzed by the same techniques to give results for a *Mott barrier*. A similar analysis applied to the Schottky ohmic contact introduces the *Debye length*, a characteristic measure of the extent of electric-field penetration in a region having significant free charge. Although basic Schottky theory provides much useful information about metal-semiconductor contacts, it does not successfully predict barrier heights to silicon. A major inadequacy of the theory lies in the treatment of surface effects. Surface states originate from the termination of the lattice and from imperfect and impure surfaces. Theory for metallic blocking contacts to real silicon surfaces is based upon the assumption of a thin interfacial layer of imprecise composition, but of well-specified electronic behavior. The applications of metal-semiconductor contacts in the form of Schottky diodes are widespread. Schottky diodes have special relevance to integrated circuits because they are relatively easy to obtain using standard silicon planar technology.

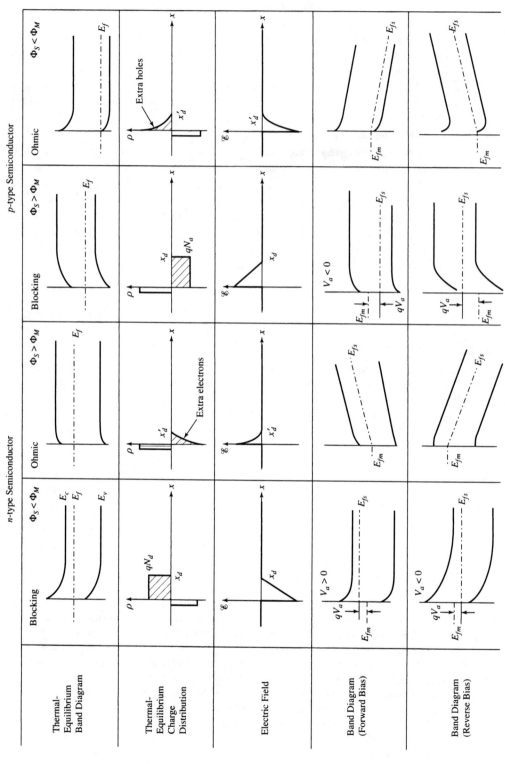

FIGURE 3.21 Diagrams for ideal metal-semiconductor Schottky diodes.

REFERENCES

1. W. SCHOTTKY, *Naturwissenschaften* **26**, 843 (1938).
2. (a) H. A. BETHE, *Theory of the Boundary Layer of Crystal Rectifiers*, MIT Radiation Laboratory Report 43–12 (1943).
   (b) S. M. SZE, *Physics of Semiconductor Devices*, 2nd Edition, Wiley-Interscience, New York (1981), pp. 255–258.
3. H. K. HENISCH, *Rectifying Semiconductor Contacts*, Oxford at the Clarendon Press (1957), p. 172.
4. N. F. MOTT, *Proceedings Cambridge Philosophical Society* **34**, 568 (1938).
5. A. ROSE, *Concepts in Photoconductivity and Allied Problems*, Wiley-Interscience, New York (1963).
6. I. TAMM, *Phys. Z. Sowjetunion* **1**, *733* (1933).
7. W. SHOCKLEY, *Phys. Rev.* **56**, 317 (1939).
8. D. PUGH, *Phys. Rev. Lett.* **12**, 390 (1964).
9. A. M. COWLEY and S. M. SZE, *J. Appl. Phys.* **36**, 3212 (1965).
10. S. M. SZE, *Physics of Semiconductor Devices*, *2nd Edition*, Wiley-Interscience, New York (1981).

BOOKS

S. M. SZE, *Physics of Semiconductor Devices*, 2nd Edition Wiley-Interscience, New York (1981).

A. G. MILNES, *Semiconductor Devices and Integrated Electronics*, Van Nostrand Reinhold, New York (1980).

PROBLEMS

3.1* Draw a diagram similar to Figure 1.11 for silicon in which the doping changes abruptly from 5×10^{18} donors cm^{-3} to 8×10^{15} donors cm^{-3}.

(a) What are the work functions associated with the two regions of the crystal?

(b) What is the potential difference between the two regions of silicon?

3.2 The energy-band diagrams for a metal and a semiconductor are sketched to the same scales in Figure P3.2a. Three possible band diagrams for a contact between these materials are sketched in Figures P3.2b to d. Assume thermal equilibrium.

(a) Explain why each figure is incorrect.

(b) Draw a correct diagram, assuming that basic Schottky theory applies.

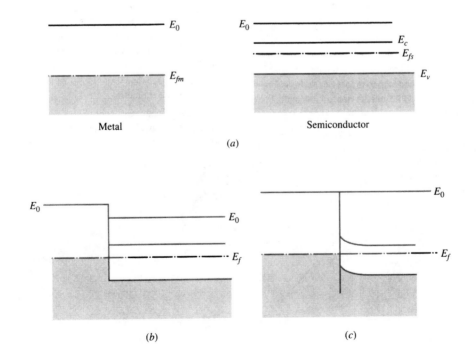

FIGURE P3.2

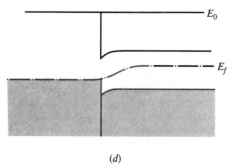

(d)

FIGURE P3.2 (continued)

3.3 Consider metal-semiconductor junctions that behave according to simple Schottky theory.

(a) Draw the theoretical energy-band diagram for copper (work function 4.5 eV) in contact with silicon having a work function of 4.25 eV.

(b) If light were to shine on this junction and create hole-electron pairs:

(i) Which way would current flow within the device if the junction were connected into a circuit?

(ii) What would be the maximum voltage that could be measured across the junction (zero output current)?

(c) Draw the energy-band diagram for copper in contact with silicon having a work function of 4.9 eV.

(d) Compare the electrical behavior of the metal-semiconductor systems described in (a) and (c).

3.4 The accompanying data were obtained on metal contacts to silicon of equal area (Figure P3.4). If Schottky theory applies, which metal probably has the higher work function? Which data were taken on 1 Ω-cm silicon and which on 5 Ω-cm silicon? Justify your answers and explain the use of "probably." (Consider Sec. 3.5.)

3.5* (a) Calculate the small-signal capacitance at zero dc bias and at 300 K for an ideal Schottky barrier between platinum (work function 5.3 eV) and silicon doped with $N_d = 10^{16}$ cm^{-3}. The area of the Schottky diode is 10^{-5} cm^2.

(b) Calculate the reverse bias at which the capacitance is reduced by 25% from its zero-bias value.

3.6† (a) Find the location x_m of the plane at which the barrier to emitted electrons [$E_2(x)$ in Figure 3.8] is a maximum and prove Equation 3.2.13.

(b) For an applied field of 10^5 V cm^{-1}, calculate x_m and $q\Delta\phi$.

3.7* Consider an aluminum Schottky barrier made to silicon having a constant donor density N_d. The barrier height $q\phi_B$ is 0.65 eV. The junction will pass high currents under reverse bias by tunneling from the metal if the barrier presented to the electrons is thin enough, as described in the following. We assume that the onset of efficient tunneling occurs when the Fermi energy in the metal is equal to the edge of the conduction band (E_c) at a distance 10 nm into the semiconductor.

(a) If this condition is reached at a total junction bias ($\phi_i - V_a$) of 5 V, what is the maximum value of N_d?

(b) What limit does this place on the resistivity of the epitaxial layers used in Schottky-clamped circuits?

(c) Draw a sketch of the energy-band diagram under the condition of efficient tunneling.

3.8† Carry through the steps needed to derive Equation 3.3.13.

3.9† Consider Equation 3.3.16 under conditions of low forward bias. Show that Equation 3.3.17 can be derived by using $(1 - V_a/\phi_i)^{1/2} = \exp[\frac{1}{2} \ln(1 - V_a/\phi_i)]$ and by approximating the resultant expression for J_S. This approach leads to Equation 3.3.17 with n = (1 + $kT/2q\phi_i$), which is generally smaller than observed values. Other effects such as rounding of the barrier contribute to values for n in Equation 3.3.17 that are somewhat higher than those found from the expression derived in this problem.

3.10 Using linear scales, plot I versus V_a for a diode that obeys the ideal-diode law (Equation 3.3.6) under the condition that:

(a) $I_0 = 1$ pA and $T = 150$ K.

(b) $I_0 = 1$ nA and $T = 300$ K.

(c) $I_0 = 1$ μA and $T = 450$ K.

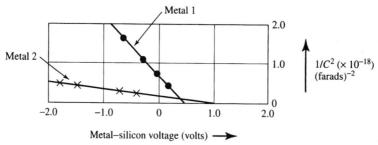

Metal–silicon voltage (volts) ⟶

FIGURE P3.4

(d) Considering the discussion in Section 3.6, state an appropriate value for the turn-on voltage V_o for each plot of I versus V_a.

For clarity in the diagrams, use a scale change at $V = 0$. Show the forward characteristic through 5 mA.

3.11[†] Use the equations in Sec. 3.4 to represent the space charge in a Schottky ohmic contact between a metal and an n-type semiconductor and set up Poisson's equation. The equation form will be $d^2\phi/dx^2 = K \exp(\phi/V_t)$, where $V_t = kT/q$. It is convenient to convert this function of voltage ϕ and position x to a function of field $\mathscr{E}$ and voltage ϕ. This can be done by making use of $d^2\phi/dx^2 = \mathscr{E} \, d\mathscr{E}/d\phi$. The resultant equation can be solved to find $\mathscr{E} = \sqrt{2n_s kT/\epsilon_s} \exp(\phi/2V_t)$.

(a) Carry through the steps which have been outlined in this problem.

(b) Derive Equations 3.4.2, 3.4.4, and 3.4.5 by continuing with this analysis.

3.12[†] Draw the band diagram for a "neutral" contact, as described in Sec. 3.4. Consider the results of Problem 1.13, which express the random thermal flux of free electrons in a semiconductor as $qn_0v_{th}/4$ where n_0 is the electron density and v_{th} is the thermal velocity. If currents drawn from the metal into the semiconductor are less than this value, the contact will not limit the flow and can be regarded as ohmic.

(a) Show that a contact is ohmic for fields in the semiconductor less than $v_{th}/4\mu_n$.

(b) Calculate the limiting ohmic current in a neutral contact made to a semiconductor having $N_d = 10^{16}$ cm^{-3} and $A = 10^{-5}$ cm^2. Take $v_{th} = 10^7$ cm s^{-1}.

(c) What is the limiting ohmic current if the bands are bent such that $q(\Phi_M - X - \phi_n) = 0.65$ eV?

3.13[†] Using Equations 3.4.2 and 3.4.5, show that half of the space charge in the Schottky ohmic contact exists within $\sqrt{2}L_D$ of the surface.

3.14 Show that the dielectric relaxation time $\tau_r = \epsilon_s/\sigma$ (discussed in Problem 1.12) can be related to the Debye length L_D by $L_D = (D\tau_r)^{1/2}$ where D is the diffusion constant in the material.

3.15 Assuming that basic Schottky theory applies, sketch the energy-band diagram for (a) an ohmic contact between p-type silicon and a metal at equilibrium, and (b) a blocking contact between p-type silicon and a metal under 2 V reverse bias.

3.16* Both Schottky-barrier diodes and ohmic contacts are to be formed by depositing a metal on a silicon-integrated circuit. The metal has a work function of 4.5 eV. For ideal Schottky behavior, find the allowable doping range for each type of contact. Consider both p- and n-doped regions and comment on the practicality of processing the integrated circuit with the required doping.

3.17[†] A back-biased Schottky diode made to silicon is to be used as a tuning element for a broadcast-band radio receiver (550 to 1650 kHz). For ease of operation, it is desirable to have the resonant frequency $(1/2\pi\sqrt{LC})$ of a tuned circuit change linearly with voltage when a dc voltage applied to the circuit changes from 0 to 5 V. If the tuning inductance L is 2 mH, we can readily calculate that the capacitance at the two extremes of bias to achieve this behavior should be 41.8 and 4.65 pF, respectively. Consider that the diode area is 10^{-3} cm^2 and find the desired dopant variation for N_d. (Calculate the numerical values and sketch a semilogarithmic plot of the results.) *Hint.* To attack this problem note that

$$\frac{df}{dV} = -\frac{1}{4\pi\sqrt{LC}} \frac{1}{C}\frac{dC}{dV} = 0.22 \text{ MHz/V}$$

from the information given. Use Equation 3.2.10 together with $C = A\epsilon_s/x_d$ to find $N(x_d)$.

pn JUNCTIONS

We saw in Chapter 3 that a system of electrons is characterized by a constant Fermi level at thermal equilibrium. This principle was initially used to deduce the energy-band diagram of a semiconductor having two doping levels. Systems that are initially not in thermal equilibrium approach equilibrium as electrons move from regions with a higher Fermi level to regions with a lower Fermi level. The transferred charge causes the buildup of barriers against further electron flow, and the potential drop across these barriers increases to a value that just equalizes the Fermi levels. These concepts were the foundation for an extensive analysis of metal-semiconductor contacts.

In this chapter we consider similar phenomena in a single crystal of semiconductor material containing regions having different dopant concentrations. We will find it useful to employ two important approximations that are frequently encountered in device analysis. One, the *depletion approximation*, has already been introduced in Chapter 3. The second, the *quasi-neutrality approximation*, serves the same simplifying purpose as the depletion approximation; it makes complicated problems tractable by focusing on the dominant physical effects in a given region of a device. The quasi-neutrality approximation is employed in a region of a semiconductor containing a slowly varying dopant concentration, while the depletion approximation is useful in the important case of a semiconductor containing adjacent *p*- and *n*-type regions.

We discuss in some detail the transition region at a *pn* junction and the barrier associated with this transition region. We next consider the influence of an applied reverse voltage on the transition region and show that changes in this applied voltage lead to capacitive behavior. We then extend the concepts to a system containing two different semiconducting materials. Limitations on the magnitude of the reverse bias imposed by two important breakdown mechanisms are then discussed. Because most semiconductor devices contain one or more *pn* junctions, our results are directly applicable to the analysis of practical devices. As examples of the use of the concepts developed for a reverse-biased *pn* junction, we conclude this chapter by discussing junction field-effect transistors.

Our focus in this chapter is on *pn* junctions at equilibrium and under reverse bias. We consider currents to be negligible except when junction fields are large enough to lead to breakdown. Current flow in *pn* junctions is discussed in Chapter 5.

4.1 GRADED IMPURITY DISTRIBUTIONS

In this section we consider equilibrium in a semiconductor with a dopant concentration that varies in an arbitrary manner with position (Figure 4.1*a*). We assume that initially the majority-carrier concentration equals the dopant concentration at every point in the material—a nonequilibrium condition. Then, we investigate the means by which the system approaches thermal equilibrium. From Equation 1.2.17 we note that a gradient in the

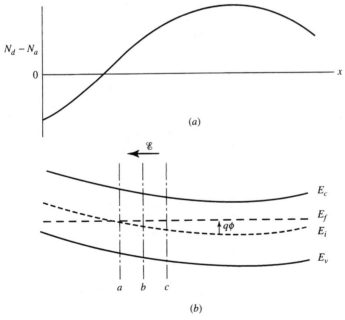

FIGURE 4.1 (*a*) Net dopant concentration as a function of position in an arbitrarily doped semiconductor. (*b*) Corresponding energy-band diagram versus position, indicating the potential ϕ. Locations *a*, *b*, and *c* are discussed in the text.

mobile carrier concentration leads to diffusion of carriers from regions of higher concentration to regions of lower concentration. As the carriers move from their initial locations, they leave behind uncompensated, oppositely charged dopant ions. This separation of positive and negative charges creates a field that opposes the diffusion flow. Equilibrium is eventually reached when the tendency of the carriers to diffuse to regions of lower density is exactly balanced by their tendency to move in the opposite direction because of the electric field created by the charge separation. Thus, the equilibrium situation is characterized by a mobile carrier distribution that does not coincide exactly with the fixed dopant distribution, and also by a built-in electric field that keeps the two charge distributions from separating further. The space charge resulting from the mechanism just described is typically a small fraction of the dopant density,* but the field arising from it can significantly influence device behavior.

We next consider how the built-in electric field affects the energy-band diagram. Because the system is at thermal equilibrium, the Fermi level is constant throughout the system. However, the variation of the dopant density and carrier concentration with position causes the separation between the Fermi level and the valence- and conduction-band edges to vary with position. Figure 4.1*b* shows the energy-band diagram corresponding to the dopant distribution of Figure 4.1*a*. The separation between the Fermi level and the band edge is less in regions of high carrier density than in regions of lower density, and the intrinsic Fermi level E_i crosses the Fermi level E_f where the net dopant concentration $N_d - N_a$ is zero.

Potential. The presence of an electric field can be seen directly from this energy-band diagram, as well as from the particle model discussed above. Since Figure 4.1*b* represents the energy-band diagram of an electron, the energy of an electron is measured by its distance above the Fermi level on the band diagram. The separation of the conduction-band edge from the Fermi level represents the potential energy of an electron while the energy above the conduction-band edge represents kinetic energy. Because the electric potential ϕ at any point is related to the potential energy by the charge $-q$, the potential can be written

$$\phi_c = -\frac{1}{q}(E_c - E_f) = \frac{1}{q}(E_f - E_c) \tag{4.1.1}$$

where the subscript c implies reference to the conduction-band energy. The reference for potential energy is arbitrary, however, and we may shift it from E_c to E_i. Because E_i is usually used for the reference, we will not subscript the symbol ϕ for potential which now is written

$$\phi = -\frac{1}{q}(E_i - E_f) = \frac{1}{q}(E_f - E_i) \tag{4.1.2}$$

as shown in Figure 4.1*b*. According to this definition the potential is positive for an *n*-type semiconductor ($E_f > E_i$) and negative for *p*-type material ($E_f < E_i$).

Field. Because the electric field is the negative of the spatial gradient of the potential, the field $\mathscr{E}_x$ is found from Equation 4.1.2 to be

$$\mathscr{E}_x = -\frac{d\phi}{dx} = \frac{1}{q}\frac{dE_i}{dx} \tag{4.1.3}$$

* This assertion is considered further in an example in Section 4.2.

Thus, a spatial variation of the band edges (and the intrinsic Fermi level) implies that a non-zero electric field exists in the semiconductor. At point b of Figure 4.1b, dE_i/dx is negative, and the field is directed toward the left. The resulting force on negatively charged electrons is toward the right; consequently, the field provides a force that opposes the tendency of electrons to diffuse from the high-concentration region at c to the low-concentration region at a. The situation in a p-type semiconductor is analogous, with the proper changes in signs and notation.

We now try to relate the electric field to the graded impurity distribution. Once the system is in thermal equilibrium, no current flows at any point in the semiconductor. In addition, because thermal equilibrium requires that every process and its inverse are in balance, the electron current and the hole current must each be zero at thermal equilibrium. In Sec. 1.2 we found that the total electron current is given by

$$J_n = q\mu_n n\mathcal{E}_x + qD_n\frac{dn}{dx} \tag{4.1.4}$$

This expression is applicable both in n-type material, where the electrons are majority carriers, and in a p-type semiconductor, where they are minority carriers.

The first term of Equation 4.1.4 represents the drift current, and the second, the diffusion current. When the total electron current is zero, the two terms are exactly balanced. No current actually flows; the drift tendency balances the diffusion tendency at each point. Because $J_n = 0$, we can solve for the field in terms of the electron concentration and its gradient

$$\mathcal{E}_x = -\frac{D_n}{\mu_n}\frac{1}{n}\frac{dn}{dx} = -\frac{kT}{q}\frac{1}{n}\frac{dn}{dx} \tag{4.1.5}$$

where we have used the Einstein relation defined in Equation 1.2.20. Similarly, the field can be expressed in terms of the hole concentration by either considering the expression for hole current (Equation 1.2.22) or using the mass-action law (Equation 1.1.13) in Equation 4.1.5:

$$\mathcal{E}_x = \frac{kT}{q}\frac{1}{p}\frac{dp}{dx} \tag{4.1.6}$$

Equations 4.1.5 and 4.1.6 show that, if we can find the mobile carrier concentrations and their gradients, we know the fields in the semiconductor.

In developing this concept we can gain some physical insight by considering the relation between the electron density and the position of the band edge (or equivalently the intrinsic Fermi level) with respect to the Fermi level. Consider an electron at x_2 in Figure 4.2a with an energy E. A portion $E - E_c$ of this energy is kinetic energy; the remainder is potential energy. The electron can move freely in the region between x_1 and x_4 because it has energy greater than the potential energy associated with this region. The electron would require more potential energy than its total energy to enter the region to the left of x_1 or to the right of x_4. Consequently, it is classically forbidden to enter these regions, and there is a *potential barrier* to the motion of the electrons.

We can relate the number of carriers at any two points in the material to the energy-band structure. We know that the electron density at x_2 is less than that at x_3 because the separation between the conduction-band edge and the Fermi level is greater at x_2. We can relate the carrier densities to the potential ϕ through the use of Equation 4.1.5. Because $\mathcal{E}_x = -d\phi/dx$, we have

$$d\phi = \frac{kT}{q}\frac{dn}{n} \tag{4.1.7}$$

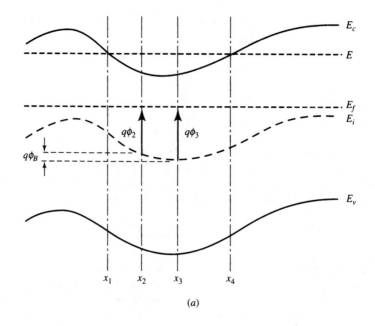

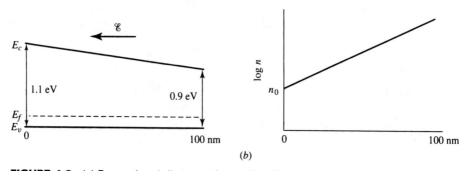

FIGURE 4.2 (a) Energy-band diagram of an arbitrarily doped semiconductor, showing that an electron with energy E is constrained to remain in the region between x_1 and x_4 where $E > E_c$. (b) Energy-band diagram for a *p*-type material with position-dependent bandgap E_g and corresponding electron concentration.

at any point in the semiconductor. Integrating this equation between any two points—for example, from x_2 to x_3—we obtain

$$\phi_3 - \phi_2 = \frac{kT}{q} \ln \frac{n_3}{n_2} \tag{4.1.8}$$

Rewriting Equation 4.1.8 in exponential form, we find

$$\frac{n_3}{n_2} = \exp\left[\frac{q}{kT}(\phi_3 - \phi_2)\right] \tag{4.1.9}$$

The ratio of the carrier densities depends on the potential difference $\phi_B = \phi_3 - \phi_2$ between the two points. Physically, we can consider that a fraction $\exp(-q\phi_B/kT)$ of the electrons at x_3 has enough energy to reach x_2.

Poisson's Equation. Equation 4.1.9 is often useful when the variation of carrier concentrations with position must be found. As is often the case, we start such an analysis by writing Poisson's equation:*

$$\frac{d^2\phi}{dx^2} = -\frac{\rho}{\epsilon_s} = -\frac{q}{\epsilon_s}(p - n + N_d - N_a) \tag{4.1.10}$$

where ρ is the space-charge density and the dopant atoms are assumed to be completely ionized. Using our definition of potential (Equation 4.1.2) in Equation 1.1.26, we can relate the carrier concentration n to the potential function ϕ:

$$n = n_i \exp\left(\frac{q\phi}{kT}\right) \tag{4.1.11}$$

Poisson's equation can then be rewritten in the form

$$\frac{d^2\phi}{dx^2} = \frac{q}{\epsilon_s}\left(2n_i \sinh\frac{q\phi}{kT} + N_a - N_d\right) \tag{4.1.12}$$

Equation 4.1.12 is the differential equation for the potential distribution in an arbitrarily doped semiconductor. Unfortunately, this equation cannot be solved in the general case, and approximations must be made to obtain analytical solutions appropriate to specific situations; alternatively, numerical techniques can be used.

To proceed, we consider two special cases. In the first, the dopant concentration varies gradually with position as, for example, the donor distribution within a diffused n-type region. The second case is just the opposite and involves abrupt spatial variations of dopant concentration as, for example, in the junction between p-type and n-type semiconductor regions.

Quasi-Neutrality. For the gradual-variation case we consider n-type silicon in which the dopant concentration varies from about 10^{18} to 10^{16} cm^{-3} within several hundred nanometers, as for a typical dopant diffusion. This change in dopant concentration corresponds to a potential difference of about 0.1 V and a field (Equation 4.1.5) of the order of 10^4 V cm^{-1} or less. If we take a specific case in which the field varies from zero to 10^4 V cm^{-1} in 0.5 μm, the average field gradient is 2×10^8 V cm^{-2}. Using this value in Poisson's equation (Equation 4.1.10) and neglecting the minority-carrier density p, we find that the difference between n and N_d must be less than about 10^{15} cm^{-3}. Because this number is only a small fraction of the donor concentration over most of the region under consideration, it is reasonable to approximate n by N_d in order to proceed with the analysis. In essence the approximation means that the majority-carrier distribution does not differ much from the donor distribution so that the semiconductor region is nearly neutral or *quasi-neutral*. This *quasi-neutrality approximation* is more valid for slowly varying dopant densities. Under the assumption of quasi-neutrality, the field in the n-type semiconductor is found directly from the donor concentration by using Equation 4.1.5:

$$\mathscr{E}_x = -\frac{kT}{q}\frac{1}{N_d}\frac{dN_d}{dx} \tag{4.1.13}$$

In a p-type semiconductor under quasi-neutral conditions, the field is similarly

$$\mathscr{E}_x = \frac{kT}{q}\frac{1}{N_a}\frac{dN_a}{dx} \tag{4.1.14}$$

---

* Because much of semiconductor device analysis is concerned with the spatial variations of carriers and potentials from one region of a device to another, Poisson's equation is often encountered, as we have already seen in Chapter 3. Together with approximations that place it in mathematically tractable form, it is one of the most useful principles in semiconductor device analysis.

Because Equations 4.1.13 and 4.1.14 depend on the quasi-neutrality approximation, they are not valid if the dopant concentration has a steep gradient.

One frequently considered case in which the quasi-neutrality approximation is useful is that of an exponential dopant distribution

$$N_d = N_0 \exp\left(\frac{-x}{\lambda}\right) \tag{4.1.15}$$

where λ is the characteristic length describing the decrease of donor atoms away from the semiconductor surface at $x = 0$. Typical Gaussian or complementary-error-function diffusion profiles (Sec. 2.5) are often approximated by exponential distributions for mathematical simplicity. Because of the relationship of an exponential function and its derivative, the field has a constant value $kT/q\lambda$ throughout the region of exponential doping. As we will see in Chapter 6, the approximation of a constant field simplifies some useful cases of device analysis. The exponential approximation may, however, obscure important implications of real diffusions that become apparent when more detailed dopant distributions are considered.

EXAMPLE   *The Quasi-Neutrality Approximation*

Investigate the assumption of *quasi-neutrality* for *n*-type silicon with nonuniform doping by considering the density of space charge present in a region where the dopant density $N_d(x)$ changes from 10^{16} to 10^{18} cm^{-3} over a length $\lambda = 1$ μm. Assume that the dopant varies as

$$N_d(x) = 10^{16} \times \exp\left\{ \ln(100)\left[\frac{x}{\lambda} - \frac{1}{2\pi}\sin\left(\frac{2\pi}{\lambda}x\right)\right]\right\} \qquad \left(0 < \frac{x}{\lambda} < 1\right)$$

(This mathematical form is a smooth differentiable function with the proper end values.) [1]

Calculate the field and charge distribution in the region $0 < \dfrac{x}{\lambda} < 1$.

Solution   Because the donor-dopant density increases with increasing x, we expect the free-electron density gradient to be positive. Hence, electrons tend to diffuse in the negative x direction. At thermal equilibrium, this diffusion tendency must be balanced by an electron drift tendency toward positive x, requiring a built-in field in the negative x direction. Associated with this field is a space charge so that the graded-dopant region is not truly charge neutral. In this example we calculate the density of charge present to investigate quantitatively the departure from charge neutrality.

Using Equation 4.1.8, we calculate the total potential difference across the region $\Delta\phi$ as

$$\Delta\phi = |\mathscr{E}_{avg}|\lambda = \frac{kT}{q}\ln(100) = 0.12 \text{ V}$$

where $\mathscr{E}_{avg} = -1.19 \times 10^3$ V cm^{-1} is the average field in the variable doping region. The field as a function of x is (from Equation 4.1.13)

$$\mathscr{E} = -\frac{kT}{q}\frac{1}{N_d}\frac{dN_d}{dx} = \mathscr{E}_{avg}\left(1 - \cos\frac{2\pi}{\lambda}x\right)$$

To find the charge density we use Poisson's equation $\rho(x) = \epsilon_s\, d\mathscr{E}/dx$. Normalizing $\rho(x)$ by the electronic charge, we have

$$\frac{\rho(x)}{q} = \frac{\epsilon_s\mathscr{E}_{avg}}{q\lambda}\left[2\pi\sin\left(\frac{2\pi}{\lambda}x\right)\right]$$

The maximum charge density ($|\rho/q|$), occurring at $x/\lambda = 1/4$, and $x/\lambda = 3/4$, is 4.8×10^{14} cm^{-3}, which is appreciably smaller than the minimum dopant density ($N_d = 10^{16}$ cm^{-3}). Hence, we

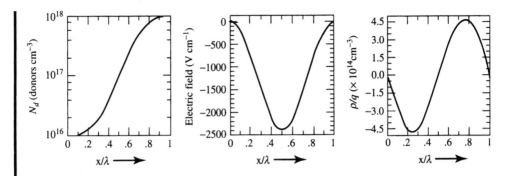

conclude that *quasi-neutrality* is a reasonable approximation in the region over which the dopant density varies. The accompanying figures show the dopant density, field, and charge density in the region of the doping gradient. ∎

Heterogeneous Material.

Up to this point, our analysis has assumed that the basic properties of the semiconductor material itself do not vary with position. However, in some cases the composition (and bandgap) of the semiconductor can be purposely varied with position to enhance device performance. Even when the dopant concentrations are constant throughout the system, the variation of the bandgap creates an electric field that can aid or retard movement of free carriers.

As an illustration, we consider a *p*-type semiconductor with a bandgap that varies linearly from 1.1 eV to 0.9 eV over a distance of 100 nm, as shown in Figure 4.2*b*. The hole and electron concentrations can be written as

$$p = N_v \exp\frac{-(E_f - E_v)}{kT} \tag{4.1.16}$$

and

$$n = N_c \exp\frac{-(E_c - E_f)}{kT} \tag{4.1.17}$$

where N_v and N_c (Equations 1.1.23 and 1.1.24) can be functions of position. However, if the band structures of the materials involved are not too different from each other, N_v and N_c do not vary significantly with composition or position.

For *p*-type material, $p \approx N_a$. Because $E_f - E_v \approx (kT/q) \ln(N_a/N_v)$, the valence band edge is approximately parallel to the constant Fermi level. However, the conduction band edge and, consequently, the electron concentration, varies with position because of the varying bandgap. Using Equations 4.1.16 and 4.1.17 in our example with a bandgap variation of 0.2 eV, we find that

$$n \approx n_0 \exp\left(\frac{0.2 \text{ eV}}{0.026 \text{ eV}} \frac{x}{100 \text{ nm}}\right) = n_0 \exp\left(\frac{x}{13 \text{ nm}}\right) \tag{4.1.18}$$

where n_0 is the electron concentration at the point where $E_g = 1.1$ eV. As expected, the electron concentration is higher where the bandgap is smaller.

Analogous to doping gradients, the bandgap gradient introduces an electric field that balances the tendency of the electrons to diffuse from regions of higher electron concentration to regions of lower concentration. The field is in the direction that pushes electrons back toward regions of higher concentration and therefore must be negative. This electric field can have important effects on any excess electrons injected into the material. Because the field is negative, an excess electron injected at the left is accelerated by the electric field. It traverses the region of graded composition more rapidly than it would travel through a region of uniform composition. The implications of this accelerating field on the speed of transistors will be discussed in Sec. 7.6.

4.2   THE *pn* JUNCTION

In the analysis of the previous section we restricted our discussion to material of one conductivity type with carrier and dopant concentrations that varied gradually with position. These limitations permitted a solution for the electric field in a quasi-neutral region having a spatially varying dopant concentration or composition. Now, we consider the other extreme: a semiconductor with a dopant concentration that has a steep gradient. In this case there can be significant departures from charge neutrality in localized regions of the semiconductor. We consider the junction between a *p*-type and an *n*-type semiconductor and find that we can treat the transition region as if it were depleted of mobile carriers. This *depletion approximation* is the opposite extreme of the *quasi-neutrality* approximation. When analyzing device structures, it is frequently useful to divide them into regions assumed to be quasi-neutral and other regions considered to be completely depleted of mobile carriers. Although an idealization, this simplification is adequate for many calculations. For more accurate analysis that avoids making this assumption, numerical techniques are generally required.

To build a model for the *pn* junction, we begin by considering initially separated *n*- and *p*-type semiconductor crystals of the same material (Figure 4.3*a*). When these are brought into intimate contact as shown in Figure 4.3*b*, the large difference in electron concentrations between the two materials causes electrons to flow from the *n*-type semiconductor into the *p*-type semiconductor and holes to flow from the *p*-type region into the *n*-type region. As these mobile carriers move into the oppositely doped material, they leave behind uncompensated dopant atoms near the junction, causing an electric field. The field lines extend from the donor ions on the *n*-type side of the junction to the acceptor ions on the *p*-type side (Figure 4.3*c*). This field creates a potential barrier between the two types of material. When equilibrium is reached, the magnitude of the field is such that the tendency of electrons to diffuse from the *n*-type region into the *p*-type region is exactly balanced by the tendency of electrons to drift in the opposite direction under the influence of the built-in field.

Potential Barrier.   The magnitude of the potential barrier associated with the built-in field can be found by considering the difference in the Fermi levels of the initially separated materials (Figure 4.3*a*) as was done for the metal-semiconductor system in Chapter 3. When the combined semiconductor regions are at equilibrium, the Fermi level must be constant throughout the entire system. Consequently, the energy barrier that forms between the two materials must equal the difference between the Fermi levels in the separated pieces of semiconductor. This difference is equivalent to the difference in work functions of the separated semiconductor regions because the work function of a semiconductor is defined as

$$q\Phi_s \equiv qX + (E_c - E_f) \tag{4.2.1}$$

where qX is the electron affinity.

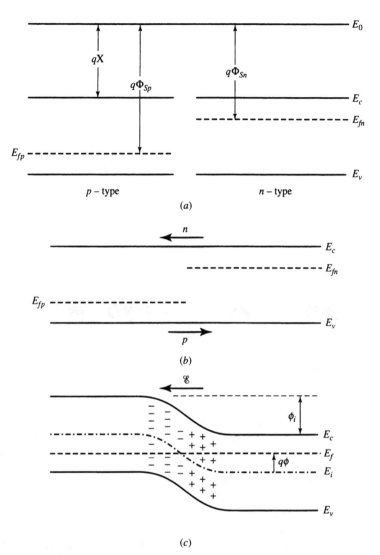

FIGURE 4.3 (*a*) *n*-type and *p*-type semiconductor regions separated and not in thermal equilibrium. (*b*) The two regions brought into intimate contact allowing diffusion of holes from the *p*-region and electrons from the *n*-region. (*c*) Transfer of free carriers leaves uncompensated dopant ions, which cause a field that opposes and balances the diffusion tendencies of holes and electrons.

Far from the junction, the carrier concentrations have the same values as in the isolated semiconductor crystals. The electron concentration in the *n*-type material is equal to the donor concentration, and the hole density is given by n_i^2/N_d. Similarly, in the *p*-type material far from the junction, the hole concentration is equal to N_a and the electron concentration is given by n_i^2/N_a. Because we know the carrier concentrations far from the junction, we can solve for the potential defined in Equation 4.1.2 by using Equation 4.1.11. Close to the junction, we cannot readily specify the free-carrier concentrations. However, Figure 4.3*c* shows that $|\phi|$ is small in this region. From Eq. 4.1.11 we see that the carrier concentrations decrease rapidly as ϕ becomes small and are, consequently, much less in the transition region than in the neutral regions. Thus, the transition region is often described as a *depletion region* in which the space charge is overwhelmingly made up of dopant ions.

In the *depletion approximation* we assume that the semiconductor can be divided into distinct zones that are either neutral or completely depleted of mobile carriers. These zones join each other at the edges of the *depletion* or *space-charge region*, where the majority-carrier density is assumed to change abruptly from the dopant concentration to zero. The depletion approximation appreciably simplifies the solution of Poisson's equation (Equation 4.1.10).

Because the carrier concentrations are assumed to be much less than the net ionized dopant density in the depletion region, the second derivative of the potential is proportional to the net dopant concentration in the depletion region:

$$\frac{d^2\phi}{dx^2} = \frac{-q}{\epsilon_s}(N_d - N_a) \tag{4.2.2}$$

In the general case, N_d and N_a may be functions of position, and we cannot solve Equation 4.2.2 explicitly. However, considering several idealized dopant configurations for which Equation 4.2.2 can be solved provides useful insight into the behavior of real *pn*-junctions. We first consider the step junction. Step junctions (sometimes called *abrupt junctions*) are characterized by a constant *n*-type dopant density on one side of the junction that changes abruptly to a constant *p*-type dopant density at a certain position. An abrupt junction can be formed, for example, by epitaxial deposition of a constant-doped *n*-type region on a *p*-type substrate, as was described in Chapter 2.

Next, we consider a *linearly graded junction*, in which the dopant concentration varies linearly with position between an *n*-type region with constant dopant concentration and a *p*-type region with constant dopant concentration. Over a limited range, some junctions formed by diffusion can be approximated as linearly graded junctions.

The final junction type we consider is a *heterojunction* between an *n*-type region of one semiconductor material and a *p*-type region of a semiconductor with a different energy gap. Heterojunctions are important in high-speed transistors and in optical devices. They can be formed between different column IV materials or between different compound semiconductor materials composed of column III and column V elements. Heterojunctions are usually formed by epitaxial deposition of one semiconductor material onto a different semiconductor material with a similar lattice constant (to allow epitaxial growth of unstrained layers), as discussed in Sec. 2.8.

Step Junction

Approximate Analysis. We can solve Equation 4.2.2 for the step junction shown in Figure 4.4, where the dopant concentration changes abruptly from N_a to N_d at $x = 0$. Using the depletion approximation, we assume that the region between $-x_p$ and x_n is totally depleted of mobile carriers, as shown in Figure 4.4b, and that the mobile majority-carrier densities abruptly become equal to the corresponding dopant concentrations at the edges of the depletion region. The charge density is, therefore, zero everywhere except in the depletion region, where it equals the ionized dopant concentration (Figure 4.4c). In the *n*-type material $(x > 0)$ Equation 4.2.2 becomes

$$\frac{d^2\phi}{dx^2} = -\frac{d\mathcal{E}}{dx} = -\frac{qN_d}{\epsilon_s} \tag{4.2.3}$$

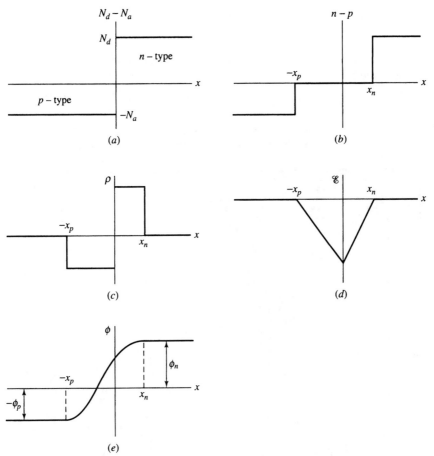

FIGURE 4.4 Properties of a step junction as functions of position using the complete-depletion approximation: (*a*) net dopant concentration, (*b*) carrer densities, (*c*) space charge used in Poisson's equation, (*d*) electric field found from first integration of Poisson's equation, and (*e*) potential obtained from second integration.

which can easily be integrated from an arbitrary point in the *n*-type depletion region to the edge of the depletion region at x_n, where the material becomes neutral and the field vanishes. Carrying through this integration, we find the field to be

$$\mathcal{E}(x) = -\frac{qN_d}{\epsilon_s}(x_n - x) \qquad 0 < x < x_n \tag{4.2.4}$$

The field is negative throughout the depletion region and varies linearly with x, having its maximum magnitude at $x = 0$ (Figure 4.4*d*). The direction of the field toward the left is physically reasonable because the force it exerts must balance the tendency of the negatively charged electrons to diffuse toward the left out of the neutral *n*-type material. The field in the *p*-type region is similarly found to be

$$\mathcal{E}(x) = -\frac{q}{\epsilon_s}N_a(x + x_p) \qquad -x_p < x < 0 \tag{4.2.5}$$

The field in the *p*-type region is also negative in order to oppose the tendency of the positively charged holes to diffuse toward the right.

At $x = 0$, the field must be continuous, so that

$$N_a x_p = N_d x_n \qquad (4.2.6)$$

Thus, the width of the depleted region on each side of the junction varies inversely with the magnitude of the dopant concentration; the higher the dopant concentration, the narrower the space-charge region. In a highly asymmetrical junction where the dopant

TABLE 4.1 Nomograph for silicon uniformly doped, one-sided, step junctions (300 K). (See Figure 4.15 to correct for junction curvature.) (*Courtesy Bell Laboratories*).

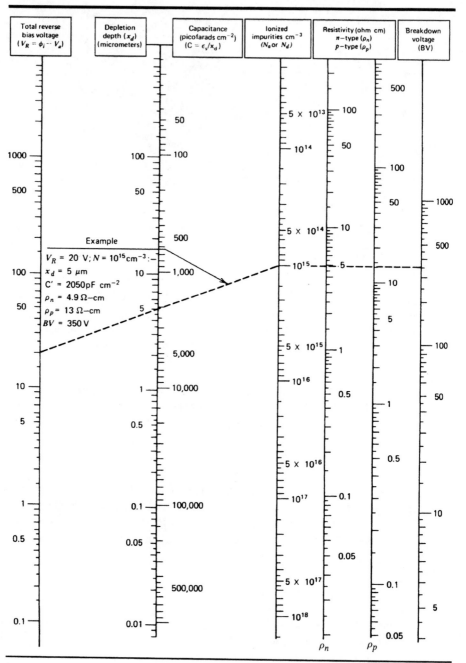

concentration on one side of the junction is much higher than that on the other side, the depletion region penetrates primarily into the lightly doped material, and the width of the depletion region in the heavily doped material can often be neglected. The charge, field, and potential at this type of junction, called a *one-sided step junction*, are identical to the results obtained for the ideal Schottky barrier, and the sketches of these quantities shown in Figure 4.4 reduce to the equivalent sketches in Figure 3.4. The properties of one-sided, abrupt, planar junctions are so frequently used that it is worthwhile to collect them as a nomograph in Table 4.1.

The expressions for the field can be integrated again to obtain the potential variation across the junction. In the *n*-type material

$$\phi(x) = \phi_n - \frac{qN_d}{2\epsilon_s}(x_n - x)^2 \qquad 0 < x < x_n \tag{4.2.7}$$

as shown in Figure 4.4e, where ϕ_n is the potential at the neutral edge of the depletion region obtained from Equation 4.1.11:

$$\phi_n = \frac{kT}{q} \ln \frac{N_d}{n_i} \tag{4.2.8}$$

Similarly, in the *p*-type material

$$\phi(x) = \phi_p + \frac{qN_a}{2\epsilon_s}(x + x_p)^2 \qquad -x_p < x < 0 \tag{4.2.9a}$$

$$\phi_p = \frac{-kT}{q} \ln \frac{N_a}{n_i} \tag{4.2.9b}$$

where $\phi_p < 0$ is the potential at the neutral edge of the depletion region in the *p*-type material.

The total potential change ϕ_i from the neutral *p*-type region to the neutral *n*-type region is $\phi_n - \phi_p (\phi_p < 0)$ and depends on the dopant concentration in each region:

$$\phi_i = \phi_n - \phi_p = \frac{kT}{q} \ln \frac{N_d}{n_i} + \frac{kT}{q} \ln \frac{N_a}{n_i} = \frac{kT}{q} \ln \frac{N_d N_a}{n_i^2} \tag{4.2.10}$$

Note that the *built-in potential* ϕ_i is positive (that is, the *n*-side is at a higher potential than the *p*-side), which is proper to obtain a balance between drift and diffusion across the junction. The major portion of the potential change occurs in the region with the lower dopant concentration, and the depletion region is wider in the same region. Note that the potential at the junction plane ($x = 0$) is not exactly zero unless the junction is symmetrical (i.e., $N_a = N_d$).

EXAMPLE   *Built-In Voltage at a pn Junction*

A lightly doped, *n*-type sample of silicon has a resistivity of 4 Ω-cm. It is used to make a *pn* junction with a *p*-region in which the dopant density N_a is 1000 times higher than that in the *n*-type silicon. What is the built-in voltage ϕ_i of the junction?

Solution   Equation 4.2.10 expresses ϕ_i as a function of the dopant densities N_a and N_d on either side of the junction. From Figure 1.15, for a resistivity of 4 Ω-cm, $N_d = 10^{15}$ cm^{-3}. Because the *p*-region dopant density is 1000 times higher, $N_a = 10^{18}$ cm^{-3}.

$$\phi_i = \frac{kT}{q} \ln\left(\frac{N_a N_d}{n_i^2}\right) = \frac{kT}{q} \ln\left(\frac{10^{18} \times 10^{15}}{(1.45 \times 10^{10})^2}\right)$$

$$\phi_i = 0.753 \text{ V}$$

This voltage, equivalent to roughly 2/3 of E_g/q, is a typical value for ϕ_i in many IC *pn* junctions. ∎

At very high dopant concentrations, Equation 4.2.10 is no longer valid because it is based on Maxwell-Boltzmann statistics (e.g., Equation 4.2.8). When the dopant density approaches the effective density of states N_c or N_v ($\approx 10^{19}$ cm^{-3}), Fermi-Dirac statistics should be used in any derivations. When calculating the potential at a *pn* junction, however, we do not need to consider Fermi-Dirac statistics in detail because at high dopant concentrations the Fermi level lies very near the band edge, and the potential in the heavily doped silicon is approximately equal to $E_g/2q$ or 0.56 V. Thus, the built-in potential across a *pn* junction composed of heavily doped *p*-type silicon (usually denoted as p^+ silicon) and lightly doped *n*-type silicon is

$$|\phi_i| = 0.56 + \frac{kT}{q}\ln\left(\frac{N_d}{n_i}\right)$$

A similar result with N_a in place of N_d applies to junctions between n^+ silicon and lightly doped *p*-type silicon.

For the step *pn* junction with arbitrary dopant concentrations, the total depletion-region width is found from Equations 4.2.6 through 4.2.10 to be

$$x_n + x_p = \left[2\frac{\epsilon_s}{q}\phi_i\left(\frac{1}{N_a} + \frac{1}{N_d}\right)\right]^{1/2} \tag{4.2.11}$$

From Equation 4.2.11, we see that the depletion-region width depends most strongly on the material with the lighter doping, and varies approximately as the inverse square root of the smaller dopant concentration.

EXAMPLE   *Reverse-Biased Step Junction*

Consider a *pn* junction with constant doping concentrations N_a on the *p*-type side and N_d on the *n*-type side.

Derive an expression for the percentage P_n of the total reverse-bias voltage that is dropped across the *n*-type region if an external voltage $V_a = -5$ V is applied. Evaluate P_n for $N_a = 10^{17}$ cm^{-3} in the following three cases: (a) $N_d = 10^{-1} N_a$, (b) $N_d = 10^{-2} N_a$, and (c) $N_d = 10^{-3} N_a$.

Use a modified form of Equation 4.2.11 to obtain the total depletion-layer width for each of these cases. Compare the calculated depletion-layer width x_d to values obtained by using Table 4.1 for a one-sided step junction having a dopant density N_d in each of these three cases.

Solution   A plot of the field for this problem is shown in Figure 4.4*d*. If we designate the maximum field by $\mathcal{E}_{max}$, the voltage V_p dropped across the *p*-region is

$$V_p = \frac{1}{2}\mathcal{E}_{max}\,x_p$$

while that dropped across the *n*-region V_n is

$$V_n = \frac{1}{2}\mathcal{E}_{max}\,x_n$$

From these two equations, we have $V_p/V_n = x_p/x_n$. The total reverse-bias voltage is $V_R = V_p + V_n$. Therefore, the required percentage P_n is

$$P_n = \frac{V_n}{V_n + V_p} \times 100 = \frac{1}{1 + V_p/V_n} \times 100$$

$$= \frac{1}{1 + x_p/x_n} \times 100$$

Using Equation 4.2.6, we obtain

$$P_n = \frac{1}{1 + N_d/N_a} \times 100$$

The percentages (P_n values) for the three junctions described above are

(a) $N_d = 10^{16}$ $P_n = 91\%$

(b) $N_d = 10^{15}$ $P_n = 99\%$

(c) $N_d = 10^{14}$ $P_n = 99.9\%$

Hence, we see that an order-of-magnitude ratio in dopant density between the two sides of a step junction results in more than 90% of the total reverse bias being dropped across the more lightly doped region. As the dopant ratio increases, the junctions becomes more and more one-sided.

In the second part of this example, we consider the total depletion-layer width when the junction is at 5 V reverse bias. To apply Equation 4.2.11 we replace ϕ_i, the built-in potential, with the total reverse voltage $V_R = \phi_i + |V_a|$ (measured in the *p*-region with respect to the *n*-region). From Equation 4.2.10, we find ϕ_i for cases (a), (b), and (c) to be

(a) $\phi_i = 0.753$ V for $N_a = 10^{17}$, $N_d = 10^{16}$ cm^{-3}

(b) $\phi_i = 0.694$ V for $N_a = 10^{17}$, $N_d = 10^{15}$ cm^{-3}

(c) $\phi_i = 0.634$ V for $N_a = 10^{17}$, $N_d = 10^{14}$ cm^{-3}

Because $V_a = -5$ V, $V_R = 5.753$, 5.694, and 5.634 V for cases (a), (b), and (c), respectively. Using these values in the modified form of Equation 4.2.11, and also in Table 4.1, we find for x_d:

	Equation 4.2.11	Table 4.1
(a)	0.91 μm	0.85 μm
(b)	2.73 μm	2.7 μm
(c)	8.54 μm	8.6 μm

From this example, it is apparent that one-sided behavior is observed in a reverse-biased *pn* junction, even with relatively small ratios of the dopant densities. Table 4.1 can be useful for rough estimates of the depletion-layer width even when dopant densities differ by only one order of magnitude. ■

The analysis of the abrupt *pn* junction has employed the depletion approximation to solve Poisson's equation and to specify the boundary conditions. Before proceeding further we briefly discuss the more exact solution and consider the consequences of the depletion approximation. We again look at a junction with a step-function change in dopant concentration at $x = 0$ (Figure 4.5a). However, we no longer assume an abrupt change from neutral regions to completely depleted regions at x_n and $-x_p$; instead, we consider transition regions that are only partially depleted near these boundaries (Figure 4.5b). Because the net charge densities in the transition regions (Figure 4.5c) are less than the dopant concentrations, the fields change more gradually than predicted by the depletion approximation. The solid line of Figure 4.5d indicates the field found from the more exact solution, while the dashed line represents the field found using the depletion approximation. In the more exact analysis the field extends further into the semiconductor interior and the space-charge region is wider.

Debye Length.[†]   Although an exact solution of the step junction can be obtained without making use of the depletion approximation, we will not carry out this more detailed analysis, but rather consider an analytical approximation for the potential near the edges

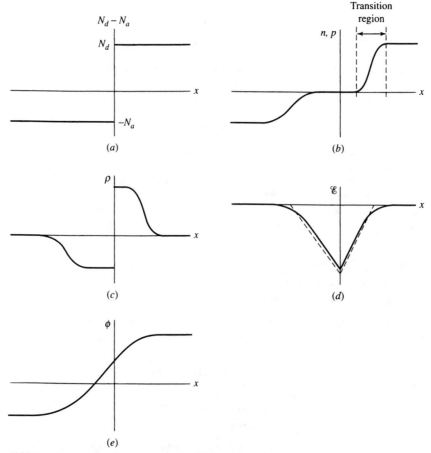

FIGURE 4.5 Properties of a step junction as functions of position considering a gradual transition between neutral and depleted regions: (*a*) net dopant concentration, (*b*) carrier densities, (*c*) space charge, (*d*) electric field, (*e*) potential. (Compare with Figure 4.4).

of the space-charge region (near x_n and $-x_p$) to investigate the validity of the depletion approximation. If we consider only small variations of potential from ϕ_n near $x = x_n$, we can rewrite Equation 4.1.10 by neglecting the minority-carrier concentration p and letting $\phi' = \phi_n - \phi$ in Equation 4.1.11:

$$\frac{d^2\phi'}{dx^2} = \frac{q}{\epsilon_s}(N_d - n) = \frac{q}{\epsilon_s}\left[N_d - n_i \exp\left(\frac{q(\phi_n - \phi')}{kT}\right)\right]$$

$$= \frac{q}{\epsilon_s}N_d\left[1 - \exp\left(-\frac{q\phi'}{kT}\right)\right] \tag{4.2.12}$$

Because we are restricting ϕ' to be small, we can expand the exponential term in Equation 4.2.12 in a Taylor series, retaining only the first two terms so that the equation reduces to

$$\frac{d^2\phi'}{dx^2} = \frac{q}{\epsilon_s}N_d\frac{q\phi'}{kT} = \frac{\phi'}{L_D^2} \tag{4.2.13}$$

where L_D, the extrinsic *Debye length* given by

$$L_D = \left[\frac{\epsilon_s kT}{q^2 N_d}\right]^{1/2} \tag{4.2.14}$$

is a characteristic length associated with the spatial variations of potential.* We met a similar form for the Debye length in Equation 3.4.3 in our discussion of Schottky ohmic contacts. The solution of Equation 4.2.13 is of the form $\phi' = B \exp(x/L_D)$ with B a constant of integration. Hence, the potential ϕ' varies exponentially with distance near the edges of the space-charge region with a characteristic length equal to the extrinsic Debye length. Because the carrier concentration itself depends exponentially on the potential, the carrier concentration changes rapidly from the dopant concentration to essentially zero within a few Debye lengths. Therefore, the depletion approximation is questionable only within a few extrinsic Debye lengths of the edges of the space-charge region, x_n and $-x_p$. For symmetric junctions with typical dopant densities of 10^{16} cm^{-3}, L_D equals 40 nm while x_n is found from simultaneous solutions of Equations 4.2.6 and 4.2.11 to be approximately 210 nm. Consequently, for this case the depletion approximation is reasonable, but clearly still an approximation.

Linearly Graded Junction

Frequently, the step or abrupt junction is not an adequate representation for a *pn* junction made by diffusion. It is especially inapplicable to most practical cases of *double-diffused junctions*, that is, junctions formed by two successive diffusions of opposite type dopant atoms. A general analytical solution of Poisson's equation (Equation 4.1.12) is not possible, but approximate analytic solutions can be obtained by making simplifying assumptions for the form of the dopant profile. If more accurate results are needed, numerical techniques are used.

One approximation to a diffused *pn* junction than can be treated exactly is the *linearly graded junction*.

In a linearly graded junction the net dopant concentration varies linearly from the *p*-type material to the *n*-type material. This type of junction is characterized by a constant a, which is the gradient of the net dopant concentration (with units of cm^{-4}). The net dopant concentration can be written

$$N_d - N_a = ax \tag{4.2.15}$$

throughout the space-charge region (Figure 4.6a). The field and potential are readily found from Poisson's equation by using the depletion approximation. Because the space

---

* More rigorously, the Debye length L_D describes the screening of electric fields by the rearrangement of mobile carriers and depends on the total (hole and electron) free-carrier concentrations in the region.

$$L_D = \left[\frac{\epsilon_s kT}{q^2 (n + p)}\right]^{1/2}$$

For an extrinsic *n*-type semiconductor, the minority-carrier density is negligible and L_D takes the form of Equation 4.2.14. The Debye length increases as carrier density decreases. The maximum length is the intrinsic Debye length L_{Di} given by the expression

$$L_{Di} = \left[\frac{\epsilon_s kT}{2q^2 n_i}\right]^{1/2}$$

At room temperature, $L_{Di} = 24$ μm.

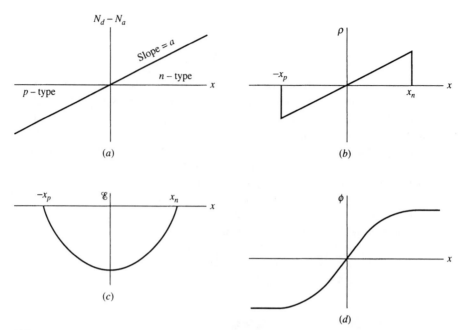

FIGURE 4.6 Properties of a linearly graded junction using the depletion approximation: (*a*) net dopant concentration: $N_d - N_a = ax$, (*b*) space charge, (*c*) electric field,
(*d*) potential.

charge varies linearly with position in the depletion region, the field varies quadratically
and the potential varies as the third power of position in the space-charge region (Figure
4.6). (The details of the analysis are considered in Problem 4.6.)

Although linearly graded junctions are not realized physically, many practical cases
can be approximated by a linearly graded junction over at least a limited voltage range.
If an abrupt junction is heated so that the dopant atoms diffuse across the junction, the
junction becomes less abrupt and can be approximated by a linearly graded junction as
long as the space-charge region is narrow compared to the diffusion length of the impurity atoms. Even diffused junctions are sometimes approximated by linearly graded junctions over a limited distance as shown in Figure 4.7*a* for an *n*-type diffusion into a *p*-type
wafer. On the other hand, a diffusion into a uniformly doped wafer can be approximated
by a one-sided step junction if the diffusion length is short compared to the width of the
space-charge region (Figure 4.7*b*) or if nonideal diffusion behavior produces a profile that
is "box-like."

Exponential Doping. Although more realistic junctions cannot be treated analytically, we can make some qualitative comments about their behavior. We can approximate either a complementary-error-function or Gaussian diffusion profile by an
exponential function over a considerable distance. Within this approximation the net
dopant concentration after an *n*-type diffusion into a uniformly doped *p*-type wafer can
be written

$$N_d - N_a = N_0 e^{-x/\lambda} - N_a \qquad (4.2.16)$$

as shown in Figure 4.8*a*. In Equation 4.2.16, λ is the characteristic length associated
with the diffusion, and N_a is the dopant concentration in the *p*-type wafer. As we saw in
Section 4.1, an electric field exists in the exponentially doped, quasi-neutral, *n*-type side

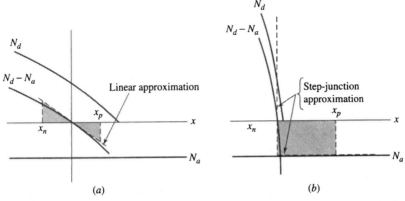

FIGURE 4.7 (*a*) A diffused junction can be approximated by a linearly graded junction if the diffusion length $2\sqrt{Dt}$ is much longer than the space-charge region. (*b*) A "one-sided" step junction is more appropriate if the space-charge region is much greater than the diffusion length $2\sqrt{Dt}$.

of the junction to balance the tendency of carriers to diffuse to regions of lower carrier density. There is no field in the uniformly doped *p*-type neutral region. The absence of mobile carriers in the depletion region on the *n*-type side of the junction requires that the field there be higher than in the quasi-neutral *n*-type region; the field throughout the structure is indicated schematically in Figure 4.8*b*. The space charge is found by differentiating the field. Because the field is constant in the quasi-neutral region, the space charge there is zero, although there is a sheet of charge at the surface (Figure 4.8*c*). The remainder of the space charge is confined to the depletion region.

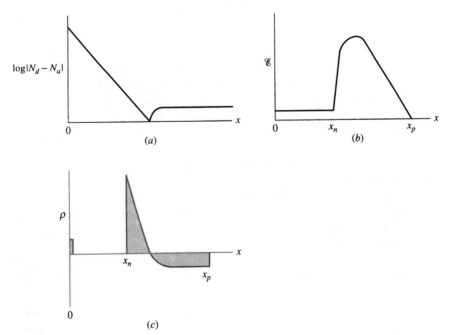

FIGURE 4.8 Properties of an exponential junction as functions of position: (*a*) net dopant concentration (semilogarithmic scale), (*b*) electric field, (*c*) space charge.

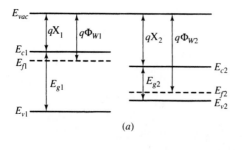

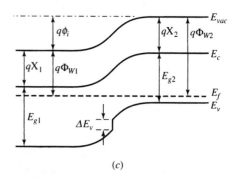

(c)

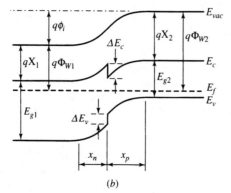

FIGURE 4.9 Energy-band diagrams for a heterojunction: (*a*) Separated regions of a wider bandgap, *n*-type material and a narrower bandgap *p*-type material. (*b*) The system after the regions are brought into intimate contact and charge flows to bring the system to equilibrium. The vacuum level remains continuous, but the conduction-band and valence-band edges can be discontinuous. (*c*) The energy bands when $X_1 = X_2$.

Heterojunction

In advanced semiconductor devices, junctions between two different semiconductor materials can be formed to improve device performance. We will briefly consider some of the characteristics of these *heterojunctions* as we continue our discussion of diodes and bipolar transistors, although a thorough analysis is beyond the scope of this book [2]. Some of the same concepts will also apply when we consider the Si–SiO$_2$ interface in Chapter 8.

For our analysis of the heterojunction system, we consider two pieces of semiconductor material 1 and 2, each with uniform composition and doping. Analogously to the *homojunction** shown in Figure 4.3, we can draw the energy-band diagram of the two isolated pieces of semiconductor (Figure 4.9*a*). In the heterojunction case, however, $E_{g1} \neq E_{g2}$, and the electron affinities X of the two materials can also be different. We consider the case where material 1 is *n*-type with a dopant density N_{d1}, a bandgap E_{g1}, and an electron affinity X_1; material 2 is *p*-type with a dopant density N_{a2}, a bandgap E_{g2}, and an electron affinity X_2.

We consider only the "straddling" band alignment shown in Figure 4.9*a*, in which the conduction-band edge of the wider-gap material is at a higher energy than that of the narrower-gap material and the valence band edge of the wider gap material is at a lower energy than that of the narrower bandgap material. Although interesting properties of an *isotype* junction (same conductivity type in the two materials) can be explored, we discuss only a *pn* heterojunction. In particular, we consider the heterojunction formed between a wide gap *n*-type material on the left and a narrower gap *p*-type material on the right. This combination of materials is especially relevant to the technologically important *heterojunction bipolar transistors* to be discussed in Chapters 6 and 7.

In our analysis we draw on our discussion of the *pn* homojunction formed between *p*- and *n*-type regions of the same semiconductor material. Now, assume that we can bring

---

* A *homojunction* contains the same material on both sides of the junction.

the two pieces of semiconductor into intimate contact with no interface states forming at the junction between the two materials. As for a homojunction, when we bring the two materials together (Figure 4.9b), charge flows to establish equilibrium, and potential barriers form to create electric fields that oppose the diffusion of carriers to regions of lower concentration. At equilibrium, the Fermi level is constant throughout the system. However, the conduction- and valence-band edges, E_c and E_v, are not necessarily continuous at the boundary between the two materials. Because $E_{g1} \neq E_{g2}$, at least one of the band edges must be discontinuous. The relative positions of the band edges across the boundary depend on the magnitude of the electron affinities and the bandgaps.

For the case we are considering, a "spike" and a "notch" appear in the conduction band. The importance of these features depends on the magnitude of ΔE_c and the doping in the two semiconductors. For example, free charge can accumulate in the notch, degrading both the dc and ac performance of a device. However, for the case shown in Figure 4.9b, the bottom of the notch is well above the Fermi level, so little charge is stored there, and we will neglect it in our discussion.

We make the usual assumption that the vacuum level is continuous across the junction. Then the change in the vacuum level, which we define as the built-in potential ϕ_i, can be found from the material properties in the neutral regions on the two sides of the junction as the difference in the work functions Φ_W of the two semiconductors:

$$\Phi_i = \Phi_{W2} - \Phi_{W1} = \left(\frac{X_2 + E_{c2} - E_f}{q} \right) - \left(\frac{X_1 + E_{c1} - E_f}{q} \right) \tag{4.2.17}$$

where

$$E_{c1} - E_f = kT \ln\left(\frac{N_{c1}}{N_{d1}} \right) \tag{4.2.18}$$

and

$$E_{c2} - E_f = E_{g2} - (E_f - E_{v2}) = E_{g2} - kT \ln\left(\frac{N_{v2}}{N_{a2}} \right) \tag{4.2.19}$$

Then,

$$\phi_i = X_2 - X_1 + \frac{E_{g2}}{q} - \frac{kT}{q} \ln\left(\frac{N_{v2}}{N_{a2}} \right) - \frac{kT}{q} \ln\left(\frac{N_{c1}}{N_{d1}} \right)$$

$$= X_2 - X_1 + \frac{E_{g2}}{q} - \frac{kT}{q} \ln\left(\frac{N_{c1} N_{v2}}{N_{d1} N_{a2}} \right) \tag{4.2.20}$$

In our analysis we need to remember that two separate material properties are changing: The band gaps (which relate minority- and majority-carrier properties in each separate piece of semiconductor) and the electron affinities (which relate properties across the junction). To help understand the influence of each of these two factors, we introduce them one at a time in the following discussion. We first consider the electron affinities to be the same in the two pieces of material ($X_1 = X_2$) so that the conduction band is continuous, as shown in Figure 4.9c. In this case the valence band must be discontinuous at the interface because $E_{g2} \neq E_{g1}$. With $X_1 = X_2$, Equation 4.2.20 becomes

$$\phi_i = \frac{E_{g2}}{q} - \frac{kT}{q} \ln \frac{N_{v2}}{N_{a2}} - \frac{kT}{q} \ln \frac{N_{c1}}{N_{d1}}$$

$$= \frac{E_{g2}}{q} - \frac{kT}{q} \ln\left(\frac{N_{c1} N_{v2}}{N_{d1} N_{a2}} \right) \tag{4.2.21}$$

Thus, the built-in potential of a heterojunction has the same general form as for a homojunction *except* that the bandgap included in the expression is that of material 2 with the smaller bandgap (for the particular case we consider here of a wider bandgap, *n*-type material and a smaller bandgap *p*-type material).

With $X_1 = X_2$, $\phi_i = \phi_{Bn}$, the barrier to electron flow from the material with the larger bandgap (material 1) into the material with the smaller bandgap (material 2). From either inspection of Figure 4.9*c* or a derivation similar to that above, we find that the barrier to hole flow from material 2 to material 1 is

$$\phi_{Bp} = \frac{E_{g1}}{q} - \frac{kT}{q} \ln\left(\frac{N_{c1}N_{v2}}{N_{d1}N_{a2}}\right) \tag{4.2.22}$$

similar to the expression for a homojunction of material 1. From Equation 4.2.22, $\phi_{Bp} = \phi_{Bn} + E_{g1} - E_{g2} = \phi_{Bn} + \Delta E_g$. The different barriers to majority-carrier and minority-carrier flow will be important when we consider current flow across a heterojunction in Chapter 5 and especially when we discuss heterojunction bipolar transistors in Chapter 6.

Now, let's relax our constraint that the two materials have the same electron affinity to see the additional effect of varying this parameter. We keep the other material properties the same as before so that we can focus on the effect of the change in electron affinity. To analyze the problem, we continue assuming that the vacuum level is continuous at the interface even though the electron affinity is not. For the vacuum level to remain continuous at the interface, the conduction band must be discontinuous there, with a discontinuity $\Delta E_c = X_2 - X_1$. The discontinuity in the valence band generally also changes. (Note that in our notation ΔE_c is the difference in electron affinities and therefore the discontinuity in the conduction-band edge at the interface. $E_{c2} - E_{c1}$ is the difference in the conduction band energies in the neutral regions far from the interface.)

The discontinuities at the interface affect the nearby space-charge regions. In the neutral region of each material, the separation between the conduction band edge and the Fermi level is determined by the doping in that material (Equations 4.2.18 and 4.2.19). Therefore, the total difference in the conduction band edge between the neutral regions in the two pieces of material does *not* depend on the difference in electron affinities and remains from Equations 4.2.18 and 4.2.19

$$E_{c2} - E_{c1} = E_{g2} - kT \ln\left(\frac{N_{c1}N_{v2}}{N_{d1}N_{a2}}\right) \tag{4.2.23}$$

even when $X_1 \neq X_2$. To retain the same value of $E_{c2} - E_{c1}$ between the neutral regions with nonzero ΔE_c at the interface, the total bending of the energy bands must increase by ΔE_c (when $\Delta E_c > 0$, as shown in Figure 4.9*b*). The change in the vacuum level (which we defined as the built-in potential ϕ_i) also increases when $\Delta E_c > 0$ because Φ_{W1} or Φ_{W2} changes. However, because of the discontinuity of the band edges in the heterojunction, it no longer equals the difference in the conduction band energies of the two materials far from the junction and is, therefore, a less useful quantity than in a homojunction.

The greater bending of the energy bands when $X_1 \neq X_2$ implies that the depletion region widens for the heterojunction shown in Figure 4.9*b*, compared to a heterojunction with $X_1 = X_2$. The width of the depletion region can be found from Poisson's equation. Within each piece of semiconductor, the analysis is identical to that for the homojunction; the differences arise from the different boundary conditions at the interface. From Gauss' law, we obtain the electric field in the depletion region within each piece of material. The solutions are identical to Equations 4.2.4 and 4.2.5 with the exception that the permittivities in the two materials can be different; we denote the permittivities as ϵ_1 and ϵ_2. At the interface the $\mathscr{D}$-field is continuous (assuming that no sheet charge exists there), and

$$\epsilon_1\mathscr{E}_1(0) = \epsilon_2\mathscr{E}_2(0) \tag{4.2.24}$$

Consequently, when $\epsilon_1 \neq \epsilon_2$, the electric field is not continuous at the interface. Let x_n and x_p be the widths of the depletion regions on the two sides of the junction. Then, using the expressions $\mathscr{E}_1(0) = (q/\epsilon_1)N_dx_n$ and $\mathscr{E}_2(0) = (q/\epsilon_2)N_ax_p$ in Equation 4.2.24, we obtain the expression $N_dx_n = N_ax_p$, as in the case of a homojunction (as required by charge neutrality).

As for the homojunction, the potentials can be obtained by integrating the fields on each side of the heterojunction and adding the potential changes on each side of the junction to obtain

$$\phi_1(-\infty) - \phi_1(x) = \frac{q}{2\epsilon_1}N_d(|x| - x_n)^2 \tag{4.2.25}$$

and

$$\phi_2(x) - \phi_2(\infty) = \frac{q}{2\epsilon_2}N_a(x_p - x)^2 \tag{4.2.26}$$

With our definition of ϕ_i, the sum of the band bending in the two materials is simply ϕ_i. From

$$\phi_i = \phi_1(-\infty) - \phi_2(\infty) \tag{4.2.27}$$

and

$$\phi_1(0) = \phi_2(0) \tag{4.2.28}$$

we find that

$$\phi_i = \frac{qN_dx_n^2}{2\epsilon_1} + \frac{qN_ax_p^2}{2\epsilon_2} \tag{4.2.29}$$

As we noted before, ϕ_i increases when $\Delta E_c > 0$, implying that the depletion region widths also increase.

Solving for the potential drop on each side of the junction, we obtain

$$\phi_{i1} = \phi_i\frac{\epsilon_2N_a}{\epsilon_1N_d + \epsilon_2N_a} \tag{4.2.30}$$

and

$$\phi_{i2} = \phi_i\frac{\epsilon_1N_d}{\epsilon_1N_d + \epsilon_2N_a} \tag{4.2.31}$$

As in a homojunction, more of the built-in voltage is dropped across the material with the lighter doping, with the exact fraction being modified by the different permittivities. However, the fraction of the built-in potential ϕ_i dropped across each semiconductor is more important for a heterojunction than for a homojunction. Because the band edges E_c and E_v can be discontinuous, irregularities, such as the spike shown in Figure 4.9b, can impede the flow of carriers across the junction. The importance of the spike as a barrier to electron flow from material 1 to material 2 depends on the fraction of the potential dropped on each side. If the acceptor concentration in material 2 is low compared to the donor concentration in region 1, the majority of the band bending occurs in

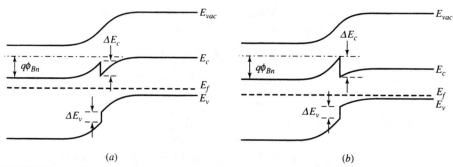

FIGURE 4.10 The barrier to electron injection from the *n*-type semiconductor to the narrower-bandgap, *p*-semiconductor depends on the discontinuity in the conduction-band edge and the doping in the *p*-type region. (a) Lightly doped *p*-type region. (b) Heavily doped *p*-type region with conduction-band spike impeding electron injection.

material 2; little occurs in material 1; and the top of the spike is lower than the conduction-band edge in the neutral region of material 2 (Figure 4.10a). The barrier ϕ_{Bn} to electron flow from material 1 into material 2 is then just the difference in the conduction band edges in the neutral regions of the two materials. However, if the doping in material 2 is high compared to that in material 1, the majority of the band bending occurs in region 1 (Figure 4.10b), and the top of the spike can be higher than the conduction-band edge in the neutral region of material 2. In this case the barrier ϕ_{Bn} to electron flow from material 1 into material 2 can be considerably greater than the difference in the conduction-band edges in the neutral regions, significantly changing the physics of any device containing such a junction. In a more detailed treatment the notch adjacent to the spike must also be considered because free carriers can accumulate there and affect the device performance.

4.3   REVERSE-BIASED *pn* JUNCTIONS

In Sec. 4.2, we used the depletion approximation to simplify our consideration of electrical effects at *pn* junctions in thermal equilibrium. To discuss the junction under bias, we again use the depletion approximation, together with several assumptions about the applied bias. We assume that ohmic contacts connect the *p*- and *n*-regions to the external voltage source so that negligible voltage is dropped at the contacts. We also consider that the voltage drops in the neutral regions are small, the *n*-region is grounded, and voltage V_a is applied to the *p*-region. With these approximations, the entire applied voltage appears across the junction. Furthermore, when these assumptions hold, the solutions of Poisson's equation that we found for thermal-equilibrium conditions in the last section also apply to the junction under bias. Only the total potential across the junction changes from the built-in voltage ϕ_i to $\phi_i - V_a$.

If V_a is positive, the barrier to majority carriers at the junction is reduced and the depletion region becomes narrower. A conceptually useful way of visualizing the reduction in depletion-layer width is that the applied voltage moves majority carriers toward the edges of the depletion region where they neutralize some of the space charge. This reduces the overall depletion-region width. The total potential barrier across the junction is $\phi_i - V_a$ with $V_a > 0$, and the junction is *forward biased*. Under forward bias, appreciable current can flow even for small values of V_a. We defer further consideration of forward bias until Chapter 5.

If negative voltage is applied to the *p*-region, the barrier to majority-carrier flow increases. Again the total potential drop can be expressed as $\phi_i - V_a$ except that now V_a is negative and the junction is *reverse biased*. Under reverse bias, majority carriers are pulled away from the edges of the depletion region, which therefore widens. Very little current flows because the bias polarity aids the transfer of electrons from the *p*-side to the *n*-side and holes from the *n*-side to the *p*-side. Because these are minority carriers in each region, their densities are low. We discuss further the small currents that do flow in a reverse-biased *pn* junction in Chapter 5.

Depletion Width, Maximum Field.   For an abrupt *pn* junction we find the depletion-region width as a function of voltage by replacing the built-in potential ϕ_i in Equation 4.2.11 by $\phi_i - V_a$ so that

$$x_d = x_n + x_p = \left[\frac{2\epsilon_s}{q} \left(\frac{1}{N_a} + \frac{1}{N_d} \right) (\phi_i - V_a) \right]^{1/2} \tag{4.3.1}$$

When the magnitude of V_a becomes considerably larger than ϕ_i, the depletion region begins to vary as the square root of the reverse bias for a step junction. Expressions for x_d for other dopant profiles are similarly modified from the thermal-equilibrium result. For example, the depletion-region width is given by

$$x_d = \left[\frac{12\epsilon_s(\phi_i - V_a)}{qa} \right]^{1/3} \tag{4.3.2}$$

in a linearly graded junction with a doping gradient a (Equation 4.2.15).

It is often important to know the maximum field at the junction $\mathscr{E}_{max}$ and its relationship to applied voltage. The step-junction case is particularly simple because the field varies linearly with distance (Equations 4.2.4 and 4.2.5). Hence, the area under the field curve that represents the potential can be written by inspection as one-half the maximum field times the depletion-layer width:

$$\tfrac{1}{2}\mathscr{E}_{max} x_d = (\phi_i - V_a)$$

so that

$$\mathscr{E}_{max} = \frac{2(\phi_i - V_a)}{x_d} \tag{4.3.3}$$

in a step junction. For the linearly graded junction, the results of Problem 4.5 can be used to find that the maximum field is

$$\mathscr{E}_{max} = \frac{3(\phi_i - V_a)}{2x_d} \tag{4.3.4}$$

Capacitance.   In Sec. 3.2, we analyzed capacitive behavior in a metal-semiconductor junction resulting from modulation of the stored charge in the junction depletion region as the applied voltage is changed. Analogous behavior occurs in *pn* junctions except that the depletion region varies with bias in two directions. In the general case for a step junction with dopant concentrations N_a and N_d on its two sides, we find the small-signal capacitance per unit area by using the expression for the charge Q_s (per unit area) in the depletion region on either side of the junction

$$Q_s = qN_d x_n = qN_a x_p \tag{4.3.5}$$

in the definition of the small-signal capacitance C (per unit area)

$$C = \frac{dQ}{dV_a} = qN_d \frac{dx_n}{dV_a} = qN_a \frac{dx_p}{dV_a} \tag{4.3.6}$$

Because $x_p = (N_d/N_a)x_n$ and $x_d = x_n + x_p$, we find from Equation 4.3.1 that

$$\frac{dx_n}{dV_a} = \frac{1}{N_d} \left[\frac{\epsilon_s}{2q\left(\dfrac{1}{N_a} + \dfrac{1}{N_d}\right)(\phi_i - V_a)} \right]^{1/2} \tag{4.3.7}$$

and

$$C = \left[\frac{q\epsilon_s}{2\left(\dfrac{1}{N_a} + \dfrac{1}{N_d}\right)(\phi_i - V_a)} \right]^{1/2} \tag{4.3.8}$$

Thus, for $|V_a|$ much greater than ϕ_i, the capacitance of a *pn* step junction varies approximately inversely with the square root of the reverse bias. Using Equation 4.3.1 in Equation 4.3.8, we obtain the expression $C = \epsilon_s/x_d$, the general relationship for small-signal capacitance.

As the dopant concentration on one side of the *pn* junction increases, the depletion region on that side of the junction narrows. At high dopant concentrations the width of the depletion region on the heavily doped side becomes a negligible part of the total space-charge-layer width, and the field- and space-charge-configurations for the *pn* junction become very similar to those at a rectifying metal-semiconductor contact.*

Although we presented a heuristic proof for the general equation $C = \epsilon_s/x_d$ in Chapter 3, the meaning of the result is important enough to be reinforced at this point by deriving the expression carefully. We consider an arbitrarily doped *pn* junction having a depletion region that extends from $-x_p$ to x_n (Figure 4.11a). The charge per unit area Q stored between a point x and the edge of the depletion region at x_n is

$$Q = q \int_x^{x_n} N \, dx \tag{4.3.9}$$

where $N = N_d - N_a$ is the net dopant density. Because $\mathscr{E}_x(x_n) = 0$, the electric field at x is found from Gauss' law to be

$$-\mathscr{E}_x(x) = \frac{1}{\epsilon_s} \int_x^{x_n} qN \, dx = \frac{Q}{\epsilon_s} \tag{4.3.10}$$

and is illustrated in Figure 4.11b. As the applied voltage V_a is changed by a small amount dV_a, the width of the *n*-type side of the depletion region changes by dx_n and the charge stored between x and the edge of the depletion region changes by

$$dQ = qN(x_n) \, dx_n \tag{4.3.11}$$

---

* Considered another way, as charge becomes more concentrated and finally approaches a δ function in space, fields are terminated more and more abruptly. Thus, the heavily doped semiconductor assumes the properties associated with idealized metals; in particular, all charge is confined to its surface.

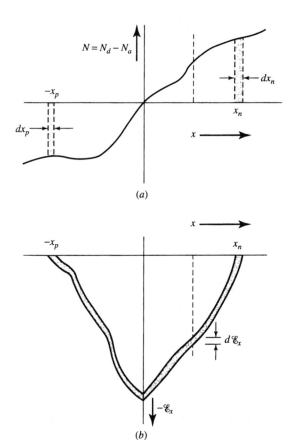

$N = N_d - N_a$

$-x_p$

dx_p

dx_n

x_n

x

(a)

x

$-x_p$

x_n

$d\mathscr{E}_x$

$-\mathscr{E}_x$

(b)

FIGURE 4.11 (*a*) Dopant concentration in an arbitrarily doped junction, showing modulation of the carrier densities at the edges of the space-charge region by an applied voltage. (*b*) Electric-field distributions for two slightly different applied voltages.

Consequently, the field at x changes by

$$-d\mathscr{E}_x = \frac{dQ}{\epsilon_s} = \frac{q}{\epsilon_s} N(x_n)\, dx_n \tag{4.3.12}$$

as the voltage changes. Because the area under the $\mathscr{E}_x$ versus x curve corresponds to the total potential $\phi_i - V_a$, the differential change in voltage corresponds to the change in area under the curve (the shaded region of Figure 4.11b)

$$dV_a \approx -x_d\, d\mathscr{E}_x = \frac{x_d}{\epsilon_s}\, dQ \tag{4.3.13}$$

Using Equation 4.3.13 in the definition of the small-signal capacitance (Equation 4.3.6), we find

$$C = \frac{dQ}{dV_a} = \frac{\epsilon_s}{x_d} \tag{4.3.14}$$

Thus, we have proven this simple relationship for an arbitrarily doped junction. The voltage dependence of the depletion-region width x_d and of the capacitance depends, of course, on the actual dopant profile.

As in the case of the metal-semiconductor junction, measurement of the variation with applied voltage of the small-signal capacitance of a *pn*-junction can be used to determine the dopant concentration as a function of position. However, an additional complication arises in the *pn* junction because the depletion region extends in both directions from the junction plane where $N = N_d - N_a = 0$. To investigate this further, we differentiate Equation 4.3.14 and write $x_d = x_n + x_p$ to find

$$\frac{dC}{dx_n} = -\frac{\epsilon_s}{(x_n + x_p)^2}\left(1 + \frac{dx_p}{dx_n}\right) \tag{4.3.15}$$

Because the changes in charge on either side of the junction are equal in magnitude,

$$|dQ| = |qN(-x_p)dx_p| = qN(x_n)dx_n \tag{4.3.16}$$

and

$$\frac{dC}{dx_n} = -\frac{C^2}{\epsilon_s}\left(1 + \frac{N(x_n)}{N(-x_p)}\right) \tag{4.3.17}$$

Combining this expression with Equation 4.3.11 and the definition of the capacitance, we obtain

$$N(x_n) = -\frac{C^3}{\epsilon_s q(dC/dV_a)}\left(1 + \frac{N(x_n)}{N(-x_p)}\right) \tag{4.3.18}$$

If the *p*-type side of the junction is much more heavily doped than the *n*-type side, the factor on the right of Equation 4.3.18 reduces to unity, and the equation simplifies to the same form as Equation 3.2.10, for the metal-semiconductor junction. Physically, this is reasonable because a depletion region scarcely extends into a highly doped semiconductor, just as it extends a negligible amount into the metal side of a metal-semiconductor junction. In the more general case, however, the presence of the factor $[1 + N(x_n)/N(-x_p)]$ in the expression for $N(x_n)$ complicates the interpretation of capacitance measurements in terms of dopant concentrations. The use of iterative techniques and measurements of other quantities are then required to obtain the dopant distributions. In practice, an additional complication is the derivative dC/dV_a in Equation 4.3.18. Differentiation of a potentially noisy experimental signal requires more complex shielding of the experimental equipment and suitable numerical averaging techniques.

For a heterojunction, the capacitance is modified from that of a homojunction by the different permittivities on the two sides of the junction and the different built-in potential. An expression for capacitance analogous to Equation 4.3.18 is

$$\frac{1}{C} = \frac{x_n}{\epsilon_1} + \frac{x_p}{\epsilon_2} = \sqrt{\frac{2\phi_i}{q}\frac{\epsilon_1 N_d + \epsilon_2 N_a}{\epsilon_1 N_d \epsilon_2 N_a}} \tag{4.3.19}$$

Photodiodes. In Sec. 1.1 we discussed photoconduction, in which the conductivity of a piece of semiconductor increases when carriers are generated by light incident on the semiconductor. When *pn* junctions are present, light interacts with semiconductors in additional ways. The electric field in the depletion region of the *pn* junction separates electron-hole pairs created there before they can recombine and accelerates the carriers toward the neutral regions, where they can contribute to the current flow. Holes are directed toward the *p*-type region, and electrons are directed toward the *n*-type region.

If no external bias is applied, the additional carriers in the neutral regions move the Fermi level toward midgap, decreasing the potential across the *pn* junction below the equilibrium built-in potential and generating a photovoltage along with the photocurrent. This *photovoltaic* mode of operation allows power to be generated and is the basis of solar cells that directly convert light to electric power.

When a reverse-biased *pn* junction is illuminated, the photogenerated carriers greatly increase the reverse current. In the dark, few carriers are present in the depletion region, and the reverse current is very low. When electron-hole pairs are generated by incident light, the number of carriers in the depletion region increases greatly, with a corresponding increase in the reverse current. The ratio of the photocurrent to the dark current in a reverse-biased junction can be many orders of magnitude. Although carriers are photo-generated throughout the structure of the *pn* diode if its depth is less than or comparable to the absorption length of the light, they are most efficiently collected when they are generated in the depletion region. The high electric field there separates the photogenerated electron-hole pairs before they can recombine and accelerates the carriers toward the neutral regions where they are majority carriers and contribute to the current. The applied bias also enlarges the depletion region, allowing the incident light to generate additional photocurrent.

To increase the efficiency, the depletion region should be made as wide as practical. In specialized photodiodes, a lightly doped, nearly intrinsic (*i*) layer is placed between the heavily doped *p*- and *n*-type regions to form a *p-i-n* diode. This *i* layer can be readily depleted by a modest reverse bias on the terminals of the diode, so that the region which leads to efficient carrier collection is larger, as is the corresponding photocurrent. Because the carriers are accelerated by the electric field within the depletion region, *pin* diodes respond rapidly to changes in the light intensity. By contrast, minority carriers generated in neutral regions travel to collecting regions by diffusion, a much slower process. Although *pin* diodes are efficient and fast, adding a lightly doped layer within the diode is not readily compatible with conventional integrated-circuit processing. Consequently, *pin* diodes are used for specialized applications, while *pn* diodes are used for more routine photo-sensing because they can be more readily integrated with other functions on an integrated circuit.

4.4 JUNCTION BREAKDOWN

In the previous sections we saw that the depletion-region width and the maximum electric field in a *pn* junction increase as reverse bias increases. Intuitively, we know that there must be physical limitations to these increases. At high voltages, some of the materials making up the device structure, such as the insulating layers of silicon dioxide or packaging materials, may rupture, or else the current through the *pn* junction itself may increase rapidly. In the first case, irreversible damage usually occurs, and the device is destroyed. The second case—breakdown of the barrier to current flow within the junction itself—is generally not destructive (unless the high currents involved melt a portion of the junction) and is of practical importance. The voltage at which breakdown occurs depends on the structure of the junction and the dopant concentrations in a reasonably well-defined manner, and junctions can be constructed with predictable breakdown characteristics.

At high fields in a semiconductor, one of two breakdown processes can occur. In one process, the field is so high that it exerts sufficient force on a covalently bound electron to free it. This creates two carriers—a hole and an electron to contribute to the current. In the energy-band picture of this breakdown process, an electron makes a transition from the valence band to the conduction band without the interaction of any other particles. This type of breakdown is called *Zener breakdown* and involves electron tunneling through energy barriers, a phenomenon introduced in our discussion of ohmic contacts in Sec. 3.4. In the other breakdown process, free carriers are able to gain enough energy from the field between collisions that they can break covalent bonds in the lattice when they collide with it. In this process, called *avalanche breakdown*, every carrier colliding with the lattice creates two additional carriers. All three carriers can then participate in further avalanching collisions, leading to a sudden multiplication of carriers in the space-charge region when the maximum field becomes large enough to cause avalanching.

One or the other of these two breakdown processes predominates in a given *pn* junction. The factors that determine which process occurs will be more evident after details of the breakdown processes have been considered. We first discuss the more common avalanche breakdown. Then we consider Zener breakdown and, finally, we contrast these two processes.

Avalanche Breakdown[†]

We consider an electron traveling in the space-charge region of a reverse-biased *pn* junction. The electron travels, on the average, a distance λ, its mean free path, before interacting with an atom in the lattice and losing energy. The energy ΔE gained from the field $\mathscr{E}$ by the moving electron between collisions is

$$\Delta E = q \int_0^\lambda \mathscr{E} \cdot d\boldsymbol{x} \tag{4.4.1}$$

The boldface letters in Equation 4.4.1 denote a vector product. If the electron gains sufficient energy from the field before colliding with an atom, it can break the bond between the atom core and one of the bound electrons during the collision so that three carriers—the initial electron and the hole and electron created by the collision—are free to leave the region of the collision. This process is indicated schematically in Figure 4.12. If all three carriers are assumed to have equal mass, conservation of energy and momentum requires that the original electron possess kinetic energy of at least $\frac{3}{2}E_g$ to break the bond (Problem 4.11).

For simplicity, we assume an abrupt junction in which the *n*-type region is much more heavily doped than the *p*-type region. In this case, as will be shown in Chapter 5, most of the carriers entering the depletion region under moderate reverse bias are electrons from the *p*-type region. The few holes entering from the *n*-type region can be neglected in our analysis. Near the edges of the space-charge region, the electric field is low and essentially no carriers can gain enough energy from the field to create a hole-electron pair before they lose their kinetic energy in a collision with the lattice. Consequently, avalanche is confined to the central portion of the space-charge region where the field is

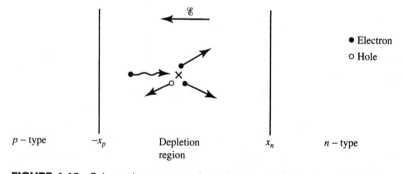

FIGURE 4.12 Schematic representation of the avalanche process. An incident electron (wavy arrow) gains enough energy from the field to excite an electron out of a silicon-silicon bond during a lattice collision, creating an additional electron-hole pair.

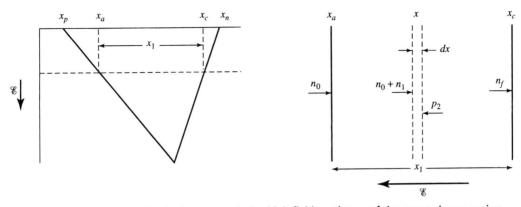

FIGURE 4.13 (a) Ionization occurs in the high-field portion x_1 of the space-charge region. (b) Carrier pairs are created in the region dx at x by electrons flowing from the left and by holes flowing from the right.

sizable. This central region is labeled x_1 in the plot of field versus distance for this case shown in Figure 4.13a.

We consider the number of carriers created by avalanche in a small volume of width dx located within x_1 at position x (Figure 4.13b). Let the density of electrons entering x_1 from the left at x_a be n_0. Avalanching increases this density between x_a and x so that the electrons entering the volume $A\,dx$ at x from the left have a density $n_0 + n_1$. The probability that electrons create electron-hole pairs while traveling through dx is given by the product of a proportionality factor α_n, called the *ionization coefficient*, times the length dx. Because the electrons gain energy more rapidly when the field is higher, we expect the ionization coefficient to be a function of electric field and, hence, of position. The added density of electrons (and the added density of holes) created in dx by the electrons entering from the left is

$$dn' = dp = \alpha_n n\,dx = \alpha_n(n_0 + n_1)\,dx \qquad (4.4.2)$$

Because we assumed that no holes enter the ionization region at x_c, any holes entering the infinitesimal volume at x from the right were created between x and x_c. We designate the density of these holes as p_2. The holes are also avalanche-multiplied within dx to create added hole and electron densities.

$$dn'' = dp = \alpha_p p\,dx = \alpha_p p_2\,dx \qquad (4.4.3)$$

where α_p is the ionization coefficient for holes. The total increase in electron density created within dx is equal to the sum $(dn' + dn'')$ (Equations 4.4.2 and 4.4.3)

$$dn = \alpha_n(n_0 + n_1)\,dx + \alpha_p p_2\,dx \qquad (4.4.4)$$

If we let n_f be the density of electrons that reaches x_c,

$$n_f = n_0 + n_1 + n_2 \qquad (4.4.5)$$

where n_2 is the density of electrons created between x and x_c; $n_2 = p_2$ because electrons and holes are created in pairs. We can then write

$$\frac{dn}{dx} = (\alpha_n - \alpha_p)(n_0 + n_1) + \alpha_p n_f \qquad (4.4.6)$$

To proceed further, we must have additional information about the ionization coefficients α_n and α_p. We can gain some useful perspective about avalanche breakdown by taking $\alpha_n = \alpha_p$, although this is only approximately true. Making that assumption and letting $\alpha \equiv \alpha_n = \alpha_p$, we can integrate Equation 4.4.6 subject to the boundary conditions $n(x_a) = n_0$ and $n(x_c) = n_f$. The result of this integration is

$$n_f - n_0 = n_f \int_{x_a}^{x_c} \alpha \, dx \tag{4.4.7}$$

We denote the ratio of the density of electrons n_f leaving the space-charge region to the density n_0 entering as the multiplication factor M:

$$M = \frac{n_f}{n_0} = \frac{1}{1 - \displaystyle\int_{x_a}^{x_c} \alpha \, dx} \tag{4.4.8}$$

As the integral in the denominator of Equation 4.4.8 approaches unity, the multiplication factor increases without bound. Hence, avalanche breakdown can be defined to occur when

$$\int_{x_a}^{x_c} \alpha \, dx = 1 \tag{4.4.9}$$

Had we not taken $\alpha_n = \alpha_p$, there would be a more complicated integral than that in Equation 4.4.9, but a form similar to Equation 4.4.8 in which the denominator vanishes could be derived.

As stated earlier, the ionization coefficient is a strong function of the electron energy which, in turn, depends on the electric field because the energy necessary for an ionizing collision is imparted to the carrier by the field. For slowly varying electric fields we can obtain a plausible expression for the ionization coefficient α by the following arguments. The density of ionizing collisions at x is proportional to n^*, the density of excited electrons arriving at x with sufficient energy to create electron-hole pairs. The density n^*, in turn, is just the total electron density n times the probability that an electron has not collided in the distance d necessary to gain adequate energy, that is,

$$n^* = n \exp\left(-\frac{d}{\lambda}\right) \tag{4.4.10}$$

where λ is the mean-free path. The length d can be found from Equation 4.4.1 by letting E_1 be the minimum energy necessary for an ionizing collision and $\mathscr{E}$ be the average field that accelerates the electron:

$$d = \frac{E_1}{q\mathscr{E}} \tag{4.4.11}$$

The number of ionizing collisions in the distance dx is also proportional to dx/d if we assume that the electron collides soon after gaining sufficient energy to ionize an atom. Then

$$dn = A'n^* \frac{dx}{d} = \frac{A'q\mathscr{E}}{E_1}\left[\exp\left(-\frac{E_1}{\lambda q\mathscr{E}}\right)\right] n \, dx \tag{4.4.12}$$

where A' is a proportionality constant. Comparing Equations 4.4.2 and 4.4.12, we obtain a reasonable expression for the ionization coefficient of the form

$$\alpha = K\mathscr{E} \exp\left(-\frac{B}{\mathscr{E}}\right) \tag{4.4.13}$$

Because the ionization coefficient depends strongly on the electric field, the multiplication factor also increases rapidly with field. Thus, a small increase in field as the voltage approaches breakdown causes a sharp increase in current, a behavior that is dramatically confirmed in practical diodes.

Not only does the ionization coefficient vary with field, and consequently with position in the space-charge region, but the width of the space-charge region also varies with voltage. Therefore, the evaluation of M in Equation 4.4.8 is difficult, and an empirical approximation of the form

$$M = \frac{1}{1 - (|V_R|/BV)^n} \qquad (2 < n < 6) \tag{4.4.14}$$

is often used, where $V_R < 0$ is the applied (reverse) bias and BV is the breakdown voltage, at which the current is observed to increase rapidly.

To relate the breakdown voltage to the material parameters, we consider a one-sided step junction with $N_a \ll N_d$ and assume that breakdown occurs when the maximum field in the junction reaches a critical value $\mathscr{E}_1$ that causes the integral in Equation 4.4.8 to approach unity. Because the maximum field is approximately given by

$$\mathscr{E}_{max} = \left(\frac{2qN_a|V_R|}{\epsilon_s}\right)^{1/2} \tag{4.4.15}$$

the breakdown voltage has the approximate form

$$BV = \frac{\epsilon_s \mathscr{E}_1^2}{2qN_a} \tag{4.4.16}$$

Equation 4.4.16 shows that the breakdown voltage decreases as the dopant concentration increases, although the decrease is not quite as rapid as indicated. In practical diodes the breakdown voltage generally varies with doping as $N^{-2/3}$. The more gradual variation with doping in the practical case is a consequence of the more efficient avalanche multiplication in junctions having wider depletion regions. At higher dopant concentrations, a slightly higher critical field $\mathscr{E}_1$ is needed (Figure 4.14).

Our comments thus far have been in terms of one-dimensional geometry. In a practical planar junction, the finite lateral dimension of the junction can significantly reduce the breakdown voltage. In Sec. 2.1 we considered a planar junction formed by diffusion through an opening in a silicon dioxide layer. The impurities diffuse laterally beneath

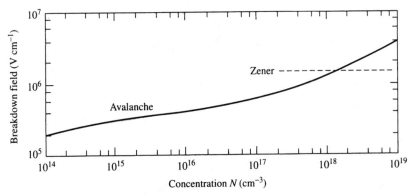

FIGURE 4.14 The critical electric fields for avalanche and Zener breakdown in silicon as functions of dopant concentration [3,4,5].

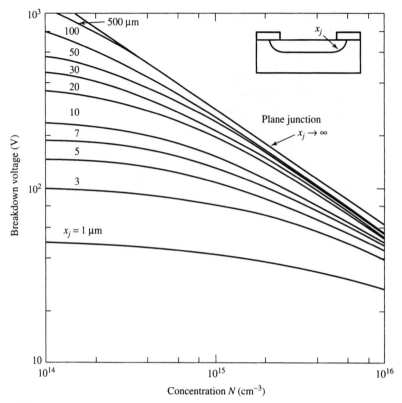

FIGURE 4.15 Breakdown voltage of one-sided, planar, silicon step junction, showing the effect of junction curvature [6,7].

the Si–SiO$_2$ interface, as well as vertically into the silicon, creating a rounded region of the junction beneath the edge of the SiO$_2$ (Figure 2.1*d*). The field in this corner region can be markedly higher than the field in the remainder of the junction, causing breakdown to occur there at unexpectedly low voltages. The reduction in breakdown voltage is especially severe for a shallow junction with a small radius of curvature. It becomes less serious as the junction depth x_j increases. The influence of junction curvature on the breakdown voltage is shown in Figure 4.15 for a one-sided silicon planar step junction.

EXAMPLE *pn-Junction-Diode Breakdown*

Diodes are to be made in a boron-doped silicon wafer that has a resistivity $\rho = 1.5$ Ω-cm. The processing sequence consists of

1. A deposition of phosphorus atoms on the surface with a concentration $N' = 5 \times 10^{15}$ cm^{-2}.

2. A drive-in diffusion at 1000°C for 27 min.

   (a) What value is expected for the breakdown voltage *BV* if the entire silicon surface is exposed to the phosphorus doping?

   (b) What value is expected for the breakdown voltage *BV* if the wafer is oxidized and *windows* are opened through the oxide to the silicon surface at selected areas before the phosphorus is deposited so that separated diode regions are formed?

Solution The process described produces *np*–junction diodes. For part (a), a large diode is formed, with the junction consisting of a single plane. For part (b), many diffused regions are created, each having a cross section similar to that shown in Figure 2.1*d*. The diodes in part (b) therefore have enhanced fields in the corner regions and we can expect them to have a lower breakdown voltage than the diode formed in part (a).

The 27 min. diffusion at 1000°C redistributes a fixed total quantity of phosphorus, and therefore results in a Gaussian distribution of the *n*-type dopant. From Figure 2.20, the diffusion coefficient of phosphorus at 1000°C is 1.6×10^{-13} cm^2 s^{-1}. Hence, $\sqrt{Dt} = 0.16\ \mu$m.

To decide whether or not the diodes can be considered to be one-sided, we determine the density of phosphorus after redistribution. The surface concentration for this Gaussian distribution is obtained from Equation 2.5.13 at $x = 0$ as $C_s = N'/(\sqrt{\pi}\sqrt{Dt}) = 1.76 \times 10^{20}$ cm^{-3}. We find the doping density N_a in the 1.5 Ω-cm wafer to be 10^{16} cm^{-3}, either from Figure 1.15 or Table 4.1. The ratio N_a/C_s is therefore 5.7×10^{-5}, and the diodes can be considered to be one-sided step junctions.

The junction depth x_j can be determined from Figure 2.21 by finding x/L such that $C/C_s = N_a/C_s$. Using the results for a Gaussian in Figure 2.21, we find $x_j = 3.1 \times 2\sqrt{Dt} = 1.0\ \mu$m.

(a) The breakdown voltage *BV* for the plane junction is now found either from Figure 4.15 or Table 4.1 to be 63 V.

(b) From Figure 4.15, the individual diffused diodes have a breakdown voltage *BV* equal to 26 V.

This example shows that field concentration at the corners of diffused junctions can have important practical consequences. From Figure 4.15 we see that the breakdown voltage increases rapidly as the junction is diffused deeper into the wafer. ∎

Because the carrier gains energy continuously as it travels between collisions, the exact expression for the ionization coefficient must consider the "nonlocal" nature of the energy transfer caused by the variation of the electric field within the distance over which the electron gains energy between collisions.

In high-performance integrated-circuit devices, the electric field can vary extremely rapidly with position, with the peak electric field extending over only a few tens of nanometers. Such rapid variations in the electric field cause impact ionization models that depend only on the local electric field to overestimate the amount of impact ionization. When the electric field increases rapidly with position, the electron gains part of its energy in lower field regions so it has less energy than that corresponding to the high local electric field. (We can say that the electron energy lags the electric field.) When the high-field region is very narrow, electrons cannot acquire sufficient energy while within this high-field region to cause impact ionization, and breakdown occurs at a higher voltage. Thus, models which assume that an electron at a position *x* has an energy corresponding to the local electric field at the same point *x* can overestimate the currents resulting from impact ionization by orders of magnitude. These effects are very significant in understanding breakdown in high-performance bipolar transistors and in estimating the substrate current in MOS transistors.

An alternate approach to modeling impact ionization calculates the average electron energy and then determines the impact ionization corresponding to the energy that the electron has gained since its last collision, rather than that corresponding to the local electric field. The electron energy at a position *x* can be found from the electric field distribution by energy-balance arguments as the convolution of the electric field and the exponential decay length λ_e [8]

$$E = \frac{2q}{5} \int_0^x \mathscr{E}(u) \exp\left(\frac{u-x}{\lambda_e}\right) du \qquad (4.4.17)$$

where λ_e is the *energy relaxation length* associated with the distance required for equilibration of the energy with the electric field. Typical values of λ_e are about 60–80 nm,

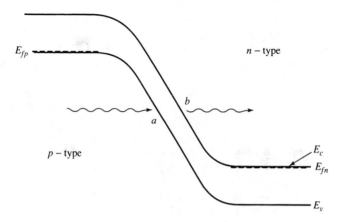

FIGURE 4.16 Energy-band diagram of a reverse-biased junction that has a high dopant concentration on both sides. Tunneling or Zener breakdown is likely in this type of junction.

corresponding to times of a fraction of a picosecond. The ionization coefficient can then be expressed as a function of this average energy, rather than as a function of the local electric field (as used in Equation 4.4.13).

Zener Breakdown†

As the dopant concentration increases, the width of the depletion region decreases and the critical field at which avalanche occurs also increases. At high dopant concentrations the field required for avalanche breakdown to occur exceeds the field necessary for Zener breakdown and the latter becomes more probable. As stated earlier, Zener breakdown occurs when the force exerted by the applied field is strong enough to rip an electron from its covalent bond to create an electron-hole pair directly. Figure 4.16 is an energy-band picture that shows Zener breakdown schematically.* As the figure indicates, a large number of electrons in the valence band on the *p*-type side of the junction are separated by the narrow depletion region from empty allowed states at the same energy in the conduction band of the *n*-type material. As the dopant concentration in the semiconductor increases, the width of the depletion region at a given reverse bias decreases, and the energy bands in the depletion region are bent more steeply. Because of the wave nature of the electron, there is a finite probability that an electron in the valence band of the *p*-type semiconductor approaching the forbidden gap can *tunnel* through the forbidden region and appear at the same energy in the conduction band of the *n*-type semiconductor. Because the probability of transmission of an electron through a barrier is a strong function of the thickness of the barrier, tunneling is only significant in highly doped material, in which the fields are high and the depletion region is narrow.

To investigate the probability of tunneling from the valence band to the conduction band, we approximate the barrier that the electron sees by a triangular barrier (Figure 4.17). The height of the energy barrier E_B decreases linearly from E_g at $x = 0$ to 0 at $x = L$, and the average field is $\mathscr{E} = E_g/qL$. The probability of tunneling Θ can be approximated by using the barrier height in the equation**

$$\Theta \approx \exp\left[-2 \int_0^L \sqrt{\frac{2m^*E_B}{\hbar^2}}\, dx \right] \tag{4.4.17}$$

---

* For this discussion it is useful to consider electrons in both the conduction band and the valence band rather than employing the concept that only holes are important in the valence band.

** This "WKB" approximation is discussed in most quantum-mechanics text books; see, for example [9].

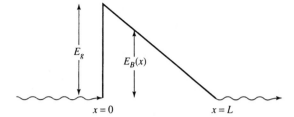

FIGURE 4.17 The probability of tunneling across the junction can be approximated by considering tunneling through a triangular barrier.

Carrying through the integration, we find the tunneling probability to be

$$\Theta = \exp\left(-\frac{B}{\mathscr{E}}\right) = \exp\left(-\frac{qBL}{E_g}\right) \tag{4.4.18}$$

where

$$B = \frac{4\sqrt{2m^*}\,E_g^{3/2}}{3q\hbar} \tag{4.4.19}$$

Thus, the probability of tunneling decreases rapidly as the electric field decreases or the tunneling distance increases.

We can roughly estimate the field necessary to obtain appreciable tunneling from the above approximations. The current is the product of the area, the electron charge, the number of valence-band electrons in the p-region arriving at the barrier per second that "see" empty states across the barrier, and the probability that each electron tunnels through the barrier. The number of electrons arriving at the barrier can be expressed as the density $\mathcal{N}$ of electrons in the valence band times their velocity v, and the current can be written as

$$I = qA\mathcal{N}v\Theta \tag{4.4.20}$$

As an indication of appreciable tunneling, we consider a current of 10 mA flowing across a junction 10^{-5} cm^2 in area. The density of electrons in the valence band that are at energies corresponding to empty allowed states in the conduction band across the barrier is comparable to the atomic density of about 10^{22} cm^{-3}. We assume that the electrons are moving with their thermal velocity of about 10^7 cm s^{-1} so that 10^{29} electrons cm^{-2} s^{-1} strike the barrier. The tunneling probability corresponding to this current is found from Equation 4.4.18 to be about 10^{-7}. Using this value and a semiconductor band gap of 1 eV, we find the corresponding tunneling distance and electric field to be about 4 nm and 10^6 V cm^{-1}, respectively. That is, for appreciable current to flow by tunneling or Zener breakdown, the effective barrier must be less than about 4 nm wide and the electric field in the depletion region must be greater than about 10^6 V cm^{-1}, corresponding to a dopant concentration of $\geq 10^{18}$ cm^{-3} on the *lightly* doped side of the *pn* junction. These values are consistent in order-of-magnitude with observations.

As the dopant concentrations decrease, the width of the space-charge region increases and the probability of tunneling decreases rapidly. Avalanche breakdown then becomes more likely than Zener breakdown. Thus, Zener breakdown is only expected for the most heavily doped junctions, while more lightly doped junctions break down by the avalanche mechanism.

Devices exhibiting Zener breakdown generally have breakdown voltages lower than those that break down by avalanching. In silicon, pure Zener breakdown is usually found in diodes having $BV < 5$ V. At higher voltages, avalanche breakdown usually predominates.

Commercially available diodes with well-defined breakdown characteristics are generally called *Zener diodes* regardless of their breakdown mechanism.

It is possible to determine whether avalanche or Zener breakdown is occurring in a junction by observing the temperature sensitivity of the breakdown voltage. Although not large, the temperature variation of the two types of breakdown is of opposite sign. In the case of Zener breakdown, the breakdown voltage decreases with increasing temperature because the flux of valence-band electrons available for tunneling increases as temperature rises. The effect of temperature on the breakdown voltage for avalanching junctions is just the opposite. The breakdown voltage increases as temperature increases because the mean-free path of energetic electrons (λ in Equation 4.4.10) decreases. For breakdown voltages in the range of ~5 to 6 V, both avalanche- and tunnel-breakdown can occur simultaneously so that the net temperature variation is very slight. This characteristic is useful for establishing a voltage reference in some integrated circuits.

4.5 DEVICE: JUNCTION FIELD-EFFECT TRANSISTORS

pn Junction Field-Effect Transistor (JFET)

We have seen that the depletion-layer width of a *pn* junction can be varied by modulating a reverse-bias voltage applied to the junction. In this section, we consider a device that makes use of this mechanism to control the current through a region bounded by two *pn* junctions. Because little current flows into a reverse-biased *pn* junction, only a small amount of power is consumed at the control electrode, while substantially more power can be delivered by the controlled current. The device, which is called a *junction field-effect transistor* (JFET), can therefore be used as a power amplifier.

Consider the structure shown in Figure 4.18, which consists of a lightly doped *n*-type layer on top of a *p*-type substrate. For the reasons outlined in Sec. 2.6, it is usual to obtain this structure by growing an *n*-type epitaxial layer on a *p*-type substrate. The *n*-type region then contains a dopant concentration that is relatively uniform through the thickness of the epitaxial layer. Alternatively, an *n*-type layer can be introduced by ion-implantation techniques.

After the uniform lightly doped *n*-type layer is formed, two heavily doped *n*-type regions (denoted n^+) are added by diffusion so that good ohmic contact can be obtained (as discussed in Sec. 3.4) to the lightly doped region known as the *channel*. Note that the channel region is similar to the resistor structure discussed in Sec. 2.10 except for the presence of the p^+ layer. The n^+ electrodes are called the *source* and *drain* electrodes.

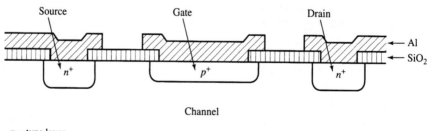

FIGURE 4.18 Basic structure of an *n*-channel, junction field-effect transistor. A lightly-doped *n*-type channel layer is bounded by the *p*-type substrate and a *p*-type gate diffusion.

The source supplies majority carriers to the channel. Consequently, conventional current flows from drain to source in an n-channel JFET. If the doping type is changed in all regions of the structure shown in Figure 4.18, the sketch would represent a p-channel JFET. In the p-channel device, conventional current flows from source to drain.

The pn junction above the channel in Figure 4.18 serves as the control element when reverse bias is applied to the p-type region which is called the *gate*. The channel is defined as the neutral region bounded from above by the depletion region at the gate and from below by the depletion region at the substrate pn junction. The substrate is usually at ground potential. If the drain of the n-channel JFET is biased positively, current flows from it to the source through the channel. If we now ground the source and apply a negative voltage to the p^+ electrode, the junction depletion region widens and the channel narrows. As the channel narrows, its resistance increases, and less current flows from drain to source. In this way, a voltage applied to the gate controls the current flowing through the channel.

With this qualitative explanation of JFET operation, we will find it straightforward to develop a quantitative theory for the device. Many of the ideas that we develop will apply directly to the GaAs metal-semiconductor field-effect transistor (MESFET) and will also be useful in our discussion of the metal-oxide-semiconductor field-effect transistor (MOSFET) (Chapters 9 and 10).

Device Analysis of the JFET.   To analyze the JFET, we first consider a small bias V_D applied to the drain electrode while the source is grounded and the voltage at the gate is V_G. With this small drain voltage, the gate–channel bias and therefore the width of the gate depletion region is uniform along the channel. An expanded view of the channel region is shown in Figure 4.19. We assume a one-dimensional structure with a gate length L between the source and drain regions and a width W perpendicular to the plane of the paper. (Usually $W \gg L$.) Drain current flows along the dimension y. We assume a one-sided step junction at the gate with N_a in the p-region much greater than N_d in the channel. Therefore, the depletion layer extends primarily into the n-channel. The distance between the p-type gate and the substrate is t, the thickness of the gate depletion region in the n-type channel is x_d, and the thickness of the neutral portion of the channel is x_w. To focus on the role of the gate, we assume that the depletion region at the substrate junction extends primarily into the substrate so that $x_w \approx (t - x_d)$. This approximation is valid when the substrate is lightly doped.

The resistance of the channel region can be written as

$$R = \frac{\rho L}{x_w W} \tag{4.5.1}$$

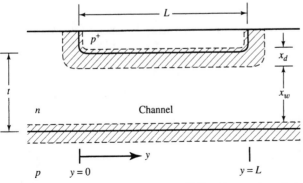

FIGURE 4.19 Channel region of a JFET with gate length L. Depletion regions are cross hatched.

where $\rho = (q\mu N_d)^{-1}$ is the resistivity of the channel. The drain current is then

$$I_D = \frac{V_D}{R} = \left(\frac{W}{L}\right)(q\mu_n N_d x_w V_D) \qquad (4.5.2)$$

The dependence on gate voltage is incorporated in Equation 4.5.2 by expressing $x_w = t - x_d$, where x_d (from Equation 4.3.1) is

$$x_d = \left[\frac{2\epsilon_s}{qN_d}(\phi_i - V_G)\right]^{1/2} \qquad (4.5.3)$$

and ϕ_i is the built-in potential. The current can now be written as a function of the gate and drain voltages.

$$I_D = \frac{W}{L}q\mu_n N_d t\left\{1 - \left[\frac{2\epsilon_s}{qN_d t^2}(\phi_i - V_G)\right]^{1/2}\right\}V_D \qquad (4.5.4)$$

The factor in front of the bracketed terms is equal to the conductance G_0 of the *n*-region when it is completely undepleted (the so-called *metallurgical channel*). Equation 4.5.4 can be rewritten in terms of G_0, as

$$I_D = G_0\left\{1 - \left[\frac{2\epsilon_s}{qN_d t^2}(\phi_i - V_G)\right]^{1/2}\right\}V_D \qquad (4.5.5)$$

At a given gate voltage, we find a linear relationship between I_D and V_D as a consequence of our assumption of a small applied drain voltage. The square-root dependence on gate voltage in Equation 4.5.5 arises from our assumption of an abrupt gate-channel junction. Equation 4.5.5 shows that the current is maximum at zero applied gate voltage and decreases as $|V_G|$ increases. The equation predicts zero current when the gate voltage is large enough to deplete the entire channel region.

EXAMPLE *Source and Drain Resistance in a JFET*

As seen in the JFET cross section in Figure 4.18, the heavily doped source and drain electrodes are typically set apart from the channel region. This separation adds undesirable resistance in series with the JFET channel. Consider an *n*-channel JFET with $L = 5$ μm and $W = 10$ μm in which the channel region is separated from the source and drain diffusions by 5 μm. The JFET channel is formed by diffusing a *p*-type gate region 0.5 μm into a 1.5-μm-thick, *n*-type, epitaxial layer doped with $N_d = 5 \times 10^{15}$ cm^{-3}. Assume that the gate diffusion forms a step junction with a dopant concentration $N_a = 1 \times 10^{19}$ cm^{-3}, and neglect the depletion layer at the junction between the channel and the substrate. Let $V_G = 0$.

Find the percentage increase in resistance between the source and drain contacts (for the JFET in the linear region of operation) caused by the separation from the channel compared to the resistance of the channel alone.

Solution   The conductance of the channel is given by Equation 4.5.4. We use Equation 4.2.10 to calculate $\phi_i = 0.854$ V. Either from Table 4.1 or from Figure 1.15, we find $\rho_n = 1$ Ω-cm in the channel region, where $\rho_n = (q\mu N_d)^{-1}$. Using these values in Equation 4.5.4, together with $W/L = 2$ and $t = 1$ μm, we calculate the channel conductance G_c

$$G_c = (2.08 \times 10^{-4})(1 - 0.472) = 1.1 \times 10^{-4}\ \Omega^{-1} = 0.11\ \text{mS}$$

where the conductance unit S is a Siemen.

The conductance of the series resistance near the source and drain regions $G_{s,d}$ is

$$G_{s,d} = \frac{W}{L}q\mu N_d t_e = 3.1 \times 10^{-4}\ \Omega^{-1} = 0.31\ \text{mS}$$

where t_e, the epitaxial layer thickness, is 1.5 μm.

The channel resistance is $1/G_c = 9.1$ kΩ. The source and drain series resistors are each $1/G_{s,d} = 3.2$ kΩ, so the total resistance is 15.5 kΩ. The percentage increase is $6.4/9.1 \times 100 = 70\%$.

The extra resistance is *parasitic*, and reduces the achievable gain of the JFET because it is connected in series with the output voltage. The parasitic resistors have a maximum effect under the conditions of this problem (maximum current with zero applied gate voltage). Their effect is reduced as the gate becomes reverse biased, causing G_c to decrease. ∎

Now that we can see the physics behind the device, we remove the restriction of small drain voltages and consider the problem for arbitrary values of V_D and V_G (with the restriction that the gate must always remain reverse biased). With V_D no longer necessarily small, the voltage between the channel and the gate is a function of position y. Consequently, the depletion-region width and the channel cross section also vary with position. The voltage across the depletion region is higher near the drain than near the source, and the depletion region is consequently wider near the drain as shown in Figure 4.20.

We now use the *gradual-channel approximation*. This approximation assumes that the channel- and depletion-layer widths vary slowly from source to drain so that the depletion region is influenced only by fields in the vertical direction and not by fields extending laterally from drain to source. In other words, the field in the y-direction is much less than that in the x-direction in the depletion regions. With this approximation we can find the depletion-region width from a one-dimensional analysis.

We write an expression for the increment of voltage across a small section of the channel of length dy at y as

$$d\phi = I_D\, dR = \frac{I_D\, dy}{Wq\mu_n N_d(t - x_d)} \tag{4.5.6}$$

The width x_d of the depletion region is now controlled by the voltage $\phi_i - V_G + \phi(y)$ where $\phi(y)$ is the potential in the channel at point y, so that

$$x_d = \left[\frac{2\epsilon_s}{qN_d}(\phi_i - V_G + \phi(y))\right]^{1/2} \tag{4.5.7}$$

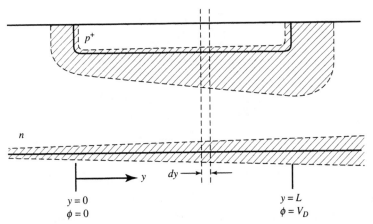

FIGURE 4.20 Channel region of a JFET showing variation of the width of depletion regions along the channel when the drain voltage is significantly higher than the source voltage.

This expression can be used in Equation 4.5.6, which is then integrated from source to drain to obtain the current-voltage relationship for the JFET

$$\frac{I_D \int_0^L dy}{Wq\mu_n N_d} = \int_0^{V_D} \left\{ t - \left[\frac{2\epsilon_s}{qN_d}(\phi_i - V_G + \phi) \right]^{1/2} \right\} d\phi \qquad (4.5.8)$$

After integrating and rearranging, we find

$$I_D = G_0 \left\{ V_D - \frac{2}{3} \left(\frac{2\epsilon_s}{qN_d t^2} \right)^{1/2} \left[(\phi_i - V_G + V_D)^{3/2} - (\phi_i - V_G)^{3/2} \right] \right\} \qquad (4.5.9)$$

At low drain voltages, Equation 4.5.9 reduces to the simpler expression of Equation 4.5.5, and the current increases linearly with drain voltage. However, at higher drain voltages, the current increases more gradually.

At large drain voltages, Equation 4.5.9 indicates that the current reaches a maximum and begins decreasing with increasing drain voltage, but this maximum corresponds to the limit of validity of our analysis. From Figure 4.21 we see that, as the drain voltage increases, the width of the conducting channel near the drain decreases, until finally the channel is completely depleted in this region (Figure 4.21*b*). When this occurs, Equation 4.5.6 becomes indeterminate ($x_d \rightarrow t$). The equations are, therefore, only valid for V_D below the

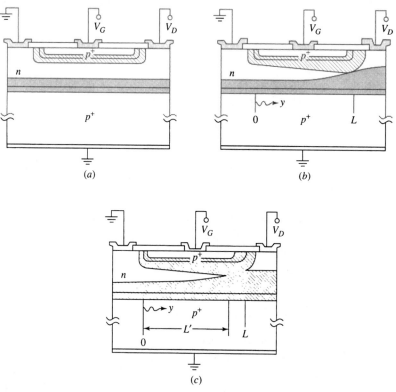

FIGURE 4.21 Behavior of the depletion regions in a JFET. (*a*) For small drain voltage, the channel is nearly an equipotential and the dimensions of the depletion regions are uniform. (*b*) When V_D is increased to V_{Dsat}, the depletion regions on both sides of the channel meet at the pinch-off point (at $y = L$). (*c*) When $V_D > V_{Dsat}$, the pinch-off point (at $y = L'$) moves slightly closer to the source [10].

drain voltage that *pinches off* the channel. Current continues to flow when the channel is pinched off because there is no barrier to the transfer of electrons traveling down the channel toward the drain. As they arrive at the edge of the pinched-off zone, they are pulled across it by the field directed from the drain toward the source. If the drain bias is increased further, any additional voltage is dropped across this depleted, high-field region near the drain electrode, and the point at which the channel is completely depleted moves slightly toward the source (Figure 4.21c). If this slight movement is neglected, the drain current remains constant (saturates) as the drain voltage is increased further, and this bias condition is called *saturation*. The drain voltage at which the channel is entirely depleted near the drain electrode (the saturation drain voltage V_{Dsat}) is found from Equation 4.5.7 to be

$$V_{Dsat} = \frac{qN_d t^2}{2\epsilon_s} - (\phi_i - V_G) \tag{4.5.10}$$

and the corresponding saturation drain current is

$$I_{Dsat} = G_0\left[\frac{qN_d t^2}{6\epsilon_s} - (\phi_i - V_G)\left\{1 - \frac{2}{3}\left[\frac{2\epsilon_s(\phi_i - V_G)}{qN_d t^2}\right]^{1/2}\right\}\right] \tag{4.5.11}$$

Based on the preceding analysis, we can divide the drain current-drain voltage characteristic into three regions (Figure 4.22): (1) the linear region at low drain voltages, (2) a region with less than linear increase of current with increasing drain voltage, and (3) a saturation region where the current remains relatively constant as the drain voltage increases further. As expected from the physics of the device, Equation 4.5.11 predicts the current to be maximum for zero gate bias and to decrease as negative gate voltage is applied. As the gate voltage becomes more negative, the drain saturation voltage and the corresponding current decrease. A family of curves can be generated (Figure 4.22), each curve showing the drain-current versus drain-voltage characteristic for a particular value of gate voltage. At a sufficiently negative value of gate voltage, the saturation drain current becomes zero. This turn-off voltage V_T is found from Equation 4.5.11 to be

$$V_T = \phi_i - \frac{qN_d t^2}{2\epsilon_s} \tag{4.5.12}$$

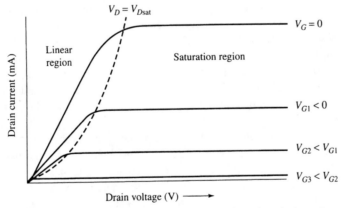

FIGURE 4.22 The output (drain-current, drain-voltage) characteristics of a JFET as a function of the gate voltage. Operation in the "linear" region (considerably to the left of the $V_D = V_{Dsat}$ curve) corresponds to Figure 4.21a. In saturation (to the right of $V_D = V_{Dsat}$), Figure 4.21c applies.

The drain current actually increases slightly as the drain voltage increases beyond V_{Dsat} because the end point for the integration in Equation 4.5.8 now becomes L' rather than L, where L' is the point at which the channel becomes completely depleted [$\phi(L') = V_{Dsat}$] (Figure 4.21). For $V_D > V_{Dsat}$, the expression for I_{Dsat} in Equation 4.5.11 is multiplied by the ratio L/L' (which is greater than unity).

Although most JFETs exhibit characteristics with well-defined saturation regions as shown in Figure 4.22, significant departures can be seen in devices with short channel lengths. In a device with a donor density of 10^{16} cm^{-3} in the channel region, an excess of 5 V beyond V_{Dsat} depletes an additional 1.0 μm. Thus, for a channel length of 8 μm, the ratio $L/L' \approx 8/7$, and the deviation from simple theory is not severe. For devices with channel lengths of the order of 2 μm, however, important deviations occur.*

Field-effect transistors are often operated in the saturation region where the output current is not appreciably affected by the output (drain) voltage but is controlled primarily by the input (gate) voltage. For this bias condition, the JFET is almost an ideal current source controlled by an input voltage. The *transconductance* g_m of the transistor expresses the effectiveness of the control of the drain current by the gate voltage. It is defined by the relation

$$g_m \equiv \left. \frac{\partial I_D}{\partial V_G} \right|_{V_D = const} \tag{4.5.13}$$

and can be found by differentiating Equation 4.5.9.

$$g_m = G_0 \left(\frac{2\epsilon_s}{qN_d t^2} \right)^{1/2} [(\phi_i - V_G + V_D)^{1/2} - (\phi_i - V_G)^{1/2}] \tag{4.5.14}$$

In the saturation region, g_m reaches a maximum value, which is found from Equation 4.5.14.

$$g_{msat} = G_0 \left[1 - \left(\frac{2\epsilon_s}{qN_d t^2}(\phi_i - V_G) \right)^{1/2} \right] \tag{4.5.15}$$

Our analysis included several simplifying assumptions. In practical devices some of these assumptions may not be sufficiently valid to obtain a good match between theory and experiment. One such assumption is that the depletion-layer width is controlled by the gate-channel junction and not by the channel-substrate junction. However, the potential across the channel-substrate junction does vary along the channel, with the maximum potential and depletion-layer thickness near the drain. Consequently, the channel becomes completely depleted at a lower drain voltage than indicated by Equation 4.5.10. A bias applied between the source and substrate is sometimes used to control the characteristics of a JFET. The effect of such a bias can be readily calculated by including the effect of substrate bias in the expression for the undepleted channel thickness in Equation 4.5.6.

The presence of lightly doped regions between the active channel and the heavily doped source and drain contacts can be more troublesome. Because of photomasking limitations and breakdown-voltage requirements, the n^+ contact diffusions are usually separated from the p-type gate diffusion. As we saw in the example, the series resistance of the intermediate regions can cause deviations from the ideal characteristics, especially at high currents, and must be considered when analyzing or designing practical devices.

---

* Channel-length modulation is discussed further in connection with metal-oxide-semiconductor field-effect transistors in Chapter 9 along with other short-channel effects.

We have seen that the characteristics of the JFET are highly sensitive to the thickness of the channel region t and to its dopant concentration. The n-type region in which the channel is formed can be made with excellent control by epitaxial deposition. The more crucial fabrication step is diffusion of the p-type gate. This diffusion can introduce troublesome variations in the effective channel thickness. To improve control, ion implantation is often used to introduce the gate impurities. JFETs can also be fabricated using two implantations, one for the n-type channel dopant and one for the p-type gate. This eliminates the need for epitaxial deposition in cases where it is not required for other devices in the same integrated circuit.

Figure 4.23 shows an integrated circuit in which several JFETs are used (type 355 operational amplifer). The interdigitated structures on the lower edge of the circuit are large input JFETs. To obtain low resistance (large W/L), the sources and drains are made in a comb-like pattern. The gate electrode snakes back and forth between the comb electrodes that contact the source and drain.

Metal-Semiconductor Field-Effect Transistor (MESFET)

In silicon integrated circuits, *pn* junctions can readily be formed by ion implantation or epitaxial deposition. We saw in the previous section that the depletion region associated with a *pn* junction can be used to modulate the thickness of a conducting region and control the current flowing in a channel, forming a useful transistor.

In other material systems, forming *pn* junctions is not always as easy. In these cases, a voltage-variable depletion region can be formed by using a Schottky barrier, rather than a *pn* junction, as shown in Figure 4.24. A voltage applied to the Schottky barrier controls the current in the channel of this *metal–semiconductor field-effect transistor* (MESFET).

We can consider the Schottky barrier to be the limiting case of a one-sided *pn* junction so that much of our analysis for the JFET applies to the MESFET. The primary modification in the analysis is contained in the built-in voltage ϕ_i. In Figure 3.4 we saw that the built-in voltage at an ideal metal–semiconductor interface is

$$\phi_i = \Phi_M - \Phi_S = \Phi_M - X - \frac{E_c - E_f}{q} \qquad (4.5.16)$$

In practice, surface states at the metal-semiconductor interface cause ϕ_i to depart from its ideal value, and it must often be determined empirically, as discussed in Sec. 3.5.

When a circuit contains a large number of MESFETs, the power consumption can be considerable, and dissipation of this power can limit the number of transistors that can be placed on the chip. To reduce the power dissipation, the threshold voltage can be adjusted so that no current flows when the gate voltage is zero. The channel thickness and dopant concentration for these *normally off* devices are chosen so that the channel is fully depleted when $V_G = 0$. To obtain $I_D = 0$ when $V_G = 0$, Equation 4.5.5 shows that

$$N_d t^2 = \frac{2\epsilon\phi_i}{q} \qquad (4.5.17)$$

A small forward gate bias (positive for an n-channel transistor) is then applied to turn on the channel and produce useful drain current. The positive gate bias must be limited to a small value (significantly less than ϕ_i) to prevent significant gate current from flowing. Because of the narrow operating range of gate voltage allowed for these devices, the thickness and doping in the channel must be very well controlled.

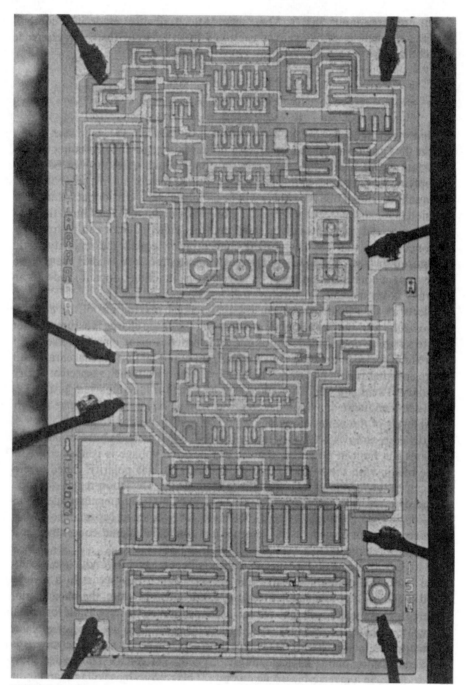

FIGURE 4.23 An integrated circuit (operational amplifier) that employs JFETs to obtain high input resistance. (*Courtesy National Semiconductor Corp.*)

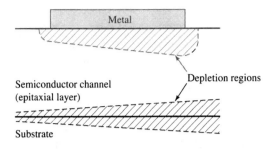

FIGURE 4.24   Schematic cross section of a metal–semiconductor field-effect transistor (MESFET), showing the depletion regions below the metal gate and at the channel–substrate interface.

MESFETs are particularly useful in GaAs integrated-circuit technology. Schottky barriers are much more readily obtained in this materials system than are *pn* junctions. The high electron mobility in GaAs makes it attractive for very-high-frequency transistors. The use of a metal as the gate electrode is also advantageous because of its low resistance, which aids high-frequency performance. Unlike silicon, GaAs can be formed with very high resistivity, so that the region under the conducting channel can have very high resistance. Consequently, the capacitance associated with the channel-substrate junction can be very low, further improving the frequency response of the transistor. In addition, the larger value of ϕ_i in GaAs than in silicon allows a wider variation of V_G in normally off MESFETs in the GaAs system.

SUMMARY

As was the case for metal-semiconductor contacts, a basic understanding of some important properties of inhomogeneously doped semiconductors can be gained by applying the principles of thermal equilibrium. If the doping in a semiconductor is nonuniform but of one type, a built-in electric field is created that balances the diffusion tendency of free carriers with an opposing drift tendency. Often, the space charge associated with this field is small and the semiconductor can be treated as quasi-neutral so that the majority-carrier concentration equals the net dopant concentration. The *quasi-neutral approximation* becomes less valid as the gradient of the dopant concentration increases. At *pn* junctions, dopant gradients are generally large and quasi-neutrality does not apply.

Effects at a *pn* junction are usually analyzed by making use of the *depletion approximation*. In the depletion approximation, the space charge is assumed to be composed of uncompensated dopant ions and its edge is assumed to be abrupt. The use of this approximation leads to predictions for field and potential that are inaccurate within a region that is roughly an extrinsic Debye length from each neutral boundary. A built-in voltage exists at a *pn* junction in equilibrium just as in a metal-semiconductor junction. In a homojunction the built-in voltage equals the difference that the Fermi levels of the *p*- and *n*-regions would have if they were isolated. In a heterojunction,

discontinuities of the band edges E_c and E_v must also be considered.

Little current flows when the voltage applied to the junction has a polarity that increases the built-in potential; e.g., a positive voltage connected to the *n*-type region with the *p*-type region grounded. This reverse bias causes the junction space-charge region to widen. Depletion-region widening can be detected readily by small-signal capacitance measurements made at different dc reverse biases. The behavior of a series of capacitance-voltage measurements can provide useful information about the dopant concentration in the junction region.

At high fields, semiconductors can suddenly become highly conductive because of the internal generation of extra free carriers. When this occurs in a reverse-biased junction the current suddenly increases. Therefore, the phenomenon is called breakdown. Either of two mechanisms can be responsible for breakdown: (1) *avalanching* or (2) *tunneling* (*Zener breakdown*). One use of reverse-biased *pn* junctions in integrated circuits is for the gate of a *junction field-effect transistor* (JFET). Operation of a JFET depends directly on the modulation of the depletion-layer width x_d in a reverse-biased *pn* junction. This reverse-bias voltage can modulate a current if that current is made to flow through a region having a cross section that depends on x_d.

REFERENCES

1. Courtesy W. G. OLDHAM.

2. R. PEOPLE, *Phys. Rev. B* **32**, 1405 (1985); R. PEOPLE, IEEE J. Quantum Electronics **QE-22**, 1696 (1986).

3. A. S. GROVE, *Physics and Technology of Semiconductor Devices*, Wiley, New York, 1967, p. 193. Reprinted by permission of the publisher.

4. A. G. CHYNOWETH, W. L. FELDMAN, C. A. LEE, R. A. LOGAN, G. L. PEARSON, and P. AIGRAIN, *Phys. Rev.* **118**, 425 (1960).

5. S. L. MILLER, *Phys. Rev.* **105**, 1246 (1957).

6. A. S. GROVE, *ibid.*, p. 197.

7. H. L. ARMSTRONG, *IRE Trans. Electron Devices* **ED-4**, 15 (1957). Reprinted by permission of the publisher.

8. J. W. SLOTBOOM, G. STREUTKER, M. J. V. DORT, P. H. WOERLEE, A. PRUIJMBOOM, and D. J. GRAVESTEIJN, Tech. Digest International Electron Devices Meeting, (Washington, DC, December 8-11, 1991), paper 5.4, pp. 127–130.

9. E. MERZBACHER, *Quantum Mechanics*, 2nd edition, Wiley, New York, 1970.

10. A. S. GROVE, *ibid.*, p. 244.

BOOKS

D. A. FRASER, *The Physics of Semiconductor Devices*, 2nd Edition, Oxford at the Clarendon Press, 1979.

G. W. NEUDECK, *The PN Junction Diode*, Vol II Modular Series on Solid-State Devices, Addison-Wesley, Reading, Mass., 1983.

PROBLEMS

4.1* An abrupt silicon *pn* junction has dopant concentrations of $N_a = 1 \times 10^{15}$ cm^{-3} and $N_d = 2 \times 10^{17}$ cm^{-3}.

(a) Evaluate the built-in potential ϕ_i at room temperature.

(b) Using the depletion approximation, calculate the width of the space-charge layer and the peak electric field for junction voltages V_a equal to 0 V and -10 V.

4.2 Consider abrupt silicon *pn* junctions that are very heavily doped on one side and have dopant concentrations of (a) 10^{15} cm^{-3}, (b) 10^{16} cm^{-3}, (c) 10^{17} cm^{-3}, and (d) 10^{18} cm^{-3} on the less heavily doped side. Find as a function of dopant concentration the length that can be depleted of mobile carriers before the maximum electric field reaches the breakdown field shown in Figure 4.14. What are the corresponding applied voltages?

4.3* Calculate the magnitude of the built-in field in the quasi-neutral region of an exponential impurity distribution:

$$N = N_0 \exp\left(-\frac{x}{\lambda}\right)$$

Let the surface dopant concentration be 10^{18} cm^{-3} and $\lambda = 0.4$ μm. Compare this field to the maximum field in the depletion region of an abrupt *pn* junction with acceptor and donor concentrations of 10^{18} cm^{-3} and 10^{15} cm^{-3}, respectively, on the two sides of the junction.

4.4 Find an expression for the potential in the example of Sec. 4.2 as a function of position within the region of variable doping. Assume that the doping is constant on either side of the transition region and make a plot of the potential and a sketch of the energy bands.

4.5 (a) Find and sketch the built-in field and potential for a silicon *pin* junction with the doping profile shown in Figure P4.5. Indicate the length of each depletion region. (The symbol *i* represents a very lightly doped or nearly intrinsic region.)

(b) Compare the maximum field to the field in a *pn* junction that contains no lightly doped intermediate region, but has the same dopant concentrations as in part (a) in the other regions.

(c) Explain physically what is happening in the intrinsic region. (That is, what does the depletion approximation mean here?)

(d) Discuss how the depletion capacitance for this structure varies with voltage, comparing it with the depletion capacitance of a structure with no intrinsic region but with the same dopant concentrations in the other regions. Sketch $1/C^2$ versus applied reverse bias for the two cases; use the same axes so that the two cases can be directly compared.

4.6 Use the *depletion approximation* to study a linearly graded junction with $(N_d - N_a) = ax$ throughout the depletion region. Assume that the equilibrium space-charge region is x_{d0} units wide. In terms of the parameters given, derive expressions for (a) the built-in potential, (b) the electric field as a function of applied voltage and distance, and (c) the depletion capacitance as a function of voltage.

4.7† We know that the capacitance of an abrupt *pn* junction varies as $V_a^{-1/2}$ for $V_a \gg \phi_i$, where ϕ_i is the built-in potential and V_a is the reverse bias applied across the junction. The capacitance of a linearly

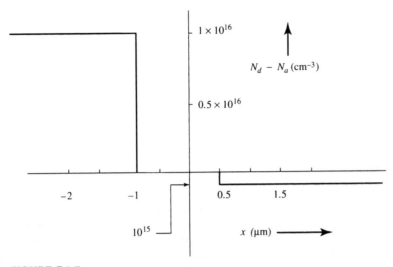

FIGURE P4.5

graded junction varies as $V_a^{-1/3}$. In a TV tuning circuit we need a capacitance that varies as V_a^{-1} for $V_a \gg \phi_i$. Qualitatively, discuss the general form of doping profile needed, indicating in each of the three cases the variation of the depletion-region width with voltage.

4.8†* Assume that the dopant distributions in a piece of silicon are as indicated in Figure P4.8.

(a) If a *pn* junction is desired at $x_0 = 1$ μm, what should be the value of the surface dopant density N_{d0}?

(b) Assume the depletion approximation and make a sketch of the space charge near the junction. Approximate this space charge as if this were a linearly graded junction. Choose an appropriate value for the doping gradient *a*.

(c) Under the approximation of part (b), take $\phi_i = 0.7$ V and use Equations 4.3.2 and 4.3.4 to calculate

$\mathscr{E}_{\max}$ at thermal equilibrium. Then, sketch the fields at thermal equilibrium throughout the regions. (Take $N_{a0} = 10^{18}$ cm^{-3}, $x_0 = 10^{-4}$ cm, $\lambda_a = 10^{-4}$ cm, and $\lambda_d = 2 \times 10^{-4}$ cm.)

4.9 The small-signal capacitance C_d of a *pn* junction diode with area 10^{-5} cm^2 is measured. A plot of $(1/C_d^2)$ *vs.* the applied voltage V_a is shown in Figure P4.9.

(a) If the diode is considered as a one-sided step junction, find the indicated doping level on the lower conductivity side (use the slope of the curve).

(b) Sketch the doping density on the low conductivity side of the junction. Calculate the location of any point at which the dopant density changes.

(c) Use the intercept on the $(1/C_d^2)$ plot to find the doping density on the highly doped side.

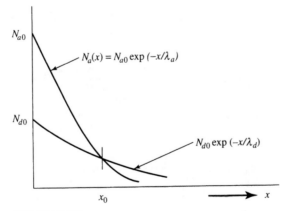

FIGURE P4.8

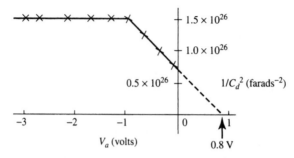

FIGURE P4.9

4.10* All parts of the following question refer to the system shown in Figure P4.10. The sketch shows the cross section of a silicon wafer fabricated by a planar process. It consists of a *p* substrate of resistivity 10 Ω-cm and an epitaxial layer 2.5 μm thick having 5×10^{15} donors cm^{-3}. At *A*, a platinum (Pt) contact is made directly to the epitaxial Si surface. At *B*, an n^+ contact is diffused 1.5 μm deep into the epitaxial material. Assume the donor density throughout the n^+ region to be 3×10^{18} cm^{-3}. Ignore all edge effects in the following analysis.

(a) What is the energy interval between the Fermi level and the mid-gap energy (E_i): (i) In the epitaxial *n*-region? (ii) In the substrate *p*-region?

(b) The barrier height (energy difference between the conduction-band edge and the Fermi level) at the Pt-Si junction is measured to be 0.85 eV.

(i) What is the built-in voltage for the Pt-Si junction?

(ii) Is the 0.85 eV barrier consistent with idealized Schottky theory? Comment intelligently.

(c) Show whether or not it would be possible to deplete fully the n^- layer below the Pt contact without reaching a breakdown field of 3×10^5 V cm^{-1}. (Consider that contact *B* is shorted to contact *C* and that both are held at ground while voltage is applied to contact *A*.) What voltage is required to accomplish this depletion?

(d) Sketch the equilibrium energy-band diagram along axis 1 − 1 (through the Pt-Si junction and into the substrate). Make the diagram qualitatively correct; show the vacuum level and assume that ideal Schottky theory *does* apply.

4.11 To treat avalanche from first principles, consider that an incident electron collides with the lattice and frees a hole-electron pair. Assume that after the interaction, the three particles each have equal kinetic energies. Also assume that all three have equal masses. Use conservation of energy and momentum principles to find that the threshold for avalanche occurs when the incident electron possesses $(3/2) E_g$ units of kinetic energy. (Despite the many approximations made in this problem, this energy is a useful first-order measure of what is found in practice.)

4.12 Is Zener breakdown more likely to occur in a reverse-biased silicon or germanium *pn*-junction diode if the peak electric field is the same in both diodes? Discuss. (Consider the size of the bandgap of each material.)

4.13† Carry through the analysis following Equation 4.4.20 and show that the tunneling distance and field derived are the proper values. The coefficient *B* given in Equation 4.4.19 is 7.87×10^7 V cm^{-1} for a bandgap of 1.1 eV if *m** is taken to be the rest mass of the

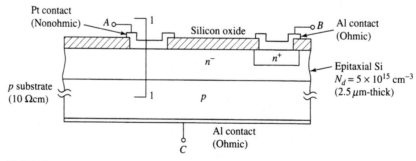

FIGURE P4.10

electron. Use the conductivity effective mass in the calculations.

4.14 For a JFET derive an expression for the major temperature variation of the conductance in the linear region at a fixed gate voltage. Assume that the mobility varies as $T^{-3/2}$.

4.15[†] For a JFET derive an expression for the drain conductance $g = \partial I_D/\partial V_D$ at a given gate voltage in the saturation region. Assume that this conductance results from the widening of the depletion region near the drain and approximate the latter by a one-dimensional step junction with very heavy doping in the drain region.

4.16[†] Figure P4.16 shows a JFET made in an annular geometry. The junctions shown can be approximated as being linearly graded with a doping gradient

$dN/dx = a$. Series resistance at the source and drain is negligible.

(a) What is the value of the gate turn-off voltage V_T in terms of the properties of the device?

(b) Write a differential equation that can be integrated to find the dependence of I_D on V_D, V_G, and device properties. Do not solve this differential equation, but place it in a form that involves definite integrals.

(c) How does the transconductance $(\partial I/\partial V_G)$ for this structure depend upon the radii r_1 and r_2? Specifically, if a device is made with $r_1 = 10$ μm and $r_2 = 40$ μm and it has a g_m value of 10 mS, what value of g_m is expected for a device made with all parameters the same except that r_2 is made 60 μm in length?

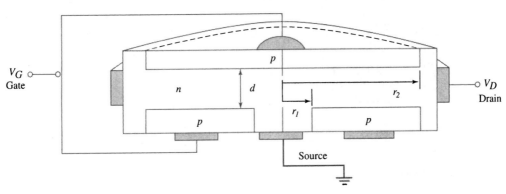

FIGURE P4.16

CURRENTS IN pn JUNCTIONS

Thus far, we have considered reverse-biased *pn* junctions, across which little current flows until the applied voltage exceeds the junction–breakdown voltage. Therefore, we have not considered current flow at low biases. Instead, we focused attention on the barrier to majority–carrier transfer at the *pn* junction and on changes in the depletion-region width as the applied voltage varied.

The major topic in this chapter is current flow across a *pn* junction under both forward and reverse biases. A detailed understanding of this topic is important not only to gain insight into junction-diode behavior but also to grasp the basis for junction-transistor operation. As a first step toward analyzing current flow, we derive a continuity equation for free carriers, an equation that considers the various mechanisms affecting the population of carriers in an infinitesimal volume inside a semiconductor. To formulate the significant terms in this equation, which account for the generation and recombination process, it will be necessary to consider several basic physical processes in more detail than was done in Chapter 1. After deriving a continuity equation that treats generation and recombination, we will be able to characterize the minority-carrier distributions in the quasi-neutral regions of a *pn* junction under bias. We will consider in detail the solutions for two especially simple cases of *pn* junctions and carry out what is known as the *ideal-diode analysis*. Then, to relate this result to real silicon diodes, we will discuss generation and recombination in the space-charge region. The physical model developed for the steady-state current-voltage relationship will then help us to consider charge storage and diode transients. Finally, we assess the role of *pn* junctions in integrated circuits and give a practical perspective to the theory in a concluding section.

5.1 CONTINUITY EQUATION

To discuss current flow in a *pn* junction, we write an equation that basically accounts for the flux of free carriers into and out of an infinitesimal volume in space. Such a *continuity equation* can be written both for majority carriers and for minority carriers in semiconductors. Solutions of the minority-carrier continuity equation in semiconductors have special importance in many practical device applications.

To derive a one-dimensional continuity equation for electrons, we consider an infinitesimal slice of thickness dx located at x (Figure 5.1). The number of electrons in the slice can increase because of net flow into the volume and from net carrier generation in the slice. The overall rate of electron increase equals the algebraic sum of

(1) the number of electrons flowing into the slice, minus

(2) the number flowing out, plus

(3) the rate at which electrons are generated, minus

(4) the rate at which they recombine.

The first two components are found by dividing the currents at each side of the slice by the charge on an electron; for the present we denote the last two by G and R, respectively. The rate of change in the number of electrons in the slice is then

$$\frac{\partial n}{\partial t} A \, dx = \left(\frac{J_n(x)}{-q} - \frac{J_n(x + dx)}{-q} \right) A + (G_n - R_n) A \, dx \qquad (5.1.1)$$

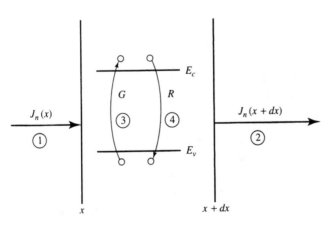

FIGURE 5.1 The increase in the electron density in an infinitesimal slice of thickness dx is related to the net flow of electrons into the slice and the excess of generation over recombination.

where A is the cross-sectional area of the slice and G_n and R_n represent the generation and recombination rates per unit volume for electrons. Expanding the second term on the right-hand side in a Taylor series,

$$J_n(x + dx) = J_n(x) + \frac{\partial J_n}{\partial x} dx + \cdots \tag{5.1.2}$$

we derive the basic continuity equation for electrons

$$\frac{\partial n}{\partial t} = \frac{1}{q} \frac{\partial J_n}{\partial x} + (G_n - R_n) \tag{5.1.3a}$$

A similar continuity equation applies to holes except that the sign of the first term on the right-hand side of Equation 5.1.3a is changed because of the charge associated with a hole

$$\frac{\partial p}{\partial t} = -\frac{1}{q} \frac{\partial J_p}{\partial x} + (G_p - R_p) \tag{5.1.3b}$$

To obtain equations that can be solved, we must relate the quantities on the right-hand side of Equations 5.1.3 to the carrier densities n and p. This is straightforward for the current terms because J_n and J_p have been written in terms of carrier densities in Equations 1.2.21 and 1.2.22. When these equations are inserted, we obtain

$$\frac{\partial n}{\partial t} = \mu_n n(x) \frac{\partial \mathscr{E}(x)}{\partial x} + \mu_n \mathscr{E}(x) \frac{\partial n(x)}{\partial x} + D_n \frac{\partial^2 n(x)}{\partial x^2} + (G_n - R_n) \tag{5.1.4a}$$

and

$$\frac{\partial p}{\partial t} = -\mu_p p(x) \frac{\partial \mathscr{E}(x)}{\partial x} - \mu_p \mathscr{E}(x) \frac{\partial p(x)}{\partial x} + D_p \frac{\partial^2 p(x)}{\partial x^2} + (G_p - R_p) \tag{5.1.4b}$$

Note that we assume that mobility μ and diffusion coefficients D are not functions of x. Although this assumption is not valid in a number of important cases, the major physical effects are included in Equations 5.1.4, and more exact formulations can be considered by numerical simulation.

If the electric field is zero or negligible in the region under consideration, the first two terms on the right-hand side of Equations 5.1.4 can be neglected, and the analysis is greatly simplified. Even if the field is not negligible, some of the terms in Equations 5.1.4 may be unimportant. For example, if the field is constant, the first term in each equation drops out. As we saw in Section 4.1, a constant field is present in an exponentially graded semiconductor. It is rarely necessary to deal with the full complexity of Equations 5.1.4.

It is worthwhile to recall some basic facts of calculus before we seek specific solutions to the continuity equations. The continuity equations (Equations 5.1.4) are partial differential equations because they are functions of time and position. Thus, they have an infinity of solutions, one of which applies to a given problem because it matches boundary and initial conditions. The continuity equations simplify to ordinary differential equations if, for example, we are interested in steady-state solutions. In that case, the time dependence on the left-hand side of the equations disappears, and only position-dependent derivatives remain.

5.2   GENERATION AND RECOMBINATION

To formulate correct expressions for generation and recombination in terms of free-carrier densities, we need to develop the physics of semiconductors beyond the discussion of Chapter 1. In Chapter 1, we noted that an electron can be excited by thermal energy from the

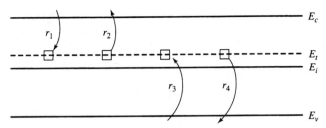

FIGURE 5.2 Free carriers can interact with localized states by four processes: r_1 electron capture, r_2 electron emission, r_3 hole capture, and r_4 hole emission. The localized state shown is acceptor type and at energy E_t within the forbidden-energy gap.

valence band to the conduction band, leaving behind a hole in the valence band. Both the hole and the electron can contribute to conduction. In thermal equilibrium, this generation rate equals the rate of the inverse process—the direct transfer of an electron into a valence-band site. These processes describe one means for free-carrier generation and recombination; there are, however, other processes by which generation-recombination can take place.

Direct transitions between the valence band and the conduction band occur in all semiconductors. Direct transitions are the most important generation-recombination mechanisms in many of the compound semiconductors, such as gallium arsenide (GaAs) and indium phosphide (InP), composed of elements from columns III and V of the periodic table. However, in silicon and germanium the details of the crystal structure make direct transitions unlikely except when very high densities of holes and electrons are present. In silicon and germanium, electrons at the lowest energy in the conduction band have nonzero "momentum." Because the holes at the valence-band maximum do have zero momentum, a direct transition that conserves both energy and momentum is impossible without a lattice (phonon) interaction occurring simultaneously. Thus, in silicon or germanium, direct transitions across the forbidden-energy gap require simultaneous interaction of three particles: the electron, the hole, and a phonon that represents the lattice interaction (refer to Section 1.2).

Three-particle interactions are far less likely than two-particle interactions, such as those between a free carrier and a phonon, that can take place if there are localized allowed energy states into which electrons or holes can make transitions. In practice, localized states at energies between E_v and E_c are always present because of lattice imperfections caused by misplaced atoms in the crystal or, more usually, because of impurity atoms. Furthermore, they are always present in sufficient numbers to dominate the generation-recombination process in silicon and germanium. These localized states act as stepping stones. In a recombination event, for example, an electron falls from the conduction band to a state that we logically call a *recombination center*,* and then it falls further into a vacant state in the valence band, thus recombining with a hole.

Localized States: Capture and Emission

The four processes through which free carriers can interact with localized states are indicated in Figure 5.2. The illustration shows a density N_t of states at an energy E_t within the forbidden gap. The states shown are acceptor type—that is, neutral when empty and negative when full—but the processes described apply also to donor-type states.

---

* Because the states behave symmetrically as interim sites either for generation or recombination of free carriers, they are properly called generation-recombination centers. For brevity, however, this is usually shortened to "recombination centers."

In the first process, *electron capture*, an electron falls from the conduction band into an empty localized state. The rate at which this process occurs is proportional to the density of electrons n in the conduction band, the density of empty localized states, and the probability that an electron passes near a state and is captured by the state. The density of empty localized states is given by their total density N_t times one minus the probability $f(E_t)$ that they are occupied. When thermal equilibrium applies, f is just f_D, the Fermi function as given by Equation 1.1.18. In the nonequilibrium case, f differs from f_D, but we need not specify it further for this discussion.

The probability per unit time that an electron is captured by a localized state is given by the product of the electron thermal velocity v_{th} and a parameter σ_n called the *capture cross section*. The capture cross section describes the effectiveness of the localized state in capturing an electron. This product $v_{th}\sigma_n$ can be visualized as the volume swept out per unit time by a particle with cross section σ_n. If the localized state lies within this volume, the electron is captured by it. The capture cross section is generally determined experimentally for a given type of localized state. A typical size for an effective recombination center is about 10^{-15} cm^2 for gold or iron [1]. An abnormally large cross section, 10^{-10} cm^2, is associated with beryllium [2]. Combining the factors discussed above, we write the total rate of capture of electrons by the localized states as

$$r_1 = n\{N_t[1 - f(E_t)]\}v_{th}\sigma_n \tag{5.2.1}$$

The second process is the inverse of electron capture: that is, *electron emission*. The emission of an electron from the localized state into the conduction band occurs at a rate given by the product of the density of states occupied by electrons $N_t f(E_t)$ times the probability e_n that the electron makes this jump.

$$r_2 = [N_t f(E_t)]e_n \tag{5.2.2}$$

The emission probability can be expressed in terms of the quantities already defined in Equation 5.2.1 by considering the capture and emission rates in the limiting case of thermal equilibrium. At thermal equilibrium, the rates of capture and emission of carriers must be equal, and the probability function $f(E)$ is given by the Fermi function $f_D(E)$ (Equation 1.1.18). Thus, we can write

$$r_1 = r_2 = nN_t[1 - f_D(E_t)]v_{th}\sigma_n = N_t f_D(E_t)e_n \tag{5.2.3}$$

and

$$e_n = v_{th}\sigma_n n_i \exp\left(\frac{E_t - E_i}{kT}\right) \tag{5.2.4}$$

The right-hand side of Equation 5.2.4 can be used to replace e_n in the general case described by Equation 5.2.2. From Equation 5.2.4, we see that electron emission from the localized state becomes more probable when its energy is closer to the conduction band because $E_t - E_i$ is then greater.

Corresponding relationships describe the interactions between the localized states and the valence band. For example, the third process, *hole capture*, is proportional to the density of localized states occupied by electrons $N_t f(E_t)$, the density of holes and a transition probability. This probability can then be described by the product of the hole thermal velocity v_{th} and the capture cross section σ_p of a hole by the localized state. Thus,

$$r_3 = [N_t f(E_t)]pv_{th}\sigma_p \tag{5.2.5}$$

The fourth process, *hole emission*, describes the excitation of an electron from the valence band into the empty localized state. By arguments similar to those for electron emission, hole emission is given by

$$r_4 = \{N_t[1 - f(E_t)]\}e_p \tag{5.2.6}$$

The emission probability e_p of a hole can be written in terms of σ_p by considering the thermal-equilibrium case, for which $r_3 = r_4$, and is given by

$$e_p = v_{th}\sigma_p n_i \exp\left(\frac{E_i - E_t}{kT}\right) \tag{5.2.7}$$

Analogously to Equation 5.2.4, the probability of emission of a hole from the localized state to the valence band becomes much greater as the energy of the state approaches the valence-band edge.

Before making use of Equations 5.2.1, 5.2.2, 5.2.5, and 5.2.6 for the kinetics of interactions between the valence and conduction bands through the localized states at E_t, it is worthwhile to consider qualitatively the physics that they represent. First, we recognize that at thermal equilibrium $r_1 = r_2$ and $r_3 = r_4$ because thermal equilibrium requires every process to be balanced by its inverse. When we have a nonequilibrium situation, $r_1 \neq r_2$ and $r_3 \neq r_4$. To gain insight about these rates, imagine specifically that the number of holes in an n-type semiconductor is suddenly increased above its thermal-equilibrium value. This causes r_3 to increase. The effect of this rate increase is to increase r_4 and r_1 (both of which eliminate holes at E_t). If most of the holes disappear from E_t via r_1, they remove electrons from the conduction band, and the localized states are effective recombination centers. If the holes are removed from the level at E_t predominantly by an increase in r_4, they return to the valence band, and the sites are effective as *hole traps*. A given localized state is generally effective in only one way: either as a *trap* or as a *recombination center*. If it is near the middle of the band gap, it is likely to be an effective recombination center; if it is closer to a band edge, it is likely to be a carrier trap rather than a recombination center. In the following section, we consider the action of recombination centers in more detail.

Shockley-Hall-Read Recombination[†]

The equations describing generation and recombination through localized states or recombination centers were originally derived by Shockley and Read [3] and by Hall [4], and the process is frequently called the Shockley-Hall-Read (or SHR) recombination. According to the SHR model, when nonequilibrium occurs in a semiconductor, the overall population of electrons and holes in the recombination centers is not greatly affected. The reason for this nearly constant population is that the recombination centers quickly capture majority carriers (there are so many of them around) but have to wait for the arrival of a minority carrier. Thus, the states are nearly always full of majority carriers, whether under thermal-equilibrium conditions or in nonequilibrium.

To illustrate this behavior, consider a typical example: acceptor-like recombination centers in an n-type semiconductor. At thermal equilibrium, the Fermi level is near E_c and, therefore, above the energy of the recombination centers. Hence, they are virtually all filled with electrons and r_1 and r_2 are both much greater than r_3 or r_4. When equilibrium is disturbed by low-level excitation, which increases the number of holes and electrons by the same amount, the electron concentration changes only by a small fraction, while the hole concentration changes by a large fraction. In this case, r_1 has to exceed r_2 by only a very slight amount to accommodate the increased rate of hole capture represented by r_3. Thus, the population of the localized states remains nearly constant, and the net rate of electron capture $r_1 - r_2$ equals the net rate of hole capture by the states. These net rates are, in turn, just the net rate of recombination that we define by the symbol U.

$$U \equiv R_{sp} - G_{sp} = r_1 - r_2 = r_3 - r_4 \qquad (5.2.8)$$

where the subscript *sp* stands for *spontaneous*, that is, recombination and generation that respond only to deviation from thermal equilibrium.* Inserting the expressions for r_1 through r_4 into Equation 5.2.8, we can eliminate f and solve for U to obtain

$$U = \frac{N_t v_{th} \sigma_n \sigma_p (pn - n_i^2)}{\sigma_p \left[p + n_i \exp\left(\dfrac{E_i - E_t}{kT}\right) \right] + \sigma_n \left[n + n_i \exp\left(\dfrac{E_t - E_i}{kT}\right) \right]} \qquad (5.2.9a)$$

$$= \frac{(pn - n_i^2)}{\tau_{no} \left[p + n_i \exp\left(\dfrac{E_i - E_t}{kT}\right) \right] + \tau_{po} \left[n + n_i \exp\left(\dfrac{E_t - E_i}{kT}\right) \right]} \qquad (5.2.9b)$$

where $\tau_{no} = (N_t v_{th} \sigma_n)^{-1}$ and $\tau_{po} = (N_t v_{th} \sigma_p)^{-1}$.

Equation 5.2.9 shows that U is positive, and there is net recombination if the *pn* product exceeds n_i^2. The sign changes and there is net generation if the *pn* product is less than n_i^2. The term $(pn - n_i^2)$ represents the restoring "force" for free-carrier populations in a nonequilibrium condition.

The dependence of U on the energy level of the recombination centers in Equation 5.2.9 can be more easily grasped by considering the case of equal electron and hole capture cross sections. For $\sigma_p = \sigma_n \equiv \sigma_0$, we can define $\tau_0 \equiv (N_t v_{th} \sigma_o)^{-1}$ and, therefore,

$$U = \frac{(pn - n_i^2)}{\left[p + n + 2n_i \cosh\left(\dfrac{E_t - E_i}{kT}\right) \right] \tau_0} \qquad (5.2.10)$$

The dependence on the energy level of the recombination center is contained in the hyperbolic cosine term in Equation 5.2.10. This term is symmetric around $E_t = E_i$, reflecting a symmetry in the capture of holes and electrons by the center. The denominator has its minimum value at $E_t = E_i$, so that U is maximum for recombination centers having energies near the middle of the gap. In Figure 5.3 (solid curve), U normalized to its maximum value has been plotted versus $(E_t - E_i)/kT$ from Equation 5.2.10 for a reasonable case of recombination in an *n*-type semiconductor. The conditions used for the figure are: $p < n$, $n = 10^{16}$ cm^{-3}, $(pn - n_i^2) = 1.5 \times 10^{31}$ cm^{-6}, and $\tau_0 = 10^{-7}$ s. We will find shortly that these values are appropriate in the quasi-neutral region near a forward-biased *pn* junction. As a second illustration of Equation 5.2.10, Figure 5.3 (dashed curve) shows the dependence of U (normalized to its maximum value) on $(E_t - E_i)/kT$ for a case involving generation. Again, reasonable values have been used for the other terms in Equation 5.2.10: $(pn - n_i^2) = -2.1 \times 10^{20}$ cm^{-6}, p and n being much less than n_i, and $\tau_0 = 10^{-7}$ s. These conditions might fit the region near the center of a *pn* junction depletion region under reverse bias. The results sketched in Figure 5.3 show that the dependence of the rate U on recombination-center energy is more pronounced for the case of generation in a depleted semiconductor than for recombination in an undepleted region. This behavior occurs because the energy level plays a greater role in equalizing the rates of transfer between the recombination center and the conduction and valence bands when all carrier

---

* In contrast to the spontaneous recombination and generation transitions are those caused by stimulation, for example, by a radiative source.

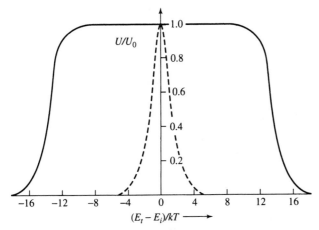

FIGURE 5.3 Recombination rate (solid curve) and generation rate (dashed curve) as functions of the difference between the energy of the recombination center E_t and the intrinsic Fermi energy E_i. The curves are normalized to the rates for $E_t = E_i$, and have been drawn using Equation 5.2.10 with values listed in the text.

densities are small. However, in either generation or recombination, the most effective recombination centers have E_t close to E_i. As practical examples, gold and copper give rise to two effective recombination centers. The values of $(E_t - E_i)$ in silicon for these elements are 0.03 and 0.01 eV, respectively [5]. Energy levels for a number of elements in silicon are given in Table 1.4.

Equations 5.2.9 and 5.2.10 are the major results of the SHR recombination analysis. They show that U, the net recombination rate through a recombination center, is a function of the free-carrier densities as well as the specific properties of the recombination site. The equations usually can be simplified for a particular problem. The predictions of material and device behavior based on these equations have been generally corroborated by experiments.

Excess-Carrier Lifetime

To understand the physical significance of the net recombination rate U, we consider a semiconductor with no current flow, in which thermal equilibrium is disturbed by the sudden creation of equal numbers of excess electrons and holes. These excess carriers then decay spontaneously as the semiconductor returns to thermal equilibrium. Solutions of the continuity equations (Equations 5.1.3) for this case give the excess electron density as a function of time. We consider this problem under an assumption that is frequently satisfied in practice: that the disturbance of equilibrium corresponds to *low-level injection*. In this condition the external disturbance does not appreciably change the total free-carrier density from its equilibrium value. If we call the extra injected electron density n' and the extra hole density p', then low-level injection implies that n' and p' are both much less than $(n_o + p_o)$ where n_o and p_o represent the thermal-equilibrium densities of carriers in the semiconductor. From these definitions, $n' \equiv n - n_o$ and $p' \equiv p - p_o$, where $n' = p'$.*

---

* Because electron and hole densities increase or decrease at the same rate, the excess densii :s n' and p' are equal to one another.

If $\sigma_n = \sigma_p$, then the recombination rate U is given by Equation 5.2.10 and the continuity equation 5.1.3a can be written

$$\frac{dn'}{dt} = G - R = -U = \frac{-(n_o + p_o)n'}{\left(n_o + p_o + 2n_i \cosh\left[\dfrac{E_t - E_i}{kT}\right]\right)\tau_0} \qquad (5.2.11)$$

Solving for n', we find that the excess carrier density decays exponentially with time:

$$n'(t) = n'(0)\exp\left(-t/\tau_n\right) \qquad (5.2.12)$$

where the *lifetime* τ_n is given by

$$\tau_n = \left[\frac{n_o + p_o + 2n_i \cosh\left(\dfrac{E_t - E_i}{kT}\right)}{(n_o + p_o)}\right]\tau_0 \qquad (5.2.13)$$

As we noted in the previous section, for recombination centers to be effective, the term $(E_t - E_i)$ is relatively small, and, therefore, the third term in the numerator of Equation 5.2.13 is negligible compared to the sum of the first two terms. Equation 5.2.13 then reduces to

$$\tau_n = \tau_0 = \frac{1}{N_t v_{th}\sigma_o} \qquad (5.2.14)$$

and

$$U = \frac{n'}{\tau_n} \qquad (5.2.15)$$

Equation 5.2.14 shows that the excess-carrier lifetime is independent of the majority-carrier concentration for recombination through recombination centers under low-level injection. We can understand this behavior physically by considering the kinetics of the recombination process. For example, in a *p*-type semiconductor, most of the recombination centers are empty of electrons because $E_f < E_t$ (assuming traps near midgap). The recombination process is therefore limited by the capture of electrons from the conduction band. Once an electron is captured by a recombination center, one of the many holes in the valence band is quickly captured. Thus, the rate-limiting step in the recombination process is the capture of a minority carrier by the recombination center; this is insensitive to the majority-carrier population.

Minority-carrier lifetimes can vary widely, depending on the density and type of recombination centers in the semiconductor. For devices such as radiation detectors that require long lifetimes for minority carriers, special care can yield silicon having millisecond lifetimes or greater. For integrated circuits, typical values range from a fraction of a microsecond to hundreds of microseconds.

Auger Recombination.[†]   In the previous section, we considered Shockley-Hall-Read recombination, in which excess carriers recombine by interacting through intermediate centers. An excess carrier is trapped at this center until a carrier of the opposite conductivity type interacts with the center and the two carriers recombine. The SHR recombination centers dominate at low-to-moderate carrier concentrations because carriers are more likely to interact with these centers than with the low concentrations of mobile carriers.

However, at high carrier concentrations, direct interaction of electrons and holes can lead to *Auger* recombination.* In Auger recombination, an electron in the conduction band falls into an empty state (hole) in the valence band. The energy emitted by this transition is absorbed by another carrier, which also helps conserve momentum. Auger recombination is the inverse of the avalanche pair–production process discussed in Sec. 4.4, in which the energy and momentum of an incoming carrier create a hole and an electron.

Auger recombination requires three carriers: the recombining electron and hole, and the carrier to which energy is transferred. For Auger recombination to occur, the probability of three carriers interacting must be reasonably high; therefore, it is only likely in highly doped material or when a very large number of excess carriers is present. In *n*-type material, two electrons and one hole interact, while in *p*-type material, two holes and one electron interact. Because two majority carriers are involved, the recombination rate U is proportional to the square of the majority carrier concentration. The Auger recombination rate U_A is given by the expression

$$U_A = R_A - G_A = c_n n(pn - n_i^2) + c_p p(pn - n_i^2) \qquad (5.2.16)$$

where c_n and c_p are the Auger recombination coefficients.

The reciprocal of the Auger lifetime can be written as

$$\frac{1}{\tau_A} = c_n N_a^2 \qquad (5.2.17)$$

for electron recombination in heavily doped *p*-type material. The coefficient c_n for electrons in silicon is approximately 1×10^{-31} cm^{-6} s^{-1}. The Auger lifetime for holes in *n*-type material is about one–half to one–third that for electrons in *p*-type material. The effective lifetime τ, which considers both SHR and Auger recombination, is the sum of the recombination rates of the two processes so that

$$\frac{1}{\tau} = \frac{1}{\tau_{SHR}} + \frac{1}{\tau_A} \qquad (5.2.18)$$

Because *n*-type regions are generally more heavily doped (e.g., the emitter of a bipolar transistor or the source- and drain-regions of an *n*-channel MOSFET) than *p*-type regions, Auger recombination is usually more important in *n*-type material.

Surface Recombination.[†]   Thus far, we have considered generation-recombination centers that are uniformly distributed throughout the bulk of the semiconductor material. As discussed in Section 3.5, a semiconductor surface can be the location of an abundance of extra localized states having energies within the forbidden gap. The presence of a passivating layer of silicon dioxide over the semiconductor surface, as is usual in devices made by the planar process, ties up many of the bonds that would otherwise contribute to surface states and protects the surface from foreign atoms. A passivating oxide can reduce the density of surface states from about 10^{15} cm^{-2} to less than 10^{11} cm^{-2}. Even with passivated surfaces, however, surface states provide generation-recombination centers in addition to those present in the bulk. Because the properties of many practical semiconductor devices are affected by generation and recombination at the surface, a brief discussion is appropriate.

The kinetics of generation-recombination at the surface are similar to those considered for bulk centers with one significant exception. While we considered the volume

* Named for physicist P. Auger and pronounced "oh-zhay."

density $N_t(\text{cm}^{-3})$ of bulk centers, we must discuss the area density $N_{st}(\text{cm}^{-2})$ of surface centers.* Although the N_{st} surface centers can be distributed over a thickness of several atomic layers, the poorly defined microscopic structure near the semiconductor surface makes useful a description in terms of an equivalent number of states located at the surface. We can write an expression for the recombination rate U per unit area at the surface analogous to Equation 5.2.9:

$$U_s = \frac{N_{st} v_{th} \sigma_n \sigma_p (p_s n_s - n_i^2)}{\sigma_p \left[p_s + n_i \exp\left(\dfrac{E_i - E_{st}}{kT} \right) \right] + \sigma_n \left[n_s + n_i \exp\left(\dfrac{E_{st} - E_i}{kT} \right) \right]} \qquad (5.2.19)$$

where the subscript s denotes concentrations and conditions near the surface and E_{st} is the energy of the surface generation-recombination centers. To stress the physical significance of surface recombination, we again simplify the mathematics by considering the most efficient centers, which are located near midgap, and equal capture cross sections for electrons and holes. With these assumptions, Equation 5.2.19 reduces to

$$U_s = N_{st} v_{th} \sigma \, \frac{(p_s n_s - n_i^2)}{p_s + n_s + 2n_i \cosh\left(\dfrac{E_{st} - E_i}{kT} \right)} \qquad (5.2.20)$$

We saw in Chapter 3 that the surface of a semiconductor is often at a different potential than the bulk so that surface carrier concentrations can differ from their values in the neutral bulk region. Even in the case of oxide-passivated surfaces, a space-charge region generally forms near the surface of the semiconductor, as shown in Figure 5.4 for p-type silicon. As described in Chapter 2, dopant segregation at the oxide-silicon interface causes silicon surfaces to be less strongly p-type or more strongly n-type than the bulk. If we assume that the pn product remains constant throughout the space-charge region, the product at the surface $p_s n_s$ can be expressed in terms of quantities at the neutral edge of the space-charge region:

$$p_s n_s = p_p(x_d) n_p(x_d) \approx N_a n_p(x_d) \qquad (5.2.21)$$

in a p-type semiconductor.** Equation 5.2.20 can then be written

$$U_s = N_{st} v_{th} \sigma \, \frac{N_a [n_p(x_d) - n_{po}]}{(p_s + n_s + 2n_i)} = N_{st} v_{th} \sigma \, \frac{N_a}{(p_s + n_s + 2n_i)} \, n_p'(x_d) \qquad (5.2.22)$$

where we have assumed $E_{st} \approx E_i$. In Equation 5.2.22 we have expressed the surface recombination rate U_s in terms of the deviation n_p' of the minority-carrier (electron) concentration from its equilibrium value at the interior boundary of the surface space-charge region.

The coefficient of n_p' on the right side of Equation 5.2.22 is usually defined as a parameter s, which describes the characteristics of the surface recombination process:

$$s = N_{st} v_{th} \sigma \, \frac{N_a}{(p_s + n_s + 2n_i)} \qquad (5.2.23)$$

The value of s depends on the physical nature and density of the surface generation-recombination centers as well as on the potential at the surface. If the surface region is

---

* The density N_{st} represents interface trapping states that are active as generation-recombination sites. These states are discussed further in Section 8.5.

** The subscripts p and n are used generally to denote carrier concentrations in p- and n-type material, respectively.

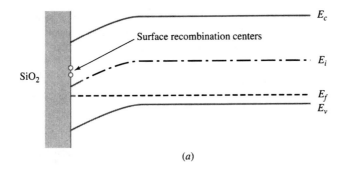

(a)

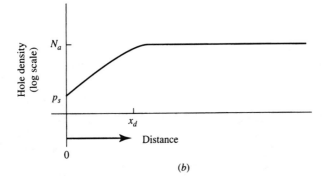

(b)

FIGURE 5.4 (a) Sketch of the energy-band diagram near the surface of p-type silicon covered with a passivating oxide. (b) Hole density near the surface.

depleted of mobile carriers, n_s and p_s are small and s is large. If the surface is neutral, $p_s \sim N_a$; s is small and is given by

$$s = s_o = N_{st} v_{th} \sigma \qquad (5.2.24)$$

where the subscript o denotes that the surface and bulk are at the same potential; that is, the surface region is neutral. The dependence of s on surface potential is important in silicon integrated circuits.

The dimensions of s are cm s^{-1}, and s is consequently called the *surface recombination velocity*, although it is not directly related to an actual velocity. A physical interpretation of s can be obtained by comparing Equation 5.2.24 to Equation 5.2.14 for the minority-carrier lifetime; s is related to the rate at which excess carriers recombine at the surface, just as $1/\tau$ is related to the rate at which they recombine in the bulk.

EXAMPLE *Surface Recombination Velocity*

An *n*-type silicon wafer with resistivity $\rho_n = 0.025$ Ω-cm at room temperature is illuminated through a passivating layer of SiO$_2$ so that the hole and electron densities exceed their thermal-equilibrium values. Assume that the surface electron density $n_s = 10^{16}$ cm^{-3} at thermal equilibrium, the photo-generation rate of carriers is 10^{14} cm^{-2} s^{-1}, and the surface hole density under illumination increases to $p_s = 10^{10}$ cm^{-3}.

(a) Find a value for the surface recombination velocity s such that 50% of the excess carriers recombine at the surface.

(b) If the surface recombination is due to the presence of a density of surface recombination centers $N_{st} = 10^{11}$ cm^{-2}, determine the interaction cross section σ giving rise to the recombination rate described in part (a).

Solution We first find $N_d = 10^{18}$ cm^{-3} for $\rho_n = 0.025$ Ω-cm using either Figure 1.15 or Table 4.1. Because $n_s = 10^{16}$ cm^{-3}, the surface of the wafer is slightly depleted of majority carriers even under illumination. Using Equation 1.1.13, we find the thermal-equilibrium density of holes at the surface

$$p_s = n_i^2/n_s = 2.1 \times 10^4 \text{ cm}^{-3}$$

Therefore, the excess density of holes at the surface

$$p_s' = 10^{10} - 2.1 \times 10^4 \approx 10^{10} \text{ cm}^{-3}$$

The rate of recombination at the surface is just half the generation rate so that

$$U_s = sp_s' = s \times 10^{10} = 0.5 \times 10^{14} \text{ cm}^{-2}\text{ s}^{-1}$$

Therefore, $s = 5000$ cm s^{-1}, which is the answer required in part (a). For part (b), we use a form of Equation 5.2.23 appropriate for *n*-type silicon together with Equation 5.2.24 to calculate

$$s_o = s\frac{(p_s + n_s + 2n_i)}{N_d} \approx 5000\frac{10^{16}}{10^{18}} = 50 \text{ cm s}^{-1}$$

Taking $N_{st} = 10^{11}$ cm^{-2} and $v_{th} \approx 10^7$ cm s^{-1} (Section 1.2), we find $\sigma = 5 \times 10^{-17}$ cm^2. ∎

5.3 CURRENT-VOLTAGE CHARACTERISTICS OF *pn* JUNCTIONS

With the continuity equations (Equations 5.1.3) and the concept of excess-carrier lifetime as derived from the SHR generation-recombination model, we are in a position to find expressions for current in a *pn* junction under bias. Solutions of the continuity equations in the quasi-neutral regions give carrier densities in terms of position and time. Expressions for current are then obtained directly by using Equations 1.2.21 and 1.2.22, which define carrier flow in terms of carrier densities. The total current generally consists of the sum of four components: hole and electron drift currents and hole and electron diffusion currents.

We consider a *pn*-junction diode connected to a voltage source with the *n*-region grounded and the *p*-region at V_a volts relative to ground. The diode structure has constant cross-sectional area A and the longitudinal dimensions sketched in Figure 5.5.

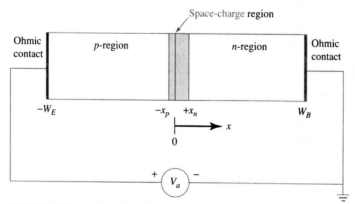

FIGURE 5.5 *pn*-junction diode structure used in the discussion of currents. The sketch shows the dimensions and the bias convention. The cross-sectional area A is assumed to be uniform.

The junction is not illuminated, and carrier densities within the diode are only influenced by the applied voltage. The applied voltage V_a is dropped partially across the quasi-neutral regions and partially across the junction itself. Because the voltage drops across the quasi-neutral regions are ohmic (current times resistance), they are typically small in IC devices at low and moderate currents. However, for small device cross sections, such as those used in some ULSI circuits, ohmic voltage drops can limit device performance.

For our analysis we neglect ohmic drops and assume that V_a appears entirely across the junction. Then, the total junction voltage is $\phi_i - V_a$ under bias, where ϕ_i is the built-in voltage. If V_a is positive (*forward bias*), the applied voltage reduces the barrier to the diffusion flow of majority carriers at the junction. The reduced barrier, in turn, permits a net transfer of holes from the *p*-side into the *n*-side and of electrons from the *n*-side into the *p*-side. When these transferred carriers enter the quasi-neutral regions, they are minority carriers and they are quickly neutralized by majority carriers that enter the quasi-neutral regions from the ohmic contacts at the ends. This neutralization of injected carriers is just the dielectric-relaxation process that we considered in Chapter 1. Once the minority carriers are injected across the space-charge region, they tend to diffuse away from the junction into the neutral region.

If V_a is negative (*reverse bias*), the barrier height to diffusing majority carriers increases. Equilibrium is disturbed, and minority carriers near the junction space-charge region tend to be depleted. The majority-carrier concentration is similarly reduced by the process of dielectric relaxation.

These brief comments suggest that the minority-carrier densities deserve special attention because they determine what currents flow in a *pn* junction. The majority carriers act only to supply the injected minority-carrier current or to neutralize charge in the quasi-neutral regions. It is helpful to consider the majority carriers as willing "slaves" of the minority carriers. Accordingly, we will seek solutions of the continuity equations for the minority-carrier densities in each of the quasi-neutral regions.

Boundary Values of Minority-Carrier Densities

To write these solutions in a useful fashion, it is necessary to relate the boundary values of the minority-carrier densities to the applied voltage V_a. The most straightforward way of doing this is to make two additional assumptions: first, that the applied bias leads to low-level injection and, second, that the applied bias is small enough so that the detailed balance between majority and minority populations across the junction regions is not appreciably disturbed. The first of these assumptions, low-level injection, was considered in Sec. 5.2. Briefly, it implies that the majority-carrier populations are not significantly changed at the edges of the quasi-neutral regions by the applied bias. The assumption that detailed balance nearly applies allows the use of Equation 4.1.9 across the junction where the potential difference is known to be $\phi_i - V_a$.

Both of these assumptions are certainly valid when V_a is small ($|V_a| \ll \phi_i$). Their validity at higher biases deserves more careful consideration, but we defer this consideration until the analysis is completed.

By the assumption of low-level injection, the electron density at the quasi-neutral boundary in the *n*-region next to the *pn* junction is equal to the dopant density whether at equilibrium or under bias. As in Chapter 4, we call this position x_n (Figure 5.5), and we denote the thermal equilibrium density with an extra subscript *o*.

Likewise, the boundary of the quasi-neutral *p*-region is at $-x_p$ and the hole density there is equal to the acceptor dopant density at equilibrium, as well as under bias.

Summarizing these statements with equations, we have

$$n_{po}(-x_p) = n_{no}(x_n) \exp\left(\frac{-q\phi_i}{kT}\right)$$

$$= N_d(x_n) \exp\left(\frac{-q\phi_i}{kT}\right) \tag{5.3.1}$$

$$p_{no}(x_n) = p_{po}(-x_p) \exp\left(\frac{-q\phi_i}{kT}\right)$$

$$= N_a(-x_p) \exp\left(\frac{-q\phi_i}{kT}\right) \tag{5.3.2}$$

$$n_p(-x_p) = N_d(x_n) \exp\left[\frac{-q(\phi_i - V_a)}{kT}\right] \tag{5.3.3}$$

and

$$p_n(x_n) = N_a(-x_p) \exp\left[\frac{-q(\phi_i - V_a)}{kT}\right] \tag{5.3.4}$$

These four equations can be combined to express the excess minority-carrier densities at the boundaries in terms of their thermal-equilibrium values. We define the excess densities by

$$n' \equiv n - n_o \tag{5.3.5}$$

and

$$p' \equiv p - p_o \tag{5.3.6}$$

Then

$$n_p'(-x_p) = n_{po}(-x_p)\left[\exp\left(\frac{qV_a}{kT}\right) - 1\right] \tag{5.3.7}$$

and

$$p_n'(x_n) = p_{no}(x_n)\left[\exp\left(\frac{qV_a}{kT}\right) - 1\right] \tag{5.3.8}$$

Equations 5.3.7 and 5.3.8 are extremely important results that we will use to obtain specific solutions for the continuity equations for minority carriers in the quasi-neutral regions near a *pn* junction. The equations show that the minority-carrier density depends exponentially on applied bias while the majority-carrier density is assumed to be insensitive to it (to first order). Because the minority-carrier density at thermal equilibrium is typically 11 or 12 orders of magnitude below the majority-carrier density, Equations 5.3.7 and 5.3.8 do not conflict with the low-level injection assumption until the exponential factor is typically of the order 10^{10} or 10^{11}. We will consider the second assumption (quasi-equilibrium of carrier fluxes) after we obtain the relationships describing the voltage dependence of the currents.

Ideal-Diode Analysis

In the discussion above we reviewed the reasons for our focus on the minority carriers and obtained the bias dependence of excess minority carriers. With this information we are ready to consider solutions of the continuity equations (Equation 5.1.4) in the quasi-neutral regions. We first solve the continuity equations under a series of idealizations that comprise the *ideal-diode analysis*.

First, we consider excess holes injected into the *n*-regions, where bulk recombination through generation-recombination centers dominates. Thus, the term $(G_p - R_p)$ in Equation 5.1.4 can be expressed through Equation 5.2.9. Because of our assumption of low-level injection, the arguments used in the discussion of excess-carrier lifetime that led to Equation 5.2.15 also apply to this case. Therefore, the continuity equation for holes in the quasi-neutral approximation discussed in Sec. 5.1 becomes

$$\frac{\partial p_n}{\partial t} = D_p \frac{\partial^2 p_n}{\partial x^2} - \frac{p_n - p_{no}}{\tau_p} \qquad (5.3.9)$$

where the subscripts *n* emphasize that the holes are in the *n*-region.

Fortunately, the simplest case of impurity doping, that of a constant donor density along *x*, is frequently encountered in practice. Taking this case and considering steady state $(\partial p / \partial t = 0)$, we can rewrite Equation 5.3.9 as a total differential equation in terms of the excess density p', which was defined in Equation 5.3.6. This equation

$$0 = D_p \frac{d^2 p'_n}{dx^2} - \frac{p'_n}{\tau_p} \qquad (5.3.10)$$

has the simple exponential solution

$$p'_n(x) = A \exp\left(-\frac{x - x_n}{\sqrt{D_p \tau_p}}\right) + B \exp\left(\frac{x - x_n}{\sqrt{D_p \tau_p}}\right) \qquad (5.3.11)$$

where *A* and *B* are constants determined by the boundary conditions of the problem. The characteristic length $\sqrt{D_p \tau_p}$ in Equation 5.3.11 is called the diffusion length and denoted by L_p. (The diffusion length of an electron in a *p*-type region is denoted by L_n.) For specific applications of Equation 5.3.11, we consider two limiting cases based on the length W_B of the *n*-region from the junction to the ohmic contact (Figure 5.5).

Long-Base Diode. If W_B is long compared to the diffusion length L_p, essentially all of the injected holes recombine before traveling completely across W_B. This case is known as the *long-base diode*. For the long-base diode, L_p is the average distance traveled in the neutral region before an injected hole recombines (Problem 5.8). Because p'_n must decrease with increasing *x*, the constant *B* in Equation 5.3.11 must be zero. The constant *A* in the solution is determined by applying Equation 5.3.8, which specifies $p'_n(x_n)$ as a function of applied voltage. The complete solution is therefore

$$p'_n(x) = p_{no}(e^{qV_a/kT} - 1) \exp\left(-\frac{x - x_n}{L_p}\right) \qquad (5.3.12)$$

as shown in Figure 5.6. Using the expression for the excess hole density in Equation 5.3.12, we obtain in a straightforward manner an expression for the hole current. The hole current flows only by diffusion, because we assumed that the field is negligible in the neutral regions; therefore, from Equation 1.2.22

$$J_p(x) = -qD_p \frac{dp_n}{dx} = qD_p \frac{p_{no}}{L_p}(e^{qV_a/kT} - 1) \exp\left(-\frac{x - x_n}{L_p}\right)$$

$$= qD_p \frac{n_i^2}{N_d L_p}(e^{qV_a/kT} - 1) \exp\left(-\frac{x - x_n}{L_p}\right) \qquad (5.3.13)$$

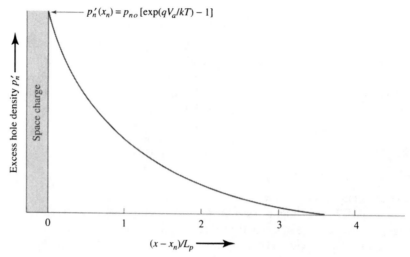

FIGURE 5.6 Spatial variation of holes in the quasi-neutral *n*-region of a long-base diode under forward bias V_a. The excess density p'_n is calculated from Equation 5.3.12.

The hole current is therefore maximum at $x = x_n$ and decreases away from the junction (Figure 5.7) because the hole gradient decreases as carriers are lost by recombination. Because the total current must remain constant with distance from the junction in the steady-state case, the electron current must increase as we move away from the junction. This electron current supplies the electrons with which the holes recombine.

The total current is entirely carried by electrons at the ohmic contact at W_B. Moving toward the junction, the electron current decreases as electrons recombine with the injected holes. At the junction, the only electron current flowing is the one injected across the junction into the *p*-region. The electrons injected into the *p*-region constitute the

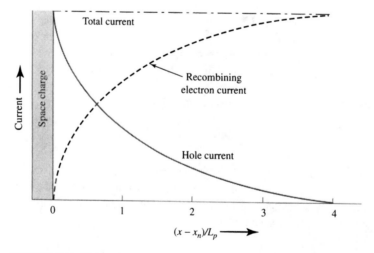

FIGURE 5.7 Hole current (solid line) and recombining electron current (dashed line) in the quasi-neutral *n*-region of the long-base diode of Figure 5.5. The sum of the two currents J (dot-dash line) is constant. The hole current is calculated from Equation 5.3.13.

minority-carrier current there. Thus, the total current is obtained by summing the two minority-carrier injection currents: holes into the n-side plus electrons into the p-side.

The minority-carrier electron current that is injected into the p-region is found by a treatment analogous to the analysis used to obtain Equation 5.3.13. If the ohmic contact is at $-W_E$ where $W_E \gg L_n \equiv \sqrt{D_n \tau_n}$, then

$$J_n = qD_n \frac{n_i^2}{N_a L_n}(e^{qV_a/kT} - 1) \exp\left(\frac{x + x_p}{L_n}\right) \tag{5.3.14}$$

Because we chose the origin for x at the physical junction (Figure 5.5), x is a negative number throughout the p-region. Therefore, J_n decreases away from the junction as did J_p in the n-region. To obtain an expression for the total current J_t, we sum the minority-carrier components at $-x_p$ and $+x_n$, as expressed in Equations 5.3.13 and 5.3.14:

$$J_t = J_p(x_n) + J_n(-x_p) = qn_i^2\left(\frac{D_p}{N_d L_p} + \frac{D_n}{N_a L_n}\right)(e^{qV_a/kT} - 1) \tag{5.3.15}$$

$$= J_0(e^{qV_a/kT} - 1)$$

where J_0 is the magnitude of the *saturation current density* predicted by this theory when a negative bias equal to a few kT/q volts is applied. Equation 5.3.15 is identical in form to Equation 3.3.6, which was derived for a metal-semiconductor Schottky-barrier diode. The similar dependence on voltage arises from the application in both cases of quasi-equilibrium assumptions, which led to Equations 5.3.7 and 5.3.8.

Short-Base Diode.   A second limiting case occurs when the lengths W_B and W_E of the n- and p-type regions are much shorter than the diffusion lengths L_p and L_n. In this case, little recombination occurs in the bulk of the quasi-neutral regions. In the limit, all injected minority carriers recombine at the ohmic contacts at either end of the diode structure. We can obtain solutions most easily in this case by approximating the exponentials in Equation 5.3.11 by the first two terms of a Taylor series expansion so that

$$p_n'(x) = A' + B'\frac{(x - x_n)}{L_p} \tag{5.3.16}$$

Because of the ohmic contact at $x = W_B$, $p_n'(W_B) = 0$. The boundary condition at $x = x_n$ is given by Equation 5.3.8 as before, so that the solution for the excess hole density in the n-region becomes

$$p_n'(x) = p_{no}(e^{qV_a/kT} - 1)\left(1 - \frac{x - x_n}{W_B'}\right) \tag{5.3.17}$$

where $W_B' = W_B - x_n$ is the length of the quasi-neutral n-type region (Figure 5.5). From Equation 5.3.17, we see that the excess hole density decreases linearly with distance across the n-type region (Figure 5.8). The approximation $L_p \gg W_B$ is equivalent to stating that all the injected holes diffuse across the n-type region before recombining. The assumption that no recombination occurs in the n-type region is equivalent to letting the lifetime τ_p approach infinity in Equation 5.3.10. The differential equation that results has a linear solution.

A linearly varying concentration indicates that the hole current remains constant throughout the n-type region, and that no electron current is needed to compensate for recombining holes. Therefore,

$$J_p = -qD_p\frac{dp}{dx} = qD_p\frac{p_{no}}{W_B'}(e^{qV_a/kT} - 1) = qD_p\frac{n_i^2}{N_d W_B'}(e^{qV_a/kT} - 1) \tag{5.3.18}$$

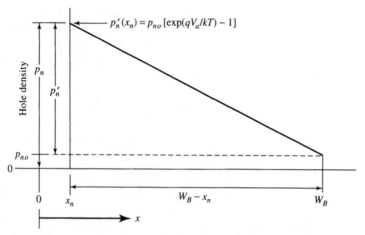

FIGURE 5.8 Hole density in the quasi-neutral *n*-region of an ideal short-base diode under forward bias of V_a volts. The excess hole density p'_n is calculated from Equation 5.3.17.

Comparing Equations 5.3.18 and 5.3.13 for the hole currents in short- and long-base diodes, we see that the solutions obtained are similar except for the characteristic length associated with each geometry. In the long-base diode, the characteristic length is the minority-carrier diffusion length. In the short-base diode, it is the length of the quasi-neutral region. As in the long-base diode, the total current in the short-base diode is made up of electrons injected into the *p*-region and of holes injected into the *n*-region. Hence, for the short-base diode with $W_E \ll L_n = \sqrt{D_n \tau_n}$ and $W_B \ll L_p = \sqrt{D_p \tau_p}$,

$$J_t = qn_i^2 \left(\frac{D_p}{N_d W_B} + \frac{D_n}{N_a W_E'} \right) (e^{qV_a/kT} - 1) \tag{5.3.19}$$

Of course, a given diode can be approximated by a combination of these two limiting cases; that is, it may be short-base in the *p*-region and long-base in the *n*-region or vice versa. Handling these cases is straightforward.

Both diode cases (Equations 5.3.15 and 5.3.19) predict a large current flow under forward bias and a small saturation current under reverse bias. Physically, this marked asymmetry results because forward bias aids the injection of majority carriers from each region across the junction. These carriers are replaced at the end ohmic contacts. Under reverse bias, the net flow across the junction is composed of minority carriers from each region. These are few in number and not replaced at the end contacts. Hence, only a small saturation current flows under reverse bias (Problem 5.7).

We see from Equations 5.3.15 and 5.3.19 that the most lightly doped side of the junction determines the reverse saturation current. For example, if the dopant concentration on the *n*-side of the junction is much less than that on the *p*-side, the hole current injected across the junction into the *n*-side is much greater than the electron current injected into the *p*-region. This situation characterizes a heavy *p*-type diffusion into a lightly doped *n*-type wafer.

EXAMPLE   *Long- and Short-Base Diodes*

(a) A silicon diode is formed by diffusing a high concentration of boron into a 700-μm-thick, phosphorus-doped wafer having resistivity $\rho = 4.5$ Ω-cm. Contacts are made to the diffused boron region and to the backside of the wafer. The *pn* junction, 0.5 μm below the surface with a plane area $= 10^{-4}$ cm², can be considered to be a one-sided step junction. Find the saturation current in the diode if the hole lifetime τ_{ps} in the substrate is 1 μs.

(b) In a different process, a similar diffusion schedule is used to form a p^+n junction in a 3-μm-thick, phosphorus-doped, epitaxial layer having a resistivity $\rho = 4.5$ Ω-cm. The epitaxial layer is deposited on a heavily doped ($N_d > 10^{19}$ cm^{-3}) *n*-type substrate. The lifetime τ_{pe} in the epitaxial layer is 1 μs. In the substrate, the lifetime is short ($\tau_{ps} \approx 100$ ps). The *pn* junction, formed 0.5 μm below the surface with a plane area $= 10^{-4}$ cm^2, can be considered to be a one-sided step junction. Find the saturation current in the diode.

Solution

(a) Either from Figure 1.15 or Table 4.1, we find that for 4.5 Ω-cm *n*-type material, $N_d = 1 \times 10^{15}$ cm^{-3}. The diffusion coefficient D_p is found from Figure 1.16 to be 12 cm^2 s^{-1}. Therefore, the diffusion length $L_p = \sqrt{D_p \tau_p}$ is 34.6 μm. Because L_n is much less than the thickness of the silicon wafer (700 μm), the device is a long-base diode. The saturation-current density J_0 is therefore given by Equation 5.3.15, and the saturation current $I_0 = J_0 A = 1.2 \times 10^{-14}$ A.

(b) In process b, the *pn* junction is formed in an epitaxial layer having a thickness much smaller than L_p. Therefore, almost all injected holes diffuse to the substrate where they rapidly recombine. This structure is therefore a short-base diode, and J_0 is given by Equation 5.3.18. The effective neutral base width is 1.5 μm (the thickness of the epitaxial layer minus the junction depth minus the ≈ 1 μm depletion region thickness in the lightly doped *n*-type region). Using Equation 5.3.18, we calculate the diode saturation current $I_0 = J_0 A = 2.7 \times 10^{-13}$ A.*

From this example we see that the short base increases the saturation current because the gradient of the minority-carrier density increases. ∎

Validity of Approximations. Before taking a more detailed look at the physics of *pn*-junction diodes, we return to consider two of the initial assumptions used to carry out the analysis. These assumptions are, first, that ohmic voltage drops in the quasi-neutral regions are small so that V_a is sustained entirely across the junction space-charge region and, second, that applied bias does not greatly alter the detailed balance between drift and diffusion tendencies that exists at thermal equilibrium.

To consider these assumptions, we consider the fairly typical case of $N_d = 5 \times 10^{15}$ and $N_a = 5 \times 10^{18}$ cm^{-3}. If we first take a short-base diode with $W_B = W_E = 3$ μm, then from Equation 5.3.19 we calculate a saturation current of approximately 10^{-9} A cm^{-2}. For a typical cross-sectional area of 10^{-5} cm^2 and a typical forward-bias voltage of ~0.65 V, this leads to a current of ~1 mA or 100 A cm^{-2}. This current consists almost entirely of holes injected into the *n*-side because the first term in the saturation current of Equation 5.3.19 far exceeds the second. Thus, the ohmic voltage drop only exists in the highly conducting *p*-region that has a resistivity of 0.03 Ω-cm. The field in this region is just 3 V cm^{-1} and, hence, the voltage dropped in the 3 μm length is approximately 1 mV, which is certainly negligible compared to the applied 0.65 volts. Note that there is essentially no ohmic voltage drop in the *n*-region for this case because there is only a negligible flow of electron current.

If we consider a long-base diode with lower doping on either side, the assumption is on shakier grounds. To see this, assume that $N_a = N_d = 10^{16}$ cm^{-3} and that we are considering a long-base diode of cross-sectional area 10^{-5} cm^2 with 0.65 V applied bias. With $L_n \approx L_p = 30$ μm, Equation 5.3.15 shows that such a diode passes approximately 32×10^{-6} A. If the *p* and *n* regions are each 100 μm long, the ohmic voltage drops are about 0.03 V in the

---

* More precisely, the reverse current for the short-base diode increases as the applied reverse bias increases because the depletion region of the *pn* junction extends farther into the neutral region in the lightly doped epitaxial layer. The significantly narrower neutral region increases the minority-carrier gradient and the reverse current of the diode.

p-region and 0.02 V in the *n*-region. To first order these can again be considered negligible compared to the 0.65 V applied. These results are typical of many practical cases, and we conclude that it is reasonable to assume that the entire applied voltage changes the height of the potential barrier at the *pn* junction for low-to-moderate current densities.

If the applied forward voltage is nearly as large as the built-in potential, however, the barrier to majority-carrier flow is substantially reduced and large currents can flow. In that case a significant portion of the applied voltage is dropped across the neutral regions, and the series resistance of the neutral regions must be considered. The effect of series resistance can easily be included using circuit analysis, and we do not consider it further here.

To determine the validity of the quasi-equilibrium assumption, we compare the magnitude of typical currents to the balanced drift and diffusion tendencies at thermal equilibrium. For a typical integrated-circuit *pn* junction, in which the hole concentration changes from 10^{18} to 10^4 cm^{-3} across a depletion region about 10^{-5} cm wide, the hole diffusion current for the average gradient is of the order of 10^5 A cm^{-2}. As we saw in Chapter 4, at thermal equilibrium this tendency of holes to diffuse is exactly balanced by an opposite tendency to drift under the influence of the electric field in the depletion region. We already argued that typical forward-bias diode currents are roughly 10^2 A cm^{-2}—only 0.1% of the two current tendencies in balance at thermal equilibrium. Thus, it is reasonable to treat the case of small and moderate biases by considering only slight deviations from thermal equilibrium. This means that we can relate the carrier concentrations on either side of the junction space-charge region by considering the effective barrier height to be $(\phi_i - V_a)$ and by using Equations 5.3.7 and 5.3.8. These two equations also depend on the validity of the low-level injection assumption. The current tendencies at *pn* junctions that are in detailed balance at thermal equilibrium are so large relative to currents that flow in practice that a more general relationship than Equations 5.3.7 and 5.3.8 is valid.

The more general relationship can be deduced by referring to the quasi-Fermi levels defined in Equations 1.1.28 and 1.1.29. The quasi-Fermi levels (ϕ_{fn} and ϕ_{fp}) express the densities of free carriers under nonequilibrium conditions. At thermal equilibrium, there is a single Fermi level that defines both electron and hole densities, while under nonequilibrium conditions, each carrier density is represented by a different quasi-Fermi level.

From our discussion of the *pn* junction under forward bias, we expect that the two quasi-Fermi levels are separated in the region near the biased junction where injection markedly changes the minority-carrier densities, while far from the junction, ϕ_{fn} and ϕ_{fp} merge with one another. Because the majority-carrier densities are only negligibly affected (for the low- and moderate-bias condition discussed so far), the quasi-Fermi levels for the majority-carrier densities are essentially unchanged, while those for the minority-carrier densities are modified.* The applied bias V_a is therefore equal to the offset $\phi_{fp}(x = -x_p) - \phi_{fn}(x = +x_n)$, that is, the difference between the quasi-Fermi level for holes on the *p*–side and that for electrons on the *n*–side. Applying Equation 1.1.30 to the space-charge region of a *pn* junction $(-x_p \le x \le +x_n)$, we find

$$pn = n_i^2 \exp\left[\frac{q(\phi_p - \phi_n)}{kT}\right] = n_i^2 \exp\left(\frac{qV_a}{kT}\right) \qquad (5.3.20)$$

---

* Another way to visualize the near constancy of the quasi-Fermi levels for the majority carriers is through Equations 1.2.25 and 1.2.26, which represent total currents in terms of the products of the carrier densities and the gradients of the quasi-Fermi levels. For a given current, the higher the carrier density, the lower is the necessary gradient—hence, the gradients of the quasi-Fermi levels for minority carriers are much larger than those for majority carriers.

Equation 5.3.20 is consistent with the boundary values for low-level injection derived for minority-carrier densities in Equations 5.3.3 and 5.3.4 and is valid within the space-charge region [6]. It is also useful in cases where injected carrier densities approach the thermal-equilibrium concentrations of majority carriers (Chapter 7).

Space-Charge-Region Currents[†]

The analysis of *pn* junctions that led to the diode equations (Equations 5.3.15 and 5.3.19) was based on events in the quasi-neutral regions. The space-charge region was treated solely as a barrier to the diffusion of majority carriers, and it played a role only in establishing minority-carrier densities at its boundaries (Equations 5.3.7 and 5.3.8). This is a reasonable first-order description of events, and the equations derived from it (Equations 5.3.15 and 5.3.19) are called the *ideal-diode* equations. Over a significant range of useful biases, however, the ideal-diode equations are inaccurate, especially for silicon *pn* junctions, and we must consider corrections to these equations that arise from events occurring in the space-charge region at the junction.

As we noted in Chapter 4, the space-charge region typically has a length of the order of 10^{-4} cm. Like the quasi-neutral regions of the diode, it contains generation-recombination centers; unlike the quasi-neutral zones, it is a region of steep impurity gradients and rapidly changing populations of holes and electrons. Because the carriers injected under forward bias must pass through this region, some carriers are lost by recombination. Conversely, under reverse bias, generation of carriers in the space-charge region leads to excess current above the saturation value predicted by the ideal-diode equations.

To find expressions for generation-recombination in the space-charge region, we use the Shockley–Hall–Read theory. For simplicity, we take the case of equal hole and electron capture cross sections (Equation 5.2.10) and evaluate the expression in the space-charge region with an applied bias V_a. The *pn* product there is given by Equation 5.3.20, and thus the overall recombination-generation rate $U = -dn/dt = -dp/dt$ is

$$U = \frac{n_i^2(e^{qV_a/kT} - 1)}{\left(p + n + 2n_i \cosh\dfrac{(E_t - E_i)}{kT}\right)\tau_0} \tag{5.3.21}$$

The recombination rate is thus positive for forward bias and negative under reverse bias (i.e., carriers are generated under reverse bias). The total current arising from generation and recombination in the space-charge region is given by the integral of the recombination rate across it:

$$J_r = q \int_{-x_p}^{x_n} U \, dx \tag{5.3.22}$$

Although this integral is not easily evaluated, a qualitative discussion permits us to determine the dependence of current on voltage. First, consider forward bias and, as in Sec. 5.2, note that recombination centers located near the middle of the forbidden gap are the most effective (Figure 5.3). Thus, we consider $E_t \approx E_i$ in Equation 5.3.21. From this equation, we see that the recombination rate is maximum when the sum $p + n$ is minimum. If we consider this sum as a function of p and n with the constraint that the product of p and n is given by Equation 5.3.20, we can easily show (Problem 5.15) that U is maximum when

$$p = n = n_i \exp\left(\frac{qV_a}{2kT}\right) \tag{5.3.23}$$

For typical forward-bias conditions, the sum $p + n$ is much greater than n_i within the space-charge region. If the dominant contribution to the integral in Equation 5.3.22 is given by this maximum value of U extending across a portion x' of the space-charge region, the recombination current can be expressed as

$$J_r = \frac{qx'n_i^2(e^{qV_a/kT} - 1)}{2n_i(e^{qV_a/2kT} + 1)\tau_0}$$

$$\approx \frac{qx'n_i}{2\tau_0}\exp\left(\frac{qV_a}{2kT}\right) \quad (5.3.24)$$

where $\tau_0 = 1/N_t\sigma v_{th}$ is again the lifetime associated with the recombination of excess carriers in a region with a density N_t of recombination centers. Unlike current arising from carrier recombination in the quasi-neutral regions, the current resulting from recombination in the space-charge region varies with applied voltage as $\exp(qV_a/2kT)$ (assuming x' is only a weak function of voltage). This different exponential behavior can be observed in real diodes, especially at low currents. Because terms like τ_0 are not known with high precision, space-charge-region recombination is usually not given more elaborate analysis than our treatment and x', in particular, is often approximated by the entire space-charge-region width x_d. If we make this approximation and express the ratio between the ideal-diode current J_t (Equation 5.3.15) and the space-charge-region recombination current J_r under forward bias, we find

$$\frac{J_t}{J_r} = \frac{2n_i}{x_d}\left[\frac{L_n}{N_a} + \frac{L_p}{N_d}\right]\exp\left(\frac{qV_a}{2kT}\right) \quad (5.3.25)$$

Thus, space-charge recombination current J_r becomes less significant relative to ideal-diode current as bias increases. Also, as the defect density decreases, the diffusion length increases, and the relative importance of J_r compared to J_t decreases. Taking values for typical (one-sided) silicon diodes of $L_n = 60$ μm, $x_d = 0.25$ μm, and $N_a = 10^{16}$ cm^{-3}, we see that J_t exceeds J_r for V_a greater than about 0.375 V.

Under reverse bias, the numerator of Equation 5.3.21 approaches $-n_i^2$. Therefore U is negative, indicating net generation in the space-charge region. From Equation 5.3.20 the pn product becomes vanishingly small, and the maximum value of U again occurs when both p and n are equal; in this case both are much less than n_i. More detailed analysis than we carry out here shows that p and n are both substantially less than n_i over a portion of the space-charge region x_i bounded by the points at which the intrinsic Fermi level E_i crosses the quasi-Fermi levels [7]. This region can be considerably smaller than the total space-charge region width x_d. Outside x_i either p or n is greater than n_i, and the generation rate falls rapidly. The net generation rate in the space-charge region can be approximated by the product of the maximum generation rate and the width x_i:

$$J_g = \frac{qn_ix_i}{2\tau_0} \quad (5.3.26)$$

where we have again assumed that the most effective centers are located at E_i. For a one-sided pn junction with a heavily doped p-type side, virtually all of the space-charge region extends into the lightly doped n-type semiconductor, and we can solve for the width of the total space-charge region x_d and for the region x_i with maximum generation rate:

$$x_d = \left[\frac{2\epsilon_s}{qN_d}(\phi_i - V_a)\right]^{1/2} \quad (5.3.27)$$

$$= \left[\frac{2\epsilon_s kT}{q^2N_d}\left(\ln\frac{N_dN_a}{n_i^2} - \frac{qV_a}{kT}\right)\right]^{1/2}$$

and

$$x_i = \left(\frac{2\epsilon_s kT}{q^2 N_d}\right)^{1/2}\left[\left(\ln\frac{N_d}{n_i} - \frac{qV_a}{kT}\right)^{1/2} - \left(\ln\frac{N_d}{n_i}\right)^{1/2}\right]$$

(5.3.28)

where V_a is the applied voltage ($V_a < 0$ for reverse bias). Both x_d and x_i depend on the square root of the applied voltage for large reverse bias, and the difference between the two quantities becomes small at high reverse bias. Figure 5.9 shows the ratio x_i/x_d as a function of voltage for several donor concentrations in a one-sided step junction. Because the density of recombination centers in practical diodes can vary with position and is generally not well known, it is often not worthwhile to distinguish between x_i and x_d, and the latter is frequently used in Equation 5.3.26.

From Equations 5.3.26 and 5.3.28, we see that the generation current in the space-charge region is only a weak function of the reverse bias, varying roughly as the square root of applied voltage. We can estimate the relative importance of the contributions to the reverse current from the quasi-neutral regions and from the space-charge region for a long-base diode by taking the ratio of Equation 5.3.15 under reverse bias to Equation 5.3.26:

$$\frac{J_t}{J_g} = \frac{2n_i}{x_i}\left[\frac{L_n}{N_a} + \frac{L_p}{N_d}\right]$$

(5.3.29)

Values of the parameters for practical diodes are such that J_t/J_g is much less than unity, and the current in reverse-biased silicon diodes is generated primarily in the space-charge region. Generation-recombination centers in the space-charge region are also responsible for excess currents in reverse-biased Schottky diodes (Chapter 3). In that case, however, the dominant currents are associated with recombination centers at the metal-semiconductor interface.

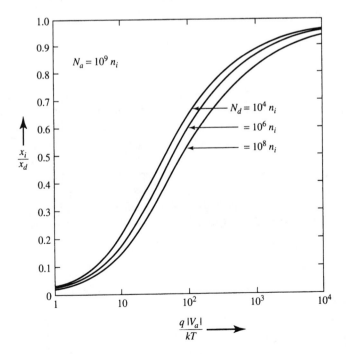

FIGURE 5.9 The ratio of generation-region width x_i to space–charge–region width x_d as a function of reverse voltage for several donor concentrations in a one-sided step junction [7].

EXAMPLE   *Generation Currents in a Reverse-Biased Diode*

(a) Compare the importance of neutral-region generation and space-charge-region generation in a reverse-biased, diffused silicon diode at room temperature. Assume that the diode is formed by introducing a high concentration of boron atoms into a thick, 5 Ω-cm, *n*-type wafer. The lifetime in both the neutral and space-charge regions is 10 μs, and 5 V of reverse bias is applied.

(b) Determine the relative importance of the two current components at 125°C. Assume that the lifetime and diffusion coefficient vary with temperature as $T^{-1/2}$ and $T^{-2.2}$ (Figure 1.17), respectively.

Solution

(a) Either from Figure 1.15 or Table 4.1, we find that $N_d = 9 \times 10^{14}$ cm^{-3} for 5 Ω-cm *n*–type silicon. Using Figure 1.16, we find $D_p = 12$ cm^2 s^{-1}. Hence, at 300 K the diffusion length $L_p = \sqrt{D_p \tau_p} = 110$ μm. For a $p^+ n$ junction, the generation current in the charge-neutral region J_t is (from Equation 5.3.15)

$$J_t \approx q n_i^2 \frac{D_p}{N_d L_p} = 4.1 \times 10^{-11} \text{ A cm}^{-2} = 41 \text{ pA cm}^{-2}.$$

To calculate the current due to generation in the space-charge region, we first use Equation 5.3.28 to determine the width x_i over which generation takes place. For the values given, we find $x_i = 2.12$ μm. From Equation 5.3.26 we calculate $J_g = 24.6$ nA cm^{-2}. Comparing these results, we see that almost 600 times as much current is produced by generation in the space-charge region as is produced in the charge-neutral regions.

(b) At 125°C, the diffusion coefficient D_p is reduced:

$$D_p(398 \text{ K}) = D_p(300 \text{ K}) \times \left(\frac{300}{398}\right)^{2.2} = 12 \times 0.534 = 6.44 \text{ cm}^2 \text{ s}^{-1}.$$

The lifetime is also reduced to 8.68 μs. Using the formula in Table 1.4, we calculate $n_i = 6.5 \times 10^{12}$ cm^{-3} at 125°C. With these values, we find $L_p = 75$ μm and $J_t = 6.45$ μA cm^{-2} for generation in the charge-neutral region. The generation width in the space-charge region x_i is 2.24 μm, and J_g is 13.4 μA cm^{-2}. Hence, at 125°C, the ratio of the two currents $J_g/J_t = 2.1$. The two currents are almost equal at the higher temperature because one increases as n_i^2 while the other varies linearly with n_i. ∎

In our discussion of carrier generation in the space-charge region, we assumed uniform properties across the entire area of the diode. However, *pn* junction diodes are usually bounded by insulating regions of silicon dioxide. The discontinuities in the crystal structure of the silicon at this boundary can create high concentrations of generation-recombination centers. In practice, carrier generation in the space-charge region where it intersects the edge of the diode usually dominates the leakage current in modern integrated-circuit diodes. Careful control of the processing is necessary to reduce the edge component of the space-charge leakage current when low leakage currents are critical to device operation. This surface (or interface) generation will be discussed in more detail in Chapter 8.

Summary.   Our analysis of current flow in *pn* junctions touched on several points and was rather lengthy. It is worthwhile to summarize the results at this point and to make several comments about them.

We considered solutions of the continuity equation for two cases of uniformly doped step-junction diodes. Both of these led to a dependence of current on voltage (Equations 5.3.15 and 5.3.19) that is the same as found earlier for Schottky-barrier diodes.

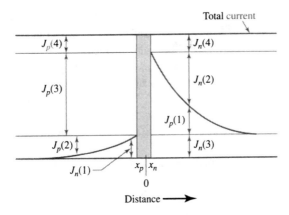

FIGURE 5.10 The current components in the quasi-neutral regions of a long-base diode under moderate forward bias: $J(1)$ injected minority-carrier current, $J(2)$ majority-carrier current recombining with $J(1)$, $J(3)$ majority-carrier current injected across the junction, $J(4)$ space-charge-region recombination current.

This behavior is generally known as the ideal-diode law [$J = J_0 \left(\exp\left(qV_a/kT\right) - 1\right)$]. In practical diodes, currents arising from events in the space-charge region must be added to the ideal behavior to obtain the overall current-voltage relationship. These added currents dominate in silicon diodes under reverse bias and under low forward bias. A sketch of the currents flowing in a long-base diode at moderate forward bias is shown in Figure 5.10. The total current close to the junction in each of the quasi-neutral regions consists of (1) injected minority carriers diffusing away from the junction, (2) majority carriers drifting toward the junction to recombine with minority carriers injected into the quasi-neutral regions, (3) majority carriers drifting toward the junction to be injected into the opposite quasi-neutral region, (4) majority carriers drifting toward the junction space-charge region where they recombine with injected carriers entering the junction region from the opposite side. Far from the junction in either region, the entire current is carried by drifting majority carriers.

A next logical step is to analyze *pn* junctions in which the dopant densities are nonuniform in the quasi-neutral regions. We expect nonuniform dopant distributions in diffused junctions and in implanted junctions. Nonuniform doping is, however, best considered by using numerical techniques, as we will discuss in Sec. 5.5, or by using approximation techniques, which we will develop in Chapter 6 when we discuss transistors. In Chapter 6 we will see that the ideal-diode law is again obtained in practical cases of nonuniform doping, but that the expression for the saturation current J_0 is different. As in the case of uniformly doped diodes, generation-recombination in the space-charge regions leads to added current components. The observed steady-state dependence of current on voltage is obtained by summing the ideal-diode current and these added components.

Heterojunctions[†]

In the previous section, we saw that the ratio of electron injection to hole injection across a silicon *pn* homojunction with current flow depends inversely on the dopant concentrations. To minimize the charge injection on one side of the junction, the dopant concentration there must be much higher than that on the other side of the junction. If the dopant concentration is limited by other constraints, the opportunity for controlling the ratio of injected currents, which often determines the gain of a transistor, is restricted.

However, if we can make the semiconductor materials on the two sides of the junction from materials with different electronic properties, we have another method of controlling the injected charge. The possibility of unequal barriers for electron injection ϕ_{Bn} and for hole injection ϕ_{Bp} across a heterojunction provides an additional parameter that we can use to control the behavior of diodes and, more importantly, transistors, as we will discuss in Chapters 6 and 7.

We saw in Chapter 4 that for a homojunction the barriers ϕ_{Bn} and ϕ_{Bp} are equal and simply the built-in potential ϕ_i; ϕ_i is determined by the difference in work functions across the junction or, equivalently, by the bandgap of the semiconductor and the doping. In a heterojunction (Figures 4.9 and 4.10), the barriers again depend on the difference in work functions. However, we now have two bandgaps to consider, so the barriers are influenced by both bandgaps, as well as by the doping, and also possibly by discontinuities in the conduction- and valence-band edges across the junction [8, 9]. In general, $\phi_{Bn} \neq \phi_{Bp}$. With these additional factors, the built-in potential ϕ_i is a less directly useful quantity, and we focus on ϕ_{Bn} and ϕ_{Bp}.

To understand the current flow across the junction, we consider two cases: First, we discuss the case where the current flow is related to the minority–carrier densities and their distributions at the neutral edges of the space-charge region surrounding the heterojunction and no other barrier limits the current flow (Figure 5.11a). This conduction mechanism corresponds to the *diffusion current* we discussed for the homojunction (Equations 5.3.13 and 5.3.18), but the magnitudes of the electron- and hole-currents are modified. The second case we discuss corresponds to one with an additional barrier that limits the current flow (Figure 5.11c). Conduction then occurs by *thermionic emission* over the additional barrier (or possibly by tunneling through the barrier).

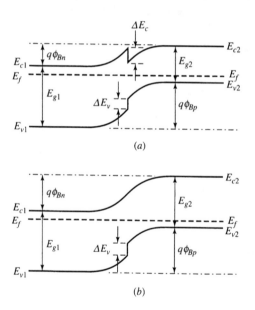

(a)

(b)

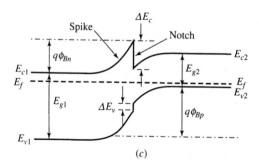

(c)

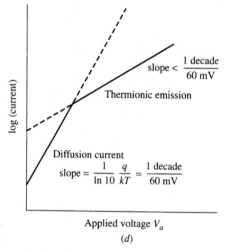

(d)

FIGURE 5.11   (a) Conduction- and valence-band edges for a heterojunction with different bandgaps and different electron affinities X on the two sides of the junction; the top of the conduction-band spike near the junction is lower than the conduction-band edge in material 2. (b) Energy bands with $X_1 = X_2$. (c) Energy bands with $X_1 \neq X_2$ and the top of the conduction-band spike higher than the conduction band in material 2. (d) (Adapted from [8].) Current-voltage characteristic for a diode near the boundary between (a) and (c), showing diffusion current at lower voltages and a transition to thermionic-emission current at higher biases.

We start our discussion with the first case (Figure 5.11*a*), where diffusion current dominates and the conduction-band spike (resulting from the curvature of E_c in the *n*-type material) does not limit carrier transport across the junction. In this case, the barriers are determined by the properties in the neutral regions of the two semiconductors (i.e., at thermal equilibrium the Fermi level must be constant throughout the system), as we saw in Sec. 4.2. For electrons

$$\phi_{Bn} = \frac{E_{c1} - E_{c2}}{q} \tag{5.3.30}$$

and for holes

$$\phi_{Bp} = \frac{E_{c1} - E_{c2} + \Delta E_g}{q} \tag{5.3.31}$$

where E_{c1} and E_{c2} are the energies of the conduction-band edges in the neutral regions of materials 1 and 2, respectively.

We can also express the barriers in terms of the built-in potential by considering the algebraic sum of the band bending in the two pieces of semiconductor (the built-in potential ϕ_i) and the discontinuities in the conduction- and valence-band edges at the interface ΔE_c and ΔE_v, respectively. Thus, $\phi_{Bn} = \phi_i - \Delta E_c/q$ and $\phi_{Bp} = \phi_i + \Delta E_v/q$. However, these expressions are somewhat misleading because ϕ_i itself depends on ΔE_c. Although the barriers appear to depend on the individual band-edge discontinuities, Equation 4.2.20 showed that the built-in potential increases by the conduction-band discontinuity, so $\phi_{Bn} = \phi_i - \Delta E_c/q$ does not depend on ΔE_c, and $\phi_{Bp} = \phi_i + \Delta E_v/q$ depends on the total bandgap discontinuity ΔE_g.

To find expressions for the diffusion currents, we need to relate the minority-carrier concentrations in the neutral regions to the material properties, especially to the bandgaps of the two semiconductors. As in Sec. 4.2, we again consider the particularly important case of an *n*-type material 1 with a bandgap E_{g1} and a *p*-type material 2 with a smaller bandgap E_{g2}, as shown in Figure 5.11*a*. We can relate the carrier concentrations on the two sides of the junction in terms of the barrier heights or directly in terms of the material properties and doping concentrations.

First, we consider the electron concentrations with no voltage applied across the junction. The majority-carrier electron concentration in the neutral region on the *n*-side of the junction in material 1 is

$$N_d = n_{no1} = N_{c1} \exp\left(-\frac{E_{c1} - E_f}{kT}\right) \tag{5.3.32}$$

and the minority-carrier electron concentration in the neutral region on the *p*-side of the junction in material 2 is

$$n_{po2} = N_{c2} \exp\left(-\frac{E_{c2} - E_f}{kT}\right) \tag{5.3.33}$$

The majority- and minority-electron concentrations across the junction are related by the barrier height ϕ_{Bn} (Equation 5.3.30).

We use the expressions for the minority-carrier concentrations to find the current flowing across the *pn* heterojunction when a forward bias V_a is applied. We assume that little voltage is dropped in the neutral regions of the semiconductors, so that the barrier heights ϕ_{Bn} and ϕ_{Bp} are each reduced by the total applied bias V_a. The minority-carrier concentrations at the edges of the neutral region are then enhanced by the factor $\exp(qV_a/kT)$, as in a homojunction (Equation 5.3.3). From the electron concentration at the edge of the neutral region in the *p*-type semiconductor, we find the minority-carrier electron current there for a long-base diode (Equation 5.3.14) to be

$$J_n = qD_n \frac{dn_{p2}}{dx} = -\frac{qD_n}{L_n} n_{po2}\left[\exp\left(\frac{qV_a}{kT}\right) - 1\right] = -\frac{qD_n}{L_n} \frac{n_{i2}^2}{N_{a2}}\left[\exp\left(\frac{qV_a}{kT}\right) - 1\right] \quad (5.3.34)$$

Thus, the injected electron current is the same as for a homojunction made from the smaller bandgap material 2. Looking ahead to our interest in heterojunction bipolar transistors, we want to express the current in terms of the intrinsic carrier density in material 1:

$$J_n = -\frac{qD_n}{L_n} \frac{n_{i1}^2}{N_{a1}} \exp\left(\frac{\Delta E_g}{kT}\right)\left[\exp\left(\frac{qV_a}{kT}\right) - 1\right] \quad (5.3.35)$$

which emphasizes that the electron current flowing across the heterojunction is greater than for a homojunction made from material 1, as we expect because of the lower barrier height ϕ_{Bn} for electron injection. For a short-base diode, we obtain a similar expression except that L_n is replaced by W'_p, the length of the quasi-neutral *p*-type region.

We can write an analogous expression for the hole current in terms of the intrinsic concentration in material 1 as

$$J_p = -qD_p \frac{dp_{n1}}{dx} = -\frac{qD_p}{L_p} p_{no1}\left[\exp\left(\frac{qV_a}{kT}\right) - 1\right] = -\frac{qD_p}{L_p} \frac{n_{i1}^2}{N_{d1}}\left[\exp\left(\frac{qV_a}{kT}\right) - 1\right] \quad (5.3.36)$$

Equation 5.3.36 shows that the hole current injected from material 2 into material 1 is the same as that in a homojunction of material 1, consistent with the barrier height ϕ_{Bp} being equal for the homojunction and the heterojunction. As expected, Equations 5.3.35 and 5.3.36 are independent of the conduction-band discontinuity and only depend on the bandgaps and doping of the two materials. Equivalently, we can express the current injection in terms of the barrier heights ϕ_{Bn} and ϕ_{Bp} for injection of electrons and holes, respectively, across the junction.

The consequences on the current of the different barriers for electron and hole injection can best be illustrated by examining a particular case. We consider a heterojunction with *n*-type silicon on one side and a *p*-type semiconductor composed of an alloy of silicon and germanium ($Si_{1-x}Ge_x$) on the other side, where *x* is the atomic fraction of germanium in the alloy. The alloy has a bandgap several tenths of an electron volt smaller than that of silicon. The electron affinity of $Si_{1-x}Ge_x$ is similar to that of silicon, so the conduction-band discontinuity is small, and we neglect it here, as shown in Figure 5.11*b*. Virtually all the bandgap difference appears as a discontinuity in the valence-band edge.

When the two materials are joined, charge transfer equalizes the Fermi levels on the two sides of the junction (Figure 5.11*b*). The Fermi level in the isolated *p*-type $Si_{1-x}Ge_x$ is closer to the vacuum level than in similarly doped Si because of its equal electron affinity and smaller bandgap. Therefore, the displacement of the Fermi level as the two pieces of semiconductor are joined is less for the heterojunction than for a similarly doped homojunction. Consequently the barrier for electrons moving from the *n*-type Si into the *p*-type $Si_{1-x}Ge_x$ is lower than its equivalent in a homojunction. However, for holes, the barrier is the same as in a homojunction; the valence-band discontinuity (by definition) identically compensates the reduced work function.

When a bias is applied, the barrier to carrier injection across the junction decreases; the electron density in the *p*-type material increases; and current flows. The smaller barrier for electrons in the heterojunction than in a homojunction allows electron current to be injected more readily, reducing the turn-on or threshold voltage at which a given current flows below that of a similarly doped Si *pn*-junction diode. The hole current, however, is similar to that in a homojunction. Thus, the important ratio of the electron- to the hole-current increases. Conversely, a given electron current is obtained at a lower applied bias, and the hole current injected into the n-type region at this bias is lower than in a homojunction.

EXAMPLE: *Current in a Heterojunction*

Consider a heterojunction composed of heavily doped n-type silicon and moderately doped p-type $Si_{1-x}Ge_x$. For simplicity, assume (somewhat incorrectly) that the bandgap of the $Si_{1-x}Ge_x$ varies linearly with composition as the germanium content increases. On a semilogarithmic plot, sketch the current flowing for $x = 0, 0.1, 0.2,$ and 0.3 (0, 10, 20, and 30% germanium in the alloy). Assume that the current for the homojunction is 1 mA at 1 V applied bias.

Solution   For the band structure shown in Figure 5.11b, diffusion current dominates. For a heterojunction with a heavily doped n-type region in the wider bandgap semiconductor, the overwhelming majority of the current is electron current. We thus find the current using Equation 5.3.35, which shows that the current increases exponentially with increasing bandgap difference:

$$\frac{I}{I_0} = \exp\left(\frac{\Delta E_g}{kT}\right)$$

The bandgap in the $Si_{1-x}Ge_x$ alloy decreases with increasing germanium content as

$$E_g = 1.12 - x(1.12 - 0.67)(eV) = 1.12 - 0.45x\,(eV)$$

and

x	$\Delta E_g(eV)$	$E_g(eV)$	I/I_0
0	0	1.12	1
0.1	0.045	1.08	5.6
0.2	0.090	1.03	32
0.3	0.135	0.99	180

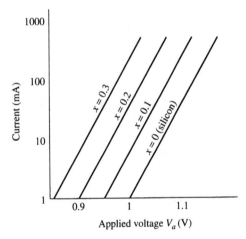

Note that the same current flows in the heterojunction diode as in the Si homojunction diode at a voltage reduced by $\Delta E_g/q$.   ∎

Now, we broaden our consideration to move toward the other limiting case of current flow. The situation of Figure 5.11b that we just considered is the optimum case. The entire applied bias is effective in reducing the barrier height and allowing carriers to flow across the junction; in addition, having $\Delta E_c = 0$ avoids difficulties with electron trapping and scattering in the notch. The situation becomes less advantageous as ΔE_c increases or the doping in the p-type region increases. (Higher doping in the p-region is needed to reduce series resistance when this region is the base of a transistor, as we will see in Chapters 6 and 7.) The consequences of electron trapping in the notch become more important as ΔE_c increases.

More seriously, when the spike in the conduction-band edge becomes higher than the conduction-band edge in the neutral region of semiconductor 2, as shown in Figure 5.11c, the analysis presented above fails. Electron transport from the n-type side of the heterojunction is now limited by thermionic emission of electrons over the energy barrier created by this spike, and $J_n \propto \exp(-q\phi_{Bn}/kT)$, where ϕ_{Bn} is now greater than $(E_{c2} - E_{c1})/q$. When a forward bias is applied, part of this voltage is dropped across the space-charge region in each piece of semiconductor. Only the portion of the applied bias dropped in material 1 reduces the barrier ϕ_{Bn} to electron flow from *n*-type material 1 to *p*-type material 2, and the current increases more slowly than $\exp(qV_a/kT)$ as the applied voltage increases. The improvement in device performance obtained by using the heterojunction then corresponds only to a fraction of the total bandgap difference.

The conditions at which the spike becomes limiting depend on the doping in the *n*- and *p*-type regions and also on the bias across the junction. As we saw in Sec. 4.2, the spike becomes more limiting when the *p*-region is heavily doped because most of the bending in the energy bands then occurs in the *n*-type material (Equations 4.2.30 and 4.2.31). Even if the spike does not limit the current flow at low forward bias, it can become limiting as the bias increases, depending on the relative decrease of the band bending on the two sides of the junction. A semilogarithmic plot of current versus voltage can show diffusion current at lower forward voltages and thermionic-emission-limited current at higher voltages [Figure 5.11*d* (adapted from [8] Figure 1.4)]. In practice, the spike and notch can be minimized at the expense of more complex fabrication by varying the composition between the two materials gradually (*grading* the junction) over an appropriate distance.

In this brief discussion of current flow across a heterojunction, we saw that the heterojunction gives us additional parameters that can be used to manipulate the ratio of electron- and hole-currents injected across the junction under forward bias. This additional control is very useful when designing heterojunction bipolar transistors, as we will discuss in Chapters 6 and 7. In particular, we will see in Sec. 7.6 that the reduction in hole current can increase the high-frequency performance of a heterojunction bipolar transistor well above that of a similarly doped homojunction bipolar transistor.

5.4 CHARGE STORAGE AND DIODE TRANSIENTS

In the previous section we saw that forward bias across a *pn* junction causes the injection of electrons from the *n*-type region into the *p*-type region and of holes in the opposite direction. After injection across the junction, these minority carriers move into the quasi-neutral regions. The resulting distribution of minority carriers leads both to current flow and to charge storage in the junction diode. In this section we consider the stored charge, its relation to the current, and its effect on the transient response of the *pn* junction to a change in diode bias.

On a fundamental level, the time-dependent behavior of minority carriers is included in the continuity equations (Equations 5.1.4). Because these partial differential equations are functions of time and position, they can be solved for various forcing functions and initial conditions to yield the transient behavior of minority carriers. This is generally not done in practice for several reasons. First, explicit solutions can be obtained only in special cases and for idealized forcing functions that can only approximate the real circuit conditions. It therefore makes little sense to carry out the exact mathematics of a solution when that solution will only apply approximately to the real problem. A second reason for not carrying out the solution of the partial differential equation is that the diode transient is not only governed by the minority charges stored in the quasi-neutral regions, but is also a function of charge storage in the depletion region. Thus, one must assess the changes in both charge stores simultaneously. The best way to look at the transient problem is to consider the physical behavior of these charges directly.

Minority-Carrier Storage

The total injected minority-carrier charge per unit area stored in the quasi-neutral n-type region can be found by integrating the excess hole distribution across the quasi-neutral region.

$$Q_p = q \int_{x_n}^{W_B} p'_n(x) \, dx \qquad (5.4.1)$$

Because the region is quasi-neutral, the n-region contains Q_p/q extra electrons per unit area above the thermal-equilibrium value to balance the $+Q_p$ charge of the holes.

We first consider the long-base diode, in which all of the injected minority carriers recombine before they reach the end of the n-type region. For simplicity, we also assume that each region of the semiconductor is uniformly doped. From Equation 5.3.12, we know that the minority-carrier distribution decays exponentially as the holes diffuse farther into the n-type region. Inserting Equation 5.3.12 into Equation 5.4.1 and integrating from the edge of the space-charge region across the quasi-neutral region, we find the charge Q_p resulting from holes stored in the n-type region to be

$$Q_p = qL_p p_{no} (e^{qV_a/kT} - 1) \qquad (5.4.2)$$

The stored charge, like the current, depends exponentially on applied bias. This behavior is expected because the exponential term arises from the magnitude of the excess minority-carrier density at the edge of the space-charge region for both the current and the stored charge. We can gain further insight into the physical behavior by expressing the stored charge in terms of the corresponding hole current. By using the expression for the hole current (Equation 5.3.13) at $x = x_n$ in Equation 5.4.2, we find the simple equation

$$Q_p = \frac{L_p^2}{D_p} J_p(x_n) = \tau_p J_p(x_n) \qquad (5.4.3)$$

Thus, the stored charge is the product of the current and the lifetime for this case of the ideal long-base diode. This relationship is reasonable because the injected carriers diffuse farther into the n-type region before recombining if their lifetime is greater; more holes are then stored.

The situation in the ideal short-base diode is slightly different. To treat it, we use the expression for the hole distribution (Equation 5.3.17) in Equation 5.4.1 to obtain the stored hole charge as

$$Q_p = \frac{q(W_B - x_n)}{2} p_{no} (e^{qV_a/kT} - 1) \qquad (5.4.4)$$

where W_B is the length of the n-type region (Figure 5.5). Again, we can express the stored hole charge in terms of the hole current (Equation 5.3.18) as

$$Q_p = \frac{(W_B - x_n)^2}{2D_p} J_p \qquad (5.4.5)$$

Although the group of constants in front of J_p in Equation 5.4.5 has the dimensions of time, it is not the lifetime, as was found in Equation 5.4.3 for the long-base diode. To see the physical significance of this group of constants, we rewrite Equation 5.4.5 as

$$\frac{Q_p}{J_p} = \frac{(W_B - x_n)^2}{2D_p} \qquad (5.4.6)$$

The amount of stored charge divided by the rate at which the charge enters or leaves the *n*-type region is just equal to the time an average carrier spends in this region. Therefore, the right-hand side of Equation 5.4.6 is equal to the average *transit time* τ_{tr} of a hole moving through the *n*-type region of the short-base diode.

Transient Behavior of Minority-Carrier Storage.[†]

We can extend the physical picture of minority-carrier storage to discuss qualitatively the nature of the transient buildup and decay of Q_p. Consider the behavior of holes in the *n*-region of an initially uncharged, ideal, long-base diode to which a positive constant-current source is suddenly applied. Before steady state can be reached, holes must be transported into the quasi-neutral *n*-region to establish a particular hole distribution. In theory, at $t = 0^+$ a large gradient of holes is possible because the *n*-region is nearly free of holes at that time. The actual gradient is set by the size of the current source. The stored hole charge Q_p increases with time as holes are supplied, and the voltage across the junction increases to reflect the boundary value of the hole density. A sketch of the transient increase of holes for this situation is shown in Figure 5.12. The time required to reach the steady-state hole distribution is given by the ratio of the steady-state stored hole charge to the size of the current source (neglecting recombination of the holes and charges stored in interface traps near the space-charge-region edges).

The turn-off time of the diode is limited by the speed with which stored holes can be removed from the quasi-neutral region. When a reverse bias is suddenly applied across the forward-biased junction, the current can reverse direction quickly because the gradient near the edge of the space-charge region can change with only a small change in the number of stored holes (curve 1 of Figure 5.13*a*). As long as sufficient minority carriers are present at the edge of the space-charge region, the diode is able to conduct a large amount of current in the reverse direction. The junction remains forward biased until the injected minority carriers near the edge of the depletion region are removed. This means that a plot of current versus time (Figure 5.13*b*) is initially nearly constant (and determined by the external circuit), and remains in this condition until t_4, at which time $p_n'(x_n)$ goes to zero. Subsequently, (curves 5 and 6 in Figure 5.13*a*), the rate at

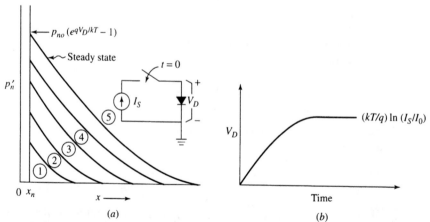

(a)   (b)

FIGURE 5.12 (*a*) Transient increase of excess stored holes in a long-base ideal diode for a constant current drive applied at time zero with the diode initially unbiased. Note the constant gradient at $x = x_n$ as time increases from (1) through (5), which indicates a constant injected hole current. (Circuit shown in inset.) (*b*) Diode voltage V_D versus time.

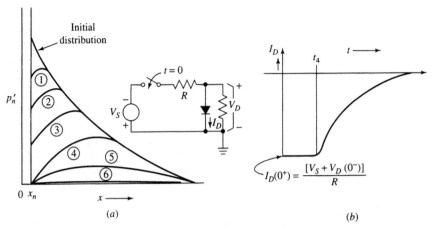

FIGURE 5.13 (a) Transient decay of excess stored holes in a long-base ideal diode. In the case shown, the initial forward bias applied through the series resistor is abruptly changed to a negative bias at time $t = 0$. (Circuit shown in inset.) (b) Diode current I_D versus time.

which holes can be delivered across the space-charge region becomes more and more limited. The current then decays on a time scale determined largely by carrier lifetime in the quasi-neutral region.

Because the switching time of a diode depends on the amount of stored charge that must be injected or removed, we can decrease the switching time by reducing the stored charge. From Equation 5.4.3 we see that this reduction in stored charge can be achieved either by restricting the forward current or by decreasing the lifetime of minority carriers. Consequently, optimal design for a given diode may not maximize the minority-carrier lifetime by forming the purest material possible. In fact, in some applications where devices must switch rapidly, recombination centers are purposely introduced into the semiconductor to reduce the lifetime so that little charge is stored. The critical concern is, of course, to introduce these "lifetime-killing" impurities in a controlled manner so that the amount of charge storage is reduced without increasing the reverse leakage current ($-J_0$ in Equation 5.3.15) to unacceptable levels. For example, gold can be diffused into a semiconductor wafer at a precisely controlled temperature.

The alternative method of reducing the switching time by restricting the forward current is more generally used in integrated circuits to avoid degrading the quality of the semiconductor material in which adjacent devices are built. The forward current through a *pn* junction can be restricted by the Schottky-diode clamping technique discussed in Sec. 3.6. Because a Schottky diode has a lower turn-on voltage than a *pn* junction, the voltage across a *pn* junction and the current through it can be limited by placing a Schottky diode in parallel with the *pn* junction. Charge storage in the *pn* junction is then decreased, and the turn-off time is reduced. Additional current flows through the Schottky diode, of course, but very little minority charge is stored there.

Thus far, we have discussed hole storage in the quasi-neutral *n*-type side of the *pn* junction. There is, of course, a similar storage of electrons in the *p*-region, and expressions analogous to Equations 5.4.2 and 5.4.4 can be derived to express this stored electron charge. The total minority-carrier charge stored in the diode is the sum of the two components.

We will find these concepts of stored minority charge, transit time, and the interrelation between them to be useful in discussing transients in *pn* junctions. To treat the transient problem, we can neglect changes in the majority-carrier concentrations because majority carriers respond to changes in fields within a dielectric relaxation time $\tau_r = \epsilon_s/\sigma$, which is of the order of 10^{-13} s (0.1 picosecond) for 0.1 Ω-cm silicon. This time is much shorter than the lifetime (Equation 5.4.3) or the transit time (Equation 5.4.6) of minority carriers. For integrated-circuit devices these times are typically between 10^{-12} and 10^{-3} s.

Total Junction Storage. In Section 4.3 we examined one mechanism of charge storage in a *pn* junction. We saw that majority carriers near the edges of the depletion region move as the depletion region expands or contracts in response to a changing reverse bias. We saw that this charge storage in the depletion region can be modeled by a small-signal capacitance. Capacitance arising from charge storage in the depletion region is usually denoted by C_j. An example is the abrupt-junction depletion capacitance derived in Equation 4.3.8.

Similarly, the variation of stored minority-carrier charge in the quasi-neutral regions under forward bias can be modeled by another small-signal capacitance. This capacitance is usually called the *diffusion capacitance* (denoted C_d), because the minority carriers move across the quasi-neutral region by diffusion. We can find the contribution to C_d from the stored holes in the *n*-region by using the definition $C_d = dQ_p/dV_a$ in Equation 5.4.2 for the long-base ideal diode or in Equation 5.4.4 for the short-base diode. (Because Q_p represents charge per unit area, C_d denotes capacitance per unit area.) Because both diodes have the same voltage dependence, we find that, in general,

$$C_d = \frac{qQ_{po}e^{qV_a/kT}}{kT} \tag{5.4.7}$$

where

$$Q_{po} = J_{p0}\tau_p = qp_{no}L_p \tag{5.4.8}$$

for the long-base diode (where $J_{p0} = qD_pp_{no}/L_p$), and

$$Q_{po} = J_{p0}\tau_{tr} = qp_{no}(W_B - x_n)/2 \tag{5.4.9}$$

for the short-base diode (where $J_{p0} = qD_pp_{no}/(W_B - x_n)$).

Again, it is straightforward to add components to C_d to represent electron storage in the *p*-region in cases where that storage is significant. As expected, Equation 5.4.7 shows that C_d is negligible under reverse bias because the minority-carrier storage is very small. Under forward bias, both Q_p and C_d increase exponentially with increasing voltage.

From the foregoing we see that the relative significance of charge storage in the space-charge region (as represented by C_j) and charge storage in the quasi-neutral regions depends strongly on the junction voltage. Under reverse bias, storage in the quasi-neutral regions is negligible and the storage represented by the junction capacitance dominates. Under forward bias, although C_j increases (because x_d decreases), the exponential factor in the formula for C_d generally makes diffusion capacitance and its associated charge storage dominant. For accurate solution of practical problems, however, it is frequently necessary to account for both types of charge storage in the forward-bias region.

EXAMPLE   *Junction and Free-Carrier Storage*

An abrupt, *pn*–junction, long-base diode in which $L_p = 20$ μm is doped with $N_a = 10^{19}$ cm^{-3} in the *p*–region and with $N_d = 10^{16}$ cm^{-3} in the *n*–region. Plot the charge Q_p as a function of applied voltage V_a for $-3 < V_a < 0.6$ V. On the same axes plot Q_V, the charge added to the *n*–region to change the depletion-region charge storage when the junction is biased.

Solution   For this p^+n diode we neglect electron injection into the heavily doped p–region and use Equation 5.4.8 to calculate $Q_{po} = 6.7 \times 10^{-18}$ C cm^{-2}. Q_p is obtained by multiplying by the diode factor

$$Q_p = Q_{po}\left(\exp\frac{qV_a}{kT} - 1\right)$$

Q_V is just the negative of the difference in the charge stored in the depletion region at $V = V_a$ and the charge stored there at $V = 0$. Hence, using Equation 4.3.1,

$$Q_V = -qN_d(x_d - x_{do}) = -\sqrt{2\epsilon_s qN_d}[(\phi_i - V_a)^{1/2} - \phi_i^{1/2}]$$

where we find $\phi_i = 0.872$ V from Equation 4.2.10.

We use these two expressions to calculate the values of Q_p and Q_V plotted in the accompanying figure. For all values of negative bias, we find that Q_p is negligible compared to Q_V. Both charge storages are negative. At a forward bias between 0.5 and 0.6 V, the two charge densities are equal. At 0.6 V, Q_p is 84.3 nC cm^{-2}, about 3.4 times as large as Q_V, and at higher voltages Q_p is far larger than Q_V. This example emphasizes that under reverse bias, charge storage at the depleted junction (Q_V) is most important, while under forward bias, minority-carrier storage (Q_p) dominates.

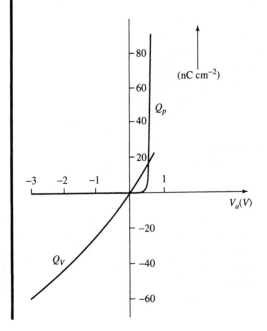

Before we conclude this section, we briefly discuss charge storage in a heterojunction. We again consider a wider bandgap n-type semiconductor and a narrower bandgap p-type semiconductor and assume that diffusion current dominates. We saw earlier in this section that the number of excess holes Q_p (per unit area) stored in the n-type neutral region of a junction under forward bias is proportional to the excess hole density at the edge of the neutral region (Equation 5.4.2). Because the excess hole density there is proportional to the current, the number of excess stored holes is also proportional to the current (Equations 5.4.3 and 5.4.5).

We saw in Sec. 5.3 that the ratio of electron current injected into the p-type region to hole current injected into the n-type region is significantly different in a heterojunction than it is in a homojunction. For a given electron current, the hole current is significantly reduced. Consequently, the number of holes (per unit area) stored in the *n*-type region of the heterojunction is markedly smaller than the number stored in a homojunction. The reduced hole storage allows faster response of the diode. The reduced charge storage and faster response of the heterojunction become especially beneficial when the heterojunction is used in a *heterojunction bipolar transistor* (HBT); we will discuss the frequency response of HBTs in Sec. 7.6.

5.5   DEVICE MODELING AND SIMULATION

The nature of charge storage at the *pn* junction strongly complicates the use of equivalent circuits for hand calculations involving transient problems. For example, the most frequent use of diodes in integrated circuits is as switches, which are sequentially forward- and reverse-biased during circuit operation. In this application, the dominant charge storage shifts during the transient itself between the diffusion and depletion-layer charges. To make hand calculations in such situations, the most successful technique is, first, to determine the total charge storage at the beginning and end of the transient. Then, the time required for switching is calculated by dividing the change in stored charge ΔQ by the driving current I that causes the diode to switch (Problem 5.18).

This technique does not, of course, give an accurate picture of the current and voltage of the diode as functions of time. For many applications, such detail is not of interest, but if it is important, the diode can be *piecewise-linearly* approximated. The phrase "piecewise linear" means that the nonlinear charge-storage effects can be approximated to first order by linear elements over a small voltage range. If the voltage increment is made small enough, this approximation is accurate, and an arbitrary precision can be obtained by joining together a sufficient number of piecewise-linear approximations to represent the entire voltage variation in a given transient problem.

Lumped-Element Model

Although we did not label them as such, the small-signal capacitances C_j and C_d in Equations 4.3.8 and 5.4.7 are piecewise-linear approximations because they represent an overall nonlinear charge-storage effect in terms of linear circuit elements (capacitors). A complete piecewise-linear model for a *pn* junction must include a conductance as well as a capacitance to represent the real current through the diode. The conductance (per unit area) in the ideal diode is (from Equation 5.3.15)

$$g_d = \frac{dJ}{dV_a} = \frac{q}{kT}J_0\left(e^{qV_a/kT}\right) = \frac{q}{kT}(J + J_0) \tag{5.5.1}$$

The total piecewise-linear circuit for the diode consists of g_d in parallel with C_j and C_d (Figure 5.14). Because we have discussed only basic junction processes, the circuit in Figure 5.14 does not include elements that are inherent to the planar technology used for IC fabrication. We will discuss these in Sec. 5.5.

For specialized applications such as the low-voltage sinusoidal excitation of a diode that is biased quiescently (at dc) at an operating point, analysis with the circuit shown in Figure 5.14 provides adequate accuracy. It is, therefore, called the *diode small-signal equivalent circuit*. The diode small-signal equivalent circuit is especially useful for analysis of a wide range of communication and linear amplification circuits.

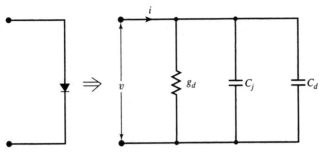

FIGURE 5.14 Diode small-signal equivalent circuit—a piece-wise–linear representation for the nonlinear current-voltage re-lation of a junction diode. Element values (per unit area) for a long-base ideal diode are given by Eqs. 4.3.8 for C_j, 5.4.7 for C_d, and 5.5.1 for g_d. This equivalent circuit is only valid for small excursions from the operating point at which the values of C_j, C_d, and g_d are calculated.

Hand calculations using the small-signal equivalent circuit in a piecewise-linear fashion to analyze a large change in voltage quickly become cumbersome. Because of the complexity, the problem is rapidly and gratefully given to a computer in all cases except exercises for university students. To carry out the analysis on a computer, an initial state is specified, for which starting values of the currents, voltages, and capacitances are found. Conditions are then allowed to change a small amount, and new values are calculated.

To illustrate this approach, consider the circuit in Figure 5.15. The initial state is taken to correspond to an initially unbiased circuit ($V_S = 0$ in Figure 5.15). At time $t = 0$, the source voltage V_S is changed. Because the voltage across a capacitor cannot change instantaneously, the size of the capacitor and the value of the current source that represents the diode can be calculated from the initial state (at $t = 0^-$). The elements in the circuit are assumed to keep these values for a time increment Δt. At the end of Δt, the voltage across the diode has changed to a new value V_D. This new voltage V_D is then used to compute new values for the components of the equivalent circuit, and the procedure is repeated in an iterative manner until the diode reaches its final state.

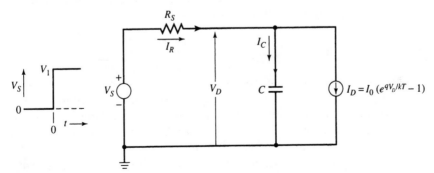

FIGURE 5.15 Circuit illustrating a typical diode switching problem. The source voltage V_S changes from zero to V_1 at $t = 0$. The analysis of this problem is discussed in the text.

If the circuit of Figure 5.15 is analyzed by a computer program in a piecewise-linear manner, the sequence of operations might be the following:

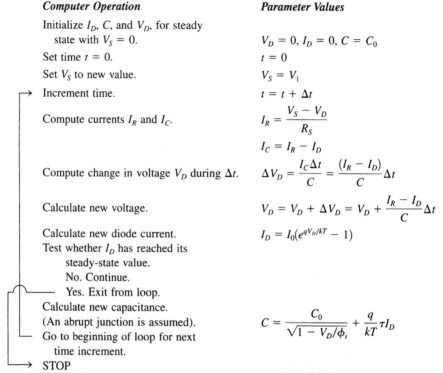

Computer Operation	Parameter Values
Initialize I_D, C, and V_D, for steady state with $V_S = 0$.	$V_D = 0$, $I_D = 0$, $C = C_0$
Set time $t = 0$.	$t = 0$
Set V_S to new value.	$V_S = V_1$
Increment time.	$t = t + \Delta t$
Compute currents I_R and I_C.	$I_R = \dfrac{V_S - V_D}{R_S}$
	$I_C = I_R - I_D$
Compute change in voltage V_D during Δt.	$\Delta V_D = \dfrac{I_C \Delta t}{C} = \dfrac{(I_R - I_D)}{C} \Delta t$
Calculate new voltage.	$V_D = V_D + \Delta V_D = V_D + \dfrac{I_R - I_D}{C} \Delta t$
Calculate new diode current.	$I_D = I_0(e^{qV_D/kT} - 1)$
Test whether I_D has reached its steady-state value.	
No. Continue.	
Yes. Exit from loop.	
Calculate new capacitance. (An abrupt junction is assumed). Go to beginning of loop for next time increment.	$C = \dfrac{C_0}{\sqrt{1 - V_D/\phi_i}} + \dfrac{q}{kT}\tau I_D$
STOP	

The analysis continues until the diode current approaches its steady-state value within some predetermined precision. The computer instructions may also include output commands so that any variable, such as voltage or current, can be plotted to give the user an indication of the circuit behavior as a function of time.

This piecewise-linear numerical approach can be utilized in any nonlinear circuit where computer simulation is the only reasonable means for studying the behavior of the total circuit. It is one of several alternative computational approaches (or algorithms) that enable the computer solution of nonlinear problems. Models for diodes and other nonlinear circuit elements based either on piecewise-linear approximations or on other techniques are typically included as subroutines in more elaborate computer-analysis programs for integrated circuits. The use of these computer-simulation programs is essential when designing and analyzing integrated circuits.

Distributed Simulation[†]

The piecewise-continuous analysis discussed above introduces the idea of calculating device behavior by iterative solution. In that case, we considered lumped elements to help visualize the concept, but more-detailed device-simulation programs that use iterative solutions to study the complex physics of electronic processes within a device can provide more detailed insight into the device operation.

To discuss device simulation we use as an example a popular device simulator, PISCES [10]. PISCES is widely available as free university software or as a commercially obtainable "industrial strength," supported-software package. Other simulators share many of the basic elements described here. PISCES solves Poisson's equation (Equation 4.1.10) and the current-continuity equations (Equations 5.1.3) for one or two carriers in two dimensions at each point of a nonuniform grid to simulate the electrical characteristics of semiconductor devices under either steady-state or transient conditions. By solving these governing equations with few approximations, the simulator is very useful in the early stages of process- and device-development to design fabrication experiments, to understand device operation, and to find potential problems [11].

Poisson's equation describes the electrostatic potential in terms of the fixed and mobile charges in the device and can be written

$$\nabla^2 \phi = -\frac{\rho}{\epsilon} = -\frac{q}{\epsilon}(p - n + N_d^+ - N_a^-) - \frac{\rho_F}{\epsilon} \qquad (5.5.2)$$

where N_d^+ and N_a^- are the ionized impurity concentrations ($N_d^+ \leq N_d$ and $N_a^- \leq N_a$), and ρ_F is the fixed charge density (especially useful when insulators are being considered).

The electron and hole continuity equations describe the carrier concentrations:

$$\frac{\partial n}{\partial t} = \frac{1}{q}\nabla J_n - U_n \qquad (5.5.3a)$$

$$\frac{\partial p}{\partial t} = -\frac{1}{q}\nabla J_p - U_p \qquad (5.5.3b)$$

The quantities U_n and U_p are the net electron- and hole-recombination rates and can include both Shockley-Hall-Read generation-recombination (Equation 5.2.9) and Auger recombination (Equation 5.2.16). The lifetimes τ_n and τ_p used in the expressions for the recombination rates can be functions of the carrier concentrations.

As for the process simulators discussed in Sec. 2.8, a grid is first established; the grid is usually nonuniform (often triangular) and includes doping profiles obtained from analytical functions or from a process simulator such as SUPREM (Sec. 2.8). Three equations are associated with each grid point: Poisson's equation and one continuity equation for each of the two carrier types. In a full solution of Poisson's equation and the two current-continuity equations, the simulator solves the three partial-differential equations self-consistently for the potential ϕ and the electron and hole concentrations n and p, throughout a structure that can contain several regions with different electronic material properties.

The continuous analytical differential equations must first be described at discrete locations in space ("discretized") by converting them to difference equations at each of the N nodes of the simulated structure. Each node represents the small volume surrounding it. The net flux of charge entering the volume is considered, along with the sources and sinks inside it, to conserve charge and current. The behavior of each node depends primarily on immediately surrounding nodes. The potential and carrier density are then calculated at each node. The current is calculated along each line segment of each triangle, and the potential gradient (electric field) along each line segment is calculated as the difference in the potentials at the two end points divided by the length of the line segment connecting them.

However, the set of 3N algebraic equations to be solved (one for Poisson's equation and two for the continuity equations) is coupled and nonlinear, and cannot be directly solved. Solutions need to be obtained by a nonlinear iteration method starting from some initial "guess." Each available numerical technique involves solving several large systems of equations, with the total number of equations approximately equal to 1–3 times the number of nodes.

In some cases, analysis can be simplified and computation time can be reduced. For example, when current is not flowing, solving only Poisson's equation is adequate (N equations). In other cases, the behavior of only one carrier needs to be obtained so that Poisson's equation can be solved together with one continuity equation (2N equations). This type of simulation can be useful for limited analysis of majority–carrier devices such as MOS transistors (MOSFETs) (Chapters 9 and 10), JFETs (Sec. 4.5), and MESFETs (Sec. 4.5), and also for devices without forward-biased junctions (e.g., charge-coupled devices and capacitors). However, minority-carrier effects are becoming increasingly important even in majority-carrier devices, such as MOSFETs. For example, substrate current and hot-carrier injection into (and degradation of) the gate oxide (Chapter 10) can greatly influence the design of MOSFETs. Therefore, for careful MOSFET analysis, both carrier types need to be considered (3N equations). For bipolar transistors (Chapters 6 and 7), both carrier types must be included.

When using iteration, the rate of convergence of the calculation is important to reduce the number of iterations. Some solutions converge at a linear rate, with the error decreasing by about the same factor for each iteration. Other methods converge more rapidly. Two common numerical techniques are Gummel's method and Newton's method.

In Gummel's method the equations are solved sequentially. Poisson's equation is solved assuming fixed carrier densities. Then the new potential is substituted into the continuity equations, which are linear and can be solved directly. The resulting carrier concentrations are then substituted back into the charge term of Poisson's equation and used in the next iteration of Poisson's equation. Because only one equation is being solved at any time, the matrix being solved contains N rows and N columns. Solving each equation separately works best when the coupling between equations is small. The most important coupling is the drift-current term. Therefore, Gummel's method converges quickly when the currents are small (e.g., in device isolation structures). It converges only slowly, if at all, when currents are large (e.g., in a resistor).

In Newton's method all the variables are allowed to change during each iteration, and the coupled algebraic equations are solved at the same time by matrix inversion. Newton's algorithm is very stable, and the convergence rate is nearly independent of the bias conditions. However, because potential, electrons, and holes are all being considered simultaneously, the matrix which must be inverted has three times as many columns and rows as for a single variable. The $3N \times 3N$ matrix can take twenty times as long to invert as an $N \times N$ matrix using the same grid. For Poisson's equation and one carrier type, a $2N \times 2N$ matrix is solved. Newton's method is preferred for solving one-carrier problems with current flow. For two-carrier problems, Newton's method is favored, but the time and required computer memory can be large for complex device structures. On the other hand, Gummel's method becomes progressively slower as the current increases and may not even converge for large currents.

Although the size of the matrix to be inverted is large, the matrix contains many zero elements. If each node depends only on immediately adjacent nodes, the matrix contains elements corresponding to coupling of each node to itself and to the four adjacent nodes (for a rectangular grid). The number of matrix elements is then 5N for Gummel's method. For Newton's method, the number of matrix elements increases by a factor of 4 (2×2) for a one-carrier solution and 9 (3×3) for a two-carrier solution. Adding more physical effects or a nonrectangular grid increases the number of matrix elements. However, the complexity of the matrix increases only linearly with the number of nodes, rather than as the square of the number of nodes because many of the nondiagonal elements are zero.

As in the case of process simulation, the time-dependent solution is obtained by solving the coupled set of equations in discrete time increments. The continuity equations (Equations 5.5.3) can be approximated by considering the incremental changes occurring during a time interval Δt_k and relating these changes to the values of the variables at only one preceding time step:

$$\frac{n_{k+1} - n_k}{\Delta t_k} = f_n(\phi_k, n_k, p_k) \tag{5.5.4}$$

and

$$\frac{p_{k+1} - p_k}{\Delta t_k} = f_p(\phi_k, n_k, p_k) \tag{5.5.5}$$

where f_n and f_p are the functions described in Equations 5.5.3a and 5.5.3b, respectively. More complex algorithms can also be used, with the increments depending on the values of the variables at more than one previous time step.

Convergence can be difficult when the parameters vary rapidly in space or time. Numerical problems often arise when applied voltages are changed abruptly; for example, by applying a step-function voltage change directly to a contact. To make the solutions converge more rapidly, the starting conditions for the calculations should be "guessed" intelligently. For example, this guess can be a previous solution for slightly different bias, an extrapolation from the previous solution, or the solution for zero bias.

When regions of dissimilar material are modeled, the boundary conditions at the interfaces must be specified. Several types of boundaries can be used, including ohmic contacts, Schottky contacts, insulating boundaries, and reflecting boundaries. At ohmic contacts the surface potential and electron and hole concentrations are fixed; the quasi-Fermi levels are fixed at the applied bias at that electrode; and no space charge can exist there. Schottky contacts are defined by the work-function difference between the electrode metal and the semiconductor and possibly by an optional surface-recombination velocity. Insulating boundaries are described by a work function, which determines the surface potential. Reflecting boundaries are used at noncontacted edges of simulated regions so that current only flows out of the device through the contacts. At the reflecting boundary the difference between the normal components of the electric displacements must be equal to the surface charge density along the interface.

As devices become smaller, second-order physical effects must be considered, as we will see in the remainder of this book. To depict the behavior of devices accurately, simulation must also incorporate models of these physical effects. In addition, as the number of devices on an integrated circuit increases, the interconnections between devices become increasingly important. For example, as interconnections are placed closer together, the capacitance between them becomes a larger fraction of the overall capacitance and cannot be neglected. In some circuits, the resistances and capacitances of the interconnections are as important as the properties of the active devices themselves. Therefore, simulation tools are available for simulating the behavior of integrated-circuit interconnection systems, in addition to programs for analyzing the active devices themselves.

As we saw in Sec. 2.8, correctly allocating the nodes (grid points) is critical in device simulation because it affects both the simulation time and the memory requirements. The number of computations increases faster than linearly with the number of grid points, often as a power in the range 1.5–2. For effective allocation of grid points, it is desirable to place more nodes in regions with rapidly varying properties, and fewer in regions with slowly varying properties. Similarly, for small and nonplanar device geometries, the grid must fit the device shape reasonably well (Fig. 2.40). An irregular grid structure permits the analysis of arbitrarily shaped devices and allows the grid to be refined in some regions without unnecessarily increasing the number of grid points in other regions.

However, an irregular grid is harder to specify than a rectangular grid, and automatic refining of the grid ("regridding") is useful in regions where key variables change rapidly. The initially specified grid can be a coarse rectangular mesh. The coarse grid is then automatically refined and perhaps converted to a triangular grid until it is fine enough to represent the structure accurately. This regridding is especially valuable when the device has a complicated doping profile. Initially, the grid can be refined in regions where the doping changes rapidly and rapid changes in the potential are expected. During the simulation, spatial variations in quantities such as charge density or potential may increase, decreasing the numerical accuracy. Advanced simulators can automatically redistribute the grid points to improve the accuracy during the simulation. For example, a coarse grid may first be refined based on abrupt changes in doping. Later in the simulation, the grid may be refined based on abrupt changes in potential.

With the increasing power of computers, a combination of process, device, and circuit simulation is increasingly being used to provide the insight needed to design advanced devices and circuits while minimizing the exorbitant processing costs of extensive experiments. In addition, very small devices and complex fabrication processes make it difficult to obtain from experimental structures physical insight and quantitative understanding of the factors governing device operation.

Because many of the elements of simulation, such as node refinement and graphical output, are common to process, device, and interconnection simulation, coupled simulators can provide efficient use of computer resources to provide an overall understanding of the factors governing device operation. A process simulator incorporates models used to find the geometrical structure and impurity distributions in the device. These outputs of the process simulator are used as the input of the device simulator, which provides information about the internal state of the device, as well as electrical terminal characteristics. The internal distributions of potentials, fields, and carrier concentrations provide understanding of the device behavior and allow improvement of the device structure (e.g., reducing peak electric fields). Device parameters extracted from the terminal characteristics and device layout are used in circuit simulators to predict circuit switching characteristics and other circuit performance parameters.

5.6   DEVICES

Integrated-Circuit Diodes

To form the simplest *pn*-junction diode using integrated-circuit technology, it is only necessary to diffuse a *p*-type region into an *n*-type wafer and make contact to the front and back of the wafer (Figure 5.16). If the diode is part of an integrated circuit, however, many *pn* junctions are typically formed on the same wafer, and simple diffusion as described above results in all of the *n*-regions being electrically connected. This, of course, would generally prevent the desired circuit operation. Unwanted dc interconnections between diodes, or more generally between devices in an integrated circuit, can be avoided by surrounding them with a combination of insulating oxide and *pn* junctions that are

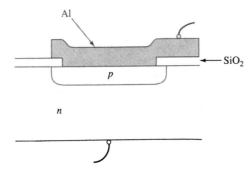

FIGURE 5.16 A diffused *pn* junction: a *p*-region is diffused into an *n*-type wafer; the surface is passivated with silicon dioxide, and ohmic contacts are made to the front and back of the wafer.

kept under reverse bias at all times. The oxide is usually used at the sides of the diode and the *pn* junction isolation is used beneath the diode to separate it from the underlying silicon substrate. This combination of oxide and junction isolation is frequently used. We illustrate some aspects of diode structures in integrated circuits by considering an array of diodes.

To achieve an isolated array of diodes, we usually start with a *p*-type wafer on which a thin *n*-type epitaxial silicon layer is grown, as shown in Figure 5.17*a*. An "epi layer" is typically 0.5–5 μm thick for a digital logic switching circuit and as thick as 5–20 μm for a linear circuit (because of the higher operating voltages). The *pn* junction at the epi-substrate interface provides electrical isolation in the vertical direction as long as it is reverse biased. The substrate is usually relatively lightly doped *p*-type material (typically 20 Ω-cm with $N_a = 7 \times 10^{14}$ cm^{-3}) to provide a wide depletion region, which reduces undesired or "parasitic" junction capacitance between the active region and the substrate and maintains a high breakdown voltage.

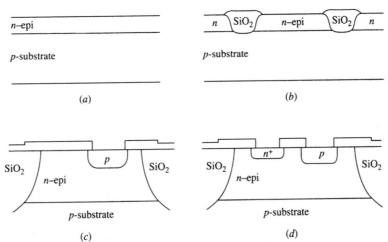

FIGURE 5.17 Planar process steps used to form an isolated junction diode in an integrated circuit: (*a*) substrate junction formed by *n*-type epitaxial layer on *p*-type substrate; (*b*) local oxidation (LOCOS) through thickness of *n*-type epitaxial layer to form lateral isolation; (*c*) shallow *p*-type diode diffusion into epitaxial layer; (*d*) highly doped, *n*-type diffusion into *n*-type epitaxial layer to provide ohmic contact to *n*-type region. (Vertical scale expanded in parts *c* and *d*.)

After the epitaxial layer is grown, the following sequence of steps includes photomasking, etching, and oxide growth to form the isolating web of oxide regions. The oxide can be formed by the local oxidation of silicon (LOCOS) process described in Sec. 2.6, perhaps combined with some etching of the silicon in the isolation regions before the oxide is grown. Because the oxide must penetrate completely through the thickness of the epitaxial layer (Figure 5.17*b*), the LOCOS process can only provide isolation for epitaxial layers up to approximately 1 μm thick. For thicker epitaxial layers, the "trench isolation" process (also described in Sec. 2.6) can be used. Trenches are etched through the entire thickness of the epitaxial layer (plus slightly into the substrate to accommodate process variations) in the isolation regions. A thin oxide is grown on the walls of the trench, and then the remainder of the trench is filled by chemical vapor deposition of either oxide or polycrystalline silicon. The excess deposited material outside the trenches is removed by polishing. This combination of oxide isolation and *pn*-junction isolation creates a "well" or "tub" in which the integrated-circuit device is then fabricated.

After isolation, the next step in producing a *pn* junction diode is to introduce a diffused *p*-region, keeping the diffusion time relatively short so that a *pn* junction is formed within the *n*-type tub (Figure 5.17*c*). It is important at this point to keep the two space-charge regions—one from the diode junction and one from the substrate isolation junction—from coming into contact. Proper design must insure that they do not touch, regardless of the voltages that will be applied on the diode and substrate junctions during "worst case" circuit operation. Because higher reverse bias widens space-charge regions, the thickness and dopant concentrations are chosen to be compatible with the highest reverse-bias conditions. If the two space-charge regions do meet, the condition is described by the term *punch-through* because the quasi-neutral *n*-region separating the two *p*-regions is "punched through." Under this condition the *p*-region for the diode and the *p*-type substrate are no longer isolated, and large currents can flow between them (Problem 5.20).

The next step is to make electrical connection to both sides of the diode from the top of the wafer. First, a highly doped n^+ contact diffusion is added to the *n*-type region (Figure 5.17*d*) to achieve good ohmic contact. (As we saw in Chapter 3, aluminum does not make ohmic contact to lightly doped *n*-type silicon.) After contact areas are opened, the diode is completed by covering the wafer with aluminum, defining the aluminum pattern by selective etching, and alloying the aluminum to make good contact with the silicon substrate. (In Figure 5.17, as in most sketches of IC cross sections, the vertical dimension has been exaggerated for clarity.)

An integrated-circuit diode made by the process outlined above suffers from a serious disadvantage, however. To obtain a relatively large current-carrying area, it is desirable to have the current flow vertically through the diode. However, this requires that majority-carrier electrons flow to the junction region through a path that includes the long, thin *n*-region between the two *pn* junctions. This is a path very like the pinch resistor described at the end of Chapter 2. For a typical epitaxial-layer resistivity of 0.5 Ω-cm and typical integrated-circuit dimensions, the series resistance introduced by this path is several kilohms. Such a high series resistance would produce sufficient ohmic drop, even at relatively low forward currents, to cause the *pn* junction bias to vary along its length. Because current depends exponentially on the junction voltage under forward bias, there would be *current crowding* toward the regions of the junction nearest the n^+ contact, as shown in Figure 5.18*a*. The effective area of the diode would be reduced from its geometric area, leading to many undesirable effects.

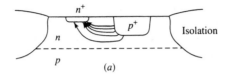

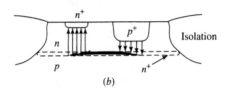

FIGURE 5.18 (a) A high series resistance in the epitaxial layer causes the current to "crowd" toward the portion of the *pn* junction nearest the n^+ ohmic contact. (b) A low-resistance *buried layer* allows uniform current flow across the entire diode area.

To eliminate this problem, a heavily doped *n*-type region is formed on the substrate before growing the epitaxial layer (Figure 5.19a). This region is usually included beneath each junction region. After the epitaxial layer is grown and the oxidation and diffusion discussed above are carried out, the structure appears as shown in Figure 5.19b. By using this *buried layer* of *n*-type material, the series resistance is reduced to a few ohms, which is generally quite acceptable, and current flows relatively uniformly across the entire diode area (Figure 5.18b). An added advantage of the buried layer is that it reduces the possibility of punch-through between the *p*-region at the surface and the substrate because the added donors in the buried layer inhibit the extension of the space-charge region into the *n*-type region. Although punch-through is inhibited, the breakdown voltage of the junction diode may be reduced by the presence of the buried layer because it confines the space-charge region and thus increases the maximum field.

Incorporating a buried layer does add some fabrication complications to integrated-circuit production. First, because the buried-layer impurities are added before the epitaxial growth, these impurities go through the temperature cycling needed for the growth and also for all the subsequent processing steps. This means that a slow-diffusing, relatively nonvolatile impurity must be used. The slow diffusion is required to maintain the desired dimensions for the devices; nonvolatility is necessary to avoid *autodoping* the epitaxial layer during its growth. In practice, relatively good results are obtained by using arsenic or antimony for the buried-layer dopant. The *p*-type diffusions are then carried out with boron, which diffuses much faster. Arsenic is used for subsequent shallow, *n*-type diffusions. The lack of a *p*-type dopant with the desirable slow-diffusing properties of arsenic or antimony makes it difficult to fabricate practical *p*-type buried layers.

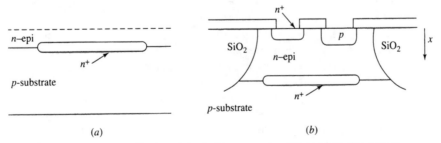

FIGURE 5.19 (a) A heavily doped *buried layer* can be diffused into the *p*-type substrate before the epitaxial layer is grown; this reduces resistance in series with the diode. (b) An integrated-circuit diode incorporating a buried layer.

Because the *n*-type buried layer narrows the space-charge region between the *n*-type region and the *p*-type substrate, it adds to the parasitic capacitance between the epitaxial layer and the substrate. This is one reason that the buried layer is usually introduced only beneath the active regions. One unavoidable disadvantage of the incorporation of a buried layer is that the strain associated with its high impurity density can reduce the lifetime and mobility in the epitaxial region above it. Because this is the critical region for device performance, the drawback can be severe, but it is less with arsenic than with antimony. Optimum design requires choosing a buried-layer doping that balances these disadvantages with the advantages that motivated the inclusion of the buried layer.

As a final topic concerned with the practical aspects of junction diodes for integrated-circuits, we give some perspective to the widespread use of tables and nomographs for design and analysis. Properties of *pn* junctions formed by (1) a Gaussian diffusion profile into a constant background doping and (2) a complementary-error function profile into a constant background doping have been published by Lawrence and Warner [12]. These authors obtained their results by finding exact solutions of the junction equations using computer routines. Their calculations permit the determination of junction capacitance, total depletion-layer thickness, and the width of the depletion layer on either side of the junction. Reference [12] presents the data in especially useful fashion. To make use of the Lawrence and Warner results, it is necessary to know the background dopant density of the region of constant doping as well as the density of the dopant at the surface from which diffusion is taking place. These data specify a family of curves that are plots of the output data as functions of total junction voltage. Because it is a closely related calculation, the Lawrence-Warner analysis also provides curves of the maximum junction field versus total voltage across the junctions. Reference to these results can provide a major, time-saving aid to the designer of an integrated circuit and can be a useful prelude to detailed numerical device simulation.

Finally we note that although the LOCOS isolation technique that we described in this section is frequently employed in IC design, newer techniques are often used for small-geometry circuits. For example, the trench isolation process described in Sec. 2.6 produces regions of insulating silicon dioxide between devices without the gradual transition of the LOCOS process. This newer technology significantly reduces the area needed for the isolation regions.

Light-Emitting Diodes

In an efficient forward-biased diode (i.e., one with little recombination in the space charge region), carriers are injected from the neutral region on one side of the diode across the narrowed space-charge region into the neutral region, where they are minority carriers. They diffuse in the neutral region and recombine with majority carriers. In an indirect-bandgap semiconductor, such as silicon, recombination of an electron and a hole requires a change in momentum, as well as a loss of energy. Therefore, direct recombination is unlikely, and the carriers recombine through midgap centers, as discussed in Sec. 5.1.

However, in direct-bandgap semiconductors, such as gallium arsenide, no change in momentum is needed and electrons and holes can recombine directly, with the electron falling from the conduction band to fill the hole in the valence band. The energy derived from this transition is often emitted as a photon of light, which can escape from a properly designed semiconductor structure. Such *light-emitting diodes* (LEDs) are widely used. Red LEDs are routinely used in indicators and in small displays in common household appliances (e.g., for clocks). The color emitted is related to the bandgap of the semiconductor, and most common direct-bandgap semiconductors emit photons corresponding to red or near infrared light ($E_g \approx$ 1.5–1.8 eV). To cover the higher-energy, blue-to-green portion of the spectrum, wider-bandgap semiconductors based on gallium nitride and related compounds have been developed. The availability of LEDs covering the entire visible spectrum has increased applications for these energy-efficient LEDs to include traffic lights. In this application, not only does the energy efficiency of LEDs reduce the cost of the electric power used but their long lifetimes (years) greatly reduce the labor needed to change lamps that fail.

EXAMPLE: *Efficiency of LEDs*

Calculate the efficiency of an LED (light power out/electrical power in) for an LED that produces 2×10^{15} photons/sec with a wavelength of 0.8 μm when a current of 1 mA is supplied at 1 V.

Solution   The energy of each photon is 1.24 eV/λ (μm) = 1.24 eV/0.8 = 1.55 eV = 2.48 $\times$ 10^{-19} joules. The corresponding power of the total light emitted is then $2.48 \times 10^{-19} \times 2 \times 10^{15}$ joules/sec = 5×10^{-4} watts. The electrical power into the LED is 10^{-3} watts, so the efficiency is 0.5 = 50%. ∎

SUMMARY

The continuity equation for free carriers, a partial differential equation that accounts for the mechanisms causing an increase or a decrease in carrier densities, is a powerful tool for semiconductor device analysis. By solving the continuity equation for minority carriers in the quasi-neutral regions near a *pn* junction, we can obtain the carrier densities. Boundary values for minority-carrier densities at the edges of the quasi-neutral regions can be written as functions of applied junction bias. Using these boundary values and the defining relationships for current as a function of free-carrier density, it is possible to obtain current-voltage characteristics in the steady state for several simple cases with important practical applications. These solutions lead to the *ideal-diode equation* $J = J_0[\exp(qV_a/kT) - 1]$.

Solutions to the time-dependent continuity equation can be carried out in simple cases to give meaning to the concept of the lifetime of excess carriers. In most cases of interest, the recombination rate, which determines the carrier *lifetime*, is a function of the properties of recombination centers. These recombination centers consist of localized electronic states having energies within the forbidden gap, typically close to the intrinsic Fermi level. The theory for recombination through localized centers is called *Shockley–Hall–Read* (SHR) recombination theory. Generation and recombination through recombination centers located inside the space-charge region are responsible for significant currents, especially under reverse and low forward bias, in silicon junction diodes. *Auger recombination* occurs when there are very high densities of carriers present. In Auger recombination the energy and momentum released by recombination are carried away by a third free particle. Because there are extra localized states at a silicon surface and also because a space-charge region is usually present, surface recombination can be an added important effect. Surface recombination is usually described by a parameter s known as the *surface-recombination velocity*.

The transient behavior of junction diodes is governed by the charge and discharge of stored minority carriers, as well as by the variation of stored depletion-layer charge. These components of charge are nonlinear functions of voltage, and a linear circuit representation is only approximately valid for small bias variations. A *small-signal equivalent circuit* can be developed with capacitors representing the charge storage. This circuit can be used in piecewise fashion to make accurate calculations for large-signal transients. Any practical problems are usually carried out by computer analysis.

The *pn*-junction diode is a basic building block for integrated circuits. There is widespread use of oxide regions as isolating elements to avoid unintentional interactions between the devices making up an integrated circuit. Series resistance between surface contacts and forward-biased *pn* junctions can be reduced by adding buried layers of donor dopant atoms below diffused junctions. Many of the properties of *pn* junctions for integrated circuits can be obtained by reference to curves of published results obtained from numerical solutions of the junction equations.

REFERENCES

1. K. THIESSEN and G. ZECH, *Phys. Stat. Sol.* (a) **10**, K 133 (1972).
2. W. FAHRNER and A. GOETZBERGER, *J. Appl. Phys.* **44**, 725 (1973).
3. W. SHOCKLEY and W. T. READ. *Phys. Rev.* **87**, 835 (1952).
4. R. N. HALL, *Phys. Rev.* **87**, 387 (1952).
5. E. M. CONWELL, *Proc. IRE* **46**, 1281 (1958).
6. J. L. MOLL, *Physics of Semiconductors*, McGraw-Hill, New York, 1964, p. 117.
7. P. U. CALZOLARI and S. GRAFFI, *Solid-State Electron.* **15**, 1003 (1972).
8. B. L. SHARMA and R. K. PUROHIT, *Semiconductor Heterojunctions*, Pergamon Press, Oxford, 1974, Sec. 1.1.
9. H. UNLU and A. NUSSBAUM, Solid-State Electron. **30**, 1095 (1987).
10. R. W. DUTTON and Z. YU, *Technology CAD: Computer Simulation of IC Processes and Devices*, Kluwer Academic Publishers, Boston, 1993.
11. K. M. CHAM, S.-Y. OH, D. CHIN, and J. L. MOLL, *Computer-Aided Design and VLSI Device Development*, Kluwer Academic Publishers, Boston, 1986.
12. H. LAWRENCE and R. M. WARNER, JR., *Bell Syst. Tec. J.* **34**, 105 (1955).
13. P. E. GRAY, D. DEWITT, A. R. BOOTHROYD, and J. F. GIBBONS, *Physical Electronics and Circuit Models of Transistors*, Wiley, New York, 1964, p. 75.

BOOKS

C. M. WOLFE, N. HOLONYAK, JR., and G. E. STILLMAN, *Physical Properties of Semiconductors*, Prentice-Hall, Englewood Cliffs, NJ, 1989, Chapter 9: Heterostructures.

H. KROEMER, Chapter 4 of *VLSI Electronics Microstructure Science, Volume 10: Surface and Interface Effects in VLSI*, ed. N. G. EINSPRUCH and R. S. BAUER, Academic Press, Orlando, 1985.

G. W. NEUDECK, *The PN-Junction Diode*, Volume II Modular Series on Solid-State Devices, Addison-Wesley Inc, Reading, MA, 1983.

PROBLEMS

5.1* A 0.6 Ω-cm, *n*-type silicon sample contains 10^{15} cm^{-3} generation-recombination centers located at the intrinsic Fermi level with $\sigma_n = \sigma_p = 10^{-15}$ cm^2. Assume $v_{th} = 10^7$ cm s^{-1}

(a) Calculate the generation rate if the region is depleted of mobile carriers.

(b) Calculate the generation rate in a region where only the minority-carrier concentration has been reduced appreciably below its equilibrium value.

5.2* Light is incident on a silicon bar doped with 10^{16} cm^{-3} donors, creating 10^{21} cm^{-3} s^{-1} electron-hole pairs uniformly throughout the sample. There are 10^{15} cm^{-3} bulk recombination centers at E_i with electron and hole capture cross sections of 10^{-14} cm^2.

(a) Calculate the steady-state hole and electron concentrations with the light turned on.

(b) At time $t = 0$ the light is turned off. Calculate the time response of the total hole density and find the lifetime. The thermal velocity is 10^7 cm s^{-1}, and there is no current flowing.

5.3 (This problem is intended to provide perspective on the concept of charge neutrality in semiconductors.) Assume that the distribution of free electrons in Figure P5.3*a* can be maintained in a block of silicon without any neutralizing charges. Consider that the electric field associated with the charge vanishes at $x = W$ and take $n_1 = 10^{18}$ cm^{-3}, $W = 1$ μm, $D_n = 7$ cm^2 s^{-1}, and $T = 300$ K.

(a) Calculate the field and current at $x = 0$ that is required to sustain this distribution of free charge. (Note that both a drift and diffusion component will be present.)

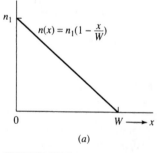

(a)

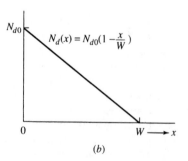

(b)

FIGURE P5.3

(b) In light of the results of (a), is the charge configuration reasonable?

(c) If a current density of approximately 10^5 A cm^{-2} can be sustained before power dissipation in the silicon causes irreversible damage, what is the maximum field that can exist in silicon having a uniform electron density equal to n_1? (The current densities in most operating IC devices are two or more orders of magnitude lower than this value.)

(d) If the compensating positive charge density shown in Figure P5.3b exists in the silicon, calculate the value of N_{d0} such that the field of part (c) will be present. Express this as a fraction of the free-electron density n_1, and note thereby how close to electrical neutrality the block of silicon is. (Use values at $x = 0$).

5.4† If recombination of electrons takes place through a donorlike recombination center, the capture cross section σ_n for the site can be crudely estimated by considering that the free electron enters a region where its thermal energy $\frac{3}{2}kT$ is less than the energy associated with Coulombic attraction; this means that the carrier is drawn to the center. The radius at which this occurs describes an area that can be taken to be equal to σ_n.

(a) Obtain an expression for σ_n based upon this model and evaluate it for silicon at 300 K.

(b) Considering the dependence on temperature of σ_n from this model together with other temperature-dependent terms, how does the electron capture rate for the center depend on temperature? (Assume that the center exists in extrinsic silicon.)

5.5† The analysis of recombination through recombination centers in Sec. 5.2 assumed low-level injection with no current flowing so that we could find the characteristic time in which excess carriers decayed. Following a similar analysis, find the "effective lifetime" in the opposite case of high-level injection in an n-type semiconductor, assuming that charge neutrality holds. Let $n' = p' \gg n_o$, p_o, and compare the decay time or effective lifetime to that found in the case of low-level injection for both $\sigma_n \neq \sigma_p$ and $\sigma_n = \sigma_p$.

5.6 Consider the continuity equation for minority-carrier holes that are injected across a pn junction into an n-type region that is short so that recombination of the holes takes place essentially only at the contact at $x = W_B$. Show by the direct use of Equation 5.3.10 that the holes are distributed linearly along x as was found in Equation 5.3.17.

5.7 An ideal pn-junction diode has long p- and n-regions and negligible generation or recombination in the space-charge region, so that the current-voltage

characteristics are described by Equation 5.3.15. Show that the limiting reverse-bias current predicted by Equation 5.3.15 corresponds to the integrated generation rate of minority carriers on either side of the pn junction. *Hint. Use Equation 5.3.12 and its equivalent for electrons and consider that* $G_p - R_p = -p_n'/\tau_p$.

5.8 Holes are injected across a forward-biased pn junction into an n-region that is much longer than L_p, the hole diffusion length. Show, by using Equation 5.3.12, that L_p is the average length that a hole diffuses before it recombines in the n-region.

5.9* For forward–biased current in an ideal pn-junction diode, show that the ratio of hole current to total current can be controlled by varying the relative doping on the two sides of the junction. If we call the ratio of hole current to total current γ, express γ as a function of N_a/N_d. Calculate γ for a silicon pn junction for which resistivity of the n-side is 0.001 Ω-cm and that of the p-side is 1 Ω-cm. Assume that $\tau_p = 0.1\ \tau_n$ and that both neutral regions are much longer than the minority-carrier diffusion lengths in them.

5.10† A pn-junction diode has the configuration shown in Figure P5.10. Assume that

1. $N_a = N_d = N_o \gg n_i$.
2. $W \ll L$, the minority-carrier diffusion length; $L_1 \gg L$.
3. All the diffusion coefficients $= D$, all lifetimes $= \tau$.
4. The space-charge-region width $\ll W$.
5. The external applied voltage is V_B, where $V_B \gg \phi_i$, the built-in voltage.
6. The recombination velocity is infinite at $x = +W$.

If a light beam that produces G_0 hole-electron pairs per unit area per unit time in a slab of negligible width along the x-direction is incident on the diode at the plane $x = -W/2$ on the p-type side

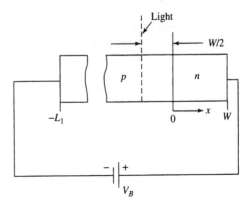

FIGURE P5.10

(a) Assuming low-level injection, find and sketch the minority-carrier concentrations in the neutral regions of the diode.

(b) Calculate the current that flows in the illuminated diode in terms of G_0 and the diode properties.

(c) In terms of the constants associated with the diode, what current flows when the light beam is removed (steady-state) [13]?

5.11* Consider an ideal, long-base, silicon abrupt *pn*-junction diode with uniform cross section and constant doping on either side of the junction. The diode is made from 1 Ω-cm *p*-type and 0.2 Ω-cm *n*-type materials in which the minority-carrier lifetimes are $\tau_n = 10^{-6}$ s and $\tau_p = 10^{-8}$ s, respectively. ("Ideal" implies that effects within the space-charge region are negligible and that minority carriers flow only by diffusion mechanisms in the charge-neutral regions.)

(a) What is the value of the built-in voltage?

(b) Calculate the density of the minority carriers at the edge of the space-charge region when the applied voltage is 0.589 V ($23 \times kT/q$).

(c) Sketch the majority- and minority-carrier currents as functions of distance from the junction on both sides of the junction, under the applied bias voltage of part (b).

(d) Calculate the location(s) of the plane or planes at which the minority-carrier and majority-carrier currents are equal in magnitude for the applied bias voltage of part (b).

5.12 Sketch the distribution of hole and electron currents in the neutral regions of a *pn*-junction diode (with forward bias) having *significant* recombination in the space-charge layer. Assume that

(a) The injected hole current is twice the injected electron current.

(b) The net rate at which pairs recombine in the space-charge layer is equal to half of the net rate at which electrons recombine in the *p*-type region.

Demonstrate that the total diode current is given by adding a term that describes the space-charge-region recombination current to the sum of the diffusion currents at the edges of the space-charge region.

5.13 Consider the integrated-circuit structures shown in cross section in Problem 4.10.

(a) Ignoring series resistance but considering all junctions, sketch the two sets of *I-V* characteristics that you expect to see for measurements of current when (*i*) voltage V_C (both polarities) is applied to contact *C* and contact *A* is grounded while contact *B* is left open, (*ii*) voltage V_B (both polarities) is applied to contact *B* and contact *A* is grounded while contact *C* is left open.

(b)† If it were possible to deplete the n^- layer fully under the conditions of part *a(i)* above, what would you expect to happen to the current flowing through contact *A*? Explain the reasons for your conclusions.

5.14† A silicon diode of junction area 10^{-5} cm^2 is made with a uniformly doped *p*-type region having 5×10^{18} acceptors cm^{-3} and a step junction to an *n*-type region doped with 10^{15} donors cm^{-3}. The diode is adequately described by the simplest theory; that is, one-dimensional flow with dominance of bulk recombination and transport of minority carriers under forward bias by pure diffusion into the neutral regions from the edge of the space-charge region. Recombination-generation in the space-charge region is negligible, and the minority-carrier lifetime in each region is 100 ns.

This junction diode is to be used in a circuit that requires a rectification ratio between forward current I_F and reverse current I_R ($|I_F/I_R|$) of 10^4 at 0.5 V and a maximum reverse saturation current of 100 nA. What is the maximum temperature at which the diode performs adequately? Consider only the major temperature dependences (i.e., neglect changes in D, μ, τ_p, N_c, and N_v with temperature).

5.15 Prove that the recombination rate in the space-charge region of a *pn* junction in which $\sigma_n = \sigma_p$ is maximum when $p = n$, as given in Equation 5.3.23.

5.16† Use the values for a silicon junction diode given in the paragraph following Equation 5.3.25 together with the equation itself to make a semilogarithmic plot of J_t/J_r versus applied voltage V_a under forward bias. The diode consists of a step junction to a highly doped *n*-region (i.e., $N_a \ll N_d$).

5.17 Consider an abrupt diode biased so that ($\phi_i - V_a$) = 5 V. The diode has the following properties: $N_a = 10^{17}$ cm^{-3}, $\tau_n = 10^{-6}$ s, $N_d = 10^{18}$ cm^{-3}, and $\tau_p = 10^{-8}$ s.

(a) Use Equation 5.3.29 to determine the ratio J_t/J_g (take $x_i = x_d$).

(b) Discuss the expected temperature dependence of this ratio.

5.18† An ideal short-base diode is built with an abrupt junction in which $N_d \gg N_a$ and $N_a = 10^{17}$ cm^{-3}. (Assume that the *n*-region is degenerately doped.) The width of the *p*-region between the edge of the space-charge region and the contact at which all recombination takes place is 3 μm. The area of the diode is 10^{-5} cm^2.

(a) Calculate the charge stored in the neutral *p*-region if 0.5 mA flows through the diode.

(b) Determine the charge stored in the narrowed space-charge layer under forward bias.

(c) How much time does it take a current source of 0.5 mA to cause the diode to switch from an "off" condition (with $V_a = 0$) to the steady state with 0.5 mA of current?

5.19 Consider a small-signal equivalent circuit for a Schottky diode under forward bias. Compare the circuit to that for a pn-junction diode as sketched in Figure 5.14. Discuss the differences and similarities in the two circuits.

5.20[†] Consider qualitatively the effects of punch-through on the structure shown in Figure 5.17d as follows: (a) sketch a band diagram at thermal equilibrium along an axis that passes through the middle of the diffused p-region and runs through to the substrate; (b) with the substrate grounded, consider that sufficient reverse bias is applied at the upper pn junction so that the junction between the epitaxial n-region and the p-substrate becomes forward biased. When this occurs, the holes in the substrate can be injected into the upper p-region. (In steady-state, a *space-charge-limited current* of holes flows.) Sketch a band diagram for this condition.

5.21* Calculate the small-signal, incremental resistance and capacitance for an ideal, long-base silicon diode using the following parameters:

$$N_d = 10^{18} \text{ cm}^{-3}, N_a = 10^{16} \text{ cm}^{-3}, \tau_n = \tau_p = 10^{-8} \text{ s},$$
$$A = 10^{-4} \text{ cm}^2, \quad T = 300 \text{ K}$$

(a) For 0.1, 0.5, and 0.7 V forward bias.

(b) For 0, 5, and 20 V reverse bias.

(c) What is the series resistance of the p-type quasi-neutral region if this region is 0.1 cm long. This resistance must be added to the elements modeling the ideal diode when considering the response of a real diode.

5.22 A pn junction is initially off. A step of current is applied to the device with a polarity that turns the device on. Explain physically why it takes much less time for the width of the space-charge region

to change than for the minority-carrier distributions to reach steady state in the quasi-neutral regions. (That is, compare the physical processes in the two situations.)

5.23 Consider the silicon integrated-circuit diode shown in Figure P5.23. Can injected holes flowing from the p-type diffusion to the n^+ buried layer best be described by the long-base or short-base approximation? What about carriers flowing sideways from the edge of the p-type diffusion to the n^+ contact region? The hole lifetime is 1 μs in the n-type epitaxial layer, and the n^+ buried layer is an effective sink for excess minority carriers. Justify your answers. (Note that the figure is not to scale.)

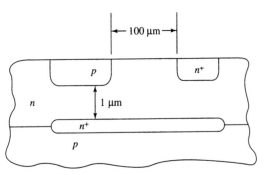

FIGURE P5.23

5.24 At high current densities, a significant fraction of the voltage applied across a diode can be dropped across the neutral regions of the device. Consider the current-voltage relation for a one-sided step junction with N_d donors on the lightly doped side of the junction. Find the current and applied voltage V_a at which 10% of V_a appears across the n-type neutral region for a typical integrated-circuit diode. Assume that the cross-sectional area is 10^{-5} cm^2, the length of the neutral region in the n-type silicon is 10 μm, $N_d = 5 \times 10^{15}$ cm^{-3}, and $\tau_p = 1$ ns.

BIPOLAR TRANSISTORS I: BASIC PROPERTIES

In **Chapter** 5, we saw that a *pn* junction, biased so that the *p*-region is positive with respect to the *n*-region conducts current because holes are injected from the *p*-region and electrons are injected from the *n*-region. The supply of these majority carriers to the junction is plentiful. Consequently, current increases rapidly as voltage increases and the barrier at the junction decreases. Currents are much smaller under reverse bias because they are carried only by minority carriers generated either in the junction space-charge region or nearby. However, the current passed by a reverse-biased junction increases if the supply of minority carriers in the vicinity of the junction is enhanced. This enhancement can, for example, result from the incident radiation of energetic particles in diode photodetectors or radiation sensors.

Another means of enhancing the minority-carrier population in the vicinity of a reverse-biased *pn* junction is to locate a forward-biased *pn* junction very close to it. This method is especially advantageous because the minority-carrier population is then under electrical control, that is, under the control of the bias applied to the nearby forward-biased junction.

The modulation of the current flow in one *pn* junction by changing the bias on a nearby junction is called *bipolar transistor action*. It is one of the most significant ideas in the history of device electronics, and the investigations that led to it were honored with a Nobel Prize in Physics for the inventors of the bipolar junction transistor (BJT), William Shockley, John Bardeen, and Walter Brattain.

In this chapter we describe the basic physical action of bipolar transistors. We will see that transistor operation can be described more simply by considering various specific bias ranges separately. We concentrate first on the *active region* in which transistors act as amplifiers, and we consider the other bias regions in terms of transistor switching. We introduce the Ebers-Moll model, which provides an extremely useful and informative means of describing basic transistor operation in each of the bias ranges. We then relate this model to our discussion of transistor action at the beginning of the chapter. We next consider the design of planar diffused bipolar amplifying and switching transistors. In the final section, we discuss heterojunction bipolar transistors (HBTs), in which different regions of the transistor are built in different semiconducting materials.

6.1   TRANSISTOR ACTION

Figure 6.1 shows a simple structure that illustrates transistor action [1]. Two *pn* junctions are spaced a distance W units apart in a single bar of semiconductor material. The bar has a uniform cross section of area A. The junctions are located "close" to one another so that electrons injected across junction 1 when V_{BE} is positive reach the vicinity of junction 2;

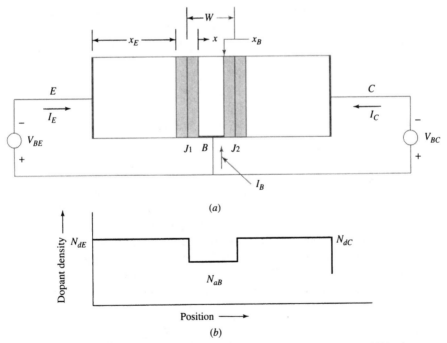

FIGURE 6.1 (*a*) Prototype transistor. Two *pn* junctions J_1 and J_2 are spaced W units apart. (*b*) Each region has a constant doping density. The doping changes abruptly at J_1 and J_2. The quasi-neutral portion of the middle *p*-region is bounded by the edges of space-charge regions at $x = 0$ and $x = x_B$, respectively. The contacts at E, B, and C are ohmic.

that is, the junction spacing W is small enough that few electrons are lost by recombination in the middle p-region. The middle region is called the *base* of the transistor.

The n-type region adjacent to the injecting (or emitting) junction is called the *emitter*, and the n-type region adjacent to the collecting junction is called the *collector*. We are especially interested in learning how electrons are injected from the emitter into the base, flow across the base, and reach the collector. For our initial discussion, we assume that electron recombination or generation in the base region is not significant; we reconsider this assumption later.

We note that in a bipolar transistor, such as that shown in Figure 6.1, there is negligible flow of holes (base majority carriers) from junction 1 to junction 2 or from junction 2 to junction 1. This is true under all bias conditions because the flow of holes from either n-region into the p-type base must be very small. If we call the longitudinal dimension x and write the expression for hole current density in the x direction (assuming no recombination) that was derived in Equation 1.2.22, we have

$$J_p = 0 = q\mu_p p \mathscr{E}_x - qD_p\frac{dp}{dx} \tag{6.1.1}$$

and

$$
\begin{aligned}
\mathscr{E}_x &= \frac{D_p}{\mu_p}\frac{1}{p}\frac{dp}{dx} \\
&= \frac{kT}{q}\frac{1}{p}\frac{dp}{dx}
\end{aligned}
\tag{6.1.2}
$$

Thus, the condition of zero hole current in the base leads to Equation 6.1.2, describing the longitudinal electric field.* The field depends both on the magnitude of the majority-carrier (hole) density and on its gradient. In contrast to the hole current, a current of *electrons* flowing between the two junctions is possible because either junction can readily supply electrons from the n-regions to the center p-region if that junction is forward biased. From Equation 1.2.21, the electron current is

$$J_n = q\mu_n n \mathscr{E} + qD_n\frac{dn}{dx} = kT\,\mu_n\frac{n}{p}\frac{dp}{dx} + qD_n\frac{dn}{dx} \tag{6.1.3}$$

Using the Einstein relation $D_n = (kT/q)\mu_n$ in Equation 6.1.3, we have

$$J_n = \frac{qD_n}{p}\left(n\frac{dp}{dx} + p\frac{dn}{dx}\right) = \frac{qD_n}{p}\frac{d(np)}{dx} \tag{6.1.4}$$

Although Equation 6.1.4 is valid for spatially varying dopant concentrations, we can gain useful insight into transistor action by considering the simple transistor structure in Figure 6.1 with abrupt junctions and constant base doping in a piece of one semiconducting material. We call this structure the *prototype transistor* because it closely resembles the device described by W. Shockley in his original discussion of the bipolar junction transistor [1].

With uniform doping in the base region, p is approximately constant. Taking the minority carrier concentration n to be small compared to the majority carrier concentration p, Equation 6.1.4 reduces to

$$J_n = qD_n\frac{dn}{dx} \tag{6.1.5}$$

---

* Strictly, Equations 6.1.1 and 6.1.2 are approximations because second-order effects to be described later can cause small hole currents in the x-direction.

We further assume that recombination of electrons in the base is small (as in the short-base diode of Sec. 5.3) so that n varies linearly across the base and

$$\frac{dn}{dx} = \frac{n_p(x_B) - n_p(0)}{x_B} \tag{6.1.6}$$

Expressing the minority-carrier electron concentrations in terms of the dopant concentration, we obtain

$$J_n = \frac{qD_n n_i^2}{x_B N_{aB}}\left[\exp\left(\frac{qV_{BC}}{kT}\right) - \exp\left(\frac{qV_{BE}}{kT}\right)\right] = J_S\left[\exp\left(\frac{qV_{BC}}{kT}\right) - \exp\left(\frac{qV_{BE}}{kT}\right)\right] \tag{6.1.7}$$

where

$$J_S = \frac{qD_n n_i^2}{x_B N_{aB}} \tag{6.1.8}$$

From Equation 6.1.7 we see that the current J_n can be switched on and off by changing the junction voltages. If both V_{BC} and V_{BE} are negative and significantly greater than kT/q, then J_n is very small. If, on the other hand, either V_{BE} or V_{CB} is positive and greater than kT/q, J_n depends strongly on the most positive voltage.

Before continuing with the analysis of the prototype transistor, we note that the base doping varies with position in most IC transistors. Equation 6.1.4 can be used in the more general case by noting that the right-hand side is a perfect differential:

$$J_n = \frac{qD_n}{p}\left(n\frac{dp}{dx} + p\frac{dn}{dx}\right) = \frac{qD_n}{p}\frac{d(pn)}{dx} \tag{6.1.9}$$

We can then write an integrated form of Equation 6.1.9 with arbitrary limits x and x' and consider recombination to be negligible so that J_n can be removed from the integral:

$$J_n\int_x^{x'}\frac{p\,dx}{q\,D_n} = \int_x^{x'}\frac{d(pn)}{dx}dx$$
$$= p(x')n(x') - p(x)n(x) \tag{6.1.10}$$

Equation 6.1.10 shows that the minority-carrier (electron) current in the base depends on the difference in the hole-electron products across a region divided by the integrated majority-carrier density in that region. If we use the junctions themselves as the two boundaries of the region, then $x = 0$ becomes the lower limit and $x' = x_B$, the upper limit for the integrals in Equation 6.1.10. From Equation 5.3.20, the pn products at the boundaries can then be related to the junction voltages:

$$p(0)n(0) = n_i^2\exp\left(\frac{qV_{BE}}{kT}\right)$$

$$p(x_B)n(x_B) = n_i^2\exp\left(\frac{qV_{BC}}{KT}\right) \tag{6.1.11}$$

Thus, the longitudinal electron current in the base can be expressed as a function of the junction voltages V_{BE} and V_{BC}:

$$J_n = \frac{qn_i^2\left[\exp\left(\frac{qV_{BC}}{kT}\right) - \exp\left(\frac{qV_{BE}}{kT}\right)\right]}{\displaystyle\int_0^{x_B}\frac{p\,dx}{D_n}} \tag{6.1.12}$$

Before we discuss the many implications of Equation 6.1.12, we modify it slightly. First, we note that D_n is often a weak function of position in the base and can be expressed by an average value $\tilde{D}_n$ and removed from the integral in the denominator of Equation 6.1.10.* With $\tilde{D}_n$ removed, the integral is just the total majority-carrier density per unit area in the base. The charge associated with this density is called Q_B

$$Q_B = q \int_0^{x_B} p\,dx \tag{6.1.13}$$

With these changes, we can express the electron current density flowing from the first junction to the second junction:

$$J_n = J_S \left[\exp\left(\frac{qV_{BC}}{kT}\right) - \exp\left(\frac{qV_{BE}}{kT}\right) \right] \tag{6.1.14}$$

where

$$J_S = \frac{q^2 n_i^2 \tilde{D}_n}{Q_B} \tag{6.1.15}$$

Prototype Transistor

Equation 6.1.12, which we derived to represent transistor action, is not a function of the base doping profile, but depends only on the integrated base majority charge. From Equation 6.1.12, we derived Equation 6.1.14, which succinctly represents the physics of transistor action in an arbitrarily doped semiconductor and can be used to describe more than just switching action. Before we explore these phenomena more fully in the general case, however, we consider the physical basis for transistor action by returning to the simpler prototype transistor structure of Figure 6.1.

In Figure 6.2 we sketch the energy-band diagrams and electron densities (base minority carriers) for the prototype transistor at equilibrium and under various bias conditions. Figure 6.2a represents the transistor under equilibrium (zero bias) conditions. Only a few electrons are in the base region, and transfer of electrons from either end region is inhibited by energy barriers. Negative bias applied to both junctions tends to increase these barriers and to deplete the base region of the few electrons present at equilibrium. The band diagram and a sketch of the electron distribution for this case are shown in Figure 6.2b. Positive bias applied to both junctions injects electrons into the base region by lowering the built-in barriers, and the resulting huge increase in the base electron density allows the ready flow of current between the two junctions (Figure 6.2c)

Active Bias. In another bias condition one junction is forward biased and the other is reverse biased. This arrangement is of greatest interest because it makes signal amplification possible. Diagrams for this bias condition, called *active bias*, are sketched in Figure 6.2d. From the figure we see that the forward-biased junction injects electrons into the base because the barrier is reduced from its equilibrium value by the forward bias. The reverse-biased junction, on the other hand, sweeps any nearby electrons out of the base into the n-region at the far end. Thus electrons (base minority carriers)

---

* $\tilde{D}_n$ is not a direct spatial average, but rather $\tilde{D}_n \equiv \int_0^{x_B} p\,dx \Big/ \int_0^{x_B} (p/D_n)\,dx$

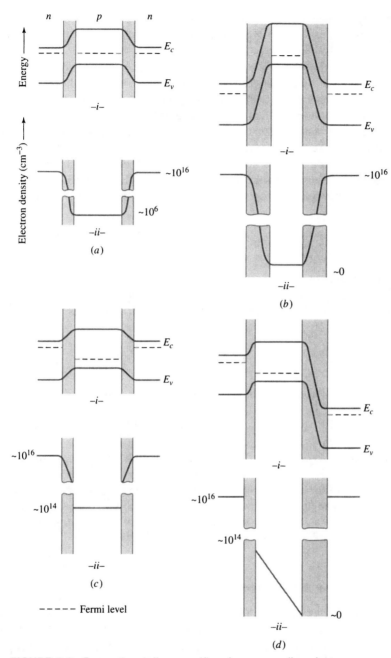

FIGURE 6.2 Energy-band diagrams (*i*) and corresponding electron-density distributions (*ii*) for the transistor of Figure 6.1. (*a*) Equilibrium, (*b*) both junctions reverse biased, (*c*) both junctions forward biased, (*d*) one junction reverse biased, and one junction forward biased. The shaded areas represent space-charge regions.

flow from the forward-biased junction, which we call the *emitter* junction, to the reverse-biased junction, known as the *collector* junction. Electrons in the base near the collector junction are swept rapidly through the space-charge region and into the *n*-type collector region.

As an introduction to transistors under active bias and to clarify some of the qualitative discussion that we have presented, we briefly consider the amplifying or active-bias operation of the prototype transistor. In Figure 6.2d we sketched the minority-carrier electron density in the base as a straight line. This electron distribution follows from our analysis of the short-base diode in Sec. 5.3 because the electrons in the base of a uniformly doped transistor are spatially distributed in the same way as in the ideal short-base diode. That is, recombination in the base is negligible, and the doping density is constant. Hence, the electron density found by solving the continuity equation is linear in x. The electron density at the emitter edge of the base region depends exponentially on the voltage V_{BE}; at the collector edge of the base, the electron density is negligible. The distribution of excess electrons n_p' is thus linear, as found in Equation 5.3.17 for the analogous case of holes. Expressed for the geometry of Figure 6.1, the excess electron density $n_p' = n_p - n_{po}$ is

$$n_p' = n_{po}\left[e^{qV_{BE}/kT}\left(1 - \frac{x}{x_B}\right) - 1\right] \qquad 0 \le x \le x_B \tag{6.1.16}$$

The current of electrons through the base for V_{BE} much greater than kT/q, as is usual for active bias, is easily found because it is purely diffusion current. Because $J_n = qD_n (dn_p/dx)$, we can write the electron current density in the base in terms of the base doping N_a as

$$J_n = \frac{-qD_n n_i^2 \exp(qV_{BE}/kT)}{N_a x_B} \tag{6.1.17}$$

To derive this result from Equation 6.1.14, we note from Equation 6.1.13 that Q_B for the prototype transistor is

$$Q_B = qN_a x_B \tag{6.1.18}$$

When this value of Q_B is used in Equation 6.1.15 and the negligible term $\exp(qV_{BC}/kT)$ in Equation 6.1.14 is dropped, we obtain the expression for current in Equation 6.1.17. The significance of the integrated equations for transistor action (Equations 6.1.12–6.1.15) lies in their ability to deal with transistors that are not uniformly doped. This approach to transistor analysis was originally developed to optimize the transistor doping profile [2]. We will consider further aspects of this approach in Chapter 7.

Transistors for Integrated Circuits

Before deriving further equations for transistors under active bias, we consider some general features of integrated-circuit transistors. We contrast IC devices with the simple prototype structure of Figure 6.1 and both develop theory and discuss practical applications at the same time. The structures used in IC transistors are considered in more detail in Sec. 6.5.

We focused on the prototype transistor in our initial discussion because it allows us to examine the physical phenomena important to general transistor operation without excessive mathematical complication. The doping profile of the prototype transistor can be achieved by the vapor-phase epitaxial techniques discussed in Sec. 2.6. [Growing a sequence of doped layers of different semiconductors is often employed to build *heterojunction bipolar transistors* (HBTs) using compound semiconductors or, at times, column IV semiconductors (e.g., alloys of germanium and silicon grown on silicon).] These HBTs are discussed further in Sec. 6.6. However, the dopant in silicon integrated-circuit

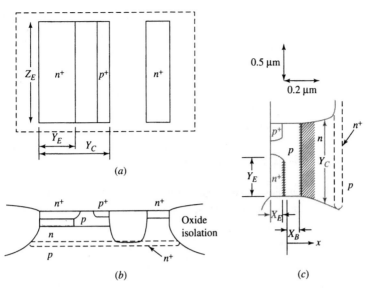

FIGURE 6.3 Top view (*a*) and cross sections (*b* and *c*) of an oxide-isolated *npn* IC transistor. The region dominating transistor action is shaded in (*b*), and the area bounded by the base diffusion is rotated 90° and expanded in (*c*); *x* and *y* scales are indicated for (*c*) only.

transistors is more commonly added by diffusion using the planar process described in Chapter 2. The structure and doping profile of IC transistors, therefore, differ considerably from those of the prototype transistor sketched in Figure 6.1.

A top view and a cross-sectional view of an oxide-isolated integrated-circuit transistor are shown in Figure 6.3. The structure is produced by a sequence of steps similar to that outlined in Sec. 5.5 where we discussed the fabrication of an array of isolated integrated-circuit diodes. To make transistors instead of diodes, an additional heavily doped *n*-region is diffused into the *p*-region. This additional diffusion produces the emitter-base junction. As with the diode array, a lightly doped *p*-type wafer is used for the substrate. A heavily doped *n*-type *buried layer* or *subcollector* region added to reduce lateral series resistance (Figure 5.18) is placed under the collector *pn* junction. In Sec. 6.5 we will discuss in more detail the dopant profiles, geometries, and tradeoffs in transistor processing. Here we focus on the basic device geometry.

The width (thickness) of the quasi-neutral base region (x_B in Figure 6.3*c*) is one of the most critical parameters in transistor design. The base width is typically of the order of 100 nm ($\sim 10^{-5}$ cm) and can be much less for very high-frequency transistors. The emitter width* Y_E is made as small as allowed by the lithography tools available and is typically a few hundred nanometers. Most of the other dimensions of the transistor are somewhat larger. Transistor action depends critically on the proximity of the two interacting *pn* junctions. Therefore, under most transistor bias conditions, efficient transistor action is confined to the shaded region within the box in the cross-sectional view in Figure 6.3*b*. This region is seen more clearly in Figure 6.3*c*. The emitter-base and base-collector space-charge regions are indicated by cross hatching in Figure 6.3*c*, and the figure is rotated by 90° to maintain the same orientation as in Figure 6.1.

---

* Note that *base* "width" is in the direction perpendicular to the surface plane while *emitter* "width" is a dimension parallel to the surface plane.

Figure 6.3c shows that transistor action is mainly confined to an area defined by Y_E, the width of the emitter quasi-neutral region. Because the emitter is usually heavily doped, the space-charge region does not extend appreciably into the emitter, and Y_E is nearly equal to the emitter stripe width. Although conventional IC processing makes the collector width Y_C much larger than the emitter width Y_E, the transistor behavior can be approximated for many purposes by considering a one-dimensional device. We use this approximation of one-dimensional current flow to specify a single area A that relates current density J to current I. For the transistor sketched in Figure 6.3, the area A is the product of Y_E and Z_E. As we develop the theory of junction transistors further, we will refer to other dimensions on Figure 6.3 and examine the constraints imposed by the IC device structure.

In Sec. 6.2 we consider transistor action under *active bias* in more detail. The concepts involved are important for understanding both amplifying and switching applications of the transistor.

6.2 ACTIVE BIAS

We described active bias for an *npn* transistor as operation with V_{BE} positive and V_{BC} zero or negative.* This bias causes electron injection at the emitter-base junction and electron collection at the base-collector junction. When V_{BC} is zero or negative and V_{BE} is substantially larger than kT/q, we see from Equation 6.1.14 that an electron current

$$J_n \approx -J_S \exp\left(\frac{qV_{BE}}{kT}\right) \tag{6.2.1}$$

flows from left to right across the collecting junction (J_2 in Figure 6.1). Using the standard convention that currents into a transistor are positive, J_n in Equation 6.2.1 is equal to $+J_C$, a positive collector-current density. Equation 6.2.1 shows that under active bias, collector current depends exponentially on emitter-base voltage.

Figure 6.4 shows experimental measurements of collector current I_C plotted on a logarithmic scale as a function of base-emitter bias voltage V_{BE}. Over nearly the entire range of voltage, these measurements confirm an exponential dependence. Furthermore, after converting base-ten logarithms to natural logarithms, the slope of the collector-current plot is just q/kT, as expected from Equation 6.2.1. This equation shows that a decade change in $|J_S/J_n|$ occurs for $\Delta V_{BE} = (kT/q) \ln(10) = 60$ mV at $T = 300$ K. Deviations from a 60 mV/decade slope indicate nonideal behavior and are carefully watched as they can indicate fabrication problems. However, *collector* current usually follows the theoretical exponential relation well enough that accurate thermometers can be made by measuring the change in collector current with temperature at a fixed emitter-base bias.

The intercept with the current axis at $V_{BE} = 0$ of an extrapolated line drawn through the J_C measurements provides a value for J_S in Equation 6.2.1. The built-in base charge in the quasi-neutral region can be obtained from Equation 6.1.15 once J_S is known:

$$Q_{B0} = \frac{q^2 n_i^2 \tilde{D}_n}{J_S} \tag{6.2.2}$$

---

* In more detail, these polarities correspond to *forward-active bias*; if the polarities of V_{BE} and V_{BC} are interchanged, the transistor is under *reverse-active bias*.

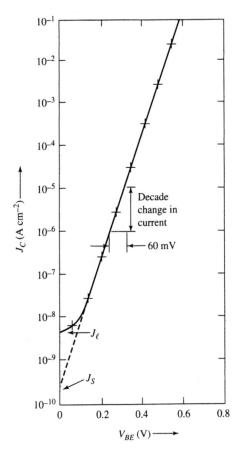

FIGURE 6.4 Semilogarithmic plot of collected current versus base-emitter voltage for an IC *npn* transistor under active bias at 300 K. Crosses designate data points. The extended straight line indicates J_S (Equation 6.2.1); J_ℓ is the leakage component of the current.

All of the other parameters in Equation 6.2.2 are well known.* We call Q_{B0} in Equation 6.2.2 the built-in base charge because, as shown in Equation 6.1.13, it represents the hole charge per unit area in the quasi-neutral base at zero base-emitter bias. This charge is determined during processing of the transistor. As we saw above, the value of Q_{B0} can be obtained from current-voltage measurements on the transistor. This behavior was first described by H. K. Gummel [3], and the number of base-dopant atoms (per cm^2) in the quasi-neutral region

$$\int_0^{x_B} N_a(x)\,dx = \frac{Q_{B0}}{q} = \frac{q n_i^2 \tilde{D}_n}{J_S} \tag{6.2.3}$$

is sometimes called the *Gummel Number GN*.

Equation 6.2.3 emphasizes that J_S, the multiplying factor for transistor current at a given bias, is inversely proportional to the Gummel Number, that is, to the total base doping. The lower the built-in base charge Q_{B0}, the higher the current at a given bias. Therefore, one might propose to design a prototype transistor with low constant base doping. A major disadvantage of this design is that even a small forward bias might make the low-injection approximation invalid near the base-emitter junction in the transistor [i.e., $n_p(0)$

---

* A possible exception is $\tilde{D}_n$, but diffusivity is not a very strong function of concentration (Figure 1.16) and thus can be approximated for analytic calculations. The local value of D_n can be used in numerical simulations to avoid this approximation.

would approach N_a]. As we will see in Chapter 7, high injection in a transistor degrades its performance. It can be avoided, and Q_{B0} can be kept small at the same time by making the base dopant profile vary from a maximum near the base-emitter junction to a lower value near the base-collector junction. The dopant density is then large in the region where the injected minority-carrier density is large and small where the minority-carrier density becomes negligible. Fortunately, IC diffusion technology automatically provides transistors with graded-base doping so that this advantage is obtained in IC transistors. Other advantages of graded-base transistors were first pointed out by Kroemer [4], and are described in Chapter 7.

Control of Q_{B0} during transistor processing is the key step in the production of integrated circuits. For high-gain transistors, in which the Gummel Number is below $10^{12} \, \text{cm}^{-2}$, Q_{B0} must be carefully controlled.

EXAMPLE   Gummel-Number Calculations

Assume that the data in Figure 6.4 was measured on a prototype *npn* transistor with a base width $x_B = 0.5 \, \mu\text{m}$. Find the Gummel Number (Equation 6.2.3) for the transistor and calculate the value of base-emitter voltage necessary to cause the electron density at the emitter edge of the base to be 1% of the base dopant density.

Solution   From Figure 6.4 we have $J_S = 2.4 \times 10^{-10} \, \text{A cm}^{-2}$. From Equation 6.2.3, the Gummel Number GN is

$$GN = \frac{q n_i^2 D_n}{J_S} = 1.4 \times 10^{11} \tilde{D}_n$$

If we assume (Figure 1.16) that $\tilde{D}_n$ is 20 cm^2s^{-1}, then $GN = 2.8 \times 10^{12}$ dopant atoms cm^{-2}, and $N_a = \dfrac{GN}{x_B} = 5.6 \times 10^{16} \, \text{cm}^{-3}$. Referring again to Figure 1.16, we see that if $N_a = 5.6 \times 10^{16}$, $\tilde{D}_n \approx 22$ instead of the value 20 which we assumed. We can redo the calculation taking $\tilde{D}_n = 22$, leading to $GN = 3.1 \times 10^{12} \, \text{cm}^{-2}$ and $N_a = 6.2 \times 10^{16} \, \text{cm}^{-3}$, which is approximately consistent with the value of $\tilde{D}_n$ used in the calculation.

The required value for the Gummel Number is, therefore, 3.1×10^{12} dopant atoms cm^{-2}.

The second part of the question requires a value for V_{BE} such that

$$n'(x = 0) = \frac{n_i^2}{N_a} \exp\left(\frac{q V_{BE}}{kT}\right) = 0.01 \times N_a = 6.2 \times 10^{14} \, \text{cm}^{-3}$$

The required value is $V_{BE} = 0.67$ V.   ∎

Current Gain

Thus far, our discussion of transistor action under active bias has considered only electrons flowing between the emitter and the collector. This represents the output current in an active-biased transistor. Collector current is an exponential function of base-emitter voltage (Equation 6.2.1) because forward bias on the base-emitter junction causes electron injection into the base to vary exponentially. Under active bias, these electrons are collected efficiently by the field in the base-collector space-charge region. The base-emitter terminals are thus the control electrodes for the collector current under active bias. The smaller the current that flows through the base terminal for any given positive V_{BE}, the more effective is the transistor as an amplifier because the input power (the product of V_{BE} and the base emitter current) is lower.

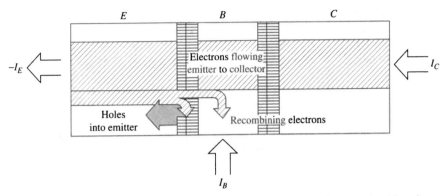

FIGURE 6.5 Terminal currents and major current components in an active-biased transistor. Not shown is the collector leakage current (J_ℓ in Figure 6.4).

Several mechanisms can cause base-emitter current in an active-biased transistor. The most straightforward of these is recombination of injected electrons with majority-carrier holes in the quasi-neutral base. Another component of base-emitter current is recombination in the base-emitter space-charge region. A third component of current flows because forward bias on the base-emitter junction not only injects electrons into the base, but it also causes holes to be injected into the emitter. These components are indicated in Figure 6.5. For a transistor to amplify effectively, all of these current components should be much smaller than the collector current.

Most silicon transistors are designed so that recombination of injected electrons in the base-emitter space-charge region is smaller than the other components mentioned except under low current conditions. (It can be much more important in transistors built from compound semiconductors.) We will consider recombination in the space-charge region separately in Chapter 7 when we discuss limitations on transistor performance.

Recombination within the quasi-neutral base itself can be expressed using the theory developed in Sec. 5.3. There we found that recombination of excess minority carriers is directly proportional to their density (in this case n' the excess base-electron density) so that the total base-region recombination current is

$$I_{rB} = qA_E \int_0^{x_B} \frac{[n - (n_i^2/N_a)]dx}{\tau_n} \tag{6.2.4}$$

where $A_E = Y_E \times Z_E$ in Figure 6.3 is the area of significant minority-carrier injection. To make I_{rB} as small as possible at a given bias, base lifetime τ_n should be maximum and base width x_B should be minimum.

Under active bias the injected excess electron density n' is much greater than the equilibrium electron density (n_i^2/N_a) over most of the base. Also, lifetime does not depend strongly on x, and Equation 6.2.4 can be simplified to

$$I_{rB} = \frac{qA_E}{\tau_n} \int_0^{x_B} n'dx \tag{6.2.5}$$

For the special case of a transistor with a uniformly doped base, such as the prototype transistor sketched in Figure 6.1, n' depends approximately linearly on x. Thus, the integration in Equation 6.2.5 is easily carried through to obtain

$$I_{rB} = \frac{qA_E n_i^2 x_B}{2N_a\tau_n}\left[\exp\left(\frac{qV_{BE}}{kT}\right) - 1\right] \tag{6.2.6}$$

Although Equation 6.2.6 was derived for a special case, the proportionality of base recombination current to $[\exp(qV_{BE}/kT) - 1]$ is obtained in general.

The loss of carriers to recombination in the base region is measured by the base transport factor, which is usually given the symbol α_T and defined by

$$\alpha_T = \frac{|I_{nE}| - |I_{rB}|}{|I_{nE}|} = 1 - \left|\frac{I_{rB}}{I_{nE}}\right| \tag{6.2.7}$$

where I_{nE} is the electron current injected from the emitter. Using Equations 6.1.17 and 6.2.6 for a transistor with a uniformly doped base, we obtain

$$\alpha_T = 1 - \frac{x_B^2}{2D_n\tau_n} = 1 - \frac{x_B^2}{2L_n^2} \tag{6.2.8}$$

Although Equation 6.2.8 is not directly applicable to IC transistors, it is sometimes used for these devices. The equation does not apply because base doping in the IC transistor is not constant. Because a graded base, as obtained by diffusion technology, increases J_S, less minority-carrier injection is needed for a given output current. Thus, a graded base improves the transport factor, and α_T calculated from Equation 6.2.8 can be regarded as a "worst-case" parameter. If we take a typical minority-carrier diffusion length of 10 μm and consider x_B to be 0.3 μm, we find $\alpha_T = 0.9996$. For the narrow base widths of high-performance IC transistors, the loss of minority carriers to recombination in the quasi-neutral base is even smaller.

The injection of base majority carriers (holes) into the emitter is the predominant cause for base current in most integrated-circuit transistors. Expressions for this current have already been derived in Sec. 5.3 because the base-emitter junction is just a forward-biased diode. For a given device it is necessary to determine where recombination of the holes injected into the emitter takes place in order to write a correct expression for the hole current. Consider first the prototype transistor shown in Figure 6.1; it has an emitter contact spaced a distance x_E from the edge of the base-emitter space-charge region. If x_E is much greater than a diffusion length for holes, virtually all injected holes recombine before reaching the ohmic contact. Therefore, by the theory developed in Sec. 5.3, the excess holes are distributed in the emitter according to an exponentially decaying function (Equation 5.3.12) as sketched in Figure 6.6a. The hole current is proportional to the hole-density gradient at the edge of the emitter quasi-neutral region.

$$I_{pE} = \frac{-qA_En_i^2D_{pE}}{N_{dE}L_{pE}}(e^{qV_{BE}/kT} - 1) \tag{6.2.9}$$

If the emitter contact is close to the base ($x_E \ll L_{pE}$), the hole concentration is a linear function of x (Figure 6.6b), and the hole current is

$$I_{pE}' = \frac{-qA_En_i^2D_{pE}}{N_{dE}x_E}(e^{qV_{BE}/kT} - 1) \tag{6.2.10}$$

For many integrated-circuit transistors, the emitter contact is closer to the base than a diffusion length. In these devices, however, emitter doping is not constant and consequently, injected holes in the emitter are influenced by a built-in electric field. Equations for this condition can be derived using a nearly analogous set of arguments to those employed in Sec. 6.1 for the transport of electrons in the base. For this case, however, a nonzero majority-carrier (electron) current is present.

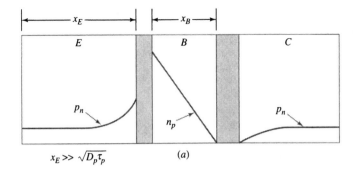

$x_E \gg \sqrt{D_p \tau_p}$ (a)

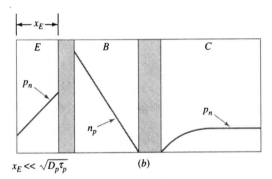

$x_E \ll \sqrt{D_p \tau_p}$ (b)

FIGURE 6.6 Minority-carrier distributions in the prototype transistor of Figure 6.1 under active bias. (a) $x_E \gg$ hole diffusion length in emitter. (b) $x_E \ll$ hole diffusion length.

Emitter Injection: Nonuniform Doping. To discuss nonuniform doping, consider the IC transistor of Figure 6.3. The impurity profile causes a built-in electric field in the emitter at thermal equilibrium. However, as we noted in Sec. 4.1, except for the junction space-charge region most nonuniformly doped regions can be treated as nearly neutral or quasi-neutral. This assumption means that $n_0(x) \approx N_d(x)$ and the built-in field at thermal equilibrium has the value obtained in Equation 4.1.13, which we repeat here for reference.

$$\mathscr{E}_0(x) = -\frac{kT}{q} \frac{1}{N_{dE}(x)} \frac{dN_{dE}(x)}{dx} \tag{6.2.11}$$

When the emitter-base junction is forward biased, we expect that the field in the quasi-neutral emitter is altered somewhat from the thermal-equilibrium value in Equation 6.2.11. For first-order analysis we express the field under bias as the sum of $\mathscr{E}_0$ plus an added term $\mathscr{E}_a$ that results from $V_{BE} \neq 0$. This is a reasonable assumption consistent with our consideration of low-level injection (i.e., $p'_n = n'_n \ll n_0$). With these considerations, the expressions for electron and hole currents in the emitter become

$$J_n = q\mu_n(n_{no} + p'_n)(\mathscr{E}_0 + \mathscr{E}_a) + qD_n\left(\frac{dn_{no}}{dx} + \frac{dp'_n}{dx}\right)$$

and

$$J_p = q\mu_p(p_{no} + p'_n)(\mathscr{E}_0 + \mathscr{E}_a) - qD_p\left(\frac{dp_{no}}{dx} + \frac{dp'_n}{dx}\right) \tag{6.2.12}$$

Because the drift and diffusion components balance at equilibrium (when $\mathscr{E} = \mathscr{E}_0$), Equations 6.2.12 can be written as

$$J_n = q\mu_n n_{no} \mathscr{E}_a + qD_n \frac{dp_n'}{dx} \tag{6.2.13}$$

and

$$J_p = q\mu_p(p_{no} + p_n')\mathscr{E}_a + q\mu_p p_n' \mathscr{E}_0 - qD_p \frac{dp_n'}{dx} \tag{6.2.14}$$

where we have neglected $p_n'(\mathscr{E}_a + \mathscr{E}_0)$ compared to $n_{no}\,\mathscr{E}_a$.

The first term in Equation 6.2.13 represents the ohmic flow of majority-carrier electrons. As with the ideal diode, we expect that only a small field $\mathscr{E}_a$ is necessary to provide this ohmic current. Typically $\mathscr{E}_a$ is less than $\mathscr{E}_0$. Making this assumption in Equation 6.2.14 for the minority-carrier hole current, we conclude that the second term is appreciably larger than the first. Thus,

$$J_p = q\mu_p p_n' \mathscr{E}_0 - qD_p \frac{dp_n'}{dx} \tag{6.2.15}$$

Using Equation 6.2.11 for the thermal-equilibrium field and combining terms, we derive

$$J_p = -\frac{qD_p}{N_d(x)} \frac{d}{dx}[p_n'(x)N_d(x)] \tag{6.2.16}$$

an equation similar to Equation 6.1.4 for electron current in the base.

Now we consider the important case of a short emitter region in which $x_E \ll L_{pE}$ so that negligible recombination of holes occurs in the bulk of the n-type region. Then the hole current is not a function of position, and Equation 6.2.16 can be integrated in the negative x direction from an arbitrary point x within the emitter to the ohmic contact where $p_n' = 0$

$$J_p \int_x^{x_E} \frac{N_d(x')dx'}{D_p} = qp_n'(x)N_d(x) \tag{6.2.17}$$

The hole current can be found by evaluating Equation 6.2.17 at the edge of the emitter-base, space-charge region $-x_n$, where the minority-carrier density is related to its thermal equilibrium value by Equation 5.3.8. The hole current in the emitter is then found to be

$$I_{pE}'' = -\frac{q\tilde{D}_p n_i^2 A_E (e^{qV_{BE}/kT} - 1)}{\displaystyle\int_{-x_n}^{x_E} N_d \, dx} \tag{6.2.18}$$

For a uniformly doped emitter, Equation 6.2.18 reduces to Equation 6.2.10.

If recombination in the quasi-neutral region is not negligible and some of the carriers recombine before traversing the emitter, Equation 6.2.16 is still valid, but it is not possible to remove J_p from the integral for an explicit solution as in Equation 6.2.18. The hole current increases above the value predicted by Equation 6.2.18; further approximations can be used to obtain a solution, but we will not consider them here.

For the two cases in which surface recombination dominates in determining I_{pE} (Equations 6.2.10 and 6.2.18), the proper value to use for A_E may not be straightforward. Surface recombination is almost always much greater at an ohmic contact than at an oxide interface. Hence, I_{pE} may be specified more accurately by using the contact area

($Y_{EM} \times Z_{EM}$ in Figure 6.3) instead of the junction area for A_E in Equations 6.2.10 and 6.2.18. We discuss this point further in Sec. 6.5 when we consider diffused planar transistors in more detail.

The effectiveness of an emitter junction in injecting electrons into the base is measured by the emitter injection efficiency, usually denoted by the symbol γ.

$$\gamma = \frac{|I_{nE}|}{|I_{nE}| + |I_{pE}|} = \frac{1}{1 + |I_{pE}/I_{nE}|} \qquad (6.2.19)$$

Because ($|I_{nE}| + |I_{pE}|$) is the total emitter current I_E, the electron current crossing the emitter-base junction I_{nE} is just γI_E. We can obtain an approximate expression for γ for the IC transistor using the approach taken previously for α_T. We apply the simple theory for the prototype transistor (Figure 6.1) and take dimensions and average doping concentrations appropriate to the IC transistor. Then using Equations 6.1.12 and 6.2.10 in Equation 6.2.19, we have

$$\gamma \approx \frac{1}{1 + \dfrac{x_B N_{aB} \tilde{D}_{pE}}{x_E N_{dE} \tilde{D}_{nB}}} = \frac{1}{1 + \dfrac{GN_B \tilde{D}_{pE}}{GN_E \tilde{D}_{nB}}} \qquad (6.2.20)$$

In the second form of the equation we introduce the Gummel Numbers GN (Equation 6.2.3) both for the base and the emitter.

Equation 6.2.20 is frequently used to represent the emitter efficiency in a *BJT*. Applying Equation 6.2.20 directly to IC transistors, however, can lead to error in predicting the emitter efficiency because two effects associated with heavy doping in the emitter were not considered in its derivation. The first effect, described in Sec. 1.1 (Figure 1.1.3 and Equation 1.1.33), is the narrowing of the bandgap and the corresponding increase in the intrinsic carrier density n_i when dopant concentrations in silicon exceed about 10^{18} cm^{-3}. Because the hole density injected into the emitter of an *npn* transistor is proportional to n_i^2 (Equation 6.2.10), bandgap narrowing in the heavily doped emitter causes increased minority-carrier injection and a corresponding reduction in the emitter efficiency. The second effect is the lifetime reduction that occurs when sufficient majority carriers are present to make Auger recombination (Equation 5.2.16) significant. The effective minority-carrier lifetime in the emitter becomes so short that recombination cannot be neglected, as we did in deriving Equation 6.2.10.

Because of these two effects, the derivation of Equation 6.2.10 for the reverse injection (of holes into the emitter) is not valid for a heavily doped emitter region. An exact analysis which includes the effects of heavy doping is not carried out here. Instead, we compensate for these effects by reducing the emitter Gummel Number in Equation 6.2.20 below the integrated dopant concentration. The reduction is sizable; if the dopant densities approach 10^{21} cm^{-3}, the effective emitter Gummel Number is only a few percent of the integrated dopant concentration in the emitter.

EXAMPLE *Emitter Efficiency of a BJT*

Use Equation 6.2.20 to calculate the emitter efficiency of an IC transistor in which the distance from the contact to the edge of the charge-neutral region in the emitter is 0.8 μm and the base Gummel Number GN_B is 3×10^{12} cm^{-2}. Assume that the emitter dopant concentration is 6×10^{20} cm^{-3} at the surface and that it decreases roughly exponentially to 5×10^{16} cm^{-3} at $x = 0.8$ μm. Assume also that heavy-doping effects reduce the effective emitter Gummel Number (GN_E) to 2% of the integrated dopant density.

Solution   We need to know the emitter Gummel Number and to estimate the ratio $\tilde{D}_{pE}/\tilde{D}_{nB}$. For the emitter, we assume an exponential variation in dopant density with a characteristic length λ

$$N_{dE} = N_{dE0} \exp\left(-\frac{x}{\lambda}\right)$$

From the values at $x = 0$ and $x = 0.8$ μm, we solve for λ

$$\lambda^{-1} = \frac{\ln(1.2 \times 10^4)}{0.8 \times 10^{-4}} \quad \text{or} \quad \lambda = 85.2 \text{ nm}$$

The integrated dopant density in the emitter is therefore

$$\int_0^{x_E} N_{dE}(x)\,dx \approx N_{dE0}\lambda = 5.1 \times 10^{15} \text{ cm}^{-2}$$

The effective emitter Gummel Number is 2% of the integrated dopant density or 1.02×10^{14} cm^{-2}. For the emitter diffusion coefficient, we find the average emitter dopant density by dividing the effective emitter Gummel Number by the emitter depth, $N_{davg} = 1.28 \times 10^{18}$ cm^{-3}. From Figure 1.16 we find $\tilde{D}_{pE} \approx 4.0$ cm^2 s^{-1}. The base diffusion coefficient $\tilde{D}_{nB}$ is approximately 22 cm^2 s^{-1} from the example considered earlier in Section 6.2. Using these numbers in Equation 6.2.20, we find

$$\gamma = \frac{1}{1 + \dfrac{3 \times 10^{12} \times 4.0}{1.02 \times 10^{14} \times 22}} = 0.9947$$

The emitter efficiency is indeed high, but not quite as high as the base transport factor α_T (0.9996) we found in the example illustrating Equation 6.2.8. This result is typical; the base-transport factor is closer to unity than is the emitter efficiency in IC transistors.

The above example relied on several approximations. Using values of D at average dopant concentrations to consider the spatial variation of D is at least plausible. However, including the effects of bandgap narrowing and Auger recombination by assuming a 50 times reduction in the emitter Gummel number seems quite arbitrary. Using numerical simulation to solve this example allows these effects to be explicitly included.   ∎

The magnitude of the ratio of the collector current I_C to the emitter current I_E under forward active bias is frequently given the symbol α_F. For our analysis, α_F is the product of the emitter injection efficiency γ and the base transport factor α_T:

$$\alpha_F = \gamma\alpha_T \tag{6.2.21}$$

Because all currents into the transistor sum to zero by Kirchhoff's current law, we have

$$I_B + I_E + I_C = 0$$

$$I_B - \frac{I_C}{\alpha_F} + I_C = 0$$

or

$$I_C = \frac{\alpha_F I_B}{(1 - \alpha_F)} = \beta_F I_B \tag{6.2.22}$$

where $\beta_F \equiv I_C/I_B$ is the current gain* for the case in which input current flows between the base and emitter and output current flows into the collector. Because α_F is nearly unity,

---

* The dc current gain is often indicated by the symbol h_{FE}.

β_F is large (typically of order 100). For the case that we took as an example, $\alpha_T = 0.9996$ and $\gamma = 0.9947$ so that $\alpha_F = 0.9943$ and $\beta_F = 174$. Small changes in α_F caused, for example, by process variations in fabricating the transistor are likewise magnified to large changes in $\beta_F [d\beta_F = d\alpha_F/(1 - \alpha_F)^2]$. This means that β_F is difficult to control precisely. Circuit designers can only be assured that β_F will be large; its value in any given process run can vary substantially.

As indicated in the example, the emitter efficiency γ is typically the factor that limits the size of the common-emitter current gain β_F in a BJT. Furthermore, as described above, both bandgap narrowing and Auger recombination limit the improvement in γ that can be obtained by increasing the emitter Gummel number.

We saw in the discussion leading to Equation 6.2.9 that the gain of a modern transistor is limited by reverse injection of holes from the base into the emitter and that this injection is proportional to the hole-density gradient at the edge of the emitter quasi-neutral region. Figure 6.6b showed that the slope of the hole density in the emitter region depends on the concentration at the surface of the emitter region when $x_E \ll L_p$ the hole diffusion length. For a typical metal contact to the emitter region, the excess hole density at the contact is virtually zero, and the hole-density gradient in the emitter is maximum, limiting the emitter injection efficiency and the transistor gain. If the hole density at the surface of the single-crystal emitter can be increased, the gradient decreases; the reverse hole injection decreases, and the transistor gain increases markedly.

The desired increase in hole density can be achieved by placing a layer of n^+ polycrystalline silicon (polysilicon—Sec. 2.6) between the metal contact and the heavily doped, single-crystal emitter region. The hole density at the single-crystal/polysilicon interface can be high, reducing the hole-density gradient in the emitter and the reverse hole injection from the base. The resulting increase in emitter injection efficiency can increase the transistor gain β_F by as much as an order of magnitude.

Several physical mechanisms have been suggested to explain this improvement. The dominant mechanism has not been firmly determined and probably depends on the details of the fabrication process. One mechanism suggests that a very thin barrier (≤ 1 nm), probably of residual silicon dioxide, blocks the flow of holes from the single-crystal emitter into the polysilicon, without severely impeding the flow of electrons from the polysilicon into the single-crystal emitter. [This mechanism relies on the lower barrier for electrons than holes ($E_{c,\text{oxide}} - E_{c,\text{Si}} < E_{v,\text{Si}} - E_{v,\text{oxide}}$)]. Another explanation relies on the shorter diffusion length of minority-carrier holes in polycrystalline silicon (with its highly imperfect crystal structure) than in single-crystal silicon. In both cases, the gradient of the minority-carrier hole density in the single-crystal emitter decreases, reducing hole injection from the base into the emitter and improving the emitter injection efficiency. The fabrication of this "polysilicon-emitter" bipolar transistor will be discussed in Sec. 6.5.

The theory that we developed in this section applies to dc bias conditions, and the equations have been derived for total currents. For application to amplifiers, we need equations that express the response to incremental changes in voltage around a dc bias point. We will consider such *small-signal* variations in Chapter 7, where we derive several equivalent circuits that describe transistor behavior in a way useful for circuit design. We will also consider frequency effects in transistors at that time. Our analysis thus far has considered only the dc (and low-frequency) case.

Before resuming our discussion of transistor action, we should point out that the analysis of hole injection into the emitter that we carried out for the case of nonuniform doping applies in general to pn junctions in integrated circuits. For example, the injected hole current given by Equation 6.2.18 is analogous to the result for electron injection into the p-type region in the diffused pn-junction diode considered in Sec. 5.6 (Figure 5.19b).

6.3 TRANSISTOR SWITCHING

Our discussion of transistor action in Sec. 6.1 emphasized that injection of electrons (base-minority carriers) into the base region is necessary for current to flow between the collector and the emitter or vice versa. Transistor switching can be understood in terms of injection, storage, extraction, and transport of electrons in the base. We can derive widely useful transistor models using this approach to transistor switching.

Regions of Operation

Figure 6.7 shows the regions of device operation in terms of the applied junction voltages V_{BE} and V_{BC} and is useful for understanding transistor switching. For example, the fourth quadrant (V_{BE} positive and V_{BC} negative) corresponds to the forward-active region that we considered in Sec. 6.2.

Reverse-active bias. In the second quadrant the polarities of both V_{BC} and V_{BE} are reversed from the forward-active biases, and the transistor is said to be biased in the reverse-active region. In this region an *npn* transistor injects electrons at the collector and collects them at the emitter. A one-to-one correspondence can be made to the parameters defined for forward-active bias. These parameters are subscripted with an R to denote a reverse measurement. For example, output current is delivered to the emitter lead and the ratio of output current to input (base) current in a reverse-active condition is defined as $\beta_R = I_E/I_B$. The distribution of minority carriers in the prototype transistor of Figure 6.1 under reverse-active bias is sketched in Figure 6.8; note its similarity to Figure 6.6a.

For the prototype transistor of Figure 6.1, the distinction between forward-active bias and reverse-active bias is not critical because the device is symmetrical; the doping is the same in the emitter and collector, and the two junctions have equal areas.

In contrast to the prototype transistor, the integrated-circuit transistor (Figure 6.3) is asymmetric both in geometry and doping, and a sketch of minority-carrier distributions under reverse-active bias differs markedly from one made in the forward-active region. First, the injection efficiency in the reverse-active mode is very much lower than in the

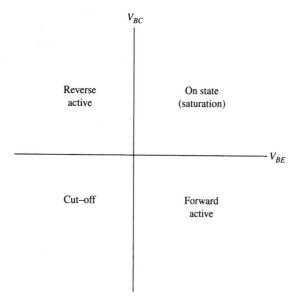

FIGURE 6.7 Regions of operation of an *npn* transistor as defined by base-emitter and base-collector biases.

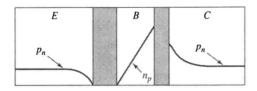

FIGURE 6.8 Minority-carrier densities in the prototype transistor of Figure 6.1 under bias in the reverse-active region. (Compare to Figure 6.6a).

forward-active region because of the ratio of dopant concentrations. Second, electrons injected from the normal collector travel against the built-in base field as they move toward the emitter. Third, the injecting junction is much larger than the collecting junction, and injection from the collector results in losses at the base contact and at the passivating oxide that are not significant in the forward-active region. These asymmetries reduce the gain for reverse-active bias below that for forward bias and have important practical consequences that we will discuss in Chapter 7.

Saturation.   The first quadrant in Figure 6.7 is defined by positive bias on both the base-emitter and base-collector junctions. This bias condition is called *saturation*.* A transistor switch in the "ON" position is biased in this region. In saturation, both junctions inject electrons, and the minority carriers are distributed throughout the prototype transistor as shown in Figure 6.9. In saturation the electron concentration is markedly increased throughout the entire base region. Comparing the minority-carrier populations in Figure 6.9 to those in Figures 6.6 and 6.8, we see that the saturation condition corresponds to the superposition of forward-active and reverse-active operation. The physical basis is that both junctions are injecting and collecting electrons *at the same time*. They inject because the built-in potential is reduced from its equilibrium value; they collect because the junction field is still of the proper polarity to sweep electrons from the base. This property of the saturated state being a superposition of forward-active and reverse-active bias conditions does not depend on transistor geometry; it is valid for the IC transistor, as well as for the prototype device.

The large amount of charge stored in the base of a saturated transistor must be removed before the transistor can be turned "OFF," reducing the switching speed of the transistor. The Schottky diode "clamp" discussed in Sec. 3.6 can be used to prevent the transistor from entering the saturation region, increasing its switching speed.

Cut-off.   When V_{BE} and V_{BC} are both negative or zero, the transistor is in the *cut-off* state. With these bias voltages applied, the base has, at most, its equilibrium population of electrons so that only very small currents can flow between collector and

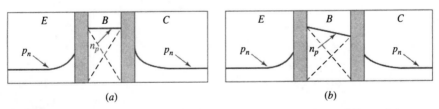

(a)   (b)

FIGURE 6.9 Minority-carrier densities in the prototype transistor of Figure 6.1 under bias in the saturation region; (a) with no current between collector and emitter, (b) with current flowing from collector to emitter.

* The term "saturation" is used because the collector current is determined by conditions in the circuit external to the transistor, rather than by the transistor itself.

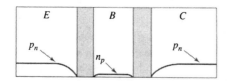

FIGURE 6.10 Minority-carrier densities in the prototype transistor of Figure 6.1 under cut-off conditions.

emitter. When negative bias is simultaneously applied to both junctions, portions of the base become depleted of even the few "built-in" minority carriers and the dc behavior of the transistor is very close to that of an open circuit. The minority-carrier populations in all three regions of a homogeneously doped transistor biased into cutoff are shown in Figure 6.10.

Although the cut-off transistor does not pass dc current, its behavior is not identical to that of an open circuit. As described in Sec. 4.3, a reverse bias on a *pn* junction exposes fixed donor and acceptor charges by removing the compensating free charges. For calculations, it is useful to make a graph showing the variation in stored junction charge, which we denote Q_V, as a function of voltage. This is readily accomplished by using the theory of Chapter 4 for step or linearly graded junctions or by using numerical simulation. A graph of this type was developed in the example of Sec. 5.4 for an abrupt *pn*–junction diode. A similar graph with Q_V plotted for both an abrupt and a linearly graded junction is given in Figure 6.11. Derivation of the graph is considered in an example and in Problem 6.7. The inset in Figure 6.11 identifies Q_V as the stored charge, and emphasizes that only the charge in excess of the junction space charge at thermal equilibrium is plotted.

To place a transistor in cut-off, the switching source must supply the junction with the incremental charge to be stored in the space-charge region; thus, the cut-off device has a capacitor-like behavior although the capacitance (dQ/dV) is not constant.

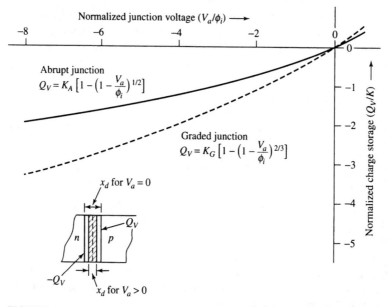

FIGURE 6.11 Normalized stored charge Q_V/K versus normalized reverse-junction bias V_a/ϕ_i in the space-charge regions of an abrupt junction and a linearly graded junction.

EXAMPLE *Stored Charge and Transistor Switching*

An *npn* prototype transistor similar to that shown in Figure 6.1 has a base doping $N_a = 10^{16}$ cm^{-3}, and emitter and collector doping $N_d = 10^{19}$ cm^{-3}. The transistor is biased in the cut-off mode with $V_C = 3$ V, $V_E = 0$ V, and $V_B = -3$ V. If the junction area is 10^{-5} cm^2, how much charge must be supplied to the base to bring the base voltage to $V = 0$ V?

Solution From Equation 4.2.10 we find $\phi_i = 0.872$ V for both the base-emitter and base-collector junctions. Using Equation 4.3.1 we find K_A in Figure 6.11 to be

$$K_A = \sqrt{2\epsilon_s q N_a \phi_i} = 53.8 \text{ nC cm}^{-2}.$$

Because the junction area is 10^{-5} cm^2, we have $K_A \times A = 0.54$ pC for the factor representing total charge at each junction. Initially, $V_a = -6$ V $= -6.88 \times \phi_i$ at the collector-base junction. From Figure 6.11, the charge stored at the collector $Q_{VC} = -1.8 \times K_A A = -0.972$ pC. When $V_B = 0$ V, $V_a = -3$ V $= -3.44 \times \phi_i$, and $Q_{VC} = -0.59$ pC. The charge supplied from the base is therefore 0.38 pC. At the base-emitter junction, the bias change is from -3 to 0 V; hence, the stored charge changes from -0.59 to 0 pC.

The total charge supplied from the base terminal is the sum of these charges or 0.97 pC. If, for example, the source switching the base voltage can supply a maximum of 1 mA of current, it requires 0.97 ns to switch the voltages as described in this example. This type of calculation is frequently carried out by designers of transistor switching circuits. ∎

A measure of the very small currents that flow in a cut-off transistor is available in the active-bias data plotted in Figure 6.4. We noted there that if the collector-base junction is reverse biased and V_{BE} is reduced nearly to zero, the collector current begins to deviate from the exponential relationship predicted by Equation 6.2.1. Instead, J_C approaches a low value J_ℓ that is independent of V_{BE}. At low base-emitter bias, the base-emitter junction injects such a small number of electrons that transistor action between the junctions is negligible. The collector-base junction behaves electrically like an isolated *pn*-junction diode under reverse bias, just as it would in cut-off, and the diode leakage current J_ℓ flows from collector to base. The small size of J_ℓ (nA cm^{-2}) indicated in Figure 6.4 is typical.

As was discussed in Chapter 5, three separate mechanisms can be responsible for the small current in a *pn* junction under reverse bias: generation of holes and electrons in the space-charge region, generation of electrons in the *p*-type base, and generation of holes in the *n*-type collector. The first of these three components, generation in the space-charge region, dominates in silicon *pn* junctions, as we discovered in evaluating Equation 5.3.29.

The space-charge region generation component I_g is approximately proportional to the width of the space-charge layer as was shown in Sec. 5.3. It depends relatively weakly on collector-base bias (Equations 5.3.26–5.3.28), and is most often written

$$I_\ell = J_\ell A_C = I_g = \frac{1}{2} \frac{q n_i x_d A_C}{\tau_0} \tag{6.3.1}$$

where x_d is the space-charge region thickness, A_C is the collector-base junction area, and τ_0 is the effective lifetime in the space-charge region. Because of its small size, I_ℓ is negligible for silicon transistors biased in the active mode. It can be important for transistor switches, however. In terms of the IC transistor structure in Figure 6.3, the area A_C is approximately $Y_C \times Z_C$ plus the vertical portions of the collector-base junction that meet the surface. There is also a contribution to I_ℓ from recombination centers at the surface (Section 5.3); with careful processing this contribution can be made very small.

We began this section with a discussion of the regions of transistor operation, which are summarized in Figure 6.7. A transistor switch is biased alternately in regions 1 and 3 (saturation and cut-off) and moves through regions 2 and 4 only during switching transients. To move from saturation to cut-off, the charges stored in and near the base of a transistor must be altered. Much of the design of transistor switching circuits consists of assuring that these charging and discharging requirements are adequately met. Models that account for the physical mechanisms we have discussed and that are useful for transistor switching calculations are presented in Chapter 7. In the next section we introduce a transistor model that has useful dc applications and that quantifies some of the physical pictures we have developed.

6.4 EBERS-MOLL MODEL

A simple and very useful model for carrier injection and extraction in bipolar transistors was developed in 1954 by J.J. Ebers and J.L. Moll [5]. More than four decades later, this model provides the basic framework for complex computer-aided models of bipolar transistors (BJTs). The *Ebers-Moll Model* is based on understanding the BJT in terms of interacting diode junctions, a viewpoint already presented in Sec. 6.1, and used to derive Equation 6.1.14. For the active-biased transistor (shown in Figure 6.5), we recognize that J_n in Equation 6.1.14 represents the electrons flowing between the emitter and the collector and linking these regions. We therefore call J_n (and its counterpart J_p for a *pnp* transistor) the *linking current*. The theory introduced in Sec. 6.1 treats only linking current, and therefore does not account for the components of base current shown in Figure 6.5.

To treat the base current, we separately consider the component flowing between the base and the emitter I_{BE}, and that flowing between the base and the collector I_{BC}. Because the base-emitter junction is a *pn* junction diode, we can express currents through it by the ideal-diode expression, denoting the saturation current as I_{0E}

$$I_{BE} = I_{0E}[\exp(qV_{BE}/kT) - 1]. \qquad (6.4.1)$$

The total current in the emitter consists of the flow to the collector (the linking current) minus the base-emitter diode current. For the linking current we use Equation 6.1.14, assuming for the present a constant area A across the base and taking $I_S = J_S A$, where J_S is given by Equation 6.1.15.* Thus, the emitter current is

$$I_E = I_S[\exp(qV_{BC}/kT) - \exp(qV_{BE}/kT)] - I_{0E}[\exp(qV_{BE}/kT) - 1] \qquad (6.4.2)$$

Similarly, for the collector current, we have

$$I_C = I_S[\exp(qV_{BE}/kT) - \exp(qV_{BC}/kT)] - I_{0C}[\exp(qV_{BC}/kT) - 1] \qquad (6.4.3)$$

where the base-collector current is

$$I_{BC} = I_{0C}[\exp(qV_{BC}/kT) - 1]. \qquad (6.4.4)$$

If we group the terms in Equations 6.4.2 and 6.4.3 according to their voltage dependences, we can write

$$I_E = -(I_S + I_{0E})[\exp(qV_{BE}/kT) - 1] + I_S[\exp(qV_{BC}/kT) - 1] \qquad (6.4.5a)$$

and

$$I_C = -(I_S + I_{0C})[\exp(qV_{BC}/kT) - 1] + I_S[\exp(qV_{BE}/kT) - 1]. \qquad (6.4.5b)$$

---

* The base area A need not be constant (it is not in an IC transistor) for the *Ebers-Moll* model to be applicable.

We now define

$$I_{ES} \equiv I_S + I_{0E}, \qquad I_{CS} \equiv I_S + I_{0C} \tag{6.4.6a}$$

and

$$\alpha_F \equiv \frac{I_S}{I_S + I_{0E}}, \qquad \alpha_R \equiv \frac{I_S}{I_S + I_{0C}}. \tag{6.4.6b}$$

In terms of these new variables, Equations 6.4.5 become

$$I_E = -I_{ES}[\exp(qV_{BE}/kT) - 1] + \alpha_R I_{CS}[\exp(qV_{BC}/kT) - 1] \tag{6.4.7a}$$

and

$$I_C = -I_{CS}[\exp(qV_{BC}/kT) - 1] + \alpha_F I_{ES}[\exp(qV_{BE}/kT) - 1] \tag{6.4.7b}$$

Equations 6.4.7 are the Ebers-Moll (E–M) equations for an *npn* transistor. In the corresponding equations for a *pnp* transistor, the current directions are changed to account for the polarity of the *pn* junctions. The diodes in a *pnp* BJT are under forward bias when V_{EB} and V_{CB} are positive (Problem 6.10).

The Ebers-Moll equations directly predict the emitter and collector currents for the transistor; in conjunction with Kirchhoff's current law (the sum of all current into a node is zero), they also specify the base current. The Ebers-Moll model has four parameters (α_F, α_R, I_{ES}, and I_{CS}). From Equations 6.4.6, however, we see that only three parameters are independent; the remaining one can be obtained from the other three by the *reciprocity relation*.

$$\alpha_F I_{ES} \equiv \alpha_R I_{CS} \equiv I_S \tag{6.4.8}$$

Equations 6.4.7 can be simplified by defining two new quantities, a diode current related to forward-active bias I_F, and one related to reverse-active bias I_R. These currents are expressed as

$$I_F = I_{ES}[\exp(qV_{BE}/kT) - 1] \tag{6.4.9a}$$

and

$$I_R = I_{CS}[\exp(qV_{BC}/kT) - 1]. \tag{6.4.9b}$$

In terms of I_F and I_R,

$$I_E = -I_F + \alpha_R I_R \tag{6.4.10a}$$

and

$$I_C = -I_R + \alpha_F I_F \tag{6.4.10b}$$

An equivalent circuit that represents Equations 6.4.10 is shown in Figure 6.12. The circuit consists of diodes and current sources connected between the base and the emitter and between the base and the collector. The current sources are needed in the circuit to represent the current components that depend on voltages across a remote junction (I_R at the emitter and I_F at the collector). Applying Kirchhoff's current law to the circuit of Figure 6.12, we solve for the base current

$$I_B = -(I_E + I_C) = I_F(1 - \alpha_F) + I_R(1 - \alpha_R) \tag{6.4.11}$$

Applications. To see how the Ebers-Moll model represents the transistor in the various regions of operation, consider first the cut-off region in which V_{BE} and V_{BC} are both negative. From Equations 6.4.9, 6.4.10, and 6.4.11, we obtain the equivalent circuit shown

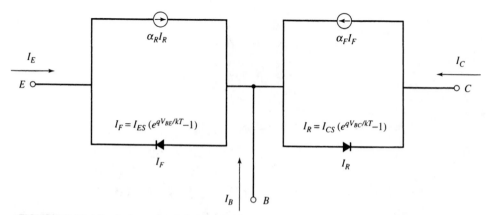

FIGURE 6.12 Equivalent circuit for the Ebers-Moll model of an *npn* transistor.

in Figure 6.13. The model reduces to two current sources that represent the reverse saturation currents of the two junctions.

In the forward-active region, the base-emitter junction is forward biased and the base-collector junction is reverse biased. We can rearrange Equations 6.4.10 to express collector current in terms of emitter current:

$$I_C = -\alpha_F I_E - I_R(1 - \alpha_F \alpha_R) \tag{6.4.12}$$

which, under active-bias conditions, becomes

$$I_C = -\alpha_F I_E + I_{CS}(1 - \alpha_F \alpha_R) \tag{6.4.13}$$

Similarly, reverse-active bias results in

$$I_E = -\alpha_R I_C + I_{ES}(1 - \alpha_F \alpha_R) \tag{6.4.14}$$

Inspection of Equations 6.4.13 and 6.4.14 shows that the parameters of the Ebers–Moll model, α_F, α_R, I_{ES}, and I_{CS}, can be obtained by measuring I_C versus I_E under forward-active bias or I_E versus I_C under reverse-active bias. If the forward-active measurements are used, a plot of I_C versus I_E should be linear with slope equal to $-\alpha_F$. The intercept with $I_E = 0$ corresponds to a measurement with open-circuited emitter, and the current flowing in this case is usually denoted I_{CB0} where

$$I_{CB0} = I_C|_{I_E=0} = I_{CS}(1 - \alpha_F \alpha_R) \tag{6.4.15}$$

[The notation indicates the current flowing between the connections denoted by the first two subscripts with the third connection (the emitter in this case) open.]

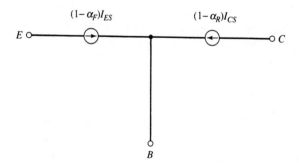

FIGURE 6.13 The Ebers-Moll representation for a transistor biased in the cut-off region.

This current can be compared to the current I_{CE0} in the collector when the base is open circuited. From Equation 6.4.13 with $I_E = -I_C$, we obtain

$$I_{CE0} = I_C|_{I_B=0} = \frac{I_{CS}(1 - \alpha_F\alpha_R)}{(1 - \alpha_F)} = \frac{I_{CB0}}{(1 - \alpha_F)} \qquad (6.4.16)$$

The difference in magnitudes between I_{CB0} and I_{CE0} can be traced to the boundary conditions determined by the bias on the emitter-base junction. When I_{CB0} flows, the emitter-base junction becomes slightly reverse biased because some electrons are extracted from the emitter without being replaced from the external circuit. In this case the collector current is carried only by electrons generated in the base and by holes generated in the collector. In the second case, I_{CE0}, the base is open circuited instead of the emitter. The base-emitter junction becomes forward biased, and the greater part of the collector current results from electrons carried across the base from the emitter. The leakage current I_{CB0} is therefore effectively multiplied by the transistor gain.

EXAMPLE *Ebers-Moll Equations*

Calculate the reverse-bias voltage present on the base-emitter junction of an *npn BJT* when the emitter is open-circuited and a reverse bias is placed on the base-collector junction. Assume $\alpha_F = 0.98$, $\alpha_R = 0.70$, $I_{CS} = 1 \times 10^{-13}$ A, $I_{ES} = 7.14 \times 10^{-14}$ A.

Solution   For this bias condition, the collector current is I_{CB0} as given by Equation 6.4.15. Because the emitter current is zero, Equation 6.4.10a establishes that

$$I_F = \alpha_R I_R \approx -\alpha_R I_{CS} = I_{ES}\left(\exp\frac{qV_{BE}}{kT} - 1\right)$$

where we have used $I_R \approx -I_{CS}$.

Using the reciprocity relationship (Equation 6.4.8), we solve for V_{BE}

$$V_{BE} = \frac{kT}{q}\ln(1 - \alpha_F) = -0.10 \text{ V}$$

at $T = 300$ K.

The reverse bias on the base-emitter junction depends only on α_F for this case of an open-circuited emitter. This dependence occurs because the bias is established by balancing the linking current with the current returned through the back-biased, base-emitter junction diode so that the emitter current is zero. ∎

It is often very useful to obtain models in which the active elements (generators) are actuated by terminal currents. This is easily done for active bias with emitter current as the variable using the equations we have developed. The equivalent circuit shown in Figure 6.14a is an emitter-actuated model that is consistent with Equation 6.4.13 if the leakage term at the collector is represented by I_{CB0} (Equation 6.4.15). It is straightforward to derive expressions for the active-biased transistor when driven from the base and to show the validity of the circuit sketched in Figure 6.14b (Problem 6.11).

In saturation, the quantity of greatest interest is $V_{CE\,sat}$, the voltage drop across the "ON-state" switch. In this case the Ebers-Moll model allows one to derive

$$V_{CE\,sat} = \frac{kT}{q}\ln\left\{\frac{\left[1 + \dfrac{I_C}{I_B}(1 - \alpha_R)\right]}{\alpha_R\left[1 - \dfrac{I_C}{I_B}\left(\dfrac{1 - \alpha_F}{\alpha_F}\right)\right]}\right\} \qquad (6.4.17)$$

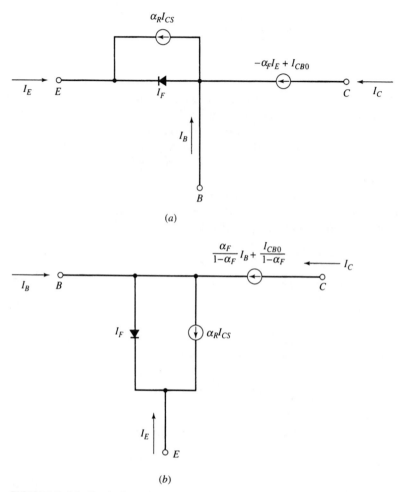

FIGURE 6.14 Equivalent circuits for *npn* transistors actuated by terminal currents. (*a*) Emitter-current actuated circuit, (*b*) base-current actuated circuit.

To obtain Equation 6.4.17, we recognize that the exponential terms in the expressions for the diode currents are large compared to unity when the transistor is saturated (Problem 6.12). Because V_{CEsat} is small, a first approximation that is often used for design purposes is to consider it negligible. When V_{CEsat} is not negligible, Equation 6.4.17 shows that V_{CEsat} changes only slowly with collector current. Hence, an equivalent circuit consisting of a voltage source from collector to emitter or, preferably of two sources connected from base to emitter and from base to collector, is appropriate. The Ebers-Moll equations do not take account of any resistance in series with the junctions. The voltage drops across series resistances, particularly in the collector regions of IC transistors, often exceed the value of V_{CEsat} predicted by Equation 6.4.17; thus the "ON-state" transistor switch is often modeled by a series voltage source and a resistor denoted by R_{Csat}.

In this section we discussed the static Ebers-Moll model, the model for dc conditions. Although this model can be modified to allow dynamic calculations (i.e., to solve for transient conditions), we find it advantageous to use another technique, known as charge-control modeling, for these calculations. The charge-control model is discussed in Chapter 7, where we also discuss a modification of the Ebers-Moll model that accounts for some important second-order effects.

6.5 DEVICES: PLANAR BIPOLAR AMPLIFYING AND SWITCHING TRANSISTORS

When discussing bipolar transistors for integrated circuits, we can consider two broad categories defined by their ultimate use: amplification and switching. Because of the higher mobility and diffusion coefficient of electrons than holes, *npn* transistors are used more frequently than are *pnp* transistors. Both amplifying and switching *npn* transistors are usually fabricated in an epitaxial layer of relatively high-resistivity, *n*-type silicon. The layer is lightly doped to allow the voltages specified by the circuit designer to be applied to the base-collector junction without junction breakdown (Sec. 4.4). The epitaxial layer is usually deposited on a *p*-type silicon substrate so that individual transistors are isolated from the substrate by *pn* junctions.

As shown in Figure 6.3*b*, contact to the collector of an integrated-circuit transistor must be made from the top surface adjacent to the active portion of the transistor. To allow lateral current flow to the metallic collector contact with minimum series resistance, a *buried layer* or *subcollector* of heavily doped *n*-type silicon is usually added between the lightly doped, *n*-type epitaxial layer and the *p*-type substrate, as discussed before (Figure 5.18*b*). The buried layer reduces series resistance from the kilohm range to a few hundred ohms in typical transistors. Adjacent device regions are laterally isolated from each other, as will be discussed in the following section. The base, emitter, and collector-contact regions are then added by successive implantations or diffusions. For amplifying applications especially, even the buried layer does not reduce the series resistance adequately; the resistance of the epitaxial layer between the buried layer and the collector contact is still substantial. In this case, an extra processing step is included in which a heavily doped *n*-type region is added under the collector contact and diffused until it reaches the buried layer. This extended diffused region is called the *collector plug*, and its inclusion reduces series resistance to the order of 10 Ω.

Typical impurity-concentration profiles perpendicular to the surface and through the emitter, base, and collector are shown in Figures 6.15 and 6.16 for switching and

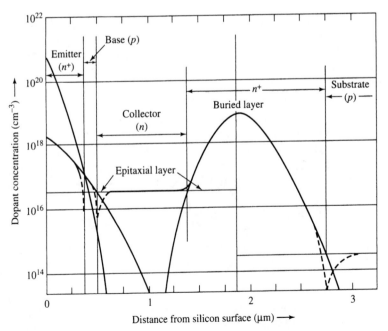

FIGURE 6.15 Diffusion profile of a switching transistor [8].

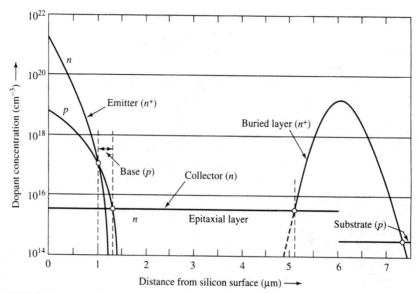

FIGURE 6.16 Diffusion profile of an amplifying transistor [8].

amplifying transistors, respectively. These figures show that the major differences in the design of switching and amplifying transistors lie in the thickness and resistivity of the epitaxial layer. Both quantities are greater in amplifying devices than in switching transistors; this results in increased breakdown voltage and reduces a parasitic effect (the Early effect) that will be described in Chapter 7. For switching devices, saturation ("ON-state") resistance must be minimized, which requires epitaxial layer thicknesses of about one micrometer with a resistivity of the order of some tenths of an Ω-cm.* If Schottky clamping is to be used on the switching transistor (as described in Chapter 3), the resistivity must be greater than 0.1 Ω-cm to obtain a good metal-semiconductor barrier. Schottky-clamped transistors differ from unclamped switching devices only in the extended finger of metal from the base contact to the lightly doped epitaxial region forming the collector, as shown in Figure 6.17. The base doping concentration should be low to maximize emitter injection efficiency γ. If the doping is too low, however, lateral series resistance in the base can limit transistor performance. If the base doping in the active portion of the transistor (the *intrinsic* base region) is fairly low to obtain a reasonable emitter injection efficiency, additional p-type doping can be added in the regions outside the active base (the *extrinsic* base region) to reduce lateral series resistance and improve the metal-semiconductor contact. This additional doping can also prevent surface inversion (described in Chapter 8), which can occur if the surface of the p-type region is lightly doped outside of the active portion of the transistor.

The emitter is usually heavily doped to increase the emitter injection efficiency. When doped above about 10^{20} cm^{-3}, however, efficiency decreases because of decreased hole lifetime in the emitter (which enhances hole injection from the base) and because of bandgap narrowing in degenerately doped silicon (Sec. 1.1) [6].

---

* When the BJT is saturated, the voltage drop between the collector and the emitter is equal to V_{CEsat} as given by Equation 6.4.17 in series with the voltage drop across series resistance in the structure. Typically the largest series resistance is in the epitaxial layer.

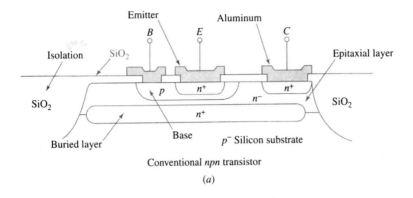

Conventional *npn* transistor

(*a*)

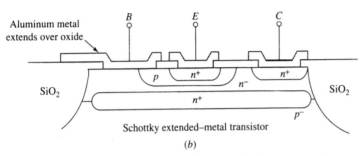

Schottky extended–metal transistor

(*b*)

FIGURE 6.17 Comparison of cross sections of (*a*) conventional *npn* transistor, (*b*) Schottky-clamped, extended-metal *npn* transistor.

As we already noted, reducing the base width and decreasing the base doping both increase gain. As the base width is reduced to submicrometer dimensions, however, the applied base-collector voltage can deplete the entire width of a moderately doped neutral base region (Figure 6.18) so that the collector-base space-charge region reaches through to the emitter-base space-charge region at higher voltages. Because the barrier at the base-emitter junction is reduced, this *punchthrough* results in a highly conductive path from emitter to collector and can lead to damaging currents in a transistor. This is one failure mode that is influenced by the collector-base bias.

The other failure mode that is sensitive to V_{CB} is avalanche breakdown of the collector junction, which was described in connection with diodes in Sec. 4.4. However, the breakdown voltage of a bipolar transistor can be markedly lower than that of the reverse-biased, collector-base diode. Consider an *npn* bipolar transistor with the base lead open. A voltage is applied between the emitter and collector with polarity such that the base-collector junction is reverse biased. Most of the applied voltage is dropped across the base-collector junction. However, to supply the reverse saturation current of the base-collector junction, a small number of holes are injected from the collector into the base, where they are majority carriers. The holes cannot accumulate within the neutral base where they are majority carriers; they must move to the edges of the neutral base region, where they neutralize a portion of the space-charge region. At the emitter-base region, the slightly narrower depletion region corresponds to a smaller barrier, which allows some electrons to be injected into the base region, where they travel to the base-collector space-charge region. The electrons entering the base-collector space-charge region are accelerated by the high electric field in this reverse-biased junction and can ionize other electrons, leading to avalanche breakdown of the junction.

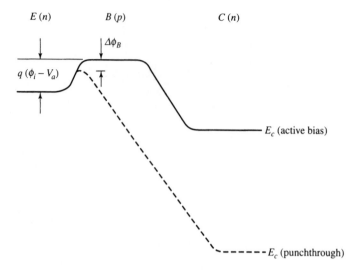

FIGURE 6.18 Conduction-band energy for an *npn* transistor under active bias with normal operating voltages (solid line). At high collector biases the entire base region can be depleted of mobile carriers, causing punchthrough (dashed curve) and a corresponding reduction in barrier height ($\Delta\phi_B$) for electron injection from the emitter.

Because the number of electrons injected at the emitter-base junction (and traveling to the base-collector junction) increases rapidly for a small decrease in the barrier height, the number of electrons entering the base-collector junction is much larger than the number corresponding to the equilibrium electron (minority-carrier) concentration in the base region. (The number of electrons of effectively amplified by the gain β of the transistor.) Because of the large number of electrons available, the avalanche breakdown process occurs at a much lower voltage than the base-collector diode breakdown voltage itself. This breakdown voltage BV_{CE0} (collector-emitter breakdown voltage with the base lead open) is often related to the base-collector diode breakdown voltage (BV_{CB0}) by the expression

$$BV_{CE0} = \frac{BV_{CB0}}{\beta^{1/m}} \tag{6.5.1}$$

where m is typically about 4.

Process Considerations

The vertical dopant profile has a major influence on the behavior of the *intrinsic* transistor, but other nearby parasitic elements often severely limit the overall transistor performance. The physical size of the regions used to bring signals into the intrinsic transistor can lead to significant *extrinsic* resistances and capacitances that limit transistor gain and frequency response.

In an integrated circuit each transistor must be totally isolated from neighboring transistors. We saw that the vertical isolation between a transistor and the substrate is provided by the *p-n* junction between the *n*-type, epitaxial collector region and the *p*-type substrate. However, lateral isolation is also needed. In older IC processes, lateral isolation was often provided by inserting a *p*-type region between the *n*-type epitaxial regions of adjacent transistors, as shown in Figures 6.19*a* and 6.20*a*. This diffused isolation region must extend through the entire thickness of the epitaxial region and intersect the *p*-type substrate to isolate the transistor completely. Because of the lateral diffusion occurring during the extended diffusion time needed to penetrate the epitaxial layer, the lateral dimensions of diffused isolation regions are large, limiting transistor density. In addition, the *p*-type base region must be spaced away from the *p*-type isolation diffusion so that

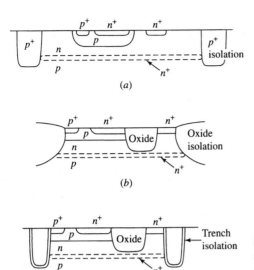

FIGURE 6.19 Cross sections of structures used to isolate adjacent BJTs. (a) pn-junction isolation; (b) LOCOS oxide isolation; (c) trench isolation.

they are separated by a neutral n-type region and also so that carriers injected from the normal base region (the emitter of the parasitic lateral bipolar transistor formed by the p-type base region, n-type collector region, and p-type isolation) are not collected by the isolation diffusion.

The lateral dimensions of the transistor can be greatly reduced by using oxide regions to laterally separate adjacent transistors. If the epitaxial region to be isolated is only

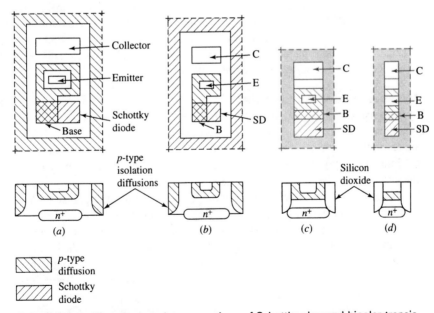

FIGURE 6.20 Plan views and cross sections of Schottky-clamped bipolar transistors: (a) junction-isolated BJT, (b) washed-emitter BJT with junction isolation, (c) oxide-isolated transistor, (d) walled-emitter BJT. The p-type diffused regions and the Schottky diode are cross hatched. The contact regions are labeled.

of the order of 1 μm thick, an oxide can be grown through the entire thickness of the epitaxial layer by the LOCOS (local oxidation of silicon) process discussed in Sec. 2.6. Somewhat thicker layers can be isolated by first etching about half the thickness of the subsequently grown oxide so that most of the grown oxide is recessed below the surface (Figure 6.19b). The isolated region then corresponds to the entire thickness of the grown oxide, rather than about half, as obtained when oxide grows on a plane surface. (The oxide grows about half above and half below the original surface.)

When thicker epitaxial layers need to be isolated, *trench* isolation can be used (Figure 6.19c). Directional, reactive ion etching is used to etch a "trench" with nearly vertical sides through the entire thickness of the epitaxial layer (plus a small amount to compensate for process variations). An oxide layer is then grown on the sides and bottom of the trench, and the remainder of the trench is filled with undoped polysilicon to provide a flat surface for further processing. (If the epitaxial layer is thin, the trench can be completely filled with insulating oxide, but the stress from an oxide-filled deep trench can cause excessive deformation of the wafer, complicating further processing.) In addition to providing flexibility, trench isolation also increases the density by avoiding the lateral oxidation at the transition between the device and isolation regions associated with LOCOS isolation.

To reduce the surface area of the transistor, the emitter contact region can be exposed without an additional mask (and the associated area needed for misalignment) by etching the thin oxide over the emitter region without a mask. Using this *washed-emitter* process (Figure 6.20b), of course, requires that thicker oxide regions cover the elements that must not be exposed. The metal emitter contact is then evaporated and covers the entire emitter diffusion; the larger contact area also reduces parasitic resistance. A base-emitter short is avoided (most of the time) because the emitter dopant diffuses laterally under the thicker oxide covering the base.

When oxide forms the lateral isolation, the base and emitter regions need not be spaced away from the isolation regions, and the area can be reduced further. The base and emitter of the BJT can be placed immediately adjacent to the isolation oxide (Figure 6.20d). Because of the lower permittivity of oxide compared to silicon, parasitic sidewall capacitance is often reduced. A narrow oxide region can also separate the emitter region and the collector contact region, saving further space.

Polysilicon-emitter process. Most advanced BJTs are currently fabricated using the polysilicon-contacted emitter introduced in Sec. 6.2. A *double polysilicon* process is outlined in Figure 6.21. A buried layer, an epitaxial layer, and LOCOS or trench isolation are first formed, as described previously. Then a first layer of polysilicon is deposited and doped heavily *p*-type. One end of lines defined in this layer can be used to make very small contacts to the base region in the single-crystal silicon; metal contacts to the other end of these lines over thick, low-permittivity oxide reduce the parasitic capacitance associated with the base contact regions. Some *p*-type dopant is diffused from the polysilicon into the *extrinsic* base contact region in the single-crystal silicon to form ohmic contact between the polysilicon and the single-crystal silicon, and an oxide is formed over the polysilicon and on its sides. The oxide is removed from the region that is to become the intrinsic base, which can be added by implantation into the exposed single-crystal silicon at this point or later.

A second layer of polysilicon is then deposited; this layer is separated laterally from the first layer by only the thickness of the oxide on the side of the first layer of polysilicon. The narrow space resulting from this *self-alignment* reduces the parasitic extrinsic base resistance. If the intrinsic base dopant has not been added previously, it

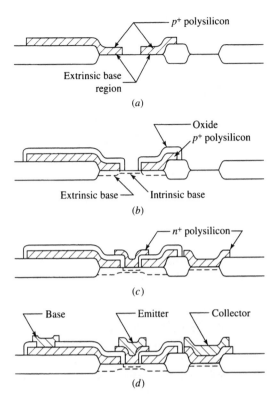

FIGURE 6.21 Two layers of polysilicon can be used in the polysilicon-emitter transistor, with an oxide on the sidewall of the first layer separating the two. (*a*) The extrinsic-base region is formed by the first polysilicon layer which extends over the field oxide. (*b*) Boron is diffused from the polysilicon into the single-crystal silicon to make contact to the intrinsic base, which can be added by implantation. The polysilicon is oxidized, and the oxide is removed from the single-crystal region which is to become the emitter. (*c*) A second layer of polysilicon is deposited, defined, and doped *n*-type. A heat treatment to form the final emitter and base dopant profiles follows. (*d*) Contact to the base region is formed over the oxide, reducing the area of the single-crystal, extrinsic-base region [14].

can be implanted into the second layer of polysilicon and diffused from the polysilicon into the single-crystal silicon. The emitter dopant is then implanted into the polysilicon and diffused into the single-crystal silicon to form a very shallow emitter region. By not implanting directly into the single-crystal silicon, crystal damage and the associated damage-enhanced dopant diffusion (*transient-enhanced diffusion*) is avoided. Base and emitter regions of the order of 50–100 nm thick can be formed with this structure. Emitter and collector contacts can be located on extensions of the *n*-type polysilicon over the oxide—just as the base contact is made to *p*-type polysilicon over the oxide—again reducing the parasitic capacitances and improving ac device performance.

General Considerations. The symbol used to represent a bipolar transistor is sketched in Figures 6.22*a* and 6.22*b*. It consists of two angled lines that identify the collector and emitter leads; these are in contact with a straight line that stands for the base. An arrow from the *p*-region to the *n*-region of the emitter-base diode differentiates an *npn* transistor (Figure 6.22*a*) from a *pnp* transistor (6.22*b*). In reality, *npn* transistors in integrated circuits are automatically coupled to a parasitic *pnp* structure with the substrate acting as a collector (Figure 6.22*c*). For forward-active bias on the *npn* transistor, the substrate *pnp* is cut off. In other modes of *npn* operation, the *pnp* can become active. It is important to be certain that the *pnp* transistor does not have appreciable gain to avoid undue parasitic loss and transient misbehavior; the buried layer is valuable here. By increasing the base charge of the parasitic *pnp* transistor and by reducing lifetime in this region, the buried layer helps to reduce α_F in the parasitic *pnp* transistor (below 0.05 in typical cases). For junction-isolated bipolar transistors, the parasitic *pnp* transistor

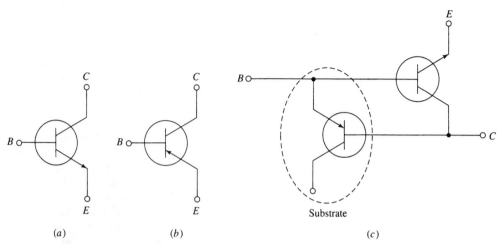

FIGURE 6.22 Standard symbols for (a) npn transistor and (b) pnp transistor. (c) The IC npn transistor automatically has a parasitic pnp transistor (dashed circle) attached to it because of its fabrication process.

also includes laterally injected holes that travel through the n-epitaxial region to be collected at the reverse-biased, p-isolation-region diffusion. In practice, it is necessary to keep the outer edge of the base appreciably separated from the isolation junction to reduce this component to an acceptable value. This constraint is removed when oxide isolation is used.

Lateral injection of electrons can also occur from the emitter into the base. We have not explicitly mentioned this injection loss in the discussion of transistor gain, but the dimensions of modern IC transistors have been so reduced that lateral injection can contribute a significant loss. The transport factor for electrons injected laterally is considerably lower than that for electrons injected downward because the effective lateral base width is large and also because some of the electrons are in the vicinity of the ohmic base contact and are lost to recombination. Lateral injection is naturally inhibited because ϕ_i increases with increasing doping (Equation 4.2.10) and electron injection at a given bias is smaller where ϕ_i is greater. It is also inhibited by the lack of any aiding field in the lateral direction across the base. The designer can keep lateral injection small by keeping the emitter shallow and the lateral base dimension large or by separating the emitter from the base contact region by a shallow oxide-isolation region.

Transistors for integrated circuits can be made in smaller surface areas than are needed for typical IC resistors and capacitors. It is thus advantageous to design circuits that use primarily transistors whenever possible, and to incorporate other devices only when absolutely necessary. If bipolar transistors are used for the active devices, they typically make up the overwhelming fraction of total devices in an integrated circuit. Transistors are very often used in place of the simpler pn-junction diodes described in Chapter 5, even though they take up slightly more surface area. This is because an IC process needed to fabricate transistors produces diodes with lower series resistances and shorter switching times when those diodes are made from the transistor structure in a diode connection. There are a total of five ways to connect a transistor as a diode because only two of three possible regions need to be accessed. Of the various diode connections, shown in Figure 6.23, the lowest series resistance and fastest switching time are obtained in the connection of Figure 6.23a, which is used most often [7].

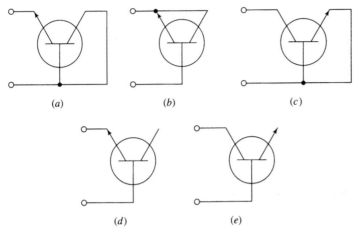

(a)   *(b)*   *(c)*

(d)   *(e)*

FIGURE 6.23 Five ways in which a transistor structure can be connected to obtain diode behavior.

6.6 DEVICES: HETEROJUNCTION BIPOLAR TRANSISTORS[†]

We saw in Sec. 6.2 that the gain of a silicon *npn* bipolar transistor is usually limited by the emitter injection efficiency γ. For an *npn* transistor, γ depends on the ratio of electron injection from the emitter into the base J_n (useful component) to hole injection from the base into the emitter J_p (wasted component) (Equation 6.2.20). Improved gain and frequency response can be achieved by reducing the base doping and making the base narrower to improve the ratio of the Gummel numbers in the emitter and base. However, good circuit performance often requires low base resistance, as well as high gain and frequency response (Sec. 7.6). If the trade-off between gain and base resistance implicit in Equation 6.2.20 can be circumvented, improved overall transistor and circuit performance can be achieved. As we saw in Chapter 5, the ratio of electron injection to hole injection across a *pn* homojunction is approximately equal to the ratio of the emitter and base Gummel numbers. On the other hand, in a heterojunction, the ratio of electron to hole injection contains an additional factor of $\exp(\Delta E_g/kT)$. We make use of this added factor in the *heterojunction bipolar transistor* HBT [10, 11].

To understand the operation of the HBT, we build on our discussion in Secs. 4.2 and 5.3, in which we considered the behavior of a single *pn* heterojunction with different semiconductor materials on the two sides of the junction. Now we use the heterojunction as the emitter-base junction of a transistor and place a second junction close to this emitter-base heterojunction. This second junction can be a homojunction or it can itself be a heterojunction. As in Secs. 4.2 and 5.3, we consider the important case of an *n*-type emitter with a wider bandgap and a *p*-type base with a narrower bandgap: $E_{gE} > E_{gB}$. Initially, we consider the base and collector to be made from the same semiconductor material, as shown in Figure 6.24a, so we are dealing with a *single-heterojunction* transistor. (The base and collector can be made of different materials, creating the *double-heterojunction* transistor shown in Figure 6.24b, which we discuss later in this section.) We focus here primarily on the important emitter-base junction which is similar in both the single- and double-heterojunction bipolar transistors.

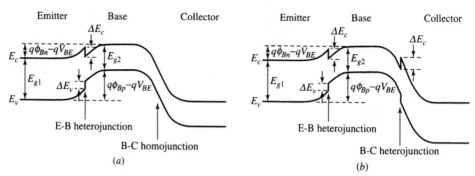

FIGURE 6.24 Energy-band diagrams for heterojunction bipolar transistors: (a) single-heterojunction transistor, (b) double-heterojunction transistor.

In the case shown in Figure 6.24a, the bending of the conduction-band edge in the base is greater than the conduction-band discontinuity at the heterojunction. Because the conduction-band edge in the quasi-neutral base is higher than the conduction-band spike at the emitter-base heterojunction, diffusion current dominates. As we saw from Equations 5.3.35 and 5.3.36, when diffusion current limits, the ratio J_n/J_p increases as $\exp(\Delta E_g/kT)$. Consequently, if the gain of the transistor is limited by the emitter injection efficiency, the gain increases exponentially with increasing total bandgap difference. For typical transistors, the gain can be enhanced by several orders of magnitude in a heterojunction bipolar transistor compared to a homojunction bipolar transistor. Alternatively, the exponential factor containing the bandgap difference allows the base doping (and Gummel number GN_B) to be increased so that adequate gain can be achieved with a much lower base resistance.

As the base doping increases or the emitter doping decreases, more of the band bending occurs in the emitter and less in the base. The band bending in the base can become smaller than ΔE_c so that the spike at the heterojunction extends above the conduction-band edge in the base, as shown in Figure 5.11c. In this case, electron injection and J_n are no longer described by diffusion current, but by thermionic emission over the conduction-band spike at the heterojunction, and the barrier ϕ_{Bn} is greater than that expected from Equation 5.3.30. Hole injection J_p is still described by the barrier ϕ_{Bp} given in Equation 5.3.31, so the ratio J_n/J_p and the gain both decrease. For a very heavily doped base with all the bending of the energy bands in the emitter, $\phi_{Bn} = X_B - X_E = \Delta E_c$ and $\phi_{Bn} = \phi_i$; $\phi_{Bp} = (\phi_i + X_E + E_{gE}) - (X_B + E_{gB}) = \phi_i - \Delta E_c + \Delta E_g$. The ratio of the electron and hole currents depends on $\phi_{Bp} - \phi_{Bn}$, which is only $\Delta E_g - \Delta E_c = \Delta E_v$ in this case, instead of the full bandgap difference ΔE_g obtained when the electron current is not limited by the conduction-band spike. Therefore, we don't achieve the entire possible benefit of the HBT. Because of the exponential dependence on energy, the gain decreases rapidly as the conduction-band spike increases. In addition, as we saw in Figure 5.11d, diffusion current can dominate at low currents while thermionic emission dominates at higher currents if the conduction band in the base is pulled below the conduction-band spike as the applied bias increases.

Thus, for the heterojunction bipolar transistor, the emitter injection efficiency is

$$\gamma = \left[1 + \frac{GN_B}{GN_E} \exp\left(\frac{-\Delta E_x}{kT} \right) \right]^{-1} \tag{6.6.1}$$

where $\Delta E_x = \Delta E_g$ or ΔE_v, depending on the details of the energy-band lineup at the emitter-base heterojunction.

EXAMPLE: *Abrupt Heterojunction Bipolar Transistor*

The most commonly discussed heterojunction bipolar transistor contains a base of GaAs and a wider bandgap emitter of an alloy of AlAs and GaAs ($Al_xGa_{1-x}As$), often with about 30% aluminum and 70% gallium ($x = 0.3$). The bandgap and other properties vary gradually between those of AlAs and those of GaAs as the composition changes, although the variation is not quite linear [12].

(a) We first consider the band diagram for moderate doping in the emitter and base: $N_{dE} = 2 \times 10^{17}$ cm^{-3} and $N_{aB} = 5 \times 10^{16}$ cm^{-3}. In this case the conduction band in the base is higher than the conduction-band spike at the heterojunction. From Equation 6.6.1 and the material properties, we can calculate the gain of the transistor at room temperature (assuming, for mathematical simplicity, similar diffusion coefficients for electrons and holes and equal thicknesses of the emitter and base regions so that $GN_B/GN_E = N_{aB}/N_{dE}$).

$$E_g(\text{GaAs}) = 1.424 \text{ eV}$$
$$E_g(\text{AlAs}) = 2.168 \text{ eV}$$
$$E_g(\text{Al}_{0.3}\text{Ga}_{0.7}\text{As}) = 1.424 + 1.247 \times 0.3 = 1.798 \text{ eV } [12]$$
$$\Delta E_g = 1.247 \times 0.3 = 0.374 \text{ eV}$$
$$\exp(\Delta E_g/kT) = \exp(0.374/0.0259) = 1.87 \times 10^6$$
$$GN_B/GN_E = 0.25$$
$$\gamma = (1 + 0.25/1.87 \times 10^6)^{-1} = (1 + 1.34 \times 10^{-7})^{-1}$$
$$\beta \approx 7.5 \times 10^6$$

However, this very high value of β is only obtained if the gain is solely limited by the ratio of the injected carriers. In practice, other factors beside the emitter injection efficiency limit the gain. Recombination in the emitter-base space-charge region is likely in transition regions of compound semiconductor systems, and recombination in the neutral base region can also limit the gain.

Without the heterojunction, the emitter injection efficiency would be about 0.8 (i.e., the reverse hole injection would be about 20% of the total current), and the gain would only be about 4.

(b) When the base doping is increased to 1×10^{19} cm^{-3} to lower the base resistance, the spike in the conduction band is higher than the conduction band in the base, as shown in Figure 5.11c. Virtually all the potential drop occurs in the emitter, and γ depends on ΔE_v instead of ΔE_g.

$$X(\text{GaAs}) = 4.07 \text{ V}$$
$$X(\text{AlAs}) = 3.5 \text{ V}$$
$$X(\text{Al}_{0.3}\text{Ga}_{0.7}\text{As}) = 4.07 - 1.1 \times 0.3 = 3.74 \text{ V } [12]$$
$$\Delta E_c = 0.33 \text{ eV}$$
$$\Delta E_v = 0.044 \text{ eV}$$
$$\exp(\Delta E_v/kT) = \exp(0.044/0.0259) = 5.5$$
$$GN_B/GN_E = 50$$
$$\gamma = (1 + 0.25 \times 50)^{-1} = (1 + 12.5)^{-1}$$
$$\beta < 1$$

and the transistor is not useful with these dopant concentrations.

The effect of the spike in the conduction band can be reduced or eliminated by grading the composition from pure GaAs to the desired composition of Al_xGa_{1-x} As over approximately the thickness of the emitter-base depletion region. This composition grading improves the device properties at the expense of more complicated materials fabrication. If the conduction-band spike is reduced to zero by grading the heterojunction material, the barriers to carrier injection are determined by the bulk material properties, and the emitter injection efficiency depends on ΔE_g. Equation 6.6.1 with the total bandgap difference ΔE_g

can then again be used to find the emitter injection efficiency:

$$\Delta E_g = 1.247 \times 0.3 = 0.374 \text{ eV}$$
$$\exp(\Delta E_g/kT) = \exp(0.374/0.0259) = 1.87 \times 10^6$$
$$GN_B/GN_E = 50$$
$$\gamma = (1 + 50/1.87 \times 10^6)^{-1} = (1 + 2.67 \times 10^{-5})^{-1}$$
$$\beta \approx 3.7 \times 10^4$$

Again, other factors probably limit the gain to less than this ideal value, but quite impressive gains can be obtained even with a high base doping. ∎

Double Heterojunction Bipolar Transistor

In the analysis of the abrupt heterojunction bipolar transistor, we focused our attention on the critical emitter-base junction. If we have a heterojunction at the base-collector junction also, we need to consider some additional effects. Again, the practical case uses a narrower bandgap in the base than in the collector. If there is a significant discontinuity in the valence band at the base-collector junction, hole injection and storage in the collector region decreases. This decrease can be important if the transistor enters the saturation region of bias where both V_{EB} and V_{CB} are positive. In the active-bias region, if there is also a discontinuity in the conduction band, electron transport from the base to the collector is impeded (Figure 6.25a). A high reverse bias at the base-collector junction can reduce this potential barrier in some circuit applications. Grading the base-collector junction can reduce the barrier, allowing better electron collection. If the base dopant diffuses out of the narrow-bandgap base material into the wider-bandgap collector material, a potential barrier forms (Figure 6.25b), again impeding electron collection from the base into the collector.

At high currents, excess free holes in the base extend the neutral base region beyond that at low currents, as we will see in Sec. 7.2. The wider base increases the transit time and the hole storage in the HBT, as in the homojunction transistor. However, additional degradation of device performance occurs in the HBT. If the neutral-base

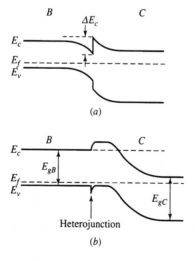

FIGURE 6.25 (a) A conduction-band offset at the base-collector heterojunction creates a barrier to electron collection from the base. (b) Extension of the neutral base (by dopant diffusion or high-current effects) also creates a barrier to electron collection.

region extends beyond the narrower-bandgap material into the wider-bandgap material, a potential barrier to electron flow forms at the base-collector junction, as shown in Figure 6.25b. We can understand the formation of this barrier by remembering that $E_f - E_v$ remains constant in the quasi-neutral base region. When the bandgap increases (as we enter the silicon), the conduction-band edge must increase in energy to keep $E_f - E_v$ constant, forming a barrier to the desired electron flow into the collector. By impeding electron flow, this barrier degrades the gain and frequency response of the transistor. The base-collector bias acts to decrease this additional barrier in some situations, but the heterojunction at the base-collector junction needs to be carefully designed.

Bandgap Grading in Quasi-Neutral Base Region

The possibility of varying the composition of the semiconductor materials in the quasi-neutral regions allows further tailoring of the device properties. Grading the bandgap of the semiconductor material in the quasi-neutral base is especially useful. The graded bandgap can create a high electric field in the base, which can increase the electron velocity and decrease the transit time of electrons through the base.

To understand the source and benefit of the field, we pursue the concept of bandgap grading that we introduced in Sec. 4.1 and Figure 4.2b, which is repeated as Figure 6.26. As described in reference [10], the forces acting on the electrons and holes in a semiconductor are equal to the slope of the edge of the band in which the carriers reside (except for the sign in the case of electrons). In homostructures the energy gap is constant; therefore, the slopes of the conduction- and valence-band edges are the same, and the forces acting on electrons and holes are equal in magnitude and opposite in sign. In a heterostructure the energy gap can vary with position; therefore, the slopes of the conduction- and valence-band edges can be different, as can the forces acting on electrons and holes. In effect, heterostructures utilize energy-gap variation, in addition to electric fields, as forces to control the distribution and flow of electrons and holes. Varying the bandgap

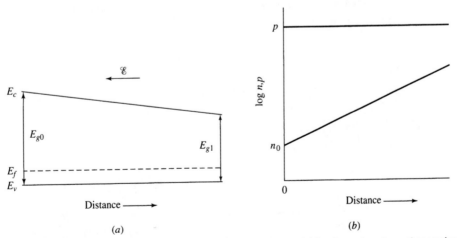

FIGURE 6.26 (a) Energy-band diagram for a p-type material having a bandgap that varies with position; (b) corresponding carrier concentrations.

provides an additional degree of freedom in device design that is especially useful when the distribution and flow of both electrons and holes must be controlled, as in a bipolar transistor.

In the present case of a bipolar transistor, the base doping is high, and the hole concentration is approximately constant, as is $E_f - E_v$. The full effect of the grading then appears in the conduction-band edge, as shown in Figure 6.26. The electric field acting on the electrons is then $dE_c/dx \approx dE_g/dx$.

A full analysis of the bandgap grading in the base of a bipolar transistor is found in reference [13], the results of which are summarized here. The key to the discussion is that the material properties, conveniently represented by the intrinsic carrier concentration, are a function of position. The conventional expressions derived earlier in this chapter can be re-derived with this added functional dependence on position. For low-level injection (low current or high base doping), the electron current is found to be [13]

$$J_n = -\frac{q \exp\left(\dfrac{qV_{BE}}{kT}\right)}{\displaystyle\int_0^{W_B} \dfrac{N_{ab}}{D_n n_i^2}\, dx} \tag{6.6.2}$$

Equation 6.6.2 is similar to the usual expression for electron current (Equation 6.1.17, which is reproduced here):

$$J_n = \frac{-qD_n n_{i0}^2}{N_{ab} W_B} \exp\left(\frac{qV_{BE}}{kT}\right) \tag{6.6.3}$$

except that n_i is now a function of position. Because n_i decreases rapidly as the bandgap increases, the major contribution to the integral in the denominator of Equation 6.6.2 comes from the portions of the base with the largest bandgap.

For the important special case of a bandgap decreasing linearly as distance increases from the emitter edge to the collector edge of the base,

$$n_i^2 = n_{i0}^2 \exp\left(\frac{\Delta E_g}{kT}\right)\frac{x}{W_B} \tag{6.6.4}$$

Equation 6.6.4 can be used in Equation 6.6.2 to find the electron current. For bandgap changes much greater than kT,

$$J_n = \frac{-qD_n n_{i0}^2}{N_{ab} W_B}\frac{\Delta E_g}{kT}\exp\left(\frac{qV_{BE}}{kT}\right) \tag{6.6.5}$$

We can compare Equation 6.6.5 to the corresponding expression for a uniform bandgap equal to that at the emitter edge of the base (Equation 6.6.3) and see that the bandgap grading increases J_n by a factor $\Delta E_g/kT$. We can also write Equation 6.6.5 in terms of the built-in field $\mathscr{E}$

$$J_n = \frac{-q\mu_n n_{i0}^2}{N_{ab}}\mathscr{E} \exp\left(\frac{qV_{BE}}{kT}\right) \tag{6.6.6}$$

Because the hole current J_p is not significantly affected by the bandgap grading, the gain also increases by $\Delta E_g/kT$. Although the increase in gain can be useful, the decrease in electron transit time through the base is much more important. We will examine the base transit time in Sec. 7.6.

EXAMPLE: *Abrupt and Graded-Bandgap HBTs*

We can compare uniform- and graded-bandgap transistors using the silicon/$Si_{1-x}Ge_x$ heterojunction transistor as an example. For a uniform bandgap base as shown in Figure 6.27a, a $Si_{1-x}Ge_x$ region with a constant germanium content is inserted between two silicon regions. Typically, the buried layer and the moderately doped silicon collector region are deposited by epitaxial deposition, as described in Sec. 2.6. The $Si_{1-x}Ge_x$ layer is next formed, also by epitaxial deposition, by adding a germanium-containing gas along with the silicon-containing gas. A layer of silicon that serves as the emitter is deposited on top of the $Si_{1-x}Ge_x$. Often, all three epitaxial layers are deposited in the same process sequence without exposing the substrate to air and the resulting native oxide formation on the semiconductor.

The majority of the thickness of the germanium layer is doped heavily with the *p*-type dopant boron to create the base region. However, undoped germanium spacer layers surround the heavily doped region. These spacers are needed so that boron does not diffuse outside the germanium layer during heat treatment after deposition of the $Si_{1-x}Ge_x$ layer. The spacer on the collector side is especially important. As we discussed above, at high currents the flowing carriers can push the neutral edge of the base region beyond the base-collector heterojunction, creating a barrier to electron flow into the collector and degrading transistor performance. Thus, the *p*-type base region must be well confined in the narrow-bandgap material.

The heterojunction at the emitter-base junction provides the desired benefits we discussed before, especially reducing reverse hole injection from the base into the emitter. The base doping can then be increased to reduce the base resistance while still maintaining a high emitter-injection efficiency and gain. The emitter doping can even be less than the doping in the base, as shown in Figure 6.27a. The region of the emitter near the surface is often heavily doped to allow low-resistance contact to the emitter.

We can contrast the uniform-base HBT to the one with a graded bandgap shown in Figure 6.27b. The Ge content increases from the emitter edge of the base toward the collector edge. An abrupt heterojunction is formed beyond the base-collection junction, so the same care is needed in designing this junction as in the uniform-base HBT. The situation near the emitter-base junction is very different, however. In this case, the barrier to reverse hole injection from the base into the emitter is only slightly greater than that in a homojunction, so the base doping cannot be greatly

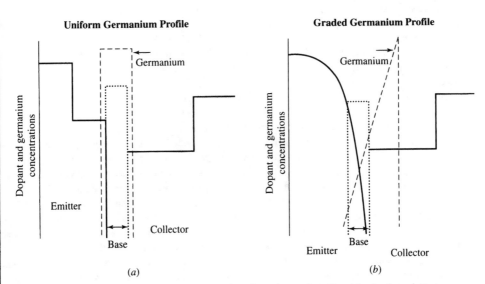

FIGURE 6.27 Uniform (*a*) and graded (*b*) bandgap heterojunction bipolar transistors. Illustrated for the silicon (emitter) – $Si_{1-x}Ge_x$ (base) – silicon (collector) HBT.

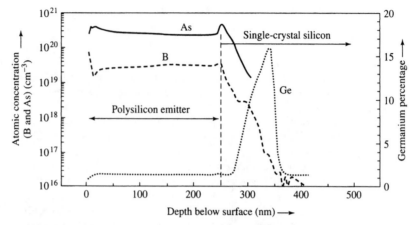

FIGURE 6.28   Germanium, arsenic, and boron profiles in a graded-bandgap $Si/Si_{1-x}Ge_x$ HBT with a polysilicon-contact region on top of a single-crystal silicon emitter [15] (© 1994 IEEE).

increased without degrading the emitter injection efficiency. The major benefit of the graded base is to provide an electric field that decreases the base transit time, as we will discuss in Sec. 7.6. Typical dopant and germanium profiles are shown in Figure 6.28.   ■

SUMMARY

An understanding of the operation of bipolar transistors can be obtained by focusing on the behavior of minority carriers at biased *pn* junctions. Because the space-charge region in a *pn* junction is a barrier to majority-carrier flow, it acts as a collector of minority carriers. Thus, when two *pn* junctions are located close together, bias on one of them can influence the minority-carrier populations in the vicinity of the second, and markedly alter its electronic behavior. By using this idea, which is called *bipolar transistor action*, we can build effective switches and amplifying devices.

Equations for transistor action can be formulated using the fact that majority-carrier currents in the transistor *base* between the two junctions must be nearly zero because a barrier to majority-carrier flow is always encountered along this path. Therefore the current linking the two junctions consists of minority carriers. The size of the linking current depends inversely on the total majority-carrier charge in the base. Currents are strongly nonlinear functions of terminal voltages in transistors, and practical design of transistor circuits is frequently aided by considering approximate equations that are valid over selected regions of base-emitter and base-collector bias. Under *active bias*, the collecting junction is continuously maintained under reverse bias while the emitting junction is continuously forward biased. Current gain in the transistor in the active-bias mode depends on the efficiency of the injection of minority carriers into the base region, from which they can be collected, as well as on their loss to recombination in the base. These mechanisms depend strongly on processing procedures and device geometry.

The *Ebers-Moll* model describes the operation of bipolar transistors over all ranges of bias. It is a very important first-order model and is the basis for more refined descriptions of the transistor. A reciprocity condition is part of the Ebers-Moll model. The reciprocity condition predicts that a transistor biased in the reverse-active mode delivers the same output current as it does when biased in the forward-active mode provided that the respective terminal voltages are interchanged. This result applies to asymmetric as well as to symmetric transistors because loss mechanisms that reduce the current gain simultaneously increase the junction-saturation currents. Provided that the losses have the same voltage dependence as does the linking current, there is an exact inverse relationship and reciprocity is valid.

Transistors in integrated circuits for both switching and amplification are typically built in an epitaxial layer with buried layers beneath the collector regions. The chief difference between devices for these two applications is in the dimensions perpendicular to the surface. Switching transistors are usually built in thinner epitaxial regions, and more emphasis is placed on reducing the effects charge storage in these devices than in amplifying transistors. The engineering trade-offs for the two classes of devices differ. In integrated circuits, conventional *npn* transistors are shunted by parasitic *pnp* transistors in which the substrate is the collector. Although the *pnp* transistor is cut off when the *npn* transistor is under active bias, it can become active during switching. Because they are smaller than other devices in bipolar integrated circuits, bipolar transistors are used by circuit designers whenever possible instead of diodes or resistors. Heterojunction bipolar transistors offer the possibility of using band discontinuities at the junctions between different semiconductors to provide additional control of transistor properties. A heterojunction at the emitter-base junction can increase current gain or can allow lower base resistance with the same current gain. The field in a graded-base HBT accelerates the carriers across the base, decreasing their transit time.

REFERENCES

1. W. SHOCKLEY, *Bell Syst. Tech. J.*, **28**, 435 (1949).
2. J. L. MOLL and I. M. ROSS, *Proc. IRE*, **44**, 72 (1956).
3. H. K. GUMMEL, *Proc. IRE*, **49**, 834 (1961).
4. H. KROEMER, *Arch. Elek. Übertrag.*, **8**, May, August, November (1954).
5. J. J. EBERS and J. L. MOLL, *Proc. IRE*, **42**, 1761 (1954).
6. J. W. SLOTBOOM, *Solid-State Electronics* **20**, 279 (1977).
7. C. S. MEYER, D. K. LYNN and D. J. HAMILTON, *Analysis and Design of Integrated Circuits*, McGraw-Hill, New York, 1968, pp. 248–258.
8. H. CAMENZIND, *Electronic Integrated Systems Design.* Copyright 1972 by Litton Educational Publishing, Inc. Reprinted by permission of Van Nostrand Reinhold Company.
9. P. E. GRAY, D. DeWITT, A. R. BOOTHROYD, and J. F. GIBBONS, *Physical Electronics and Circuit Models of Transistors*, Wiley, New York, 1964, p. 145.
10. H. KROEMER, *Proc. IEEE.* **70**, 13 (1982).
11. H. KROEMER, "Heterostructure device physics: Band discontinuities as device design parameters," Chapter 4 in *VLSI Electronics Microstructure Science*, **10**: *Surface and Interface Effects in VLSI*, eds. N. G. EINSPRUCH and R. S BAUER, Academic Press, New York, 1985.
12. Properties of AlGaAs and GaAs: S. ADACHI, *J. Appl. Phys.* **58**, pp. R1–R29 (1985).
13. H. KROEMER, *Solid-State Electronics* **28**, 1101 (1985).
14. T. KAMINS, *Polycrystalline Silicon for Integrated Circuits and Displays*, Second Edition, Kluwer Academic Publishers, Boston, 1998, p. 272, Figure 6.10.
15. D. VOOK, et al., *IEEE Trans. Electron Devices*, **41**, 1013 (1994).

BOOKS

S. M. SZE, *Semiconductor Devices: Physics and Technology*, second edition, Wiley, New York, 2002.

B. G. STREETMAN and S. BANERJEE, *Solid-State Electronic Devices*, fifth edition, Prentice-Hall, Upper Saddle River, NJ (2000).

A. SEDRA and K. SMITH, *Microelectronic Circuits*, fourth edition, Oxford Univ. Press, 1998.

A. G. MILNES, *Semiconductor Devices and Integrated Electronics*, Van Nostrand Reinhold, New York, 1980.

PROBLEMS

6.1* Show by using Equation 6.1.2 that there is a constant field in a region in which the doping density varies exponentially with distance. If a constant field of -4000 V cm^{-1} exists in the base of a transistor having a base width of 0.3 μm and the dopant density is 10^{17} cm^{-3} at the emitter-base edge, what is the dopant density at the edge of the collector-base space-charge region?

6.2 Apply Equation 6.1.10 to an *npn* transistor under active bias. Consider that $I_C = -J_n A_E$ and prove that

$$I_C = \frac{q A_E \tilde{D}_n N_a(x) n(x)}{\int_x^{x_B} N_a \, d\xi}$$

6.3[†] Use the expression derived in Problem 6.2 to find the *x* dependence of the electron density in the base of the transistor of Problem 6.1. Sketch plots of $n(x)$ versus *x* in the base under active-bias conditions for two cases of *npn* transistors that are biased so that both pass the same collector current. Use a single set of axes for the two transistors: (i) the transistor described in Problem 6.1, and (ii) a transistor having a constant base doping of 10^{17} cm^{-3} but otherwise identical to that in Problem 6.1. Note that both plots have the same gradient at the edge of the collector space-charge region. Why is this the case?

6.4[†] Derive expressions for the total stored minority charge in the base in the two transistors described in Problem 6.3. Use this result to compare the values of β_F in the two devices under the assumption that base current results only from recombination in the base.

6.5* (a) Find Q_{BO}, the base-majority charge in the quasi-neutral region for the transistor data plotted in Figure 6.4.

(b) Find the base width x_B if the average base doping is 10^{17} atoms cm^{-3}. (Assume that $\tilde{D}_n = 25$ cm^2 s^{-1}.)

6.6† Use approximation techniques to determine the net number of dopant atoms per unit area between the base-emitter junction plane and the base-collector junction plane in the switching transistor of Figure 6.15 and the amplifying transistor of Figure 6.16. Compare these values with the dopant atoms per unit area in the quasi-neutral base (Gummel Number in Equation 6.2.3) for the transistor of Problem 6.5 and explain any differences between the results.

6.7 Redo the problem considered in the example of Sec. 6.3 if both the base-emitter and base-collector junctions are linearly graded with the grading coefficient $a = 10^{22}$ cm^{-4} (where $N_d - N_a = ax$), and $\phi_i = 0.872$ V.

6.8* The accompanying table gives values of the net dopant density $(N_d - N_a)$ as a function of position relative to the junction plane between the emitter and base on a transistor with the doping profile of Figure 6.16.

Net dopant density $(N_d - N_a)$ (cm^{-3})	Distance from base-emitter junction (μm)
7.36×10^{17}	-0.125
-1.25×10^{16}	$+0.045$
-4.0×10^{16}	$+0.115$
-1.9×10^{16}	$+0.15$
-6.4×10^{15}	$+0.205$
-1.5×10^{15}	$+0.250$

Assume that the transistor has an active area of 2×10^{-6} cm^2. (a) Plot these values on a linear scale and argue that the profile can be roughly approximated by a one-sided step junction having $N_a \approx 2 \times 10^{16}$ cm^{-3}. (b) Make a plot similar to that of Figure 6.11 to show Q_{VE} versus total voltage for the junction over the bias range from breakdown to $+0.3$ V. Find values for ϕ_i and K_A as shown in Figure 6.11.

6.9 Use the result of Problem 6.1 and the data of Problem 6.8 to estimate the size of the field in the quasi-neutral base of the transistor having the doping profile sketched in Figure 6.16.

6.10 Draw an equivalent circuit similar to Figure 6.12 and write equations similar to Equations 6.4.9 and 6.4.10 to represent the Ebers–Moll model for a *pnp* transistor.

6.11 Show that Figure 6.14*b* represents the Ebers–Moll model for a transistor driven by base current.

6.12* Derive Equation 6.4.17 for the voltage drop across a saturated transistor as predicted by the Ebers–Moll model. Evaluate $V_{CE\,sat}$ for $I_C/I_B = 10$, $\alpha_F = 0.985$, and $\alpha_R = 0.72$.

6.13 Use the Ebers–Moll equations for a *pnp* transistor (Problem 6.10) to find the ratio of the two currents I_{CE0} to I_{CB0} where I_{CE0} is the current flowing in the reverse-biased collector with the base open circuited, and I_{CB0} is the current flowing in the reverse-biased collector with the emitter open circuited. Explain the cause for the difference in the currents in terms of the physical behavior of the transistor in the two situations.

6.14† Consider an *npn* transistor that is biased in the active mode. At time $t = 0$, an intense visible light is focused on the collector-base space-charge region. The light produces G hole-electron pairs per unit time inside the space-charge zone. (Consider that qG is roughly of the same order as the dc base current I_B.)

(a) If the emitter-base voltage and collector-base voltage are both held constant, what are the values of base, emitter, and collector currents for $t > 0$?

(b) Repeat part a if the base is driven by a current source so that the illumination does not alter I_B.

6.15 Near room temperature, the collector current I_C and the base current I_B are both normally positive for an *npn* transistor biased in the active mode. Assume that an *npn* transistor is under active bias and I_C is held constant while the temperature is increased. If we measure I_B, we find that its magnitude decreases and ultimately changes sign. What physical effects account for this behavior? (Consider all currents at the base lead.) [9]

6.16 Consider an *npn* transistor in which both the base and emitter regions are nonuniformly doped and in which recombination of reverse-injected holes into the emitter takes place at the emitter contact. Show that the emitter efficiency γ can be written as $\gamma = (1 + Q_{BO}\tilde{D}_{pE}/Q_{EO}\tilde{D}_{nB})^{-1}$ where Q_{EO} is the emitter dopant in the quasi-neutral region of the emitter, Q_{BO} is the base dopant in the quasi-neutral region of the base, and $\tilde{D}_{pE}$ and $\tilde{D}_{nB}$ are the effective minority-carrier diffusion coefficients in these regions. Show the consistency of this result with that given for the prototype transistor when it is subjected to the same recombination constraints (Equation 6.2.20).

6.17* An *npn* transistor has a cross-sectional area of 10^{-5} cm^2 and a quasi-neutral base that is uniformly doped with $N_a = 4 \times 10^{17}$ cm^{-3}, $D_{nB} = 18$ cm^2 s^{-1}, and $x_B = 0.5$ μm.

(a) If conditions in the emitter are similar to those described in Problem 6.16 and the emitter has total dopant (Q_{EO}/q) of 8×10^9 atoms and $\tilde{D}_{pE}$ is 2 cm^2 s^{-1}, calculate γ.

(b) Estimate α_T, the transport factor if the base lifetime is 10^{-6} s.

(c) Calculate β_F for this transistor. What is the percentage error involved in taking $\beta_F \approx Q_{EO}\tilde{D}_{nB}/Q_{BO}\tilde{D}_{pE}$? This simplified form is sometimes used to estimate β_F.

6.18† Assume that the transistor of Problem 6.17 is placed in a radiation environment where the electron lifetime in the base decays according to the equation: $\tau_n = \tau_{n0} \exp -(t/t_d)$. In this equation t is measured in days and t_d is 10 days. Plot β_F as a function of time in days and determine the time interval until β_F drops to unity. What is the base lifetime when this occurs?

6.19† Consider the transistor structure shown in Figure P6.19.

(a) Derive an expression for the electron distribution in the base as a function of x assuming that recombination takes place. (This will demand solving an equation similar to Equation 5.3.10.)

(b) The slope of the distribution at $x = 0$ is proportional to the injected electron current, and the slope at $x = x_B$ is proportional to the collected electron current. The difference represents current lost because of recombination in the base (supplied by the base lead). Find an expression for this base recombination current as a function of L_n and x_B/L_n [$L_n = (D_n\tau_n)^{1/2}$].

(c) The dopant concentrations in the emitter, base, and collector are assumed to be uniform throughout each region with values of 10^{19}, 10^{17}, and 10^{15} cm^{-3}, respectively. The hole lifetime in the emitter is 10^{-9} s and the electron lifetime in the base is 10^{-7} s. The emitter and base widths are each 1 μm. Calculate emitter efficiency and β_F. Use Figure 1.16 to obtain values for the diffusion coefficients. [Hint: For this problem, it is best to express solutions for the continuity equation as the sum of the hyperbolic functions $\cosh (x/L_p)$ and $\sinh (x/L_p)$].

6.20† By considering collector current as charge in transit ($I_C = qnv_dA$), use the known distribution of injected electrons in the base of an npn prototype transistor biased in the active mode to solve for the base transit time τ_B. That is, formulate the transit time as the integrated sum of incremental path lengths divided by velocity and then carry out the integration to prove that τ_B is given by $\tau_B = x_B^2/2D_n$ (i.e. take $\tau_B = \int_0^{x_B} dx/v$).

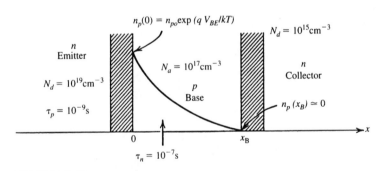

FIGURE P6.19

CHAPTER 7

BIPOLAR TRANSISTORS II: LIMITATIONS AND MODELS

Thus far, our discussion of bipolar transistors has described their fundamental operation. For example, the description of a transistor biased in the active mode emphasized that carriers are injected with high efficiency from the emitter into the base. There, the injected carriers greatly increase the minority-carrier population and flow from the base-emitter contact toward the collector. On reaching the base-collector space-charge region, these added free carriers are swept across to the quasi-neutral collector region where they are majority carriers and flow as ohmic current through the collector lead.

In real transistors this simple description is remarkably accurate over a wide range of conditions. However, we need to consider more than this basic ideal behavior. Real transistors are limited by processes that we have not yet considered. For example, as described in Chapter 6, an ideal transistor has a current-source output in the active mode; that is, output current does not depend in any way on the voltage applied between the base and collector terminals. In real transistors, however, both output currents and output voltages influence the device performance. Similarly, the theory presented so far predicts a current gain in the active mode that is insensitive to bias. But this is only a rough approximation. Therefore, a first objective of this chapter is to investigate additional mechanisms important in real transistor operation.

A more complete picture of transistor operation must also deal with time-varying effects, which are the second major topic to be developed in this chapter. A great simplification in considering time-varying effects will be obtained by modeling transistors as charge-controlled devices. This viewpoint leads very naturally into the important topic of large-signal transient models for bipolar transistors. The combination of the Ebers-Moll model with the charge-control model provides a powerful analysis technique and is the basis of much of computer-based transistor modeling. It will be straightforward to derive a small-signal ac equivalent circuit from the more general large-signal representation. This small-signal equivalent circuit, known as the *hybrid-pi* circuit, is relatively easy to characterize because it is composed of elements that relate directly to physical mechanisms in the transistor. Charge-control concepts also lead to physical understanding of the high-frequency limits of bipolar transistors.

The production of *pnp* transistors within the confines of standard IC processing is described in a final section. Two basic types of *pnp* transistors, *substrate pnp* and *lateral pnp* transistors, are typically fabricated. The substrate *pnp* transistor has only limited application because its collector is not isolated. The lateral *pnp* transistor has severely limited performance compared to *npn* transistors. Simple models for loss mechanisms in this device explain the limited performance.

7.1 EFFECTS OF COLLECTOR BIAS VARIATION (EARLY EFFECT)

When we considered bipolar transistors under active bias in Chapter 6, the function of the voltage applied across the collector-base junction was merely to insure the efficient collection of base minority carriers and their delivery to the collector region. The magnitude of the bias only limited the range of the permissible collector voltage swing, provided that it was below the breakdown value. However, as we noted in Chapter 4, the width of a reverse-biased *pn* junction is voltage dependent; in fact, this bias dependence makes possible the operation of junction field-effect transistors. In the case of bipolar transistors, a changing collector-base bias varies the space-charge layer width at the collector junction and, consequently, the width of the quasi-neutral base region. This variation results in several effects that complicate the performance of the transistor as a linear amplifier. Base-width modulation resulting from variations in collector-base bias was first analyzed by James Early [1], and the phenomenon is generally called the *Early effect*.

The dependence of collector current on collector-base bias can be formulated directly by using the integral equations for active bias (*npn* transistor) developed in Sec. 6.2. In particular, from Equations 6.2.1 and 6.1.15, we can write

$$I_C = \frac{q\tilde{D}_n n_i^2 A_E \exp(qV_{BE}/kT)}{\int_0^{x_B} p \, dx} \tag{7.1.1}$$

where the integration is performed over x_B, the width of the quasi-neutral base region and the other terms were defined in Sec. 6.1.

Variation of the base width with voltage V_{CB} causes a change in the collector current that can be written

$$\frac{\partial I_C}{\partial V_{CB}} = \frac{-q\tilde{D}_n n_i^2 A_E \exp(qV_{BE}/kT)p(x_B)}{\left[\int_0^{x_B} p\,dx\right]^2}\frac{\partial x_B}{\partial V_{CB}} \tag{7.1.2}$$

Several of the terms in Equation 7.1.2 can be combined to represent collector current itself, so that Equation 7.1.2, which represents the small-signal conductance at the collector-base junction, can be written

$$\frac{\partial I_C}{\partial V_{CB}} = -I_C p(x_B)\left[\frac{1}{\int_0^{x_B} p\,dx}\right]\left[\frac{\partial x_B}{\partial V_{CB}}\right]$$

$$= -\frac{I_C}{V_A} = \frac{I_C}{|V_A|} \tag{7.1.3}$$

Because the collector is reverse biased, the derivative in Equation 7.1.3, $\partial x_B/\partial V_{CB}$, is negative, and the Early effect results in an increase in I_C when V_{CB} is increased. This increase in I_C is evident if we examine *common-emitter* output characteristic curves such as the family shown in Figure 7.1a. These curves are actually plots of I_C versus V_{CE}, but V_{CE} varies similarly to V_{CB} for bias in the active region ($V_{CE} \approx V_{CB} + 0.7$ V).

Equation 7.1.3 reveals that the Early effect varies linearly with collector current. The reciprocal of the factor multiplying the current has the dimension of a voltage and has been defined [2] as the *Early voltage*. The Early voltage is usually given the symbol V_A.

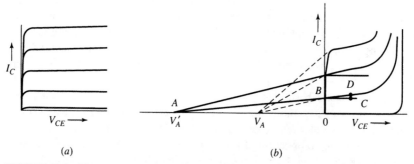

(a)   (b)

FIGURE 7.1 Measured output characteristics for an amplifying transistor: collector current versus V_{CE}. (a) Vertical 0.5 mA full scale, horizontal 5 V full scale. (b) Vertical 0.5 mA full scale, horizontal 50 V full scale; tangents to the measured curves (at the edge of saturation) are extended to the voltage axis (dashed lines) to determine the Early Voltage V_A. Extensions of lines drawn approximately tangential to the characteristics in the active region intersect the voltage axis at V_A' (solid lines).

From Equation 7.1.3, V_A for an *npn* transistor can be written as

$$V_A = \frac{\int_0^{x_B} p\, dx}{p(x_B)\partial x_B/\partial V_{CB}} \tag{7.1.4}$$

Again, the derivative in Equation 7.1.4 is negative and, thus, the Early voltage is negative for an *npn* transistor. The analogous effect of emitter-base space-charge layer widening when the transistor is biased in the reverse-active mode can be considered by using a different Early voltage, usually denoted as V_B.

Except for high-level effects, the three terms that define V_A depend only on the transistor manufacturing process and on collector-base voltage. In practice, the collector-base voltage dependence of V_A itself is usually treated as negligible and the Early voltage is approximated by its value at a single bias (often at $V_{CB} = 0$). With this condition to specify V_A, we expect that a series of tangents to curves of I_C versus V_{CB} (V_{CE} in practice), drawn at the edge of the forward-active region where V_{CB} is approximately zero, should intersect the V_{CE} axis at V_A. In Figure 7.1*b* dashed lines drawn from the point at which the transistor moves out of saturation intersect at a common point, indicating the Early voltage. For circuit design and analysis, however, interest is not in an Early voltage characterizing the edge of saturation, but rather in a parameter to use in the forward-active region. If tangents are drawn to the curves of I_C versus V_{CE} in the active region, they do not, in general, intersect each other on the voltage axis. It is usual, however, to approximate an intersection point appropriate to the range of bias of the transistor as at V_A', formed by the solid lines in Figure 7.1.

A useful alternative expression for V_A can be obtained by rearranging some terms in Equation 7.1.4. First, we use Equation 6.1.13 to express the numerator in terms of the base majority-charge density Q_B in that portion of the base where transistor action is occurring:

$$\int_0^{x_B} p\, dx = \frac{Q_B}{q} \tag{7.1.5}$$

We then recognize that the denominator of Equation 7.1.4 represents the derivative of base charge Q_B with respect to V_{CB}:

$$qp(x_B)\frac{dx_B}{dV_{CB}} = \frac{dQ_B}{dV_{CB}} \tag{7.1.6}$$

The derivative on the right-hand side of Equation 7.1.6 can be related to C_{jc}, the small-signal capacitance per unit area at the collector-base junction.

$$\left|\frac{dQ_B}{dV_{CB}}\right| = C_{jc} \tag{7.1.7}$$

Therefore, the Early voltage is

$$|V_A| = \frac{Q_B}{C_{jc}} \tag{7.1.8}$$

To reduce the influence of collector-base voltage on collector current, the magnitude of V_A should be increased. From Equation 7.1.8 we see that this can be accomplished by increasing the ratio of base majority-charge per unit area to the capacitance per unit

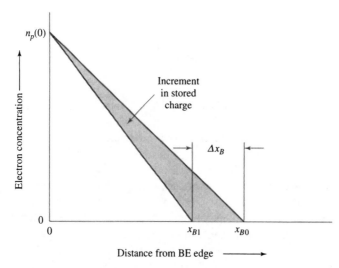

FIGURE 7.2 The influence of the Early effect on the minority–carrier distribution in the quasi-neutral base region of a prototype transistor. An increase in V_{CB} reduces the base width from x_{B0} to x_{B1}, increasing the gradient of n_p and decreasing the total minority charge in the base.

area at the collector-base junction. Physically, this reduces the movement of the base-collector boundary into the base region.

For computer modeling we can describe the widening of the base-collector space-charge layer through the Early voltage V_A, as we will see in Sec. 7.7. However, it helps conceptually to consider an alternative view of the Early effect that is physically more revealing for the prototype transistor introduced at the beginning of Chapter 6. For homogeneous doping and active-mode bias, the minority-carrier distribution in the base has the triangular shape shown in Figure 7.2. Increasing the collector-base voltage from V_{CB} to $(V_{CB} + \Delta V_{CB})$ moves the base-collector junction edge a distance Δx_B from x_{B0}. A second triangular distribution specifies the new minority-carrier profile, and the shaded area between the two distributions represents the decrease in stored base charge.

The collector current is thus increased in the ratio x_{B0}/x_{B1} by the change in V_{CB}. By looking directly at stored base charge, we also see that the Early effect reduces charge storage. This affects both the transient behavior of transistors and the dc base current because base recombination depends directly on stored base charge. We will consider these effects in more detail in Sec. 7.5 when we discuss BJT models for circuit applications.

7.2 EFFECTS AT LOW AND HIGH EMITTER BIAS

Additional physical insight can be gained by studying a transistor biased into the active region with the base and collector currents measured as functions of the base-emitter voltage. Because of the physics of transistor action, such measurements can be more easily interpreted when plotted semilogarithmically with junction voltage on the linear scale. Typical data measured on an amplifying IC *npn* transistor are given in Figure 7.3. The excellent fit of the data for I_C and I_B to straight lines over the midrange of currents indicates an exponential dependence on voltage, consistent with the theory we developed in Chapter 6. However, straight lines of constant slope do not adequately represent all of the data in Figure 7.3, for either I_C or I_B. The reason that the collector current is "ideal" over a wider voltage range than is the base current will become clear during our discussion. First,

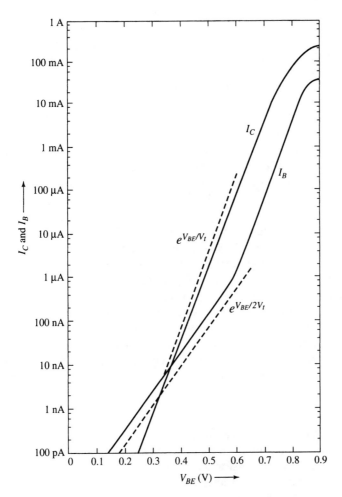

FIGURE 7.3 Typical behavior of collector and base currents as functions of base-emitter voltage for bias in the forward–active region.

however, we consider the variation in I_B at very low voltages where the theory given thus far obviously disagrees with experiment.

Currents at Low Emitter Bias

At low base-emitter voltage the slope of the log $I_B - V_{BE}$ curve decreases. The experimental data indicate that the base current asymptotically approaches a curve that can be represented by

$$I_B = I_0 \exp\left(\frac{qV_{BE}}{nkT}\right) \tag{7.2.1}$$

as V_{BE} goes to zero. In Equation 7.2.1 the parameter n is generally between one and two. Furthermore, the parameter I_0 in Equation 7.2.1 is larger than the corresponding multiplier for an exponential form that fits the data at intermediate biases.

The source of the added base-emitter current at low biases is clear from our discussion of current in *pn* junctions in Chapter 5; recombination in the base-emitter space-charge region causes the excess base-emitter current. Space-charge-region recombination current in a diode structure was considered in Sec. 5.3, where the Shockley-Hall-Read theory was used to derive Equation 5.3.24. That equation for recombination current in the space-charge region has the same voltage dependence as Equation 7.2.1 if n is taken equal to 2. Values for n between 1 and 2 can be justified by considering possible variations of parameters affecting recombination within the space-charge region. As we showed in Equation 5.3.25, the importance of recombination relative to injection into the quasi-neutral regions increases as voltage decreases. For the example used in the diode discussion that illustrated Equation 5.3.25, recombination current dominated for junction voltages less than 0.48 V.

Recombination current in the space-charge region flows only in the base and emitter leads. It does not affect the collector current, which consists almost entirely of electrons that are injected at the base-emitter junction. Hence, the collector current continues to be represented by Equation 7.1.1 as V_{BE} decreases until injection becomes so low that the collector current is dominated by generation in the space-charge region as expressed by Equation 6.3.1. Thus, at low biases the collector current is a smaller fraction of emitter current than in the intermediate bias range. This behavior is seen more clearly in a plot of $|I_C/I_B| \equiv \beta_F$, as defined in Equation 6.2.22. The data of Figure 7.3 was used to make the plot shown in Figure 7.4. The decrease in β_F as base-emitter bias decreases represents a clear limitation on the use of transistors to amplify low voltages.

Maintaining large current gains at low bias levels and, therefore, reducing space-charge recombination current is of great importance and enormous commercial significance. Applications for transistors and integrated circuits such as hearing-aid amplifiers and *pacemaker* circuits for the medical treatment of heart patients depend on achieving adequate performance at the lowest possible currents. For circuit applications such as these, processing efforts are usually concentrated on maintaining as high a lifetime as possible within the base-emitter space-charge region.

High-Level Injection

The transistor theory considered thus far has relied on the low-level injection approximation: that majority-carrier populations under bias are essentially unperturbed from their values at thermal equilibrium. As transistors are biased more heavily in the active mode, violations of the low-level injection assumption first become apparent either at locations where there is a high density of injected minority carriers or where there is a low density of dopant atoms. The first situation is expected in the region of the base nearest the emitter

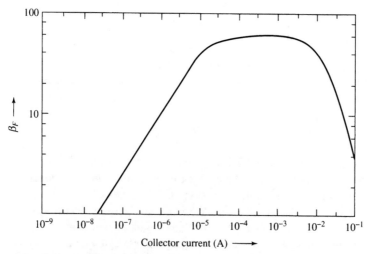

FIGURE 7.4 Current gain (β_F) as a function of collector current for the transistor of Figure 7.3.

edge, while the second can occur in an IC transistor near the base-collector junction. Both cases are found in practical IC transistors. We will consider them separately, concentrating first on high-level injection near the emitter junction.

High-Level Injection at the Base-Emitter Junction.

The basic equation for collector current under active bias (Equation 7.1.1) was derived by considering a special case of the equation for transistor action (Equation 6.1.12). The exponential dependence of collector current on voltage predicted by Equation 7.1.1 is seen in Figure 7.3 to be valid over nearly eight decades of current until it deviates with high bias on the base-emitter junction.

One cause for the deviation is high-level injection into the base. If the injection of electrons into the base is sufficient to cause a significant increase in the population of base majority carriers, then the full dependence of current on voltage is not stated explicitly in Equation 7.1.1. In this case, the integrated majority charge in the denominator depends on bias. (An associated, but second-order effect is a change in $\tilde{D}_n$.) Because of quasi-neutrality in the base, the expression for the integral of $p(x)$ should be written

$$\int_0^{x_B} p(x)dx = \int_0^{x_B} [N_a(x) + n'(x)]dx \tag{7.2.2}$$

where $n'(x)$ is the injected electron density and $N_a(x)$ represents the base doping profile. To proceed further, we need to specify the position dependences in the integral and, therefore, to consider a specific dopant profile.

To continue a qualitative discussion, however, we focus on the boundary value for the injected-electron density $n'(0)$. Because this is the maximum density, it dominates in the integral in Equation 7.2.2. The boundary value can be determined by using Equation 5.3.20 together with the quasi-neutrality assumption. Equation 5.3.20, which states that the pn product at the emitter-base boundary is related to the applied bias by the expression. $p(0)n(0) = n_i^2 \exp(qV_{BE}/kT)$, remains valid* provided that the current levels remain well below the balanced equilibrium flow tendencies (estimated at $\sim 10^5$ A cm^{-2} in Sec. 5.3). Currents of this size do not flow in practice without causing irreparable thermal damage to the junction.

The minority-carrier boundary value $n(0)$, which under active bias is almost the same as the injected electron density at the boundary $n'(0)$, is therefore

$$n(0) = \frac{N_a(0)}{2}\left[\left(1 + \frac{4n_i^2 \exp(qV_{BE}/kT)}{N_a^2(0)}\right)^{1/2} - 1\right] \tag{7.2.3}$$

At low or moderate injection levels, Equation 7.2.3 reduces to Equation 5.3.7 as expected. Under high-injection conditions, however, the second term in the square root dominates and the variation of $n(0)$ approaches $\exp(qV_{BE}/2kT)$, partially cancelling the exponential increase predicted by the numerator of Equation 7.1.1. Collector current tends likewise to become proportional to $\exp(qV_{BE}/2kT)$.

High-Injection Effects at the Collector (Kirk Effect).[†]

The introductory comments about low-level injection pointed out the weakness of this assumption in locations where the material is lightly doped. Inspection of a typical doping profile for an amplifying IC transistor (Figure 6.16) shows that the region of lowest doping is in the

---

* The validity of Equation 5.3.20 under high-level-injection conditions is also necessary to permit the use of the basic equation for transistor action (Equation 6.1.12) because Equation 5.3.20 was used in the derivation of Equation 6.1.12.

collector, close to the base region, where no additional dopant has been added to the epi-taxial material. This lightly doped collector region serves several useful purposes. It significantly increases the breakdown voltage of the collector-base junction, it reduces collector capacitance, and it reduces the widening of the base-collector space-charge layer into the base (Early effect). In this section, we will see that several complicated and undesirable effects also arise from this low collector doping.

Under amplifying conditions, the base-collector junction is reverse biased so that any minority carriers that reach its edges are quickly swept across to the opposite region. We used this physical picture to justify our assumption of essentially zero density for minority carriers at the edges of the reverse-biased collector-base space-charge region. This assumption is clearly an approximation, however; it must be violated if any carriers are to flow across the space-charge region. Because free carriers are limited in their velocities, a current density J requires a density of minority carriers at least equal to J/qv_l, where v_l is the scattering-limited velocity (Equation 1.2.12). This charge density is neutralized by majority carriers in the base region, but it adds algebraically to the fixed background charge associated with the uncompensated dopant atoms in the space-charge region. The presence of this charge component associated with collector current modifies the theory that we developed thus far when the density of the injected, current-carrying charge becomes comparable to the background density of dopant ions. As the current in a transistor is increased, this mechanism can significantly alter important transistor properties.

Qualitatively, the effect can be described as follows: the free carriers entering the base-collector space-charge region modify the background charge in that region and thus affect the electric field. For a constant collector-base voltage, the integral of the field across the space-charge region is constant. Hence, any space-charge modification must be accompanied by a change in the width of the region. We will see that the width of the space-charge region tends to decrease and, therefore, the neutral base tends to widen. This mechanism is often called the *Kirk effect* after the author of the first paper to analyze it [3].

In Figure 7.5a we show a portion of the doping profile for an IC amplifying transistor from Figure 6.16. We consider a case in which this device is biased in the active mode with 10 V applied to the collector-base junction. A plot of the electric field with little or no current flowing is sketched in Figure 7.5b; $\mathscr{E}_l$, the field at which free electrons reach their limiting drift velocity v_l (Figure 1.18), is noted on Figure 7.5b. Poisson's equation for the collector-base space-charge region of the npn transistor takes the form

$$\frac{d\mathscr{E}}{dx} = \frac{1}{\epsilon_s}\left[qN(x) - \frac{J_C}{v(x)}\right] \tag{7.2.4}$$

where $N(x)$ is the net dopant density $(N_d - N_a)$ in the space-charge region (negative acceptor ions in the diffused region near the base and positive donor ions in the high-resistivity epitaxial layer and in the heavily doped buried layer). With a constant collector-base bias V_{CB}, we obtain a second equation involving the field $\mathscr{E}$

$$V_{CB} + \phi_i = \int_{x_B}^{x_C} - \mathscr{E}\,dx \tag{7.2.5}$$

where ϕ_i is the built-in junction voltage (Equation 4.2.10). Equation 7.2.5 implicitly specifies the width of the space-charge layer. Referring again to Equation 7.2.4, we note that the second term on the right-hand side is negligible at low currents, but dominates at high currents. The critical current J_1 dividing the low- and high-current regions can be found by equating the two terms:

$$J_1 = qN(x)v(x) \tag{7.2.6}$$

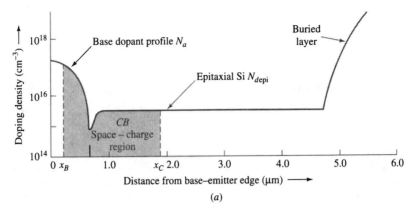

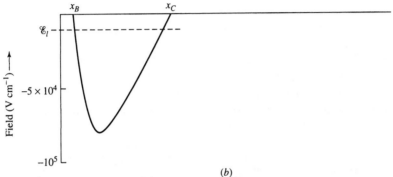

FIGURE 7.5 Space charge and field at the collector-base junction of the amplifying transistor shown in cross section in Figure 6.16. (*a*) The space-charge region for 10 V collector-base reverse bias, (*b*) the electric field in the collector-base space-charge region. The approximate field $\mathcal{E}_l$ at which free electrons reach limiting velocity is marked in (*b*).

The critical current is reached first where $N(x)$ is minimum; that is, in the epitaxial region, where for most biases $v(x)$ is at its maximum value v_l and N is N_{epi}. Thus, $J_1 = qN_{epi}v_l$. An analysis of this effect is of practical importance because in many applications the collector current density equals and exceeds the critical current J_1. For example, if $A_E \approx 10^{-7}$ cm^2 and $N_{epi} = 5 \times 10^{15}$ cm^{-3}, the critical collector current is 0.8 mA.

To carry out the mathematical treatment, we first multiply Equation 7.2.4 by x and then integrate between x_B and x_C, the boundaries of the collector-base space-charge layer.

$$\int_{x_B}^{x_C} x \frac{d\mathcal{E}}{dx} \, dx = \frac{1}{\epsilon_s} \int_{x_B}^{x_C} x \left[qN(x) - \frac{J_C}{v(x)} \right] dx. \tag{7.2.7}$$

The left-hand side of Equation 7.2.7 can be integrated by parts and used with Equation 7.2.5 to yield

$$\int_{x_B}^{x_C} x \, d\mathcal{E} = - \int_{x_B}^{x_C} \mathcal{E} \, dx = V_{CB} + \phi_i \tag{7.2.8}$$

where $\mathcal{E}(x_C)$ and $\mathcal{E}(x_B)$ are taken to be zero. Thus, the collector voltage can be expressed as

$$V_{CB} = \frac{1}{\epsilon_s} \int_{x_B}^{x_C} x \left[qN(x) - \frac{J_C}{v(x)} \right] dx - \phi_i \tag{7.2.9}$$

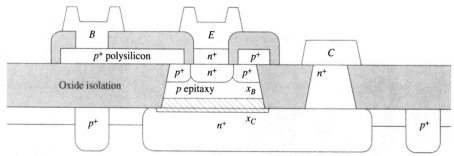

FIGURE 7.6 Cross section of an oxide-isolated *npn* transistor made using a *p*-type epitaxial layer and polysilicon connections to the base and emitter. The collector-base space-charge layer extending from x_B to x_C is shown by the (left-slanting) cross-hatching.

Applying Equation 7.2.9 to the transistor represented in Figure 7.5 is not straightforward because of the shape of the doping profile. Mathematical complications can be avoided and a worthwhile physical picture of the Kirk effect can be obtained by considering first the less common structure shown in Figure 7.6. For this "epitaxial-base" structure, the epitaxial region is a high-resistivity *p*-type layer, instead of the more commonly used *n*-type material. For an *npn* transistor made in a *p*-type epitaxial layer, the base-collector space-charge region extends outward from the n^+ buried layer, and the point x_C is relatively insensitive to bias changes because of the high density of donors in the buried layer. We therefore consider x_C as constant and solve Equation 7.2.9 to find the variation with current density of $x_{CB} = x_C - x_B$, the width of the collector-base space-charge region. For this case of constant acceptor doping, we take the charge $qN(x)$ in Equation 7.2.9 to be a constant equal to $-qN_{epi}$ up to the buried layer, and assume $v = v_l$; then

$$V_{CB} = \frac{1}{2\epsilon_s}\left[qN_{epi} + \frac{J_C}{v_l}\right]x_{CB}^2 - \phi_i$$

$$= \frac{qN_{epi}}{2\epsilon_s}\left[1 + \frac{J_C}{J_1}\right]x_{CB}^2 - \phi_i \qquad (7.2.10)$$

which can solved for x_{CB}.

At zero current the collector space-charge region has width x_{CO} where

$$x_{CO} = \left[\frac{2\epsilon_s(V_{CB} + \phi_i)}{qN_{epi}}\right]^{1/2} \qquad (7.2.11)$$

The variation with current of the collector-base space-charge region x_{CB} can then be expressed as

$$x_{CB} = \frac{x_{CO}}{(1 + J_C/J_1)^{1/2}} \qquad (7.2.12)$$

This variation is shown in Figure 7.7, where we see that the Kirk effect results in a widening of the charge-neutral base region as currents approach and exceed the critical value J_1. This base-layer widening decreases the transistor current gain at high currents and also degrades the frequency response of the device. We consider the latter effect in later sections, but first we return to the case of the Kirk effect in *npn* transistors with an *n*-type epitaxial layer, the most widely used structure.

The complication in the case of an *n*-type epitaxial layer arises because the dopant concentration varies considerably through the collector-base space-charge region, going

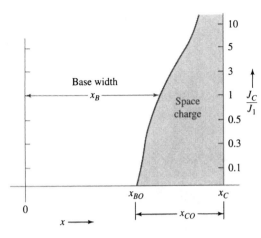

FIGURE 7.7 Base widening resulting from the Kirk effect in a transistor having the cross section sketched in Figure 7.6. The variation shown has been calculated using Equation 7.2.12.

from $-qN_a(x_B)$, in the tail of the p-type base diffusion, to $+qN_{epi}$, the fixed-donor density in the epitaxial region and finally to $+qN_d(x)$ at the outer edge of the out-diffusing buried layer. In this case, as the current level increases above $J_1 = qN_{epi}v_l$, the sign of the space charge in the epitaxial layer changes from plus to minus. When current densities exceed J_1, this sign reversal causes the electric field in the epitaxial layer to vary with position in the opposite sense from the zero-current case. Electric flux lines that end on the negative acceptor ions in the base when currents are very low become terminated instead on the current-carrying free electrons when the Kirk effect dominates. The entire space-charge region is then pushed toward the heavily doped collector region, moving from an initial position at the pn junction formed by the base acceptor diffusion to a final position at the edge of the buried layer at high currents. Under this high-current condition, the electric field increases linearly with distance, and the space-charge region is compressed against the highly doped collector edge (Problem 7.10). Computer calculations of the Kirk effect for this case have been carried out [4] and several of the results obtained are shown in Figure 7.8.

The doping profile of the transistor is shown in Figure 7.8a, while the field and electron densities at various current levels are indicated in Figures 7.8b and 7.8c, respectively. At low currents, the field in the collector-base region has the roughly triangular shape sketched in Figure 7.5b. (Note that the *negative* of the field is plotted in Figure 7.8b.) The critical current density J_1 for the structure of Figure 7.8 is of the order of 500 A cm^{-2}. Near this current density, the high-field region is seen in Figure 7.8b to migrate toward the highly doped buried layer. At still higher currents, the field becomes compressed against the boundary between the epitaxial region and the buried layer. The plot of electron concentration in Figure 7.8c has a steep gradient near the emitter to overcome the built-in opposing field in this region. The extension of the base region with increasing current is evident from this figure. Note that at higher currents, the electron density has a nearly constant gradient over most of the epitaxial region.

Base Resistance

Because current gain β_F in amplifying transistors is typically so high, we may be tempted to assume that base current is negligible and therefore that resistance in the base region is of little consequence to transistor operation. However, even small voltage differences in the base region are magnified by the exponential diode factor, and thus can cause much larger differences in current densities across the emitter-base pn junction.

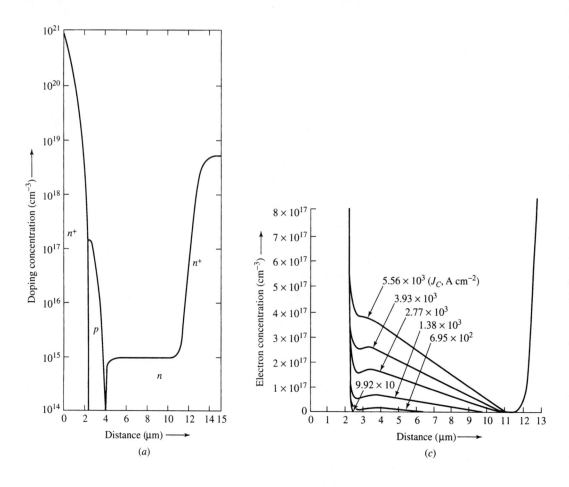

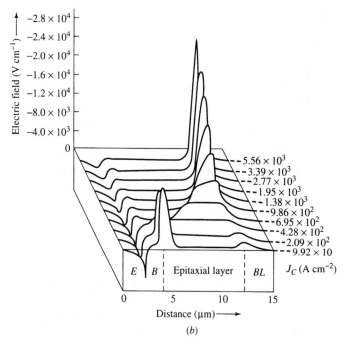

FIGURE 7.8 (a) Doping profile in an *npn* transistor for which a computer analysis of the Kirk effect has been carried out [4]. (b) Electric field as a function of distance for various collector currents in the transistor of (a). (c) Base minority-carrier (electron) density as a function of distance for various collector currents in the transistor of (a).

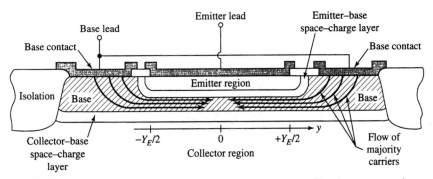

FIGURE 7.9 Cross section of a transistor under active bias. The base current is supplied from two side base contacts and flows toward the center of the emitter, causing the base-emitter voltage to vary with position.

Consider the cross section of the transistor shown in Figure 7.9. As V_{BE} increases from zero, injection of electrons from the emitter region is greatest at that part of the junction where the base doping N_a is the smallest. Because the base is diffused, this is across the bottom plane of the emitter diffusion. Majority–carrier base current flows into this region to supply carriers for recombination and to be injected into the emitter. A typical base thickness, however, is much less than one micrometer and therefore a sizable resistance is generally present between the base electrode and the active area of the transistor. Because the emitter-base current is spread over the active region of the emitter (Figure 7.9), the base current decreases continuously as we move toward the center line of the emitter. Therefore, we cannot calculate a resistance value that accounts in a straightforward way for the ohmic drop in the base region. More important, the potential dropped along the base region decreases the base-emitter bias away from the base contact. The density of injected-electron current therefore decreases from its highest value in the active region nearest the base electrode to a minimum at the center of the emitter. This *current crowding* toward the perimeter of the emitter increases with bias and causes localized heating at current levels that might be tolerable if the current were distributed uniformly. The high-injection effects that we considered earlier in this section also occur at lower currents because of the uneven current density across the active region. To reduce base resistance, power transistors are made with large interdigitated, comb-like base and emitter contacts.

A plot of collector current versus base-emitter voltage for a transistor in which base resistance is significant is shown in Figure 7.10. Because base resistance reduces the junction bias, we can try to fit the measured data from Figure 7.10 by the expression

$$I_C = I_S \exp\left[\frac{q(V_{BE} - I_B R_B)}{kT} \right] \qquad (7.2.13)$$

To do so, we recognize from our discussion of current crowding that the resistance R_B in Equation 7.2.13 (called the *base spreading resistance*) must be variable. Using the measurements of I_C, I_B, and V_{BE}, we find that to correspond to Equation 7.2.13, R_B must decrease with increasing current (Figure 7.11). The initial reduction in base resistance apparent in Figure 7.11 corresponds to the decreasing path length from the base electrode to the effective region of the transistor as current increases.

If only base spreading resistance were important, an asymptotic value of R_B would be approached when all of the current had crowded toward the perimeter of the emitter and all resistance was associated with inactive regions of the base. However, the onset of the other high-level effects previously mentioned can cause R_B to decrease to a lower value.

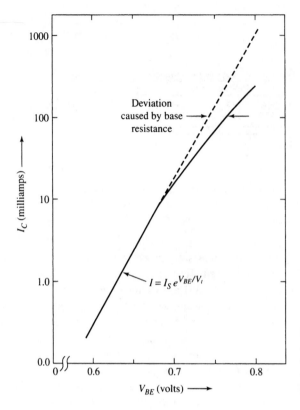

FIGURE 7.10 Collector current as a function of base-emitter voltage, showing the deviation from ideal behavior at high currents.

It is possible to analyze exactly the current-crowding effects that we described by using a spatially distributed form of the diode equation. The mathematics become relatively tedious, however, and can obscure the relevant physical mechanisms. Therefore, we consider an approximate analysis in which the transistor is divided into sections. Each section is assumed to have the same current gain as the original device and to follow the ideal transistor equation, that is, to have negligible base resistance. Each section is characterized by its proportional share of the overall saturation current (I_S in Equation 7.2.13) and is separated from the adjacent section by a resistance corresponding to part

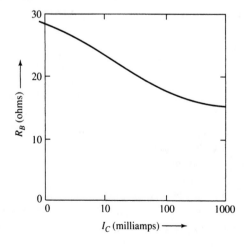

FIGURE 7.11 Base resistance in the transistor of Figure 7.10. Values of R_B are calculated using measured data and Equation 7.2.13.

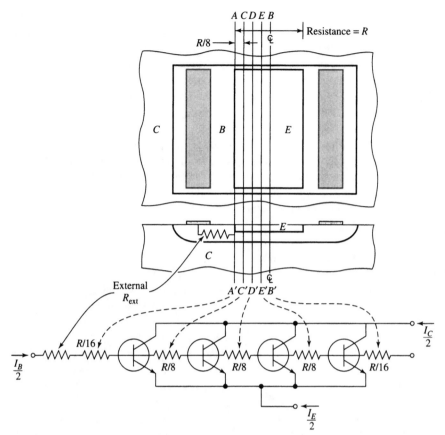

FIGURE 7.12 Modeling the effect of base-spreading resistance. The base region of an IC transistor is divided into eight sections (four on each half) and total built-in resistance between the base contacts is divided between eight ideal transistors [5]. Network analysis allows us to determine the effective resistance seen at the base-emitter junction as bias varies (Problem 7.12).

of the physical resistance along the base region. By increasing the number of pieces into which the transistor is sliced, we can increase the accuracy of the analysis and ultimately approach the exact distributed case. We use the four-section structure in Figure 7.12 to illustrate both the physical mechanism of current crowding and the method of solution. The calculation of base resistance from the network sketched in Figure 7.12 can be accomplished by assuming a current in the last leg of the ladder and then applying network laws to calculate voltage and current at the input terminal. Base resistance is then obtained by using Equation 7.2.13 at each value of input current I_B and voltage V_{BE}. Further consideration of base resistance is found in the problems.

In this section we considered several bias-dependent physical effects that were not included in the basic theory of junction transistors. The effects described provide limits on the useful range of bias for an amplifying bipolar transistor. Analysis of these effects shows that the equation for transistor action (Equation 6.1.12) adequately describes high-level injection at the base-emitter junction provided that the total majority charge in the base is properly specified. The basic theory of Chapter 6 must be extended, however, to treat low-level injection, the Kirk effect, or base resistance. Further consequences of these effects are described in the discussion of transistor speed and transistor models that follows.

To illustrate the concepts involved, our discussion of high-level injection treated effects one-by-one. In practice, of course, they occur simultaneously. Some interactions between effects, such as a reduction in base resistance because of high-level injection at the base-emitter junction, can be surmised; others are more subtle. Computer-aided transistor analysis treats high-level injection from first principles, providing a comprehensive analysis. The computer analysis [6] shows that the Kirk effect, sometimes called *base push-out*, is the dominant high-level injection effect in a well-designed IC transistor.

The list of limiting effects we treated is not all-inclusive. The mechanisms described are concerned with internal transistor operation. Additional performance limits are caused by parasitic elements introduced by the IC fabrication process. Limiting effects resulting from these elements can usually be treated by using circuit analysis, as illustrated in the discussion of lateral *pnp* transistors in Sec. 7.8.

7.3 BASE TRANSIT TIME

The transit time of minority carriers across the quasi-neutral base of a transistor under active bias is a significant parameter. The delay associated with this transit time was the dominant limitation on the transient behavior of the first transistors. The search for methods of reducing base transit time was responsible for many advances in semiconductor processing technology. In the present design of bipolar transistors for integrated circuits, the base transit time has been so greatly reduced that it is just one among several important time limitations. Considering the base transit time, however, illustrates the interrelationships between transistor design and performance.

We first consider the magnitude of the excess minority charge in the quasi-neutral base of a transistor under active bias. This charge, which we designate Q_{nB}, can be expressed as*

$$Q_{nB} = \int_0^{x_B} qA_E n'(x)\, dx \tag{7.3.1}$$

The injected charge Q_{nB} carries the collector current of Equation 7.1.1 that flows because of transistor action between the emitter and collector. Accordingly, a characteristic *base transit time* τ_B for the transfer of minority carriers across the base is given by the ratio of Q_{nB} to I_C

$$\tau_B = \frac{Q_{nB}}{I_C} \tag{7.3.2}$$

Calculation of the base transit time in the prototype transistor sketched in Figure 6.1 is particularly simple. Because $n'(x)$ is linear (Equation 6.1.16), $Q_{nB} = \frac{1}{2} q n'(0) x_B A_E$ and I_C is given by Equation 6.1.17. Hence, for this case

$$\tau_B = \frac{x_B^2}{2\tilde{D}_n} \tag{7.3.3}$$

Equation 7.3.3 is a measure of the time required for diffusion transport across a region. Some appreciation of the low speed of diffusion transport relative to drift of carriers in an electric field can be gained by calculating τ_B for a lightly doped base ($N_a \approx 10^{16}$ cm^{-3})

---

* A similar definition for excess hole charge in an *n*–type region in a diode was made in Equation 5.4.1.

having a width x_B of 1 μm. Using these values in Equation 7.3.3 gives $\tau_B \simeq 144$ ps. On the other hand, when a field is present in the base, the carrier drift transit time is $x_B/\mu_n\mathcal{E}$ or x_B^2/μ_nV for a constant field. The drift transit time is smaller than the diffusion time for any voltage drop greater than about 50 mV.

For an arbitrarily doped transistor, an expression for τ_B can be written using Equations 7.1.1 and 7.3.1 in Equation 7.3.2:

$$\tau_B = \frac{\int_0^{x_B} p\,dx \int_0^{x_B} n'\,dx}{\tilde{D}_n n_i^2 \exp(qV_{BE}/kT)} \tag{7.3.4}$$

Equation 7.3.4 is particularly useful when considering the effects of various base dopant profiles on τ_B.

Applying Equation 7.3.4 to the prototype transistor under high-level injection conditions is also enlightening. For the prototype transistor under active bias, the carrier densities vary linearly with position, and the integrals in Equation 7.3.4 are particularly simple to evaluate. A plot of the minority-carrier density in the base is triangular and, because of charge neutrality, the majority carriers have a trapezoidal distribution, decreasing away from the edge of the base-emitter space-charge region. The carrier densities are sketched in Figure 7.13. Under these conditions, the integrals in Equation 7.3.4 can be evaluated, and the base transit time can be written as

$$\tau_B = \frac{[\frac{1}{2}n'(0)x_B][\frac{1}{2}(n'(0) + 2N_a)x_B]}{\tilde{D}_n[n'(0)(n'(0) + N_a)]}$$

$$= \frac{x_B^2}{4\tilde{D}_n}\left[1 + \frac{N_a}{n'(0) + N_a}\right] \tag{7.3.5}$$

Under low- or moderate–injection conditions, for which $n'(0)$ is much less than N_a, Equation 7.3.5 shows that τ_B approaches $x_B^2/2\tilde{D}_n$. High-bias conditions reduce the base transit time, with τ_B asymptotically approaching $x_B^2/4\tilde{D}_n$. This factor of two reduction in base transit time for homogeneously doped transistors under high-level injection was first predicted by W. M. Webster [7] and is frequently called the *Webster effect*. The physical source for the reduction in τ_B is the field associated with the increased density of majority carriers

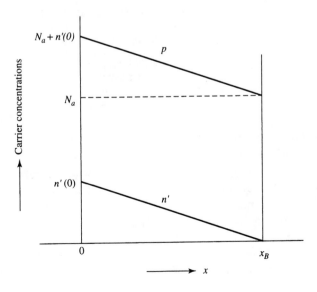

FIGURE 7.13 Free-carrier densities in the base of a homogeneously doped transistor. Charge neutrality causes the hole density to increase above the dopant concentration N_a.

that exists throughout the base under high-level injection conditions. To balance a diffusion tendency of these majority carriers resulting from their nonuniform distribution, an electric field must exist between the emitter and collector. This electric field balances the diffusion tendency and prevents majority-carrier flow toward the collector. At the same time, the field aids the flow of minority carriers and acts to reduce their transit time across the base. In the case of uniform base doping, the transport of minority carriers changes from pure diffusion at low biases to drift-aided diffusion at higher biases.

Although IC transistors are typically not made with uniformly doped base regions, we saw in Sec. 7.2 that high-level injection conditions extend the base into the uniformly doped epitaxial region (Kirk effect). For this reason, high-level injection analyses often take τ_B to be $x_B^2/4\tilde{D}_n$ [3, 4].

An alternative form for τ_B can be derived from Equations 6.1.10 and 7.3.4. If we take the density of electrons at the collector boundary $n(x_B)$ to be zero, then Equation 6.1.10 shows that the minority-carrier density as a function of x is

$$n(x) = -\frac{J_n}{q\tilde{D}_n p(x)} \int_x^{x_B} p(\xi)\, d\xi \tag{7.3.6}$$

Under the usual active-bias approximation that $n \approx n'$, Equation 7.3.6 can be combined with Equation 7.3.4 to give

$$\tau_B = \frac{1}{\tilde{D}_n} \int_0^{x_B} \frac{1}{p(x)} \left[\int_x^{x_B} p(\xi)\, d\xi \right] dx \tag{7.3.7}$$

If the distance variable x in Equation 7.3.7 is normalized to x_B by introducing $y = x/x_B$, then Equation 7.3.7 becomes

$$\tau_B = \frac{x_B^2}{\tilde{D}_n} \left\{ \int_0^1 \frac{1}{p(y)} \left[\int_y^1 p(\xi)\, d\xi \right] dy \right\}$$

$$= \frac{x_B^2}{\nu \tilde{D}_n} \tag{7.3.8}$$

where the factor ν represents the reciprocal of the nested definite integral in the first form of Equation 7.3.8. Equation 7.3.8 shows that the effect of an inhomogeneous base doping on τ_B can be incorporated into a parameter ν.

For constant base doping $p(x) = N_a$, it is easy to use Equation 7.3.8 to show that $\nu = 2$. When graded profiles are used, τ_B can be reduced from $x_B^2/2\tilde{D}_n$ by about an order of magnitude. Limits on reducing τ_B are imposed by the need to maintain acceptable injection efficiency [which constrains $N_a(0)$ to be appreciably lower than the emitter doping] and the requirement of an extrinsic p-type doped material at $x = x_B$ [which forces $N_a(x_B)$ to be greater than N_{depi}].

7.4 CHARGE-CONTROL MODEL

The concept introduced in Equation 7.3.1 of determining time variations by relating current to stored charge can be extended beyond calculating the base transit time. The charge-control model can be used for more general time-dependent analysis [8]. In this model the controlled variable is not current or voltage; instead, equations are expressed in terms of controlled charges within regions of the device. In this section, we develop the full charge-control model and illustrate its use. In Sec. 7.6 we use the charge-control concept to gain insight into the physical mechanisms limiting the frequency response of bipolar transistors.

A typical charge-control relationship for a transistor under active bias was derived in the previous section in Equation 7.3.2. This equation, $I_C = Q_{nB}/\tau_B$, relates the minority charge stored in the quasi-neutral base Q_{nB} to the current I_C carried by transistor action between the emitter and the collector. The charge and current are linearly related with τ_B, the transit time in the quasi-neutral base, as the proportionality factor. Because Equation 7.3.2 represents only minority-carrier transport across the base, however, it is just one portion of the charge-control model for a transistor.

An amplifying *npn* transistor is controlled by the bias on the base-emitter junction. This bias affects not only Q_{nB}, but other charge components as well. The major additional components to be considered are the charges represented by holes injected into the emitter, which we designate as Q_{pE}, and the charges stored on the base-emitter and base-collector depletion capacitances, which are given the symbols Q_{VE} and Q_{VC}, respectively. Figure 7.14 shows these components for a prototype transistor. We first discuss the two injection components, Q_{nB} and Q_{pE} that are responsible for steady-state base current. The other charge components shown in Figure 7.14 influence the time-varying behavior of the BJT and will be considered later. Because both Q_{nB} and Q_{pE} increase when the base-emitter voltage increases, their sum ($Q_{nB} + Q_{pE}$) is called Q_F (because Q_F increases when the BJT is under *forward*-active bias). The sign of Q_F is the sign of the controlling (base majority-carrier) charge, positive for an *npn* transistor and negative for a *pnp* transistor. The steady-state collector current can be written in terms of Q_F if a characteristic time τ_F is introduced. The equation for current (in analogy to Equation 7.3.2) is

$$I_C = \frac{Q_F}{\tau_F} \tag{7.4.1}$$

Note that all that is formally required to make Equation 7.4.1 accurate is that I_C must be linearly related to Q_F (so that τ_F is constant).

Because Q_F represents the sum of the magnitudes of the excess minority charge in the quasi-neutral emitter and base regions, its voltage dependence is generally that of a diode, and we can write

$$Q_F = Q_{FO}\left[\exp\left(\frac{qV_{BE}}{KT}\right) - 1\right] \tag{7.4.2}$$

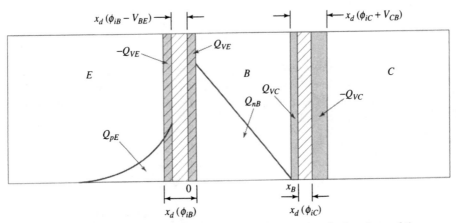

FIGURE 7.14 Cross section of a prototype transistor showing the locations of the charge components used in charge-control modeling. The charges Q_V represent storage at the edges of the space-charge regions. The cross-hatching indicates the two junction space-charge regions at thermal equilibrium.

where Q_{FO} is a function of the dopant profiles and device geometry. Because Q_F is at least somewhat greater than Q_{nB}, τ_F must be greater than the base transit time τ_B (Equation 7.3.2). We will discuss τ_F further after the charge-control model is developed more fully.

The steady-state current flowing in the base lead is proportional to the rate at which Q_{nB} recombines in the quasi-neutral base plus the rate at which holes are injected into the emitter to replenish Q_{pE}. These two rates are proportional to the diode factor $[\exp(qV_{BE}/kT) - 1]$, and therefore (by Equation 7.4.2) proportional to Q_F. It is therefore possible to write a charge-control expression for the input (base) current of the transistor

$$I_B = \frac{Q_F}{\tau_{BF}} \tag{7.4.3}$$

A considerable amount of physical analysis involving emitter efficiency and the recombination processes for excess carriers in both the emitter and base is required to express τ_{BF} or τ_F in terms of more fundamental parameters, and this analysis is not pursued here.

Equations 7.4.1 and 7.4.3 can be used to show that the steady-state current gain is simply the ratio of the two characteristic times:

$$\frac{I_C}{I_B} = \beta_F = \frac{\tau_{BF}}{\tau_F} \tag{7.4.4}$$

For example, Equations 7.4.1, 7.4.3, and 7.4.4 can be applied to the prototype transistor of Figure 6.1. If the emitter efficiency is very high in the prototype device, then $Q_F \approx Q_{nB}$ with $Q_{nB} = \frac{1}{2} q n'(0) x_B A_E$. Because only base recombination is significant for this case, $\tau_{BF} = \tau_n$ and $\tau_F = x_B^2/2\tilde{D}_n$ as derived in Equation 7.3.3. Therefore, from Equation 7.4.4 β_F is $2L_n^2/x_B^2$ where $L_n = \sqrt{\tilde{D}_n \tau_n}$. This result for dc current gain can be compared with earlier analysis of the same problem in Sec. 6.2. There, α_F was derived as the base transport factor to be $[1 - (x_B^2/2L_n^2)]$ in Equation 6.2.8. If we use this result for α_F in the equation $\beta_F = \alpha_F/(1 - \alpha_F)$, we find β_F to be identical to the expression obtained from the charge-control analysis.

A full charge-control model for the bipolar transistor is derived by adding terms to represent the currents that flow because of time variations in stored charge. Clearly, if Q_F increases with time, there is a component of base current equal to dQ_F/dt. Likewise, changes in the charge stored at the base-emitter and base-collector junctions (Q_{VE} and Q_{VC}) result in added base current. An overall expression for the base current is, therefore,

$$i_B = \frac{Q_F}{\tau_{BF}} + \frac{dQ_F}{dt} + \frac{dQ_{VE}}{dt} + \frac{dQ_{VC}}{dt} \tag{7.4.5}$$

The first three components of current in Equation 7.4.5 flow from the base to the emitter; the last flows from the base to the collector. Combining Equations 7.4.1 and 7.4.5 and using Kirchhoff's current law, we obtain a set of charge-control equations for the transistor under active bias:

$$i_C = \frac{Q_F}{\tau_F} - \frac{dQ_{VC}}{dt}$$

$$i_B = \frac{Q_F}{\tau_{BF}} + \frac{dQ_F}{dt} + \frac{dQ_{VE}}{dt} + \frac{dQ_{VC}}{dt} \tag{7.4.6}$$

$$i_E = -Q_F\left(\frac{1}{\tau_F} + \frac{1}{\tau_{BF}}\right) - \frac{dQ_F}{dt} - \frac{dQ_{VE}}{dt}$$

We have thus derived a set of *linear* equations relating currents and charges in a bipolar transistor. The linearity of these equations contrasts with the nonlinear expressions that relate currents and voltages in the transistor.

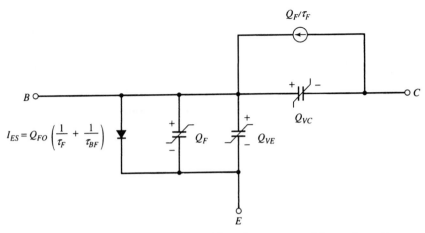

FIGURE 7.15 Charge-control representation of an *npn* transistor under active bias with junction-charge storage and injected-base charge included.

The circuit diagram of Figure 7.15 represents the various terms in Equation 7.4.6. The diode from the base to the emitter passes the steady-state current and has a saturation current $I_{ES} = Q_{FO}[(1/\tau_F) + (1/\tau_{BF})]$ as indicated by Equation 7.4.2. The elements storing the charges Q_F, Q_{VE}, and Q_{VC} are shown as capacitors with a line across them to indicate that they store charge (like capacitors) but depend on voltage. Now that we have a basic set of charge-control equations for a *BJT*, we are better able to gain a perspective on the useful range and the limitations of this viewpoint.

The basic premise underlying charge-control analysis is the existence of a constant proportionality between amount of charge and current. Another way of stating this is that the characteristic times in the charge-control equations must not themselves be functions of charge or of bias voltages. With this premise, the characteristic times can be derived for dc conditions and then used to express dynamic conditions. Strictly this premise is not correct; the characteristic times are functions of the charge. For example, to affect current at the collector, minority carriers must disperse through the base region after being injected at the edge of the base-emitter junction. Thus, during transient conditions, collector current does not have the same ratio to base charge as it does in steady state.

If a dynamic problem is analyzed with the charge-control model, the solutions for charge are constrained to be a time sequence of differing "steady-state" solutions. Hence, these solutions are sometimes called *quasi-static approximations*. For the greater part of the transient, however, the charge-control solution is usually a fair representation of the more exact result. The time scale over which there is significant error in the charge-control solution is of the order of the transit time in the base, that is, the order of τ_F. For most applications the time scale of interest is appreciably greater than τ_F, and solution by means of the charge-control model is adequate. A specific example in the next section shows the value and limitations of the charge-control approach.

Applications of the Charge-Control Model

Before discussing the bipolar charge-control model further, we illustrate the use of Equations 7.4.6 in an application. A simple circuit appropriate for this purpose is shown in Figure 7.16*a*, and the circuit plus charge-control model is sketched in Figure 7.16*b*. In this analysis we want to find the collector-current response of an *npn* transistor under active bias when driven by a current source at the base. For hand calculations, several

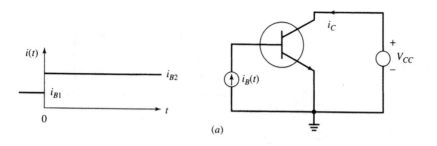

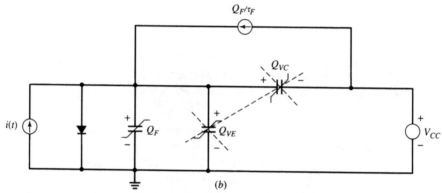

FIGURE 7.16 (a) Simple circuit for illustration of charge-control model. (b) Equivalent-circuit model. The cancelled elements carry negligible currents.

simplifications are appropriate. First, the current dQ_{VE}/dt can be considered negligible because the base-emitter voltage varies only slightly under forward bias. This simplification is usually made for active-bias operation. Because we have chosen a simple circuit in which the collector voltage is constant, it is also possible to treat the current dQ_{VC}/dt as negligible.

With these approximations the equation for the base current has only one unknown, the controlled charge Q_F, and takes the form

$$i_B = \frac{Q_F}{\tau_{BF}} + \frac{dQ_F}{dt} \tag{7.4.7}$$

Because $i_B(t)$ is specified by

$$i_B = i_{B1} \ (t < 0)$$
$$= i_{B2} \ (t > 0)$$

the solution for Q_F contains terms associated with both the homogeneous and particular forms of Equation 7.4.7. When the boundary values $Q_F(t = 0) = i_{B1}\tau_{BF}$ and $Q_F(t \rightarrow \infty) = i_{B2}\tau_{BF}$ are used, the solution becomes

$$Q_F = \tau_{BF}[i_{B2} + (i_{B1} - i_{B2}) \exp(-t/\tau_{BF})] \tag{7.4.8}$$

With the approximations that we made, the collector current is given by Q_F/τ_F so that its time dependence becomes (using Equation 7.4.4)

$$i_C = \beta_F[i_{B2} + (i_{B1} - i_{B2}) \exp(-t/\tau_{BF})] \tag{7.4.9}$$

The collector current thus changes from its initial to its final value following an exponential function with a characteristic time constant equal to τ_{BF} (Figure 7.17).

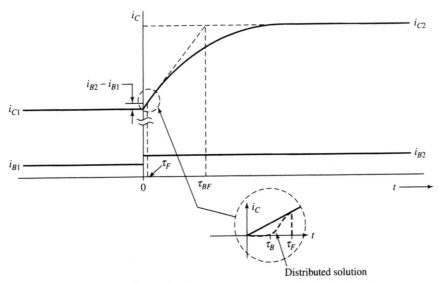

FIGURE 7.17 Time variation of collector current in the circuit of Figure 7.16 as calculated from the charge-control model.

As mentioned in the previous section, the transient solution given in Equation 7.4.9 is not accurate for small values of t. At zero time, for example, Equation 7.4.9 predicts an abrupt change in the slope of the collector current equal to $(i_{B2} - i_{B1})/\tau_F$, but the collector current actually does not change until the extra electrons injected at the emitter side of the base reach the collector. A more complete analysis that considers the distributed nature of base charging predicts that i_C does not change at all for times close to $t = 0$; initially, the increments in the base and emitter currents are equal. The collector current first begins to change when t reaches the base transit time τ_B; i_C then rapidly approaches the transient solution predicted by the charge-control model and essentially matches that result for times larger than $t = \tau_F$. (See inset in Figure 7.17.)

The foregoing example may seem artificial because of the number of simplifications that we imposed, first in the choice of circuit and second in the approximations made. These physical simplifications, however, avoided mathematical complications that tend to obscure the use of the model.

For example, a complication arises if the collector in Figure 7.16a is not connected to an ac ground, but rather is connected to the source through a load resistor R_L. Because V_{CB} varies in this case, the current required to charge Q_{VC} is not negligible. The solution for i_C can be obtained readily if we define an effective capacitance C_{jC} equal to the average of dQ_{VC}/dV_{CB} over the collector voltage interval. The solution for i_C is then equal to Equation 7.4.9 provided that the time constant τ_{BF} in Equation 7.4.9 is replaced by [Problem 7.19]:

$$\tau'_{BF} = \tau_{BF}\left(1 + \frac{R_L C_{jC}}{\tau_F}\right) \tag{7.4.10}$$

To derive Equation 7.4.10, we note that the presence of the collector capacitance affects the base current much more than it does the collector current. Hence, we can still approximate $i_C = Q_F/\tau_F$, but i_B must now include the transient charge storage at the collector junction dQ_{VC}/dt.

This result shows that a collector resistor lengthens the time of the current transient by the binomial factor in Equation 7.4.10.

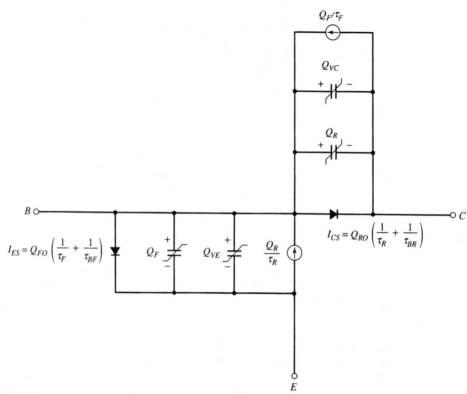

FIGURE 7.18 Complete bipolar charge-control model for large-signal applications.

Large-Signal Model. The most extensive use of charge-control equations is for the solution of large-signal transient problems, and they are especially useful when dealing with switching between regions of operation, typically between cut-off and saturation. Writing charge-control equations for large-signal switching is straightforward because all regions of operation have in common the injection and extraction of charge at the two junctions. These injection and extraction processes can be represented by adding together two sets of charge-control equations: one set that characterizes forward-active operation (Equations 7.4.6) and a companion set having the same form to represent operation under reverse-active bias. The total controlled charge is a superposition of charges representing forward-active bias Q_F and reverse-active bias Q_R.

The full set of equations for an *npn* transistor is*

$$i_E = -\frac{dQ_F}{dt} - Q_F\left(\frac{1}{\tau_F} + \frac{1}{\tau_{BF}}\right) + \frac{Q_R}{\tau_R} - \frac{dQ_{VE}}{dt}$$

$$i_C = \frac{Q_F}{\tau_F} - \frac{dQ_R}{dt} - Q_R\left(\frac{1}{\tau_R} + \frac{1}{\tau_{BR}}\right) - \frac{dQ_{VC}}{dt} \qquad (7.4.11)$$

$$i_B = \frac{dQ_F}{dt} + \frac{Q_F}{\tau_{BF}} + \frac{dQ_R}{dt} + \frac{Q_R}{\tau_{BR}} + \frac{dQ_{VE}}{dt} + \frac{dQ_{VC}}{dt}$$

A circuit model corresponding to these equations is shown in Figure 7.18. Its form clearly indicates the superposed forward-active and reverse-active charge-control models. The dc

---

* Charge-control equations for a *pnp* transistor are given in Problem 7.17.

components in the circuit of Figure 7.18 have a one-to-one correspondence to the components of the Ebers-Moll large-signal representation sketched in Figure 6.12. The Ebers-Moll model is also built up from superposed forward-active and reverse-active equivalent circuits.

In Equations 7.4.11, parameters associated with the reverse-active region of bias are defined analogously to those representing forward bias. For example, the controlled charge Q_R is related to V_{BC} by

$$Q_R = Q_{RO}\left[\exp\left(\frac{qV_{BC}}{kT}\right) - 1\right] \tag{7.4.12}$$

and this charge represents storage in both the base and collector quasi-neutral regions.

The density of dopant atoms in the collector region of an IC transistor is normally of the same order of magnitude or lower than the density in the base near the collector junction. Hence, when a transistor is under reverse-active bias, the efficiency of electron injection into the base is typically quite low (approximately 60 to 85%), and a relatively large amount of charge is stored in the collector. In fact, because the base is so narrow, this charge is typically the dominating component of Q_R. For example, Figure 7.19 is a sketch of the components of Q_F and Q_R for two cases of saturated transistors. Figure 7.19a shows the case of a homogeneously doped transistor with a lightly doped collector region and Figure 7.19b shows a typical IC transistor. The large amount of collector charge Q_C in either transistor when it is saturated must be removed to bring it out of saturation. This can cause considerable delay. One means of speeding its removal—the incorporation of

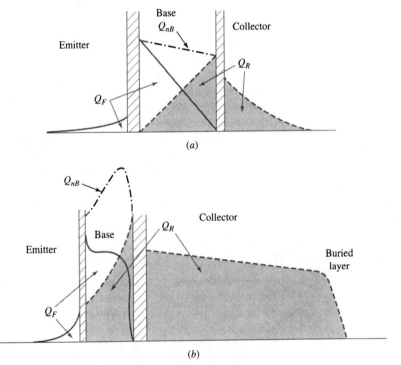

FIGURE 7.19 The locations of Q_F and Q_R (dashed lines) for saturated conditions: (a) in a homogeneously doped transistor with a lightly doped collector region and (b) in an epitaxial-collector diffused IC transistor. The dot-dash lines represent the total base charge Q_{nB}.

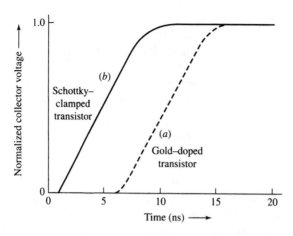

FIGURE 7.20 Comparison of turn-off times of (a) gold-doped and (b) Schottky-clamped transistors. Switching of the gold-doped device is delayed 7 ns as recombination occurs before the transistor begins to change state [9].

a large number of recombination centers, usually by doping the collector region with gold—was extensively employed in early designs of fast switching integrated circuits. Superior switching results can be obtained by preventing the transistor from entering the saturated mode of operation. An effective method of accomplishing this was described in Sec. 3.6 where Schottky clamping was discussed. An IC realization of a Schottky-clamped switching transistor was shown in Figure 6.17. Figure 7.20 shows the comparative switching times of a gold-doped switching transistor and a Schottky-clamped transistor having the same dimensions. The impressive reduction in the time delay before the output voltage rises makes clear why Schottky clamping has become so important for fast switching transistors.

Saturation Transient.[†] As a final application of charge-control modeling, we consider the solution for the transient behavior of a bipolar transistor within the saturated region of operation (for a transistor that has not been Schottky clamped). For this case all terms in Equations 7.4.11 need to be retained.

The circuit is shown in Figure 7.21. We assume that $V_S \gg V_o$, the "turn-on voltage" of the base-emitter diode that was introduced in Sec. 3.6. We also take V_C to be much greater than the drop across the transistor in saturation (Equation 6.4.13) and assume that the base drive is strong enough to saturate the transistor fully; that is, that $\beta_F V_S / R_S \gg V_C/R_L$. After the switch is closed at $t = 0$, the transistor enters the active-bias region and current increases in the collector circuit according to Equation 7.4.9 with the modification of the time constant given in Equation 7.4.10. The final value of current $\beta_F i_2$ is not reached because the transistor enters saturation when V_{BC} becomes roughly 0.5 V. The

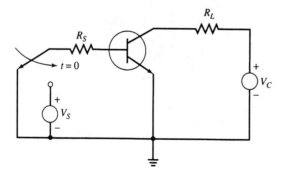

FIGURE 7.21 Switching circuit in which a transistor undergoes transient behavior from cut-off to saturation. The source voltage V_S is taken to be much greater than the "ON voltage" V_o of the base-emitter diode. The base drive is taken to be sufficient to cause transistor saturation $(\beta_F V_S / R_S \gg V_C/R_L)$.

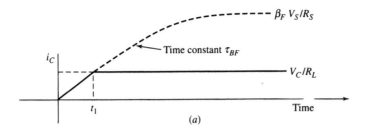

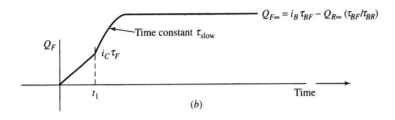

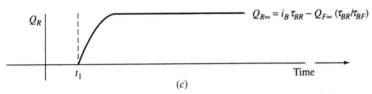

FIGURE 7.22 Transient behavior of (a) i_C, (b) Q_F, and (c) Q_R for the switching circuit in Figure 7.21.

collector current is then limited to V_C/R_L as shown in Figure 7.22a. Before the transistor saturates, the stored charge Q_F increases with a time dependence similar to that of i_C (Figure 7.22b). Although the collector current stabilizes after the transistor saturates, the charges Q_F and Q_R continue changing with time, and the transistor is not in a steady-state "ON" condition until these quantities become constant. Before saturation is reached, the collector-base junction is back biased, and Q_R is essentially zero. After saturation Q_R increases, as does Q_F.

To analyze the saturated behavior, a set of independent simultaneous equations for Q_F and Q_R must be solved. The set consists of equations for i_B and i_C:

$$i_B = \frac{Q_F}{\tau_{BF}} + \frac{dQ_F}{dt} + \frac{Q_R}{\tau_{BR}} + \frac{dQ_R}{dt}$$

$$i_C = \frac{Q_F}{\tau_F} - Q_R\left(\frac{1}{\tau_R} + \frac{1}{\tau_{BR}}\right) - \frac{dQ_R}{dt} \quad (7.4.13)$$

In writing these equations, the terms representing charging currents for Q_{VC} and Q_{VE} are ignored as negligible because V_{BC} and V_{BE} are roughly constant in saturation.

Equations 7.4.13 represent simultaneous differential equations for which the natural frequencies are solutions for s that satisfy the equation:

$$\left(s + \frac{1}{\tau_{BF}}\right)\left(s + \frac{1}{\tau_R} + \frac{1}{\tau_{BR}}\right) + \left(s + \frac{1}{\tau_{BR}}\right)\frac{1}{\tau_F} = 0 \quad (7.4.14)$$

The roots of this quadratic equation are approximately given by

$$|s_1| = \frac{1}{\tau_{FAST}} = \left(\frac{1}{\tau_F} + \frac{1}{\tau_R} + \frac{1}{\tau_{BR}} + \frac{1}{\tau_{BF}} \right)$$

and

$$|s_2| = \frac{1}{\tau_{SLOW}}$$

$$= \tau_{FAST} \left(\frac{1}{\tau_F \tau_{BR}} + \frac{1}{\tau_R \tau_{BF}} + \frac{1}{\tau_{BF} \tau_{BR}} \right) \tag{7.4.15}$$

These roots represent normal modes of the transient solution for Q_F and Q_R in saturation. As the names imply, the root s_1 is responsible for a transient solution that is completed long before the steady state is reached, while the root s_2 leads to the transient behavior of Q_F and Q_R that dominates while the transistor remains saturated. Hence, a good approximation to the transient behavior of Q_F and Q_R in saturation (denoted by Q_{FS} and Q_{RS}) is

$$Q_{FS} = (Q_{F\infty} - Q_{F1}) \left[1 - \exp\left(\frac{-(t - t_1)}{\tau_{SLOW}} \right) \right] + Q_{F1}$$

$$Q_{RS} = (Q_{R\infty} - Q_{R1}) \left[1 - \exp\left(\frac{-(t - t_1)}{\tau_{SLOW}} \right) \right] + Q_{R1} \tag{7.4.16}$$

where t_1 is the time at which saturation begins, $Q_{F\infty}$ and $Q_{R\infty}$ are the final values, and Q_{R1} and Q_{F1} are the values of the two controlled charges at $t = t_1$.

The total saturation charge $Q_{ST} = (Q_{FS} + Q_{RS})$ is sometimes considered in terms of the forward charge $Q_F = I_C \tau_F$ at the edge of saturation and an added portion that enters the channel after the transistor saturates.

$$Q_{ST} = I_C \tau_F + Q_S \left[1 - \exp\left(\frac{-(t - t_1)}{\tau_{SLOW}} \right) \right] \tag{7.4.17}$$

where

$$Q_S = Q_{F\infty} + Q_{R\infty} - I_C \tau_F$$

For the prototype transistor in which essentially all charge is stored in the base, these definitions aid in understanding the physical mechanisms. They can be simply represented as shown in Figure 7.23. The important feature, emphasized by Figure 7.23, is that transistor switching is not completed when the collector current reaches a steady value (which occurs at the time of saturation), but that the steady state is only reached when Q_R and Q_F are at their final values.

We have only considered the "turn-on" transient. To calculate the "turn-off" transient, a similar two-part analysis is carried out. We first find solutions in the saturated region while Q_R and Q_F decrease to their values at the edge of active-mode bias and then solve for the transient through the active region to transistor cut-off.

The above discussion helps show the physical basis of the charges in saturated bipolar transistors and illustrates the use of charge-control analysis. The charge-control equations are especially useful because they provide a set of simultaneous linear equations that are readily solved by computer-aided analysis. Further applications of the model to switching circuits are extensively described in references [10] and [11]. In the following section, we will use the charge-control relations to develop a model useful for small-signal transistor analysis.

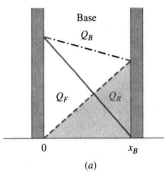 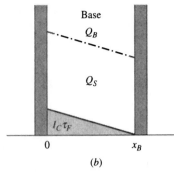

(a)   (b)

FIGURE 7.23 Two representations for total charge at steady state in a saturated, homogeneously doped transistor with high emitter and collector doping. (a) The two charge-control variables Q_F and Q_R are preserved. (b) The total charge is divided into $I_C\tau_F$, which is present at the onset of saturation, and Q_S, which is added during the saturation transient.

Equivalence between Models.   Two models representing bipolar transistors have now been discussed: The Ebers-Moll model in Sec. 6.4 and the charge-control model in this section. Because these models represent the same physical devices, we expect the parameters of the two models to be related. For example, in Problem 7.20 we find that the Ebers-Moll parameters α_F and α_R are related to the charge-control parameters τ_F and τ_{BF} by

$$\alpha_F = \frac{\tau_{BF}}{\tau_F + \tau_{BF}} \tag{7.4.18}$$

and

$$\alpha_R = \frac{\tau_{BR}}{\tau_R + \tau_{BR}} \tag{7.4.19}$$

respectively. In the dc Ebers-Moll representation, the six parameters of the basic charge-control model (Q_{FO}, Q_{RO}, τ_F, τ_{BF}, τ_R, and τ_{BR}) are reduced to only four parameters (α_F, α_R, I_{ES}, and I_{CS}).

Relationships such as Equation 7.4.19 are useful to obtain model parameters by measurements and to establish properties of the models. For example, the reciprocity condition of the Ebers-Moll model $\alpha_F I_{ES} = \alpha_R I_{CS}$ (Equation 6.4.8) can be used with the charge-control model to derive

$$\frac{Q_{FO}}{\tau_F} = \frac{Q_{RO}}{\tau_R} \tag{7.4.20}$$

7.5 SMALL-SIGNAL TRANSISTOR MODEL

Bipolar transistors are often used to amplify small signals, especially at high frequencies. In this section we focus on the small-signal behavior of the transistor and its frequency response. In the following section, we use the charge-control model to find the highest useful frequencies at which a transistor can operate.

When transistors are biased in the active region and used for amplification, it is often useful to approximate their behavior under conditions of small voltage variations at the base-emitter junction. If these variations are smaller than the thermal voltage $V_t = (kT/q)$, the transistor can be represented by a linear equivalent circuit. This representation greatly aids the design of amplifying circuits and is called the small-signal transistor model. It can be readily derived by making use of several of the charge-control relationships that were discussed in Sec. 7.4.

When a transistor is biased in the active mode, collector current is related to base-emitter voltage by Equation 6.2.1, which is repeated here for convenient reference.

$$I_C = I_S \exp\left(\frac{qV_{BE}}{kT}\right) = I_S \exp\left(\frac{V_{BE}}{V_t}\right) \tag{7.5.1}$$

Hence, if V_{BE} varies incrementally, the variation of I_C is given by

$$\frac{\partial I_C}{\partial V_{BE}} = \frac{I_S}{V_t} \exp\left(\frac{V_{BE}}{V_t}\right) = \frac{I_C}{V_t} \equiv g_m \tag{7.5.2}$$

This derivative is recognized as the *transconductance* defined in Equation 4.5.13 and is given the usual symbol g_m. Note that g_m is directly proportional to the bias current in the transistor. The variation of base current with base-emitter voltage can be found most directly by using the charge-control expressions for I_C and I_B (Equations 7.4.1 and 7.4.3)

$$\frac{\partial I_B}{\partial V_{BE}} = \frac{\partial(Q_F/\tau_{BF})}{\partial V_{BE}} = \frac{\partial(I_C\tau_F/\tau_{BF})}{\partial V_{BE}} = \frac{\tau_F\, g_m}{\tau_{BF}} = \frac{g_m}{\beta_F} = \delta g_m \tag{7.5.3}$$

(The ratio of τ_F to τ_{BF} is called the *defect factor* δ. From Equation 7.4.4 we see that δ is equal to β_F^{-1}.)

The base minority charge Q_F varies with base-emitter voltage according to

$$\frac{\partial Q_F}{\partial V_{BE}} = \frac{\partial(I_C\tau_F)}{\partial V_{BE}} = g_m\tau_F \equiv C_D \tag{7.5.4}$$

where the symbol C_D (often called the *diffusion capacitance*) represents the capacitance associated with incremental changes in the injected minority-carrier charge.

If the incremental voltages and currents in Equations 7.5.2, 7.5.3, and 7.5.4 are identified with ac signals, then the system of equations can be represented by the equivalent circuit shown in Figure 7.24. The small-signal incremental currents and voltages are denoted by lowercase symbols. The base-emitter input circuit is a parallel RC network having a time constant $\beta_F \tau_F$, which is just the base time constant τ_{BF}.

The collector-emitter output circuit consists of a current source activated by the input voltage. The output current for a given v_{BE} is proportional to g_m, and therefore depends

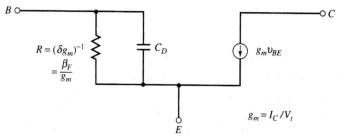

FIGURE 7.24 Equivalent circuit representing small-signal active bias for a bipolar junction transistor. Only first–order effects are included.

on the dc bias I_C as shown in Equation 7.5.1. The equivalent circuit of Figure. 7.24 emphasizes that, to first order, the input is decoupled from the output and the output is insensitive to collector-base voltage variations. From the analysis of Sec. 7.1, we know that the voltage across the collector-base junction does influence collector current, chiefly as a result of the Early effect. The variation of I_C with V_{CB} was shown in Sec. 7.1 to be the ratio of the collector current I_C to the Early voltage V_A. In terms of the small-signal parameters,

$$\left| \frac{\partial I_C}{\partial V_{CB}} \right| \equiv \frac{I_C}{|V_A|} = \frac{g_m V_t}{|V_A|} = \eta g_m \tag{7.5.5}$$

where a new parameter $\eta \equiv V_t/|V_A|$ is introduced to represent the ratio of the change in I_C when V_{CB} is varied to the change in I_C when V_{BE} is varied.

Variation in I_C with V_{CB} must also cause a change in the controlled charge Q_F with V_{CB}. This can be calculated from the charge-control relationship of Equation 7.4.1.

$$\left| \frac{\partial Q_F}{\partial V_{CB}} \right| = \left| \frac{\partial (I_C \tau_F)}{\partial V_{CB}} \right| = \tau_F \eta g_m = \eta C_D \tag{7.5.6}$$

Any change in base minority charge results in a change in base current, as well as in collector current. Thus, varying V_{CB} causes a change in I_B given by

$$\left| \frac{\partial I_B}{\partial V_{CB}} \right| = \left| \frac{\partial (Q_F/\tau_{BF})}{\partial V_{CB}} \right| = \frac{\eta g_m \tau_F}{\tau_{BF}} = \eta \delta g_m \tag{7.5.7}$$

The variations calculated in Equations 7.5.5, 7.5.6, and 7.5.7 can be incorporated into the linear equivalent circuit of Figure 7.24 by adding three elements as shown in Figure 7.25. The variation in I_C calculated in Equation 7.5.5 represents a change in the current flowing from collector to emitter in response to a change in collector-base voltage. It is thus modeled as a current generator activated by collector-base voltage. The current generator is directed from collector to emitter because an increase in V_{CB} increases I_C as described in Sec. 7.1. The change in injected charge storage in response to a change in collector-base voltage calculated in Equation 7.5.6 is modeled as a capacitor from collector to base. The variation in current calculated in Equation 7.5.7 flows between the base and the emitter and is caused by a changing collector-base voltage. It is thus modeled as a current generator directed from the emitter to the base. This direction is consistent with a reduced base current as a consequence of a reduction in Q_F.

The equivalent circuit sketched in Figure 7.25 can be simplified by employing two procedures common in circuit analysis. First, the generators involving v_{CB} can be drawn

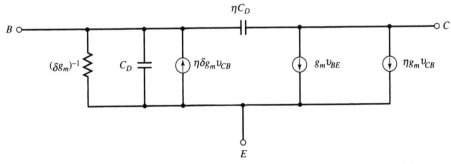

FIGURE 7.25 Small-signal equivalent circuit for a bipolar junction transistor with Early-effect elements.

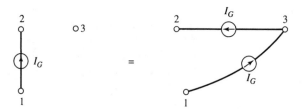

FIGURE 7.26 Two equivalent circuits representing the same current flowing between nodes.

between the collector and base nodes by using the equivalence illustrated in Figure 7.26. With this change some generators are activated by the voltage that appears across their terminals and can be replaced by passive elements. The second procedure involves re-expressing the activating voltages for several generators by making use of the identity: $v_{CE} \equiv v_{CB} + v_{BE}$. When these operations are performed and the admittance of elements in parallel is summed, the circuit can be simplified to the form shown in Figure 7.27. This form of the small-signal equivalent circuit is called the *hybrid-pi circuit* for the transistor. The term *hybrid* is used because the generator is a voltage-activated current source and therefore relates quantities of differing dimensions. The reference to *pi* denotes the general geometric shape of the circuit in the form of a Greek letter Π.

For accurate representation of the transistor small-signal behavior, two additional effects remain to be considered. The first of these is base resistance, discussed in Sec. 7.2. There we saw that current crowding causes an overall dc base resistance R_B (defined implicitly in Equation 7.2.13) that is a function of collector current. The dependence of R_B on I_C for a typical *npn* transistor is shown in Figure 7.11. To incorporate this variation into a small-signal equivalent circuit, the interdependences of the small-signal variations in voltage and current must be considered. To do this properly, we consider the functional relationships between V_{BE}, I_B, I_C, and R_B. Total differentiation of V_{BE} with respect to I_C, for example, results in three terms:

$$\frac{dV_{BE}}{dI_C} = \left.\frac{\partial V_{BE}}{\partial I_C}\right|_{I_B,R_B} + \left.\frac{\partial V_{BE}}{\partial I_B}\right|_{I_C,R_B}\left(\frac{dI_B}{dI_C}\right) + \left.\frac{\partial V_{BE}}{\partial R_B}\right|_{I_C,I_B}\left(\frac{dR_B}{dI_C}\right) \tag{7.5.8}$$

From Equation 7.2.13 we have

$$V_{BE} = I_B R_B + V_t \ln\left(\frac{I_C}{I_S}\right) \tag{7.5.9}$$

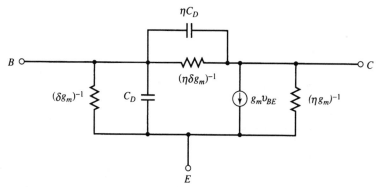

FIGURE 7.27 Simplified small-signal equivalent circuit including Early-effect elements. In general, $\delta \ll 1$ and $\eta \ll 1$.

Therefore, using Equation 7.5.8, we obtain

$$\frac{dV_{BE}}{dI_C} = \frac{V_t}{I_C} + R_B \frac{dI_B}{dI_C} + I_B \frac{dR_B}{dI_C} = \frac{1}{g_m} + \delta R_B + I_B \frac{dR_B}{dI_C} \tag{7.5.10}$$

where we have used the previously defined symbols g_m and δ while retaining the term involving the derivative of R_B. In practice this derivative can be obtained from a plot similar to that in Figure 7.11.

To obtain an equivalent-circuit representation from Equation 7.5.10, we solve for the base input resistance R_I where

$$R_I = \frac{dV_{BE}}{dI_B} = \frac{dV_{BE}}{dI_C} \cdot \frac{\partial I_C}{\partial I_B} = \frac{dV_{BE}}{dI_C} \cdot \frac{1}{\delta} \tag{7.5.11}$$

Using Equation 7.5.10 in 7.5.11, we find

$$R_I = \frac{1}{\delta g_m} + \left(R_B + \frac{I_B}{\delta} \frac{dR_B}{dI_C} \right) \tag{7.5.12}$$

Thus, to account for base resistance, a resistor of value

$$r_b = R_B + I_C \frac{dR_B}{dI_C} \tag{7.5.13}$$

must be added in series with the base-emitter resistance $(\delta g_m)^{-1}$ that was previously determined. This added resistance is included in the low–frequency circuit sketched in Figure 7.28. In Figure 7.28 we simplified the equivalent circuit shown in Figure 7.27 by omitting capacitors (valid at low frequency) and large resistors in order to focus our attention on the effects of base resistance. As apparent from this circuit, when base resistance is taken into account, the current generator in the output circuit is no longer actuated by the applied base-emitter voltage, but is rather a function of an internal node voltage. It is left as a problem to show that the current gain in the presence of base resistance is also properly modeled when the circuit of Figure 7.28 is used.

Our consideration of base resistance has been limited to dc and low-frequency effects. Base resistance can lead to more complex behavior at high frequencies because the base resistance and junction capacitances behave like distributed transmission lines. These effects can be modeled with fair accuracy by using an equivalent shunt RC network in place of r_b. A full discussion of this topic is provided in reference [5].

The last complication that we add to the hybrid-pi circuit is the capacitance associated with the junction space-charge regions. This capacitance is in parallel with the base-emitter and base-collector capacitors shown in Figure 7.27. The added capacitance is usually denoted by C_{jE} and C_{jC}. It is calculated from the junction-capacitance equations already derived in Chapter 4 and discussed in terms of the charge-control model in Sec. 7.4. The

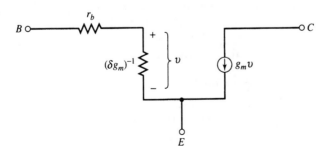

FIGURE 7.28 Low frequency, small-signal equivalent circuit including base resistance.

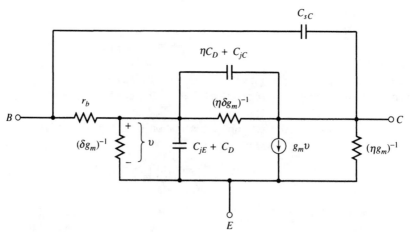

FIGURE 7.29 Hybrid–pi circuit with space-charge capacitances and base resistance.

emitter-base junction capacitance is $C_{jE} = dQ_{VE}/dV_{EB}$, and the collector-base junction capacitance is $C_{jC} = dQ_{VC}/dV_{CB}$. For increased accuracy, it is sometimes necessary to divide the total collector junction capacitance into a portion across the base impedance element (referred to as C_{sC}) and a portion C_{jC} that is returned to the intrinsic transistor base node (the node connected to δg_m). This division implies that some parts of the collector capacitance are not charged through the base resistance. The overall equivalent circuit including these effects is shown in Figure 7.29.

The circuit taken in its entirety appears formidable. Fortunately, it is seldom necessary to deal directly with the overall hybrid-pi circuit in hand calculations. Either several elements in the circuit have negligible effect under given conditions, allowing the circuit to be simplified, or else the calculations are carried out by computer-aided analysis.

Frequency Response. In this section, we developed a small-signal, hybrid-pi equivalent circuit useful in active bias by considering some of the physical effects discussed earlier. We can use this circuit to study the frequency response of the transistor, and, in particular, the highest frequencies at which it can operate.

We start our discussion by relating the small-signal, hybrid-pi equivalent circuit to the charge-control model and see how measuring the short-circuit current gain as a function of frequency can determine the charge-control parameter τ_F. In the next section, we extend the charge-control model to consider additional charges that must be moved within the transistor and the delays associated with their movement. These considerations provide a more detailed and physical measure of the frequency limits of the transistor.

We consider that only dc bias is applied to the collector so that the ac small-signal equivalent circuit has the collector shorted to ground. Under this condition, it is usually a good approximation to neglect the Early-effect elements in the circuit of Figure 7.29 and also to consider base resistance as insignificant. The circuit of Figure 7.29 can then be simplified to the form shown in Figure 7.30. The current gain i_C/i_B in this circuit is

$$\frac{i_C}{i_B} = \frac{(1/\delta)(1 - j\omega C_{jC}/g_m)}{1 + j\omega[(C_{jE} + C_{jC})/g_m\delta + \tau_F/\delta]}$$

$$\approx \left(\frac{\tau_{BF}}{\tau_F}\right)\left[1 + j\omega\left(\tau_{BF} + \frac{(C_{jE} + C_{jC})\tau_{BF}}{g_m\tau_F}\right)\right]^{-1} \tag{7.5.14}$$

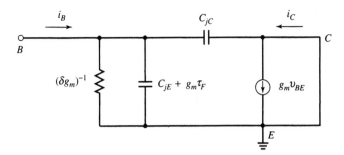

FIGURE 7.30 Equivalent circuit for obtaining the interrelationship between f_T and τ_F.

where we used Equation 7.4.4 and omitted the frequency-dependent term in the numerator. This numerator term is only important at frequencies appreciably above those at which the imaginary term in the denominator dominates.

As the frequency increases, the current gain decreases and, from Equation 7.5.14, the magnitude of the current gain is unity at a frequency f_T which is approximately given by

$$f_T = \frac{1}{2\pi\left(1 + \dfrac{C_{jE} + C_{jC}}{g_m \tau_F}\right)\tau_F} \tag{7.5.15}$$

Solving this equation for τ_F, we have

$$\tau_F = \frac{1}{2\pi f_T} - \frac{(C_{jE} + C_{jC})}{g_m} \tag{7.5.16}$$

Thus, measuring the short-circuit gain (i.e., current gain with the collector in an ac short-circuit connection) as a function of frequency provides a means of obtaining the charge-control parameter τ_F. The parameter f_T can be obtained by extrapolating a logarithmic plot of gain versus frequency to unity gain. A value for τ_F is then obtained by using f_T in Equation 7.5.16. The low-frequency gain β_F can then be used to calculate $\tau_{BF} = \beta_F \tau_F$ (Equation 7.4.4). If these measurements are repeated for the transistor under reverse-active bias, the parameters β_R and τ_{BR} can be obtained in a similar manner.

7.6 FREQUENCY LIMITS OF BIPOLAR TRANSISTORS

As we saw in the previous section, at higher frequencies the gain of a transistor decreases as the frequency increases because of the finite time needed to move charges to various parts of the transistor. For amplifying circuits, which usually operate in the forward-active-bias region, the small-signal gain is generally the most important parameter describing the transistor. The frequency dependence of the gain is described by two parameters: f_T (the short-circuit cut-off or transition frequency introduced in Sec. 7.5)—the frequency at which the gain of the transistor decreases to one when the output is short circuited—and f_{max} (the maximum oscillation frequency)—a parameter that also includes the effect of the base resistance.

Useful expressions for these parameters can be obtained using charge-control analysis; that is, by considering movement of charges into various regions of the transistor and the rearrangement of the charges. The associated delay times are found by dividing the charge that must be moved by the current that moves it. The total delay time $\tau_{EC} = 1/(2\pi f_T)$ to charge the transistor can be written as the sum of several components associated with the collector, base, and emitter regions of the transistor, respectively [12–15]

$$\frac{1}{2\pi f_T} = \tau_{EC} = \tau_C + \tau_B + \tau_E \tag{7.6.1}$$

Other components involving RC time constants corresponding to the resistance (both real and "dynamic") of each region and the capacitances of the junctions must also be added. Each of these components will be briefly discussed in the following paragraphs. Additional terms representing movement of other charges can be added for more accurate analysis.

The first-order terms to consider are the base transit time τ_B (discussed in Sec. 7.3), which describes the time for a minority carrier to move from the emitter edge of the quasi-neutral base to the collector edge, and the times associated with moving charge at the edges of the depletion regions surrounding the neutral base in response to the changing voltage (i.e., the time to charge the emitter-base and base-collector capacitances).

For a uniform base region of an npn transistor, current flows primarily by diffusion across the quasi-neutral base, and the base transit time is given by Equation 7.3.3, which is repeated here,

$$\tau_B = \frac{Q_B}{I_C} = \frac{q x_B n_{po}/2}{q D_{nB} n_{po}/x_B} = \frac{x_B^2}{2 D_{nB}} \tag{7.6.2}$$

where x_B is the width of the quasi-neutral base region and D_{nB} is the diffusion coefficient of the minority-carrier electrons in the base. In a nonuniform base, the electric field associated with the nonuniformity can greatly accelerate the electrons across the base, so that drift is the most important transport mechanism. When drift dominates, the base transit time is given by

$$\tau_B = \int_0^{x_B} \frac{dQ}{I_C} = \int_0^{x_B} \frac{q n_p(x) dx}{q n_p(x) \mu_n \mathscr{E}} = \frac{x_B}{\mu_n \mathscr{E}} \tag{7.6.3}$$

where μ_n is the mobility of the minority-carrier electrons in the base region and $\mathscr{E}$ is the electric field in the base region (assumed to be constant in this derivation).

The field can be created either by a doping gradient or by a change in the bandgap of the semiconductor across the base region. For an exponentially decreasing dopant concentration in the base, Equations 4.1.14 and 4.1.15 show that the field is constant and equal to

$$\mathscr{E} = \frac{kT}{q x_B} \ln\left(\frac{N_{BE}}{N_{BC}}\right) \tag{7.6.4}$$

where N_{BE} and N_{BC} are the dopant concentrations at the emitter and collector edges of the quasi-neutral base, respectively.

For the highest field, the dopant concentration at the emitter edge of the quasi-neutral base should be as high as possible, and the concentration at the collector edge should be as low as possible. The base doping near the emitter edge is limited by the need to keep the emitter injection efficiency high and is typically in the mid-10^{18} cm^{-3} range. The base doping near the collector must be high enough that the base-collector depletion region extends primarily into the collector, rather than into the base, to reduce the Early effect. For a typical collector doping $N_{dC} = 10^{16}$ cm^{-3}, the base doping near the collector should be at least 10^{17} cm^{-3}. Thus, the maximum electric field that can be obtained by grading the dopant concentration is limited by constraints imposed by the dopant concentrations in the adjacent emitter and collector regions. For a base width of 100 nm and the dopant concentrations described above, the maximum electric field is $\lesssim 1 \times 10^4$ V/cm. Higher electric fields can be obtained by varying the bandgap across the base region, as shown in Figure 4.2b. For a linear variation of the bandgap $\Delta E_g = 0.2$ eV across a 100-nm-wide base region, the electric field $\mathscr{E} = \Delta E_g/(qx_B)$ is 2×10^4 V/cm. The field increases significantly as the thickness of the base region decreases.

With very thin bases, the base transit time is small enough that other delay times significantly limit the speed of the transistor. First, carriers spend a finite length of time traveling through the depletion region of width x_{dC} separating the quasi-neutral base and collector regions. With proper design and bias, the carriers travel at their limiting velocity v_l (also called the *saturation* velocity v_{sat}). The transit time of carriers through the base-collector depletion region can then be written

$$\tau_C = \frac{x_{dC}}{2v_{sat}} \tag{7.6.5}$$

(The factor of two in the denominator arises from an approximation that suggests that roughly half of the charge is supplied through the base lead [16, 17].) τ_C can be reduced by increasing the donor concentration in the collector region; however, a higher collector doping decreases the breakdown voltage, requiring a trade-off between frequency response and breakdown voltage.

In our discussion of the charge-control model in Sec. 7.4, we saw that the base transit time can be viewed as the delay time to change the distribution of charge injected into the base region and is given by the charge Q_B divided by the electron current I_{nB}. An analogous delay time is associated with rearranging holes injected from the base into the emitter region. This unwanted component of base current (discussed in Sec. 6.2) degrades the frequency response of the transistor, as well as reducing its gain. The delay time associated with charge storage in the quasi-neutral emitter region can be written as the hole charge stored in the emitter divided by the charging current. Recognizing that the current is mainly electron current in a reasonably efficient transistor and that $J_{nB} \approx \beta J_{pE}$, we find

$$\tau_E = \frac{Q_{pE}}{\beta J_{pE}} = \frac{x_E^2}{2D_{pE}} \frac{GN_B}{GN_E} \tag{7.6.6}$$

where x_E is the thickness of the quasi-neutral emitter region and GN is the Gummel number defined in Equation 6.2.3. We have also assumed that $\beta \approx GN_E/GN_B$ and that the minority carrier diffusion length L_{pE} is much greater than x_E. (A more thorough analysis shows that τ_E is increased by bandgap narrowing in the emitter resulting from the heavy doping there.)

In addition to the components τ_C, τ_B, and τ_E, we also need to consider the delays associated with various RC products. We can group these products into two terms, one associated with each space-charge junction capacitance:

$$\tau_1 = r_e C_{jE} \tag{7.6.7}$$

where r_e is the "dynamic" emitter resistance $(dI_C/dV_{BE})^{-1} = kT/qI_C = 1/g_m$, and

$$\tau_2 = (r_e + R_E + R_C)C_{jC} \tag{7.6.8}$$

where R_E and R_C are the geometrical resistances of the neutral emitter and collector regions, respectively, plus the associated contact resistances. Then,

$$\frac{1}{2\pi f_T} = \tau_{EC} = \tau_C + \tau_B + \tau_E + \tau_1 + \tau_2 \tag{7.6.9}$$

At low currents, r_e is large and limits f_T. r_e decreases with increasing current until other terms dominate. At high currents, the Kirk effect widens the base, increasing τ_B and decreasing f_T; other high-current effects can also limit f_T. Some physical insight into the mechanism limiting the frequency response can be obtained by plotting $1/f_T$ as a function of $1/I_C$ to find the region where r_e limits f_T, and the region where other effects dominate.

Although f_T is widely used to describe the frequency response of a bipolar transistor, it does not consider the limitations caused by the physical base resistance R_B. A second parameter f_{max}, the *maximum oscillation frequency* (the frequency at which the power gain equals one), includes the effect of the base resistance R_B, along with the parameters in f_T. Thus, f_{max} is a more comprehensive measure of the frequency performance of a transistor.

$$f_{max} = \sqrt{\frac{f_T}{8\pi R_B C_{jC}}} \tag{7.6.10}$$

Equations 5.4.2 and 5.4.4 and the relation between minority and majority carriers $p_{no} = n_i^2/N_{dE}$ show that the number of holes injected from the base across the junction into the emitter depends on the dopant concentration in the base. For the lowest base resistance, the dopant concentration should be high. However, a high base dopant concentration increases the hole injection into the emitter and the component τ_E of the transition time τ_{EC}, as well as decreasing the emitter injection efficiency and the transistor gain.

In a heterojunction bipolar transistor, a bandgap discontinuity at the emitter-base junction reduces the need for a trade-off between base resistance and frequency response. For the same collector current ($\sim$electrons injected across the emitter-base junction), the applied voltage V_{BE} is lower for the heterojunction because of the reduced barrier to electron injection. This lower operating voltage reduces the hole current injected into the emitter and the charge stored there. The net effect is that the discontinuity in the bands impedes hole injection from the base into the neutral emitter region by a factor $\exp(\Delta E_g/kT)$ so that the emitter component τ_E of the transition time decreases from the value found in Equation 7.6.6 to

$$\tau_E = \frac{x_E^2}{2D_{pE}} \frac{GN_B}{GN_E} \exp\left(\frac{-\Delta E_g}{kT}\right) \tag{7.6.11}$$

Thus, the discontinuity in the bandgap at the emitter-base heterojunction eases the trade-off between τ_E and base resistance inherent in a homojunction bipolar transistor. The heterojunction allows us to reduce the delay time for a fixed dopant concentration in the base; alternatively, the base doping can be significantly increased to improve f_{max} while keeping the same delay time and f_T. By varying the parameters appropriately, optimum values of f_T and f_{max} can be obtained. (Note that τ_C is not significantly different for an HBT compared to a homojunction transistor. The collector depletion region extends primarily into the moderately doped collector in either case, and the doping there is still limited by the need for an adequate breakdown voltage.)

EXAMPLE *Frequency Response of Bipolar Transistors*

Using reasonable values of material parameters, calculate f_T and f_{max} both for a silicon homojunction bipolar transistor and for a heterojunction bipolar transistor with a valence-band discontinuity of 0.1 eV at the emitter base junction (using silicon materials properties for the other parameters). The gain β of the transistor should be 200, and the emitter-collector breakdown voltage BV_{CE0} should be 7 V. The transistor is biased with a collector current of 10 mA at $V_{CB} = 3$ V.

Solution   With these design criteria, the base-collector junction breakdown voltage $BV_{CB0} = BV_{CE0}\beta^{1/m}$ [where $m \sim 4$ (Equation 6.5.1)] must be about 26 V, corresponding to a collector doping N_{dC} of 3×10^{16} cm^{-3}. For a 3 V base-collector voltage, the base-collector depletion region width is 0.4μm, and from Equation 7.6.5, $\tau_C = 2.0$ ps.

To reduce the base transit time in the homojunction transistor, the dopant concentration in the base region is graded. For a graded base of width 100 nm, an electron mobility of 400 cm^2/V-s, and a field of 10 kV/cm, the base transit time τ_B is 2.5 ps. (The transit time across a uniformly doped base of the same width is almost twice as long.)

For a neutral emitter region $x_E = 0.25$ μm deep, a hole diffusion coefficient of 3 cm^2/s, and a ratio GN_E/GN_B of 200 (assuming the gain is determined by the ratio of the emitter and base Gummel numbers), the emitter delay τ_E is calculated from Equation 7.6.6 to be 0.5 ps. For a current of 10 mA, the dynamic emitter resistance r_e is 2.6 Ω.

The other parameters in the expression for f_T depend on the geometry of the transistor. As shown in the figure, let's consider a multi-emitter transistor with 5 emitter stripes, each 20 μm long and 1 μm wide for a total emitter area of 100 μm^2. With these dimensions, a reasonable value for the emitter resistance R_E (geometrical emitter resistance in the silicon plus the contact resistance) is about 1 Ω.

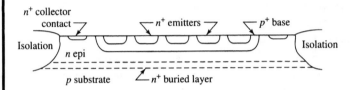

The emitter-base depletion-region capacitance C_{jE} is of the order of 200 fF (1 fF $= 10^{-15}$ F). [C_{jE} is actually significantly higher (perhaps twice as high) when the perimeter capacitance is included.] The collector-base depletion-region capacitance C_{jC} can be estimated from the collector depletion region width of 0.4 μm found above, and an area we assume to be 3 times as large as the emitter area; with this assumption, C_{jC} is 80 fF.

The collector resistance depends on the details of the layout and method of making contact to the buried layer. It's composed of the undepleted vertical portion of the epitaxial layer, the lateral resistance in the buried layer (suitably weighted for the varying distance from each location to the collector contacts), the vertical resistance between the buried layer and the collector contact, and the collector contact resistance. For our purposes, we estimate the collector resistance as 7 Ω (2 Ω for the buried layer and 5 Ω for the contact and vertical connection from the contact to the buried layer).

With these values, we can estimate f_T from Equations 7.6.1, 7.6.7, and 7.6.8. (all times in ps)

$$
\begin{aligned}
\tau_{EC} &= \tau_C + \tau_B + \tau_E + \quad (r_e + R_E + R_C)C_{jC} \quad + \quad r_e C_{jE} \\
&= 2.0 + 2.5 + 0.5 + (2.6 + 1 + 7) \times 80 \times 10^{-15} + 2.6 \times 210 \times 10^{-15} \\
&= 2.0 + 2.5 + 0.5 + \quad\quad 0.9 \quad\quad + \quad\quad 0.5 \\
\tau_{EC} &= 6.4 \text{ ps}
\end{aligned}
$$

The corresponding value of f_T is $(2\pi\tau_{EC})^{-1} = 2.5 \times 10^{10}$ Hz = 25 GHz. For this design, the dominant delays are in the collector depletion region and in the base. The former is limited by the requirement for an adequate breakdown voltage, and the latter is limited by the need for a low base resistance to achieve a high value of f_{max}.

To calculate f_{max} we also need to know the base resistance, which is composed of the *intrinsic* base resistance under the emitters, the *extrinsic* base resistance between the active regions and the base contact, and the base contact resistance. Reasonable values for these parameters with the geometry being used are 8 Ω for the intrinsic base resistance, 2 Ω for the extrinsic base resistance, and 2 Ω for the contact resistance, for a total base resistance of 12 Ω. Using this value of base resistance and the other parameters determined above in Equation 7.6.10, we find f_{max} to be 32 GHz.

For a heterojunction bipolar transistor, we can focus on improving either f_T or f_{max}, depending on which is most critical for the circuit application. As an example, to obtain some increase in f_T and a larger increase in f_{max}, we can decrease the base width by a factor of 2 to improve the base transit time and f_T while we increase the base dopant concentration by 10 times to reduce the base resistance and improve f_{max}. The heterojunction barrier at the emitter-base junction allows us to increase the base doping without injecting a significant number of holes into the emitter and degrading the emitter delay τ_E. In this example, the base Gummel number increases by a factor of 5, but the 0.1 eV barrier decreases the injection by a factor of 47, so τ_E decreases by almost 10 times to a negligible value of 0.05 ps.

With these modified numbers, $\tau_{EC} = 2.0 + 1.25 + 0.05 + 0.9 + 0.5 = 4.7$ ps, and $f_T = 34$ GHz, an appreciable improvement from the homojunction transistor value of 25 GHz. The improvement in f_{max} is even more significant. The intrinsic component of the base resistance decreases by almost 5 times while we assume the extrinsic base resistance and contact resistance are unchanged, so that the total base resistance decreases from 12 Ω to 5.6 Ω. Again, using Equation 7.6.10 with the new values of f_T and R_B, we find f_{max} to be 55 GHz, a marked increase from $f_{max} = 32$ GHz in the homojunction transistor. Different choices of parameters allow the majority of the improvement to be included in f_T, rather than in f_{max}, as in this example. ∎

7.7 BIPOLAR TRANSISTOR MODEL FOR COMPUTER SIMULATION[†]

For computer simulation of transistors, precision takes precedence over conceptual or computational simplicity. To maximize the usefulness of computer programs, models for transistors should be accurate for both large- and small-signal applications and they should also be readily characterized by parameters that are relatively easy to obtain and to verify. These requirements have been met most successfully by simulations based on the Ebers-Moll equations, which were introduced in Sec. 6.4 [18].

The starting point for our discussion is the so-called "transport version" of the Ebers-Moll equations. This consists of Equations 6.4.2 and 6.4.3, which specify transistor currents in terms of the linking current between the emitter and the collector and additional base-emitter and base-collector diode components.

In Equation 6.1.14 we derived an expression for the linking-current density J_n, which we write here in terms of total current I_n.

$$I_n = I_S \left[\exp\left(\frac{V_{BC}}{V_t}\right) - \exp\left(\frac{V_{BE}}{V_t}\right) \right] \tag{7.7.1}$$

This equation and the forms derived in Equations 6.4.7a and 6.4.7b allow the Ebers-Moll equations to be written in the following form:

$$I_C = -I_n - \frac{I_S}{\beta_R} \left[\exp\left(\frac{V_{BC}}{V_t}\right) - 1 \right]$$

$$I_E = I_n - \frac{I_S}{\beta_F} \left[\exp\left(\frac{V_{BE}}{V_t}\right) - 1 \right]$$

$$I_B = \frac{I_S}{\beta_F} \left[\exp\left(\frac{V_{BE}}{V_t}\right) - 1 \right] + \frac{I_S}{\beta_R} \left[\exp\left(\frac{V_{BC}}{V_t}\right) - 1 \right] \tag{7.7.2}$$

In this formulation, the three parameters I_S, β_F, and β_R characterize the basic Ebers-Moll relationships. Additional terms must be added to the equations in this set to represent effects not included in the Ebers-Moll model, such as the phenomena described earlier in this chapter. Gummel and Poon [19] showed relatively straightforward methods by which Equations 7.7.2 can be modified to incorporate three important second-order effects: (1) recombination in the emitter-base space-charge region at low emitter-base bias, (2) current-gain decrease under high-current conditions, and (3) effects of space-charge-layer widening (Early effect) on the linking current between the emitter and the collector. The consequences of these second-order effects lead to the deviations from

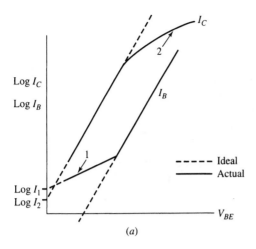

(a)

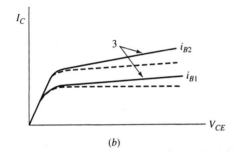

(b)

FIGURE 7.31 The results of second-order effects on bipolar transistor characteristics in the active mode. The numbers on the figures refer to the effects numbered in the text. The base current extrapolated to zero base-emitter voltage is I_1 in Equation 7.7.3.

ideal performance that are seen in the sketches in Figures 7.31a and 7.31b. Inclusion of these effects leads to the *Gummel-Poon model*, which is useful for computer simulation.

Recombination in the Space-Charge Regions. As we saw in Chapter 5, recombination in the space-charge region leads to modified diode relationships for the junction currents. These can be modeled by adding four parameters to the Ebers-Moll model to define base current in terms of a superposition of ideal and nonideal-diode components.

$$I_B = \frac{I_S}{\beta_F}\left[\exp\left(\frac{V_{BE}}{V_t}\right) - 1\right] + I_1\left[\exp\left(\frac{V_{BE}}{n_e V_t}\right) - 1\right]$$
$$+ \frac{I_S}{\beta_R}\left[\exp\left(\frac{V_{BC}}{V_t}\right) - 1\right] + I_2\left[\exp\left(\frac{V_{BC}}{n_c V_t}\right) - 1\right] \quad (7.7.3)$$

The new parameters I_1, I_2, n_e, and n_c are found in practice by measurements made at low base-emitter biases. For example, I_1 is obtained from the intercept of a plot of log I_B versus V_{BE} extrapolated to $V_{BE} = 0$ (Figure 7.31).

Early Effect and High-Level Operation. Both the high-current effect (2) and the Early effect (3) can be incorporated by modifying the value of I_S, the multiplier for the linking current between the emitter and the collector. In Sec. 6.1 we showed that J_S depends inversely on the base majority-charge density Q_B. If we rewrite Equations 6.1.13 and 6.1.15 to represent the total base charge Q_{BT} and total saturation current I_S, we have

$$I_S = J_S A_E = \frac{q^2 A_E^2 n_i^2 \tilde{D}_n}{Q_{BT}} \quad (7.7.4)$$

where

$$Q_{BT} = qA_E \int_0^{x_B} p(x) \, dx \qquad (7.7.5)$$

In the Gummel-Poon model, Q_{BT} is represented by components having a bias dependence that can be readily calculated. First, there is the "built-in" base charge Q_{BO} where

$$Q_{BO} = qA_E \int_0^{x_B} N_a(x) \, dx \qquad (7.7.6)$$

In addition to this term, there are emitter and collector charge-storage contributions (Q_{VE} and Q_{VC}) plus the charge associated with forward and reverse injection of base-minority carriers. These are all summed to represent Q_{BT} by the equation:

$$Q_{BT} = Q_{BO} + C_{jE} V_{BE} + C_{jC} V_{BC} \frac{A_E}{A_C} + \frac{Q_{BO}}{Q_{BT}} \tau_F I_S \left[\exp\left(\frac{V_{BE}}{V_t}\right) - 1 \right]$$

$$+ \frac{Q_{BO}}{Q_{BT}} \tau_R I_S \left[\exp\left(\frac{V_{BC}}{V_t}\right) - 1 \right] \qquad (7.7.7)$$

By defining several parameters, Equation 7.7.7 can be put into a more manageable format.

$$q_b \equiv \frac{Q_{BT}}{Q_{BO}}; \qquad I_{KF} \equiv \frac{Q_{BO}}{\tau_F};$$

$$I_{KR} \equiv \frac{Q_{BO}}{\tau_R}; \qquad |V_A| \equiv \frac{Q_{BO}}{C_{jC}} \frac{A_C}{A_E} \qquad (7.7.8)$$

$$|V_B| \equiv \frac{Q_{BO}}{C_{jE}}$$

The key variable, total base charge Q_{BT}, is normalized in Equation 7.7.8 to Q_{BO}, and its dimensionless counterpart is designated as q_b. The two-charge control time constants τ_F and τ_R together with Q_{BO} define "knee currents" I_{KF} and I_{KR} having a significance that will shortly become apparent. The definition of the Early voltage V_A and the equivalent Early voltage for reverse operation V_B is the same as derived in Equation 7.1.8.

In terms of the normalized parameters, Equation 7.7.7 can be written in the following form:

$$q_b = q_1 + \frac{q_2}{q_b} \qquad (7.7.9)$$

where q_1 and q_2 are auxiliary variables defined by

$$q_1 = 1 + \frac{V_{BE}}{|V_B|} + \frac{V_{BC}}{|V_A|}$$

$$q_2 = \frac{I_S}{I_{KF}} \left[\exp\left(\frac{V_{BE}}{V_t}\right) - 1 \right] + \frac{I_S}{I_{KR}} \left[\exp\left(\frac{V_{BC}}{V_t}\right) - 1 \right] \qquad (7.7.10)$$

These new variables provide a convenient indication of the significance of the second-order effects. If the Early effect is negligible, q_1 approaches unity. If high-level injection effects are not important, q_2 is small.

Thus, base-width modulation effects have been modeled through the introduction of the two Early voltages while high-level bias effects are specified through the knee currents I_{KF} and I_{KR}. The Gummel-Poon model thus requires specifying three variables I_S, β_F, and β_R for the basic Ebers-Moll model and then adds four more, I_1, I_2, n_e, and n_c to model space-charge-region recombination effects. The Ebers-Moll parameters used should be valid in the mid-bias range, where high-level effects are not significant.

Finally, base-width and majority-charge modulation are modeled by specifying a variable q_b that depends on the values of four additional variables I_{KF}, I_{KR}, V_A, and V_B. The overall model is thus specified by 11 parameters plus the temperature (to allow calculating V_t). The collected equations making up the model for an *npn* transistor are

$$I_B = \frac{I_S}{\beta_F}\left[\exp\left(\frac{V_{BE}}{V_t}\right) - 1\right] + I_1\left[\exp\left(\frac{V_{BE}}{n_e V_t}\right) - 1\right]$$

$$+ \frac{I_S}{\beta_R}\left[\exp\left(\frac{V_{BC}}{V_t}\right) - 1\right] + I_2\left[\exp\left(\frac{V_{BC}}{n_c V_t}\right) - 1\right]$$

$$I_C = \frac{I_S[\exp(V_{BE}/V_t) - \exp(V_{BC}/V_t)]}{q_b} - \frac{I_S}{\beta_R}\left[\exp\left(\frac{V_{BC}}{V_t}\right) - 1\right]$$

$$- I_2\left[\exp\left(\frac{V_{BC}}{n_c V_t}\right) - 1\right] \qquad (7.7.11)$$

$$q_b = \frac{q_1}{2} + \frac{\sqrt{q_1^2 + 4q_2}}{2}$$

$$q_1 = 1 + \frac{V_{BE}}{|V_B|} + \frac{V_{BC}}{|V_A|}$$

$$q_2 = \frac{I_S}{I_{KF}}\left[\exp\left(\frac{V_{BE}}{V_t}\right) - 1\right] + \frac{I_S}{I_{KR}}\left[\exp\left(\frac{V_{BC}}{V_t}\right) - 1\right]$$

To illustrate the validity of this equation set we consider low-level, active-mode operation for which $q_2 \approx 0$ (because the collector current in Figure 7.32 is much less than the knee current I_{KF}). For this case,

$$I_C \approx \frac{I_S \exp(V_{BE}/V_t)}{1 + V_{BC}/|V_A|} \approx I_S \exp\left(\frac{V_{BE}}{V_t}\right)\left(1 - \frac{V_{BC}}{|V_A|}\right) \qquad (7.7.12)$$

and

$$\frac{\partial I_C}{\partial V_{CB}} = \frac{I_C}{|V_A|} \qquad (7.7.13)$$

as was derived from fundamental considerations in Equation 7.1.3.

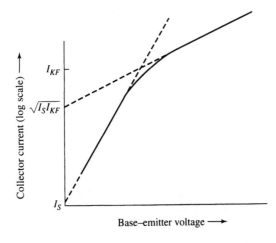

FIGURE 7.32 Logarithm of collector current versus V_{BE} in the active mode to illustrate the Gummel-Poon model for high-level effects. The asymptotes for low and high bias intersect at the "knee" current $I_C = I_{KF}$.

For operation at high-current levels, for which the Early effect is of far less consequence than high-level injection effects, we have $q_2 > q_1$. Under this condition, the normalized base charge q_b has a high-bias asymptotic behavior of the form

$$q_b = \sqrt{\frac{I_S}{I_{KF}}} \exp\left(\frac{V_{BE}}{2V_t}\right) \tag{7.7.14}$$

Thus, collector current varies as (see Figure 7.32)

$$I_C = \sqrt{I_S I_{KF}} \exp\left(\frac{V_{BE}}{2V_t}\right) \tag{7.7.15}$$

It is left as a problem to show that the intercept of the asymptotic behavior of the Gummel-Poon model at low bias with the high-bias asymptote (Equation 7.7.15) occurs at the "knee" current $I_C = I_{KF}$. The physical basis for the form represented by Equation 7.7.15 is that sufficiently high injection into the base region leads to a bias dependence for the base majority-carrier concentration, as was described in the discussion of Equation 7.2.3.

7.8 DEVICES: *pnp* TRANSISTORS

If only one type of transistor (i.e., *npn* or *pnp*) is fabricated by an integrated-circuit process, the fabrication process is simpler, with the associated lower cost and higher yield. Because of the higher mobility of electrons than holes, *npn* transistors are more frequently used than are *pnp* transistors, especially as *driver* transistors, as shown in Figure 7.33. To complete the circuit, a *load* element is needed between the collector of the *npn* transistor and the power supply V_{CC}. Although the load element can be a resistor, using a transistor as the load element is often advantageous. The transfer characteristics of a transistor used as the load element allow faster switching of the circuit. In addition, IC transistors can be more compact (and hence less costly) than resistors.

When a transistor is used as the load element, as shown in Figure 7.33c, it must be of the *complementary* type from the driver transistor—that is, it must be a *pnp* transistor. Building an optimum *pnp* transistor complicates the fabrication process significantly, but a *pnp* transistor adequate for the less-critical load element can be made without significantly modifying the IC-processing steps. Two types of *pnp* transistors for which this is possible are the *substrate pnp transistor* and the *lateral pnp transistor*.

Substrate *pnp* Transistors

One reason for the excellent performance of *npn* transistors is that transistor action takes place in a region designed to be away from the surface and uniform over the major area of the junction that is parallel to the surface plane. A *pnp* transistor with these features can be obtained using a standard planar process by making the *pnp* emitter from the *p*-type diffusion that is normally used for the *npn* base region. The structure of this *pnp* transistor is shown in cross section in Figure 7.34.

The epitaxial region serves as the *pnp* base and the grown *np* junction at the interface between the epitaxial layer and the substrate is the collector junction. Because the collector region is the substrate of the integrated circuit, it is not isolated from other *pnp* transistors formed in the same way. For this reason, these *substrate pnp transistors* can only be used in an integrated circuit when the collector junction is an ac ground as

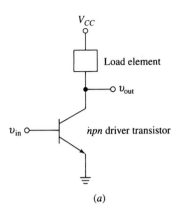

(a)

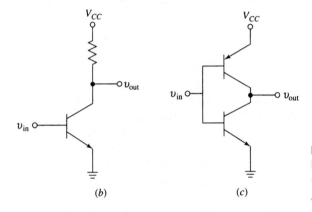

(b)   (c)

FIGURE 7.33 (a) Load element in a bipolar circuit. (b) Resistor used as load; (c) complementary *pnp* transistor used as load.

in an emitter-follower circuit. This is a valuable component for many integrated circuits, but it obviously cannot be used in every case for which a *pnp* transistor is needed. Although substrate *pnp* transistors do not have the built-in base field resulting from a graded base that is found in double-diffused *npn* transistors, they can be made with values of β up to about 100 at 1 mA. Depending on the process used, vertical *pnps* have been designed to operate in the range 1 μA to roughly 10 mA and to have values of f_T as high as 50 MHz.

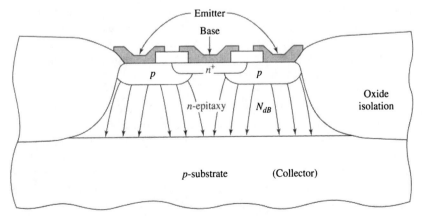

FIGURE 7.34 Cross section of a substrate *pnp* transistor. Flow lines for the linking current are sketched on the figure.

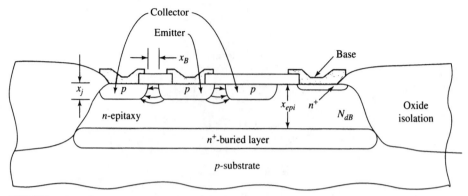

FIGURE 7.35 Cross section of a typical lateral *pnp* transistor. The diffused collector region completely surrounds the emitter. Flow lines for the linking current are sketched on the figure.

Lateral *pnp* Transistors

There is a convenient way to make a *pnp* transistor with an isolated collector using standard processing. The standard *npn* base, *p*-type diffusion can be used as both the emitter and the collector by placing two *p*-regions close together as shown in the cross-sectional view in Figure 7.35. The device formed by this construction is known as a *lateral pnp transistor* because transistor action takes place laterally—that is, parallel to the surface between the emitter and collector regions. This design sacrifices the advantages normally gained by moving transistor action away from the surface region. As a result, the performance of lateral *pnp* transistors is markedly inferior to that of standard, vertical *npn* transistors. Nevertheless, lateral *pnp* transistors are frequently used in both analog and digital integrated circuits.

The two collector regions for the lateral *pnp* transistor shown in Figure 7.35 are joined; the collector typically completely surrounds the emitter region to improve current gain in the transistor. A buried layer is also included, as shown in Figure 7.35. The buried layer improves transistor gain and frequency response in two ways: (1) by reducing base resistance and (2) by suppressing the collection of holes at the junction between the epitaxial layer and the substrate. The second improvement can be understood by noting that the built-in field resulting from the doping gradient in the buried layer repels holes approaching from the epitaxial region. Alternatively, we can consider that including a buried layer increases the base doping Q_B of the parasitic substrate *pnp* transistor. As seen in Equation 6.1.15, this reduces the loss of holes to the parasitic device, thereby improving the gain of the lateral *pnp* transistor.

Collector Current. The current linking the emitter and the collector in a lateral *pnp* transistor follows a two-dimensional path, as seen from the flow lines sketched in Figure 7.35. Because the epitaxial region is uniformly doped, the boundary value for the injected hole density in the base at a given emitter-base bias is uniform along the edge of the emitter-base junction. The carrier density gradient that causes the injected holes to diffuse toward the collector is maximum near the surface where the spacing between the junctions is minimum. Away from the surface along the emitter-base junction, the gradient in the hole density decreases slowly as the distance between the diffused *p*-regions increases. Thus, the base width for the lateral *pnp* transistor is only approximated by the spacing between the junctions at the surface (x_B in Figure 7.35), and the hole current is nonuniform along the emitter-base junction. The flow lines for the linking current are also influenced by the thickness of the epitaxial layer and by the geometry of the buried layer.

An analysis of lateral *pnp* transistors [20] considered these effects in detail and gave an empirical means of relating the linking current I_p to the flow in a one-dimensional transistor of base width x_B and emitting area $P_E x_j$ where P_E is the perimeter of the emitter and x_j is the depth of the diffused junction. The current is written

$$I_p = F \frac{q P_E x_j D_p n_i^2}{N_{dB} x_B} \exp\left(\frac{q V_{EB}}{kT}\right) \tag{7.8.1}$$

where F can be shown to depend on the two dimensionless ratios: x_{epi}/x_j and x_B/x_j [20]. As seen in Figure 7.35, x_j is the depth of the emitter and collector diffusions, x_{epi} is the thickness of the epitaxial region up to the edge of the buried layer, and x_B is the separation of the emitter and collector at the oxide-silicon interface. Curves showing $F(x_{epi}/x_j, x_B/x_j)$ from [20] are given in Figure 7.36. A typical value for both of these arguments is 2, in which case F from Figure 7.36 is roughly 1.8. Thus, the total current is nearly doubled from that predicted by a one-dimensional analysis.

The spacing x_B is determined by the separation of the emitter and collector windows in the photolithographic mask minus the lateral diffusion under the oxide toward one another by the acceptors forming the emitter and the collector. The photolithographic spacing is typically limited to about 0.3 μm, and the sideways diffusion is somewhat less than the diffusion in the vertical direction. Thus, x_B is generally of the order of 0.2 μm, although by diffusing deeper one can attain smaller spacings. By

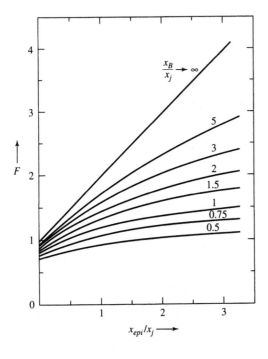

FIGURE 7.36 Geometry-dependent factor F relating actual collector current to that obtained from a one-dimensional model [20].

reducing x_B, gain and frequency response improve, but reproducibility and reliability degrade.

The epitaxial layer is typically lightly doped ($\sim 10^{16}$ donors cm^{-3}). Hence, only moderate forward bias at the emitter-base junction causes high-level injection in the base (Problem 7.9). Chou [20] has shown that high-level injection decreases the gain of lateral *pnp* transistors in three ways: (1) a lessened dependence of $p'(0)$—the excess hole density at the emitter-base junction—on emitter-base voltage, which can be described by writing an equation analogous to Equation 7.2.3 for holes, (2) an effective variation in the diffusion coefficient D_p because of the *Webster effect* (Sec. 7.3), and (3) voltage drops in series with the applied emitter-base bias because of resistances in the base and emitter.

Base Current.   When we discussed the *npn* transistor, we only needed to consider three components of base current to obtain an accurate device model. The three components were caused by injection of minority-carrier holes into the emitter (the dominant source for base current under most conditions of bias for the *npn* transistor), recombination of injected electrons in the base, and recombination of injected electrons in the emitter-base depletion region (important at low emitter currents). Currents analogous to these three components, in which the roles of electrons and holes are interchanged, are also important in lateral *pnp* transistors. However, because of the presence of the oxide-silicon interface and the lateral geometry of the device, two other significant components of base current are present in the lateral *pnp* transistor. These other components are caused by extra recombination at the oxide-silicon interface and in the neighborhood of the buried layer. An adequate representation of base current in the lateral *pnp* transistor can be obtained when these five current components are considered. These five components of base current are indicated schematically in Figure 7.37.

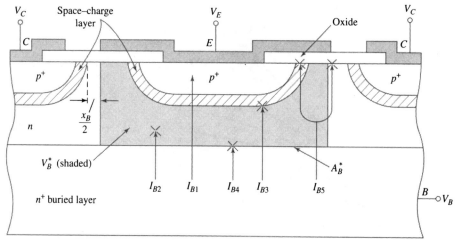

FIGURE 7.37 Schematic illustration of the physical origins of the base current in a lateral *pnp* transistor [20]. I_{B1} represents electron injection into the emitter, I_{B2} represents base recombination, I_{B3} represents space-charge region recombination away from the surface, I_{B4} accounts for recombination at the buried layer and collection by the substrate, and I_{B5} represents surface recombination of holes.

The three contributions to I_B that are analogous to those important in *npn* transistors can generally be expressed by equations similar to those already presented. An exception is the equation analogous to Equation 6.2.4 for base recombination. The two-dimensional flow pattern in the lateral *pnp* transistor complicates the expression for the volume of the base. Chou [20] has shown that base recombination can be written

$$I_{B2} = \frac{q n_i^2 [\exp(q V_{EB}/kT) - 1]}{N_{dB} \tau_p} V_B^* \tag{7.8.2}$$

where τ_p is the lifetime of holes in the epitaxial region and V_B^* (shaded region in Figure 7.37) represents the volume defined by a surface that bisects the base width x_B.

It is convenient to account for the flow of holes to the buried layer (I_{B4}) by assigning a recombination velocity s_{nn^+} to the interface between the undoped epitaxial region and the buried layer. As introduced in Equation 5.2.23, the recombination velocity is multiplied by the incident excess carrier density to express the total recombination rate at a surface. Assuming a relatively long lifetime in the epitaxial layer, we can use the boundary value for excess holes at the base-emitter junction to write

$$I_{B4} = q s_{nn^+} A_B^* \frac{n_i^2}{N_{dB}} \left[\exp\left(\frac{q V_{EB}}{kT}\right) - 1 \right] \tag{7.8.3}$$

where A_B^* is the area at the lower surface of V_B^* (Figure 7.37). In the lateral *pnp* transistor, there is really no plane having the recombination velocity s_{nn^+}. Rather, the recombination velocity is an effective parameter that can be used to account for all the hole current flowing into the buried layer. The incident holes can recombine at the interface, recombine within the buried layer, or else be collected across the junction between the buried layer and the substrate. Recombination at the oxide-silicon interface is also treated by defining a recombination velocity s_{os} and expressing current by equations similar to Equation 7.8.3.

After all the expressions for base current are written, the current itself can be calculated if appropriate values for s_{nn^+}, s_{os}, and the hole lifetimes in the bulk and space-charge regions can be obtained. In general, special test structures are needed to determine these parameters [20]. Experiments on lateral *pnp* transistors have shown that for typically low values of s_{os} ($s_{os} \sim 1-5$ cm s^{-1}), the recombination current at the oxide-silicon surface I_{B5} is generally negligible. All of the other components are significant in one or another range of useful emitter-base bias. The recombination velocity at the buried layer s_{nn^+} varies over a fairly wide range. Values between 10 and 2000 cm s^{-1} have been reported. This variation apparently arises from differing dopant densities and gradients, buried-layer geometries, and processing schedules. When s_{nn^+} is $\gtrsim 100$, the vertical hole current is generally important.

The current gain β of lateral *pnp* transistors is substantially lower than that of *npn* transistors. Values of 20 or less are common, although with care, values of β up to about 100 have been achieved. In the low microampere range, β is typically lower than one. It increases with increasing current until I_C is roughly 100 μA (for an emitter area $\sim 10^{-7}$ cm^2). At higher currents, β decreases strongly as collector current continues increasing. The increasing β at low biases corresponds to the decreasing importance of recombination in the emitter-base space-charge region. The decrease in β at higher currents accompanies the onset of high-level injection effects.

Figure 7.38 is a composite micrograph (enlarged ~ 1500 times) of a cross section through a lateral *pnp* transistor. It was made by using angle-lapping and staining techniques that delineate diffused regions in integrated circuits. Such micrographs are of great value in assessing IC process control and in locating defects.

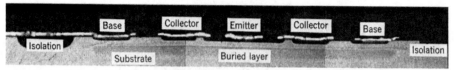

FIGURE 7.38 Stained cross section of a lateral *pnp* transistor. (*Courtesy Signetics Corporation.*)

SUMMARY

The basic theory of transistor action introduced in Chapter 6 must be augmented by considering several important physical effects to obtain the accuracy needed for design. For transistors biased in the active mode, one such effect is the variation in collector current resulting from variation in collector-base bias. The phenomenon, called the *Early effect*, can be conveniently described by the introduction of an *Early voltage*, which indicates the output current variation at a given bias point. Other effects limit the use of transistors in different bias ranges. The operation of bipolar transistors at low currents is generally limited by recombination within the base-emitter space-charge region. This recombination reduces the injection of minority carriers into the base and decreases the current gain (β_F) as the dc bias of the device is reduced.

When transistors are biased at high currents, several additional effects become important. One of these is a reduction in emitter efficiency when the base minority-carrier density begins to approach the dopant density at the edge of the base-emitter space-charge region. Another high-current effect is the modification of the space-charge configuration within the collector-base depletion region. When this high–current condition is reached, the boundaries of the quasi-neutral base region are affected; in general, the base region widens. To analyze this *Kirk effect*, we must consider in detail the doping profile of the transistor and obtain a consistent set of solutions of Poisson's equation and the equations expressing mobile space charge in terms of bias currents. Base spreading resistance also degrades transistor performance at higher currents. The base spreading resistance reduces the bias of those parts of the base-emitter junction far from the base contact because of the ohmic voltage drop associated with the base majority-carrier current in the active base region. In a given device all the high-current effects can be important simultaneously. Understanding them in detail is more important in transistor design than in the use of the transistors in circuits. For circuit design, an empirical model of high-level performance is usually all that is necessary.

An equation for the transport of injected carriers across the quasi-neutral base of a transistor (the *base transit time*) can be written for an arbitrary base doping. An understanding of this equation and of the steps in its derivation is useful in developing the charge-control representation of the transistor. Considering the effect of high-level operation on the base transit time leads to an understanding of the *Webster effect*, in which minority-carrier transit time across the base is reduced by the field associated with excess majority carriers introduced into the base.

The charge-control model of the transistor comprises a set of linear differential equations that provide an extremely useful description of the transistor for circuit-design purposes. The model is useful when considering the device from its terminals outward; that is, the charge-control model does not give accurate information about any of the distributed effects within the device. When incremental voltages are considered, the charge-control model can be used to derive a small-signal equivalent circuit (*hybrid-pi model*) that is especially helpful for the design of amplifying circuits.

Interrelations between the parameters of the charge-control model, the Ebers-Moll model, and the hybrid-pi model are useful both in gaining an understanding about the various models and in designing experiments to obtain model parameters. The *Gummel-Poon* model successfully represents most of the significant physical effects in bipolar transistors and is suitable for computer analysis of circuit behavior. In this model, several parameters are added to the basic Ebers-Moll transistor representation. The charge-control model can be used to study the high-frequency performance of bipolar transistors and to determine two useful parameters: f_T, the *cut-off* or *transition* frequency; and f_{max}, the maximum oscillation frequency. A heterojunction bipolar transistor provides additional design flexibility by reducing some of the conflicting constraints found in a homojunction bipolar transistor and can also provide an electric field that accelerates carriers across the base region. Because of the processing requirements for integrated circuits, lateral *pnp* transistors—that is, transistors in which the separation between the emitter and collector is parallel to the surface—are often employed. The performance of lateral *pnp* transistors is frequently dominated by effects different from those important in *npn* transistors.

REFERENCES

1. J. M. Early, *Proc. IRE*, **40**, 1401 (1952).

2. F. A. Lindholm and D. J. Hamilton, *Proc. IEEE*, **59**, 1377 (1971).

3. C. T. Kirk, *IRE Trans. Electron Devices*, **ED-9**, 164 (1962).

4. H. C. Poon, H. K. Gummel, and D. L. Scharfetter, *IEEE Trans. Electron Devices*, **ED-16**, 455 (1969): Reprinted by permission.

5. P. E. Gray, D. DeWitt, A. R. Boothroyd, and J. F. Gibbons, *Physical Electronics and Models*, SEEC Volume II, Wiley, New York, 1964.

6. O. Manck, H. H. Heimeier, and W. L. Engl, *IEEE Trans. Electron Devices*, **ED-21**, 403 (1974).

7. W. M. Webster, *Proc. IRE*, **42**, 914 (1954).

8. R. Beaufoy and J. J. Sparkes, *Automat. Teleph. Elect. J.*, **13**, Reprint 112 (1957).

9. R. N. Noyce et al, *Electronics*, July 21, 1969, p. 74.

10. P. E. Gray and C. L. Searle, *Electronic Principles: Physics, Models and Circuits*, Wiley, New York, 1969.

11. A. Sedra and K. Smith, *Microelectronic Circuits*, Fourth Edition, Oxford Univ. Press, 1998.

12. H. D. Barber, *Can. J. Phys.* **63**, 683 (1985).

13. L. H. Camnitz and N. Moll in *Compound Semiconductor Transistors: Physics and Technology*, ed. S. Tiwari, IEEE Press, New York, 1993, p. 21.

14. D. J. Roulston and F. Hebert, *IEEE Electron Device Lett.*, **EDL-7**, 461 (1986).

15. S.-Y. Chiang, D. Pettengill, and P. vande Voorde, IEEE 1990 Bipolar Circuits and Technology Meeting, paper 8.1, p. 172.

16. R. G. Meyer and R. S. Muller, *IEEE Trans. Electron Devices*, **ED-34**, 450 (1987).

17. S. E. Laux and W. Lee, *IEEE Elect. Dev. Lett.*, **11**, 174 (1990).

18. C. C. McAndrew, J. A Seitchik, D. F. Bowers, M. Dunn, M. Foisy, I. Getreu, M. McSwain, S. Moinian, J. Parker, D. J. Roulston, M. Schroter, P. van Wijnen, and L. F. Wagner, *IEEE J. Solid-State Circuits*, **31**, 1476 (1996).

19. H. K. Gummel and H. C. Poon, *Bell Syst. Tech. J.*, **49**, 827 (1970).

20. S. Chou, *Solid-State Electron.*, **14**, 811 (1971).

21. A. S. Grove, *Physics and Technology of Semiconductor Devices*, Wiley, New York, 1967, pp. 229, 240.

22. J. Logan, *Bell System Tech. J.*, **50**, 1105 (April 1971).

23. A. Bar-Lev, *Semiconductor and Electronic Devices*, Prentice–Hall International, Englewood Cliffs, N.J., 1984.

24. I. Getreu, *Modeling the Bipolar Transistor*, Tektronix, Inc., Beaverton, OR 97077, 1976.

PROBLEMS

7.1 Show that the relationship for the Early voltage V_A derived in Equation 7.1.3 properly specifies $\partial I_C / \partial V_{CB}$ for the prototype transistor by considering collector current to be carried purely by diffusion between the emitter and the collector.

7.2 Calculate the value of V_A at $V_{CB} = 0$ for a prototype transistor in which the base is doped with 10^{17} atoms cm^{-3} of boron and the collector is doped with 10^{16} atoms cm^{-3} of phosphorus if the neutral base width is 2.5 μm. Consider the junction to be a step between the two concentrations. What is the indicated slope of I_C resulting from the Early effect?

7.3* Compare the Early voltages for the two transistors considered in Problem 6.3. Assume that $\partial x_B / \partial V_{CB}$ is approximately equal for both devices.

7.4† Consider an *npn* transistor in which the base doping varies *linearly* across the quasi-neutral region, going from 10^{17} cm^{-3} at the emitter side to 10^{16} at the collector side. The base width is 1 μm and both emitter and collector have $N_d = 10^{19}$ cm^{-3}.

(a) Sketch the minority-carrier densities (i) at thermal equilibrium, and (ii) under low-level, active-bias conditions.

(b) Sketch the "built-in" electric field in the base.

(c) Derive an expression for the field and give its maximum value.

(d) Determine the approximate ratio between the two Early voltages (i) V_A encountered in forward-active bias, and (ii) V_B that characterizes reverse-active bias.

7.5 Using Equation 7.1.4, discuss qualitatively the dependence of V_A on collector-base bias for (a) the prototype transistor, and (b) an IC amplifying transistor.

7.6 One criterion for the onset of current crowding is a drop in the transverse base voltage exceeding kT/q. Estimate the corresponding collector current for a transistor that has $\beta_F = 50$, the impurity distribution shown in Figure P7.6, and a stripe geometry with $Z_E = 0.1$ cm and $Y_E = 2 \times 10^{-3}$ cm (see Figure 6.3). It can be shown that, for the stripe geometry, the base spreading resistance is

$$R_B = \frac{\bar{\rho}_B Y_E}{6 x_B Z_E}$$

The average resistivity of the base region $\bar{\rho}_B$ is given by

$$\bar{\rho}_B \simeq \frac{A_E x_B}{\mu_p Q_{BO}}$$

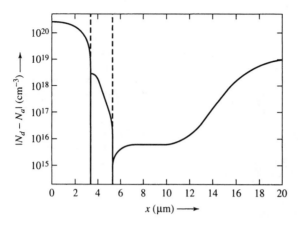

FIGURE P7.6

where μ_p is the mobility that corresponds to $N_a = Q_{BO}/qA_E x_B$ [21].

7.7* Consider a *pn* junction having the properties described in Problem 5.12. If the junction is to be used as the emitter-base junction of a *pnp* transistor, what is the maximum value of β_F that can be attained?

7.8† Reconsider Problems 6.17 and 6.18 in which β_F was to be found in a transistor and degradation due to radiation was considered. In the present case, assume that the transistor is to be used at a voltage such that $J_t/J_r = 10$ in Equation 5.3.25.

(a) Compute β_F for this case.

(b) If lifetime in the space-charge region varies in the same way as base minority lifetime resulting from the radiation damage, calculate the information required in Problem 6.18.

7.9* Derive an expression for and plot β_F/β_{FO} versus collector current for an *npn* transistor in which high-level injection effects at the emitter edge of the base (as described in Sec. 7.2) degrade current gain. The current gain β_{FO} is the value obtained in the middle bias range. Consider a prototype transistor with $A_E = 5 \times 10^{-5}$ cm^2, $N_a = 5 \times 10^{16}$ cm^{-3}, $D_n = 20$ cm^2 s^{-1}, and $x_B = 5$ μm. Consider that the base current is dominated by reverse injection into the emitter. Under this condition, base current remains proportional to exp (V_{BE}/V_t). Thus, beta falls off as $n(0)$ begins to depart from its mid-range dependence on V_{BE}.

7.10† Consider the extreme case of the Kirk effect (described in Sec. 7.2) in which the collector space-charge region is moved to the edge of the n^+ buried layer. Take the case that the negative space charge is completely due to electrons in transit and the positive space charge is provided by a very high donor concentration that starts abruptly at the edge of the buried layer. Calculate the field and space-charge-layer

width as functions of current assuming that the electron velocity v is given by (a) $v = \mu\mathscr{E}$ or (b) $v = v_l$. These are two cases of space-charge-limited currents in solids.

7.11 (a) Show that R_B in Equation 7.2.13 can be obtained from a plot of I_C versus V_{BE} such as in Figure 7.10. In particular, if I_{CA} is the *actual* collector current and I_{CI} is the collector current that would flow in the absence of base resistance (the *ideal* current), show that

$$R_B = \frac{V_t \beta_F}{I_{CA}} \ln\left(\frac{I_{CI}}{I_{CA}}\right)$$

(b) Use the following data for the transistor of Figure 7.10 to plot R_B versus I_{CA}. $I_S = 3 \times 10^{-14}$ A, $V_t = 0.0252$ V, $\beta_F = 100$ (assumed constant).

$V_{BE}(V)$	$I_{CA}(mA)$
0.70	11.25
0.72	22.4
0.75	56.2
0.80	200

7.12† If the total resistance R in the transistor shown in Figure 7.12 is 150 Ω (0.75 square at 200 Ω/square), and the external resistor is 20 Ω, use the network in the inset of the figure to investigate the crowding of current due to base resistance. Assume that each of the transistor segments has $\frac{1}{8}$ of the I_S given in Problem 7.11, but that each has $\beta_F = 100$.

(a) Assume currents of 1 and 10 mA in the innermost transistor segment. Work through the network to calculate the total current flowing in the overall transistor.

(b) Use the results of Problem 7.11 to calculate R_B and the total applied bias between the external base and emitter leads.

(c) Assuming that base resistance is the only important high-current effect, what value does R_B approach at the highest current levels?

(d) Show that R_B approaches the sum of the external resistance plus $11R/128$ at very low currents.

(You will need to maintain about four significant figures for V_{BE} to derive accurate results.)

7.13 Prove the statements made in the paragraph containing Equation 7.3.3.

7.14[†] Use Equation 7.3.8 to calculate τ_B for a transistor having a constant (built-in) base field resulting from an exponential variation in base doping. Specifically, take the case that the built-in voltage drop between $x = 0$ at the emitter side and $x = x_B$ at the collector side of the base is κV_t where V_t is the thermal voltage kT/q.

(a) Show that $\nu = \kappa^2/(\kappa - 1 + e^{-\kappa})$ and that ν behaves properly as $\kappa \to 0$.

(b) Calculate τ_B for $\kappa = 20$, $x_B = 0.5$ μm, $D_n = 20$ cm^2 s^{-1}, and explain why this value of κ is about as large as can be realized practically.

7.15[†] Consider the influence of the Kirk-effect results sketched in Figure 7.8 on τ_B. Use Equation 7.3.8 to make a semiquantitative plot of τ_B versus collector current. The indicated fall-off in high-frequency performance is an important consequence of the Kirk effect [3].

7.16 Find an expression for τ_{BF} (as introduced in Equation 7.4.3) in terms of transistor geometry and minority-carrier lifetime in the case of a prototype transistor for which the emitter efficiency is given by Equation 6.2.20. Formulate the conditions required in order that τ_{BF} becomes nearly equal to τ_n, the electron lifetime in the transistor base.

7.17 Show that the set of charge-control equations for a *pnp* transistor is of the form:

$$i_C = -\frac{Q_F}{\tau_F} + \frac{dQ_R}{dt} + Q_R\left(\frac{1}{\tau_R} + \frac{1}{\tau_{BR}}\right) + \frac{dQ_{VC}}{dt}$$

$$i_E = \frac{dQ_F}{dt} + Q_F\left(\frac{1}{\tau_F} + \frac{1}{\tau_{BF}}\right) - \frac{Q_R}{\tau_R} + \frac{dQ_{VE}}{dt}$$

$$i_B = -\frac{dQ_F}{dt} - \frac{Q_F}{\tau_{BF}} - \frac{dQ_R}{dt} - \frac{Q_R}{\tau_{BR}}$$
$$- \frac{dQ_{VE}}{dt} - \frac{dQ_{VC}}{dt}$$

7.18 Consider the physical nature of the space charges represented by Q_{VE} and Q_{VC} to argue that the signs for the derivatives of these quantities are correct in Equation 7.4.11 and in the equation set of the preceding problem (for a *pnp* transistor). *Hint: Consider the sign of the charges supplying the base current and*

the sign of the Q_V terms. Reference to Figure 7.14 may be helpful.

7.19 Carry through the steps necessary to show the validity of Equation 7.4.10.

7.20 Analyze the circuit shown in Figure 7.18 (which corresponds to Equation 7.4.11) under dc conditions. Show that if we define

$$\alpha_F = \frac{\tau_{BF}}{\tau_F + \tau_{BF}}$$

$$\alpha_R = \frac{\tau_{BR}}{\tau_R + \tau_{BR}}$$

and

$$I_{ES} = Q_{FO}\left(\frac{1}{\tau_F} + \frac{1}{\tau_{BF}}\right)$$

$$I_{CS} = Q_{RO}\left(\frac{1}{\tau_R} + \frac{1}{\tau_{BR}}\right)$$

Equations 7.4.11 reduce to the Ebers-Moll equations (Equations 6.4.10).

7.21[†] Show that τ_{SLOW} in Equation 7.4.15 can be expressed by

$$\tau_{SLOW} = \frac{(\beta_F + 1)\tau_{BR} + (\beta_R + 1)\tau_{BF}}{1 + \beta_F + \beta_R}$$

This form for τ_{SLOW} is frequently used in practice.

7.22 (a) Show that Q_{FO} as defined in Equation 7.4.2 is linearly related to $n_{po}(0)$ and write its value for a prototype transistor. *Hint: Consider Equations 7.1.1 and 7.4.1*

(b) Show that V_{CE} for a saturated transistor is given by

$$V_{CE} = \frac{kT}{q}\ln\left[\frac{Q_{RO}(Q_F + Q_{FO})}{Q_{FO}(Q_R + Q_{RO})}\right]$$

7.23* A transistor has the following charge-control parameters:

$$\tau_F = 12 \text{ ns}, \quad \beta_F = 100, \quad \tau_R = 36 \text{ ns}, \quad \beta_R = 10$$

(a) Evaluate the forward stored charge Q_F if the collector current $I_C = 2$ mA and the transistor operates just on the boundary of the saturation region with $V_{CB} = 0$.

(b) Determine the base-charge components Q_F and Q_R if the base current is now changed to $I_B = 0.5$ mA with I_C remaining at 2 mA.

(c) Compare the charge stored for cases (a) and (b).

7.24 For a calculated variation in charge with voltage to be modeled by a capacitance, not only must the variation be activated by the terminal voltage associated with the charge, but also the sign of the charge variation must be consistent with that of a capacitor. Show that this condition is fulfilled when the variation in Q_F (as calculated in Equation 7.5.6) is modeled in the capacitor ηC_D of Figure 7.27.

7.25 Show by considering the signs of v_{BE} and the generator that the equivalent circuit shown in Figure 7.27 is valid either for *npn* or for *pnp* transistors.

7.26 Discuss the limitation of the small-signal equivalent circuit to variations in base-emitter voltage that are less than V_t. What steps in the derivation of the circuit require this limitation?

7.27 Carry out the steps necessary to reduce the equivalent circuit of Figure 7.25 to the form sketched in Figure 7.27. (Make use of the fact that $\delta \ll 1$ and $\eta \ll 1$.)

7.28 An increment of emitter current dI_E is applied to a transistor under active bias with quiescent current I_E and base-emitter voltage V_{BE}. Assume that the simplified hybrid-pi circuit of Figure 7.30 (with C_{jC} negligibly small) is applicable. Show that a time

$$\tau_E \simeq \tau_F + (C_{jE}V_t/I_C)$$

is required in order to bring the base-emitter voltage to a new steady-state value. This is one of the delays that affects the measured f_T in a transistor.

7.29* (a) Obtain the hybrid-pi representations for the two transistors described in Problems 6.1 and 6.3. Take $I_C = 2$ mA and $\phi_i + V_{CB} = 10$V for both transistors. In the uniform base transistor, take $\tau_n = 100$ ns.

(b) Comment on the relative performance of the two transistors when they are used as small-signal amplifiers.

7.30[†] Consider an *npn* transistor biased in the active mode and illuminated in the collector-base space-charge

region. The radiation produces hole-electron pairs at a rate r pairs per unit time. Consider r to be a sinusoidal function of time.

(a) Briefly indicate the flow of the generated carriers.

(b) Indicate how the effects of radiation might be incorporated in the low-frequency hybrid-pi circuit (Figure 7.28) (take the Early effect to be negligible).

(c) Use the circuit of part b to compute i_C produced by the radiation if v_{BE} is zero (base-emitter ac shorted).

(d) Repeat part c if the base-emitter is ac open-circuited (i.e., $i_B = 0$).

7.31 Show that a symmetrical equivalent circuit diagram like that in Figure P7.31 represents the "transport version" of the Ebers-Moll equations (Equations 6.4.2 and 6.4.3), provided that we define

$$I_{AA} = I'_{ES}\left[\exp\left(\frac{qV_{BE}}{kT}\right) - 1\right]$$

$$I_{BB} = I'_{CS}\left[\exp\left(\frac{qV_{BC}}{kT}\right) - 1\right]$$

where I'_{ES} and I'_{CS} differ from I_{ES} and I_{CS} in the original Ebers–Moll equations. Advantages for this representation over that of the more conventional Ebers–Moll equivalent circuit (Figure 6.12) are discussed by J. Logan [22].

7.32[†] Derive Equations 7.7.2.

7.33[†] Derive the solution for q_b in Equation 7.7.9.

7.34[†] Using the Gummel-Poon equations derived and discussed in Sec. 7.7, consider active-bias and low-level conditions to show that a decrease in $\beta_F(I_C/I_B)$ is predicted as current decreases. Find the variation in β_F as a function of I_C at low levels. Show that the parameter n_e can be obtained from the behavior noted and plot a reasonable sketch for β_F in a transistor having $\beta_F = 100$ at currents above 0.9 mA with a dropoff asymptote that intersects the midrange β_F at $I_C = 0.5$ mA. How might the curve of β_F be used to obtain a value for I_1?

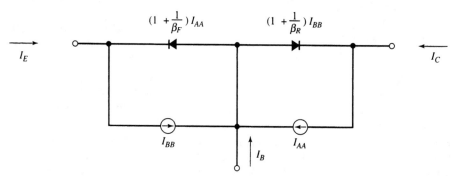

FIGURE P7.31

7.35[†] (a) Show that the intercept of the two asymptotic forms expressing collector current as a function of base-emitter bias in the Gummel-Poon model occurs at the "knee" current $I_C = I_{KF}$ (Figure 7.32).

(b) Show that the Gummel-Poon model predicts that, in the high-current region, β_F becomes proportional to I_C^{-1}.

7.36 Using reasonable values of material parameters, calculate the base delay time τ_B: (1) for a silicon bipolar transistor with a uniformly doped base, (2) for a silicon bipolar transistor with an exponentially graded dopant concentration in the base, and (3) for a heterojunction bipolar transistor with a bandgap change of 0.2 eV across the base region (using silicon values for the other parameters).

PROPERTIES OF THE METAL-OXIDE-SILICON SYSTEM

In **our** discussion of the electronics and technology of materials in Chapters 1 and 2, we saw that silicon has become the dominant semiconductor material because of its special properties. The chief reason for its prominence among competitive semiconductor materials is the ability to produce by compatible technologies both a semiconductor (single-crystal-silicon) and an insulator (amorphous silicon dioxide SiO_2) that have superb electrical and mechanical properties. This ability made planar technology possible, and, in turn, led to the reliable production of large-scale integrated circuits. Properties of the oxide-silicon system are fundamental to the performance of integrated-circuit devices. Knowledge of these properties and their control has been responsible for many advances in device design and performance. Despite years of work in the area, research on the oxide-silicon system is still ongoing, and a better understanding is continually being developed. A useful starting point for the consideration of the oxide-silicon system is the construction of an energy-band diagram. The usefulness of the band diagram is considerably enhanced by adding to the figure a third material: a metal layer above the oxide. The metal provides an electrode at which the voltage can be fixed, and the resulting three-component, metal-oxide-silicon (MOS) system is useful in understanding several important integrated-circuit structures: most notably the metal-oxide-silicon field-effect transistor (MOSFET), sometimes called the insulated-gate field-effect transistor (IGFET).*

The MOSFET is a device of such major importance that the following two chapters are devoted to it. The development of dense, very-large-scale integrated circuits that can accommodate tens of millions of MOSFETs on a single chip has made the MOSFET more important than the bipolar transistor among the devices used in integrated circuits. The discussion of the MOS system in this chapter will greatly help us understand the physical electronics that underlie operation of the MOSFET.

Acceptor-doped (*p*-type) silicon is considered throughout this chapter for uniformity in the presentation and because of its relevance to the dominant type of MOSFET. The oxide-silicon system with donor-doped (*n*-type) silicon is discussed as a natural extension and is also the subject of several problems. The results for both systems are included in a summary of important equations at the end of the chapter.

Throughout the discussion that follows, we describe the characteristics of the silicon dioxide-silicon system. However, as MOS technology develops and the oxides required for high-performance ICs become thinner, insulating materials other than pure silicon dioxide are being considered. For example, combinations of silicon dioxide and silicon nitride, as well as insulators with high *permittivities* (or *dielectric constants*), such as tantalum pentoxide, can be useful, despite having a less perfect insulator-silicon interface than that between silicon dioxide and silicon. Most of our remarks about the oxide-silicon system can be applied to these more complex systems.

8.1 THE IDEAL MOS STRUCTURE

To derive an energy-band diagram for the metal-oxide-silicon system, we apply the basic principles that we previously used in studying systems of metals and semiconductors, and of *p*-type and *n*-type silicon. Our starting point is recognition that the Fermi energy is constant throughout a system at thermal equilibrium. In this case of a three-component system,

---

* More precisely, IGFET denotes a broader class of devices than those made from the metal-oxide-silicon system, although in most cases IGFET and MOSFET are used interchangeably. "MOSFET" can also be interpreted more generally as a metal-oxide-*semiconductor* field-effect transistor.

the Fermi level is constant throughout all three materials: the metal, the oxide, and the silicon. For the present, we idealize the MOS system by considering that both the oxide and the interfaces between the materials are free of charges.

Thermal-Equilibrium Energy-Band Diagram

The Fermi levels in the different materials are equalized by the transfer of negative charge from materials with higher Fermi levels (smaller work functions) across the interfaces to materials with lower Fermi levels (higher work functions). As discussed in Chapter 3, the vacuum level is a continuous function of position. Therefore, knowing the electron affinities of the insulator and semiconductor and the work functions of the semiconductor and metal permits the construction of a unique energy-band diagram. In Figure 8.1, aluminum (work function = 4.1 V), silicon dioxide (electron affinity ~0.95 V), and uniformly doped p-type silicon (electron affinity = 4.05 V) having a work function of 4.9 V are considered. The vacuum level is designated E_0, and the various energies when the materials are separated are indicated on the figure.

When the materials are brought together, negative charge is transferred from the aluminum to the silicon to bring the system to equilibrium because the work function of the metal is 0.8 V less than the work function of the silicon. The insulator is incapable of transferring charge because it ideally possesses zero mobile charge, so a voltage drop appears across it because of the charge stored on either side. There is a thin sheet of positive charge (a plane of charge in the ideal case of a perfect conductor) at the surface of the metal and negatively charged acceptors extending into the semiconductor from its

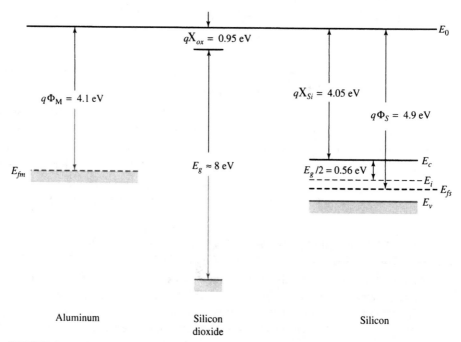

FIGURE 8.1 Energy levels in three separated materials that form an MOS system: aluminum, thermally grown silicon dioxide, and p-type silicon containing $N_a \approx 1.1 \times 10^{15}$ cm^{-3}. (Note that there is considerable variation in tabulated values for work functions and electron affinities. Commonly used values are indicated.)

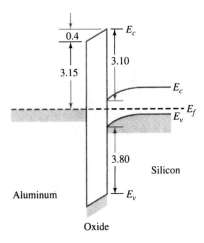

FIGURE 8.2 Energy-band diagram at thermal equilibrium for an ideal MOS system composed of the materials indicated in Figure 8.1. The oxide and Si–SiO$_2$ interface are assumed to be ideal and free of charges.

surface. The voltage corresponding to the energy difference divides across the oxide and the space-charge region near the surface of the silicon.

It may be puzzling to consider charge transfer in the MOS system with reference to an oxide that is an almost perfect insulating material. In fact, if the system being considered were fabricated without any path for charge flow between the metal and the silicon other than through an ideal oxide, the materials could exist in a condition of non-equilibrium (i.e., with unequal Fermi levels) for long periods.* However, nearly every MOS system of interest has some alternative path for the transfer of charge that is much more transmissive to charge flow than is the oxide; e.g., the aluminum gate electrode and the silicon substrate may be connected together, or an ohmic conducting path may exist between them. Hence, we assume that thermal equilibrium exists between the metal and the semiconductor. With these assumptions the band diagram for the MOS system formed with the materials of Figure 8.1 is shown in Figure 8.2.

In terms of charge flow, the diagram of Figure 8.2 results from the transfer of holes from the p-type silicon to an ohmic contact (not shown) where the holes are freely converted into electrons. These electrons are then supplied from the aluminum of the MOS system. The charge imbalance in the aluminum because of the transfer of these electrons leaves a sheet of positive charge at the metal surface near the oxide (as close as it can get to the equal quantity of negative charge stored near the silicon surface). The resulting drop of 0.4 V across the oxide shown in Figure 8.2 applies to the ideal MOS system we are considering. If charges are introduced intentionally or uninten-tionally into the bulk of the oxide or at the interfaces between the various materials, as is the case for virtually all real MOS systems, the value of the energy difference be-tween the metal and the surface of the semiconductor changes, as we will see shortly in an example.

The band diagram of Figure 8.2, in which the p-type silicon is depleted of holes at its surface, can be compared to the diagram for gold and n-type silicon sketched in Figures 3.3 and 3.4. The space charge near the surface of the silicon consists of ionized dopant atoms in both cases, and the energy bands in the silicon are curved, as found by solving Poisson's equation. The presence of the oxide shown in Figure 8.2 reduces the

---

* This nonequilibrium charge retention is often used in the nonvolatile memory devices we will discuss in Chapter 10.

surface field by separating the surface charges, but there is otherwise no important difference in the band diagram within the silicon itself when the surface is depleted.

In terms of electrical characteristics, there are, of course, major differences in the two situations. In the MOS system, electrons cannot pass freely in either direction across the oxide. This distinction is evident on the band diagram by the presence of abrupt steps in the energy of the allowed states for electrons in the MOS system. From Figure 8.2 we see that electrons in the allowed states in the metal at the Fermi level are 3.15 eV lower in energy than the allowed states in the conduction band in the silicon dioxide.* Because of this separation of the allowed energies for free electrons, we can characterize the metal-oxide interface by a 3.15-eV barrier for electron emission from the metal into the oxide. By the same reasoning, a barrier of 3.10 eV exists at the oxide-silicon interface for electrons in the conduction band of the silicon, as does a barrier of 4.20 eV for electrons in the valence band of the silicon. The validity of these barrier heights has been confirmed directly by measurements of the photon energies required to inject electrons from the metal and the semiconductor into silicon dioxide.

EXAMPLE   *MOS Energy-Band Diagram*

What is the thickness of the silicon-dioxide layer for the MOS energy-band diagram sketched in Figure 8.2?

Solution   Figure 8.2 is the energy-band diagram of the MOS system at thermal equilibrium for the materials shown in Figure 8.1. A voltage difference exists between the metal and the silicon (because of the differing work functions) equal to Φ_{MS} which is (4.9–4.1) or 0.8 V. Because Figure 8.2 indicates that 0.4 V is dropped across the oxide, the voltage drop near the silicon surface is also 0.4 V.

Because there is no charge in the SiO_2, the oxide field $\mathscr{E}_{ox}$ is constant and the voltage across the oxide V_{ox} is simply $\mathscr{E}_{ox} \times x_{ox}$, where x_{ox} is the oxide thickness. Therefore, x_{ox} can be found once $\mathscr{E}_{ox}$ is known.

Because we assumed the oxide-silicon interface to be charge-free, the electric displacement D perpendicular to the interface is continuous and the field in the oxide is related to the field $\mathscr{E}_{s0}$ at the surface of the silicon by the equation

$$\mathscr{E}_{ox} = \frac{\epsilon_s}{\epsilon_{ox}} \mathscr{E}_{s0}$$

A depletion layer exists at the surface of the silicon with a constant charge density qN_a that extends a distance x_d away from the Si–SiO$_2$ interface. In this region, the field and voltage dependences on x are the same as in the Schottky diode that we considered in Section 3.2. By an analysis similar to that used to obtain Equations 3.2.2 and 3.2.3, we can write expressions for the surface field $\mathscr{E}_{s0}$ in the silicon and for the depletion-layer width x_d.

$$\mathscr{E}_{s0} = \frac{qN_a x_d}{\epsilon_s}$$

and

$$x_d = \left[\frac{2\phi_s \epsilon_s}{qN_a} \right]^{1/2}$$

The acceptor density N_a in the silicon can be obtained by using Equation 1.1.27 with $p = N_a$ and $(E_i - E_f)$ as given in Figure 8.1. From Figure 8.1, we have

$$(E_i - E_f) = (4.9 - 4.05 - 0.56) = 0.29 \text{ eV}$$

---

* As mentioned previously, there are theoretical objections to describing an amorphous material like silicon dioxide in terms of energy bands. However, within the context of our present discussion, the concept is useful.

and, from Equation 1.1.27,

$$p = N_a = n_i \exp[(E_i - E_f)/kT] = 1.1 \times 10^{15}\,\text{cm}^{-3}$$

Using the value $\phi_s = 0.4$ V, we calculate $x_d = 685$ nm, and $\mathscr{E}_{s0} = 1.17 \times 10^4\,\text{V cm}^{-1}$. Therefore,

$$\mathscr{E}_{ox} = 3.505 \times 10^4\,\text{V cm}^{-1} \quad \text{and} \quad x_{ox} = \frac{V_{ox}}{\mathscr{E}_{ox}} = 114\,\text{nm}. \qquad \blacksquare$$

Polysilicon and Metals as Gate-Electrode Materials

Because the surface condition of the silicon can be controlled by the metal electrode, the metal layer is usually called the *gate*, and the voltage on the metal is denoted by V_G. Aluminum was the predominant metal gate material for MOS technology until heavily doped polycrystalline silicon (*polysilicon*) was introduced in the late 1970s as a better alternative. Extremely high purity silicon can be conveniently deposited over the gate oxide by chemical vapor deposition (CVD), as discussed in Chapter 2, immediately after the gate oxide is grown, thus protecting the oxide from contamination in subsequent fabrication steps. Because the silicon is deposited over amorphous silicon dioxide, it is a polycrystalline film typically consisting of sub-micrometer-size crystallites. A very high concentration of either *n*-type or *p*-type dopant is subsequently introduced into the polysilicon to make it sufficiently conducting to behave electrically like a metal. One major advantage of polysilicon is its ability to withstand high-temperature thermal treatments. It can be deposited and defined, and then its edges can be used to precisely position (*self-align*) the end of the source and drain regions of an MOS transistor, reducing unwanted *overlap* capacitance and its variability.

As MOS technology entered the sub-micrometer regime in the 1990s, some of the deficiencies of polysilicon became troublesome; its primary drawback is the high resistance of even heavily doped polysilicon compared to that of a metal. Other materials, especially refractory metals such as tungsten, were explored as alternatives. However, none of these polysilicon substitutes were found to be production worthy; their use caused contamination, and they were technologically incompatibile with other process materials. Attempts to find attractive metals to replace polysilicon as the gate electrode are continuing.

The Flat-Band Voltage

We saw that for the idealized MOS system at thermal equilibrium, the metal and the semiconductor form two plates of a capacitor. The capacitor is charged to a voltage corresponding to the difference between the metal and semiconductor work functions. Applying a bias voltage between the metal and the silicon causes the system to depart from thermal equilibrium and changes the amount of charge stored on the capacitor. For the case considered in Figure 8.2, a negative voltage applied to the metal with respect to the silicon opposes the built-in voltage on the capacitor and tends to reduce the charge stored on the capacitor plates below its equilibrium value. We can think of the negative gate voltage as pulling positive holes toward the surface of the semiconductor to neutralize some of the negatively charged acceptors.

At one particular value, the applied voltage exactly compensates the difference in the work functions of the metal and the semiconductor. The stored charge on the MOS capacitor is then reduced to zero, and the fields in the oxide and the semiconductor vanish. In this situation, the energy bands in the silicon are level or flat in the surface region as

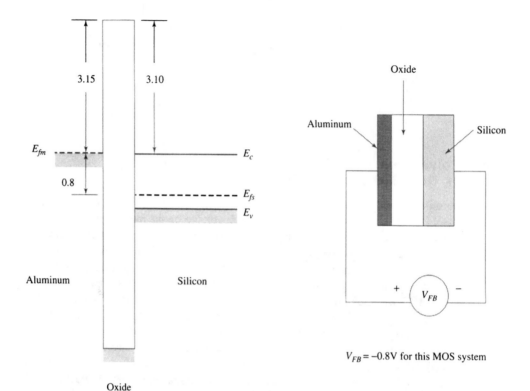

Oxide

FIGURE 8.3 Energy-band diagram of the MOS system of Figure 8.2 under flat-band conditions. An external voltage equal to $V_{FB} = -0.8$ V is applied between the metal and the silicon to achieve this condition, which does not correspond to thermal equilibrium.

well as in the bulk (Figure 8.3). Because of the effect of the applied voltage on the band diagram, the voltage that produces flat energy bands in the silicon is called the *flat-band voltage* and usually designated V_{FB}. The flat-band voltage varies with the dopant density in the silicon, as well as with the specific metal used for the MOS system. Note that the MOS system is *not* at thermal equilibrium at the flat-band condition; therefore, the Fermi energy is different in the metal and in the silicon (Figure 8.3). The voltage applied to the ideal MOS system to bring it to the flat-band condition equals the difference in the work functions of the metal and the silicon.

$$V_{FB} = \Phi_M - \Phi_S = \Phi_{MS} \tag{8.1.1}$$

The flat-band voltages for ideal MOS systems with different gate materials are shown in Table 8.1.

TABLE 8.1 Work Functions (Φ_M and Φ_S) and Flat-Band Voltages for Commonly Used Gate Materials and p-Type Silicon with $N_a = 1.1 \times 10^{15}$ cm^{-3}.

Gate material parameter	Aluminum	n^+ polysilicon	p^+ polysilicon	Tungsten
Φ_M (V)	4.1	4.05	5.17	4.61
Φ_S (V)	4.9	4.9	4.9	4.9
V_{FB} (V)	−0.8	−0.85	0.27	−0.29

8.2 ANALYSIS OF THE IDEAL MOS STRUCTURE

Qualitative Description

We continue with our consideration of the system of Figure 8.2. If the silicon is held at ground and the voltage applied to the metal is negative but increases in magnitude above V_{FB}, additional positively charged holes are attracted toward the silicon surface, and the MOS capacitor begins to store positive charge there. This positive charge is made up of an increase in the hole population at the surface, so the surface has a greater density of holes than N_a, the acceptor density. This condition is called *surface accumulation* (or simply *accumulation*), and the region at the surface containing the increased hole population is known as the *accumulation layer.*

Because the surface accumulation layer is a space-charge layer composed of free carriers, the solution for a Schottky ohmic contact that was discussed in Sec. 3.4 describes its spatial distribution. In Equation 3.4.2 we showed that half of the space charge of free carriers exists within $\sqrt{2}$ times the Debye length L_D that characterizes the surface, and the total extent of the accumulation layer is, therefore, a few times the value of L_D at the surface. To grasp the order-of-magnitude of these quantities, consider a *p*-type silicon wafer with $N_a = 10^{15}$ cm^{-3} with its surface accumulated such that the surface density is 10 times the bulk density. The Debye length at the surface for this case is about 40 nm (from Equation 3.4.3), greater than the thickness now used for most MOSFET gate oxides. Sketches of the energy-band diagram and of the charge configuration for the case of MOS surface accumulation are given in Figures 8.4*a* and 8.4*b*, respectively.

Now we consider other biases. We already saw that the MOS system of Figure 8.2 with zero voltage applied between the metal and the silicon stores negative charge at the silicon surface and positive charge on the metal. The negative charge in the silicon consists of uncompensated acceptors in a region depleted of holes. This charge configuration is consistent with a built-in positive voltage between the metal and silicon. If this built-in voltage is aided by applying a positive gate voltage between the metal and the silicon, the silicon becomes further depleted as more holes are repelled from its surface and more acceptors are exposed. Correspondingly, the positive charge on the metal

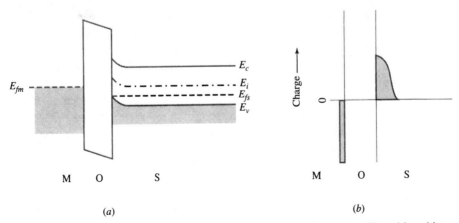

(a) (b)

FIGURE 8.4 (*a*) Energy-band diagram of an MOS system with *p*-type silicon biased into accumulation. (*b*) Charge in the same MOS system.

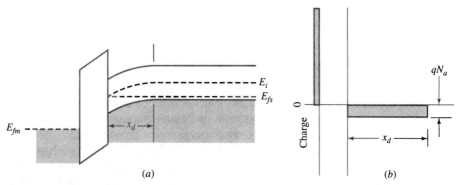

FIGURE 8.5 (*a*) Energy-band diagram of an MOS system with *p*-type silicon biased into depletion. (*b*) Charge in the same MOS system.

increases. Because mobile silicon charge is withdrawn from the surface, this condition is called *surface depletion.* The energy-band diagram and the charge configuration for depletion bias in the MOS system of Figure 8.2 are sketched in Figure 8.5. Under depletion bias, the energy-band diagram behaves very similarly to the reverse-biased Schottky-barrier diode. Figure 8.5*a* can be compared to Figure 3.5*b* to demonstrate this similarity (with due allowance for the fact that the silicon is *n*-type in Figure 3.5 and *p*-type in Figure 8.5).

If the voltage applied to the metal in the MOS system increases further, however, behavior unlike that of a metal-semiconductor diode occurs. In the MOS case, as the voltage on the metal increases, the field at the surface of the silicon increases and the energy bands bend considerably away from their levels in the bulk of the silicon. In the surface region, the majority carriers have been depleted and generation of carriers therefore exceeds recombination (Equation 5.2.9). The generated hole-electron pairs are separated by the field, the holes being swept into the bulk and the electrons moving to the oxide-silicon interface where they are confined by the conduction-band barrier. If the voltage at the metal is changed slowly enough, this generation process can bring the densities of free carriers at the silicon surface into equilibrium with the bulk densities, Then, a constant Fermi level can be drawn from the bulk to the oxide interface, and Fermi-Dirac statistics can be used to calculate the concentrations of carriers in the silicon. In fact, all of the energy-band diagrams drawn in this section assumed that equilibrium was achieved. (We consider alternative behavior later.) If the Fermi level remains constant in the silicon while the energy bands bend as the applied voltage changes, at sufficiently high voltages the intrinsic Fermi level E_i at the silicon surface crosses the Fermi level corresponding to the silicon bulk. In the silicon near the oxide-silicon interface, the Fermi level is closer to the conduction-band edge than to the valence-band edge. In terms of carrier densities, this means that the applied voltage has created an *inversion layer,* so-called because the surface contains more electrons than holes, even though the silicon was doped with acceptor impurities. The applied voltage between the metal and the silicon has induced a *pn* junction near the surface. A sketch of the energy-band diagram for the MOS system of Figure 8.2 when biased into inversion is shown in Figure 8.6*a*

When E_i is slightly below E_f at the surface, the electron density in the inversion layer is low (order n_i), and the MOS system is said to be biased in the *weak inversion* region. On the other hand, when $(E_c - E_f)$ at the surface is smaller than $(E_f - E_v)$ in the bulk, the electron density in the inversion layer is greater than hole density in the bulk, and the system is in the *strong inversion* region. The dividing point between the two cases

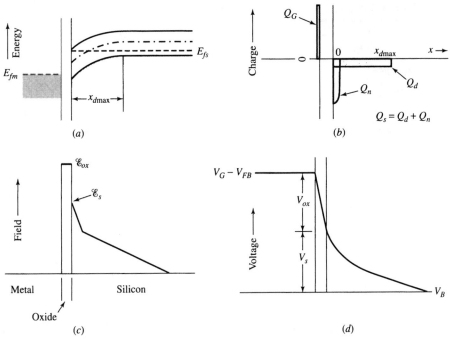

FIGURE 8.6 (a) Energy-band diagram, (b) space-charge configuration, (c) field, and (d) potential distribution for an MOS system with p-type silicon biased into inversion.

is conveniently taken when the electron density at the surface equals the acceptor concentration. While this division between weak and strong inversion is somewhat arbitrary, it is a convenient distinction for many purposes.

In the depletion and inversion regions of bias, the applied voltage gives rise to negative charge in the semiconductor by repelling holes from the surface to create the depletion region and then inducing electrons to form the inversion layer. Figure 8.6b shows the charge configuration in the MOS system of Figure 8.2 when it biased into inversion. The free-electron charge density in the inversion layer, which we call Q_n (C cm^{-2}), is located close to the surface because of the attractive force of the surface field. The acceptor charge is distributed throughout the depletion region, which extends away from the surface toward the bulk of the silicon. The depletion-charge density is called Q_d, as in Chapters 3 and 4. The sum $Q_n + Q_d$ is Q_s, the silicon space-charge density.

Once the silicon is biased into strong inversion, the free-electron density at the surface is nearly an exponential function of the potential at the surface, so the surface potential changes little with increasing gate voltage after the inversion layer forms. The charge induced by the additional gate voltage is virtually all free-electron charge just below the oxide-silicon interface. Thus, the total potential drop across the depletion region and the depletion-layer width in the silicon are both relatively constant after the surface is inverted. The maximum width of the depletion layer is usually denoted x_{dmax}, as in Figure 8.6b. The field and potential variations in the inverted MOS structure are shown in Figures 8.6c and 8.6d, respectively. The field at the interface between the oxide and the silicon is discontinuous, decreasing from $\mathscr{E}_{ox}$ to $\mathscr{E}_s$ because of the different permittivities of the two materials. The total voltage across the structure ($V_G - V_B - V_{FB}$) is composed of a drop across the oxide and a drop across the space-charge region in the silicon. Equations for these two voltages are derived in Sec. 8.3.

TABLE 8.2 MOS Surface-Charge Conditions for _p_-type Silicon

$(V_G - V_{FB})$	ϕ_s	Surface Charge Condition	Surface Carrier Density
Negative	Negative $\|\phi_s\| > \|\phi_p\|$	Accumulation	$p_s > N_a$
0	Negative $\phi_s = \phi_p$	Neutral (Flat-band)	$p_s = N_a$
Positive (small)	Negative $\|\phi_s\| < \|\phi_p\|$	Depletion	$n_i < p_s < N_a$
Positive (larger)	0	Intrinsic	$p_s = n_s = n_i$
Positive (larger)	Positive $\|\phi_s\| < \|\phi_p\|$	Weak inversion	$n_i < n_s < N_a$
Positive (larger)	Positive $\phi_s = -\phi_p$	Onset of strong inversion	$n_s = N_a$
Positive (larger)	Positive $\|\phi_s\| > \|\phi_p\|$	Strong inversion	$n_s > N_a$

This qualitative discussion of surface charge, voltages, and fields has introduced the important effect of the gate voltage applied to the overlying metal in determining the properties of the silicon surface. Although the system is basically only a capacitor, the various forms that the surface charge in the silicon can take produce very significant differences in the electrical properties of the silicon surface. For example, the surface can be highly conductive and electrically connected to free carriers in the bulk when it is biased in accumulation; it can be highly resistive when it is biased in depletion; or it can be highly conductive but disconnected from the bulk when it is biased in inversion. Table 8.2 summarizes these various MOS surface conditions.

8.3 MOS ELECTRONICS

Model for Charges in the Silicon Substrate

The previous two sections contained a qualitative introduction to the charges induced near the silicon surface by the voltage applied between the overlying gate and the substrate. The particular application of this theory to the MOSFET in the next two chapters demands a more quantitative background, especially for inversion bias. To provide this background, a simplified model for the charge distribution in the gate and bulk will be presented. The model avoids an exact representation of the carrier densities, which cannot be written in closed form. The complexity of the exact forms is not needed to develop a model of the MOS system that adequately meets our needs. The complete solutions have been published [1, 2]. The detailed treatment is also used in computer simulation for some regions of bias.

Thermal Equilibrium

As a first level of analysis, we consider the silicon surface region to be in thermal equilibrium with the bulk. The results obtained for thermal equilibrium can then be readily

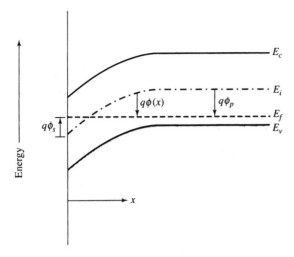

FIGURE 8.7 Energy-band diagram showing the potential as defined in Equation 8.3.1 in the vicinity of the silicon surface in an MOS system. The sketch corresponds to a positive value of surface potential (ϕ_s).

generalized for the nonequilibrium case we will discuss shortly. We define potential in the silicon as in Equation 4.1.2:

$$\phi(x) = \frac{1}{q}[E_f - E_i(x)] \tag{8.3.1}$$

Because we are considering thermal equilibrium, E_f is constant but E_i can vary with position. As shown in Figure 8.7, the potential ϕ_p in the neutral bulk of the silicon is negative because the material is p-type and E_f is less than E_i. At the surface, the potential ϕ_s is expressed as

$$\phi(0) = \phi_s = \frac{1}{q}[E_f - E_i(0)] \tag{8.3.2}$$

The carrier densities are related to $\phi(x)$ by Equations 1.1.26 and 1.1.27:

$$p = n_i \exp\left(-\frac{q\phi}{kT}\right) \quad \text{and} \quad n = n_i \exp\left(\frac{q\phi}{kT}\right) \tag{8.3.3}$$

From these equations and the definitions of ϕ_p and ϕ_s, we can express the surface free-carrier densities n_s and p_s in terms of the potential drop ($\phi_s - \phi_p$) across the depletion region at the silicon surface.

$$p_s = N_a \exp\left[\frac{q(\phi_p - \phi_s)}{kT}\right]$$

$$n_s = \frac{n_i^2}{N_a}\exp\left[\frac{q(\phi_s - \phi_p)}{kT}\right] \tag{8.3.4}$$

As we discussed in the previous section, when the MOS system is biased in accumulation or inversion, free carriers in the bulk pile up against the oxide-silicon interface. Equation 8.3.4 shows that the densities of these charges depend exponentially on the local band bending, as required by thermal statistics. As we mentioned earlier, when the surface is biased into strong inversion or accumulation, the surface potential ϕ_s remains

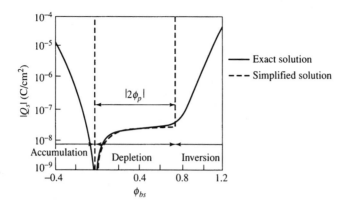

FIGURE 8.8 Total charge Q_s in the silicon substrate as a function of ϕ_{bs}, the voltage equivalent of the total bending of the energy bands from the substrate to the surface of the silicon.

relatively constant because n_s and p_s change rapidly with only a small change in ϕ_s. An exact solution for the silicon space-charge density Q_s, including both the free carriers and the depletion charge, is plotted (solid line) in Figure 8.8 as a function of the total band bending, ϕ_{bs}, from the substrate to the surface ($\phi_{bs} = |\phi_s - \phi_p|$). Because of the exponential dependence of the free charge density on the local band bending, the physical thickness (for example, as defined by the region that contains 90% of these charges) of either the inversion or the accumulation layer is only of the order of ~10 nm.

For a simplified model, we assume that the onset of *accumulation* occurs at the flat-band condition (when the total band bending in the substrate is zero) and that *inversion* starts when the surface band bending ϕ_{bs} reaches a value $2\phi_p$. In addition, we assume that the inversion or accumulation layer has zero thickness so that the free charges form a charge sheet at the silicon surface, and there is no additional band bending within this layer. When the surface potential is between the value that leads to accumulation and the value that causes inversion, the silicon is biased in *depletion*, and we assume that the only space charge that exists in the depletion region is that on ionized impurities.

With this simplified model, we can relate the surface potential ϕ_s to the width of the depletion layer x_d by using the depletion approximation, as we did for the Schottky barrier in Chapter 3 (Equation 3.2.3) to find

$$x_d = \sqrt{\frac{2\epsilon_s|\phi_s - \phi_p|}{qN_a}} \tag{8.3.5}$$

The depletion-charge density Q_d is given by

$$Q_d = -qN_a x_d$$

The maximum value that the depletion region can attain is limited by the onset of inversion (when $\phi_s = |2\phi_p|$); thus,

$$x_{d\max} = \sqrt{\frac{4\epsilon_s|\phi_p|}{qN_a}} \tag{8.3.6}$$

and

$$Q_{d\max} = -qN_a x_{d\max} = -\sqrt{4\epsilon_s qN_a|\phi_p|} \tag{8.3.7}$$

The solution for the silicon charge density Q_s as a function of the silicon surface potential ϕ_s found using this model is plotted as the dotted line in Figure 8.8.

EXAMPLE   *Potential near the Oxide-Silicon Interface*

Derive an expression for the potential distribution in an ideal MOS capacitor in the depletion condition in terms of the surface potential ϕ_s and the depletion width at the surface x_d taking the zero for potential in the silicon bulk. The silicon is doped p–type and $x = 0$ at the oxide-silicon interface.

Solution   Using the depletion approximation and Poisson's equation, the space charge and field gradient are constant and negative in the silicon for $0 \leq x \leq x_d$. Hence, if the surface field is $\mathscr{E}_s$, the field as a function of position away from the surface is

$$\mathscr{E}(x) = \mathscr{E}_s(1 - x/x_d) \qquad 0 < x < x_d$$

and

$$\phi(x) = \phi_s - \int_0^x \mathscr{E} \, dx$$

$$= \phi_s - \mathscr{E}_s x + \mathscr{E}_s x^2/2x_d$$

$$= \tfrac{1}{2}\mathscr{E}_s x_d - \mathscr{E}_s x + \mathscr{E}_s x^2/2x_d$$

$$= \frac{\mathscr{E}_s}{2x_d}(x_d - x)^2$$

so that

$$\phi(x) = \phi_s\left(1 - \frac{x}{x_d}\right)^2 \qquad 0 < x < x_d$$

is the required expression for potential as a function of x.

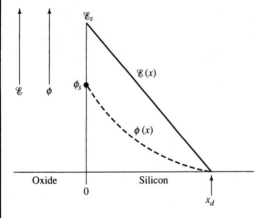

The surface potential ϕ_s corresponds to the area under the curve of $\mathscr{E}$ versus x. Therefore $\phi_s = \tfrac{1}{2}\mathscr{E}_s x_d$. ∎

Nonequilibrium

Once the MOS system is biased into inversion, a *pn* junction exists between the surface and the bulk of the silicon. If there is a nearby *n*-type diffused region that contacts the inverted surface as shown in Figure 8.9, we can apply a bias to the *pn* junction. When bias is applied, it causes a nonequilibrium condition within the silicon, and some current flows between the inversion layer at the surface and the bulk. However, in practical applications of MOS systems, the junction is reverse biased and the current is small.

An energy-band diagram for the case of bias applied to an inverted surface is described by two quasi-Fermi levels (Equations 1.1.28 and 1.1.29), one for the *p*-region and

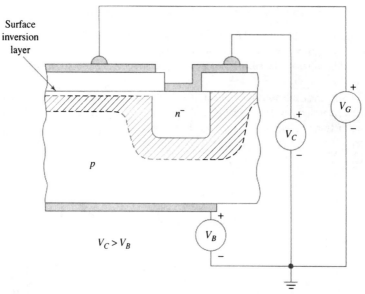

FIGURE 8.9 A diffused junction in the vicinity of an MOS capacitor can be used to bias the induced junction between the bulk of the silicon and an inversion layer formed at the oxide–silicon interface. The cross-hatching indicates the extent of the space-charge region in the depleted silicon.

one for the n-region. Just as for the pn junction under reverse bias (Chapter 4), the two quasi-Fermi levels are separated by the applied reverse bias. This situation is sketched in Figure 8.10 for a reverse bias ($V_C - V_B$) applied between the inversion layer (or *channel*) and the substrate (or *bulk*).

A reverse bias applied between the induced surface n-region and the bulk increases the charge Q_d in the depletion layer. Because the negative charge induced by $V_G - V_B$ is shared between the depletion and inversion layers, an increase of the charge in the depletion layer means that there is less charge available to form the inversion layer for a given gate voltage. Viewed another way, more gate voltage must be applied to induce the same number of electrons in the inversion layer when a reverse bias is applied across the surface-bulk pn junction. The applied voltage extends the range of gate voltages for which the surface region is depleted; consequently, $x_{d\max}$ is larger. The bias at the surface prevents an inversion layer from forming as readily as when no bias is present by draining away electrons that could form the inversion layer until the surface potential ϕ_s reaches

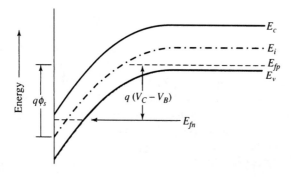

FIGURE 8.10 Energy-band diagram for an inverted surface on p-type silicon with a voltage ($V_C - V_B$) applied between the inversion layer and the substrate.

$|\phi_p| + V_C - V_B$, instead of the unbiased inversion potential $|\phi_p|$. Consequently, the change of the surface potential between flat-band and strong inversion is $2|\phi_p| + V_C - V_B$, rather than $2|\phi_p|$. The corresponding maximum depletion-region width x_{dmax} and the depletion-layer charge Q_d (per unit area) become

$$x_{dmax} = \sqrt{\frac{2\epsilon_s(2|\phi_p| + V_C - V_B)}{qN_a}} \tag{8.3.8}$$

and

$$Q_d = -\sqrt{2\epsilon_s qN_a(2|\phi_p| + V_C - V_B)} \tag{8.3.9}$$

The previous discussion about charge in the MOS system showed that flat band $(V_G - V_B = V_{FB})$ corresponds to the condition of charge neutrality in the silicon. Therefore, $(V_G - V_B) - V_{FB}$ is the effective voltage that charges the MOS capacitor. In this context, the flat-band voltage V_{FB} for the MOS system is analogous to ϕ_i, the built-in voltage for the *pn* junction; that is, both act like offsets of the zero level in equations relating stored charge to an applied bias. To express the charge in terms of applied voltage, we carry out the following analysis: The charging voltage $[(V_G - V_B) - V_{FB}]$ is the sum of a voltage drop V_{ox} across the oxide and a drop in the silicon $(\phi_s - \phi_p)$, as shown in Figure 8.6d.

$$V_G - V_B - V_{FB} = V_{ox} + \phi_s - \phi_p \tag{8.3.10}$$

The field in the insulating oxide is constant in the absence of oxide charge. In terms of the applied voltage and the oxide thickness, this field $\mathscr{E}_{ox}$ is

$$\mathscr{E}_{ox} = V_{ox}/x_{ox} = [(V_G - V_B - V_{FB}) - (\phi_s - \phi_p)]/x_{ox} \tag{8.3.11}$$

Just inside the silicon and adjacent to the oxide (before any silicon charge is encountered), the normal displacement D is constant and the field $\mathscr{E}_{s0}$ is

$$\mathscr{E}_{s0} = \frac{\epsilon_{ox}\mathscr{E}_{ox}}{\epsilon_s} \tag{8.3.12}$$

Combining Equation 8.3.12 with Equation 8.3.11 and using the definition for oxide capacitance (per unit area) $C_{ox} = \epsilon_{ox}/x_{ox}$, we find

$$\epsilon_s\mathscr{E}_{s0} = C_{ox}[(V_G - V_B - V_{FB}) - (\phi_s - \phi_p)] \tag{8.3.13}$$

Gauss' law states that the charge contained in a volume equals the permittivity times the electric field emanating from the volume. Applying Gauss' law to a volume extending from just inside the silicon at the oxide-silicon interface to the field-free bulk region, we can write

$$-\epsilon_s\mathscr{E}_{s0} = Q_s = Q_n + Q_d \tag{8.3.14}$$

where Q_s, the total charge induced in the semiconductor, is composed of the mobile electron charge Q_n and the depletion-region charge Q_d (all per unit area). These quantities are shown in Figure 8.6b. Using Equation 8.3.14 in Equation 8.3.13, we can express the mobile charge Q_n as

$$Q_n = -C_{ox}[(V_G - V_{FB} - V_B) - (\phi_s - \phi_p)] - Q_d \tag{8.3.15}$$

We insert the values of ϕ_s and Q_d into Equation 8.3.15 to relate the mobile charge Q_n to the applied voltages. In strong inversion with an applied reverse bias between the channel and the bulk, $\phi_s = |\phi_p| + (V_C - V_B)$, and Equation 8.3.15 becomes

$$Q_n = -C_{ox}(V_G - V_{FB} - V_C - 2|\phi_p|) + \sqrt{2\epsilon_s qN_a(2|\phi_p| + V_C - V_B)} \tag{8.3.16}$$

When no applied bias is applied between the channel and the bulk, Equation 8.3.16 reduces to the simpler expression

$$Q_n = -C_{ox}(V_G - V_{FB} - V_B - 2|\phi_p|) + \sqrt{4\epsilon_s q N_a |\phi_p|} \qquad (8.3.17)$$

Note that the first term on the right-hand side of Equations 8.3.16 and 8.3.17 is negative while the last term in each equation is positive. The positive terms are, however, smaller in magnitude (because $|Q_s| > |Q_d|$), so that the difference Q_n is negative, as it must be for an inversion layer on a p-type substrate.

From these equations, we can directly express the gate voltage necessary to induce a conducting channel at the surface of the semiconductor. This voltage, known as the threshold voltage V_T, is defined as the gate voltage that results in $Q_n = 0$. From Equation 8.3.16, an expression for V_T can be written as

$$V_T = V_{FB} + V_C + 2|\phi_p| + \frac{1}{C_{ox}}\sqrt{2\epsilon_s q N_a(2|\phi_p| + V_C - V_B)} \qquad (8.3.18)$$

Taking the terms in Equation 8.3.18 one by one, we can readily see that they produce the proper qualitative behavior. First, V_T contains V_{FB} because a gate voltage equal to V_{FB} is necessary to bring the silicon to a charge-neutral condition. (In the system of Figure 8.2, V_{FB} is negative, tending to reduce the size of V_T.) Second, increasing the channel voltage V_C increases the gate voltage necessary to induce a given charge near the silicon surface. Third, $2|\phi_p|$ volts must be applied to cause the silicon energy bands to be bent to an inverted condition. Finally, the square-root term accounts for the uniform distribution of uncompensated ionized acceptors in the depletion region. This term varies inversely with the oxide capacitance. It increases with increasing $V_C - V_B$ to reflect the redistribution of semiconductor charge Q_s from the inversion layer (where it contributes to Q_n) into the depletion layer (where it forms part of Q_d).

For strong inversion, the charge in the inversion layer can be expressed very simply in terms of the difference between the applied gate voltage and the threshold voltage. From Equations 8.3.16 and 8.3.18

$$Q_n = -C_{ox}(V_G - V_T) \qquad (8.3.19)$$

Equation 8.3.19 should be used with caution if V_G is approximately equal to V_T, because it is derived with the assumption that there are no electrons at the surface until the onset of strong inversion (when the surface potential $\phi_s = -\phi_p + V_C - V_B$). This is, of course, an approximation that can be removed by considering the exact solutions of Poisson's equation for the MOS system [1, 2].

The more frequently used equations that we derived for the MOS system are in Table 8.3, which is included at the end of the chapter for ready reference. Results for both n-type substrates and p-type substrates are given.

8.4 CAPACITANCE OF THE MOS SYSTEM

Analysis of the behavior of the small-signal capacitance measured between the output electrodes provides valuable insight into the electrical behavior of the MOS system, as it did for the Schottky barrier and the pn junction. In the case of the MOS system, this analysis has been central to the research leading to the present well-developed state of understanding of the oxide-silicon system and its technology. MOS capacitance-voltage (MOS C-V) measurement and analysis techniques are standard tools in the diagnosis and monitoring of technology under development or in production. In this section, we discuss the basic physics behind MOS C-V behavior assuming that the MOS system is ideal and that there are no charges in the oxide or traps at the oxide–silicon interface.

C-V Behavior of an Ideal MOS System

The small-signal capacitance (per unit area) of a two-terminal device is defined as the derivative of the charge Q on the terminals with respect to the voltage across them:

$$C = \frac{dQ}{dV} \tag{8.4.1}$$

From Equation 8.4.1, once Q is known as a function of V, C can easily be derived, as we did for the Schottky diode and the pn junction in earlier chapters. In those cases, the free electrons in the metal and the majority carriers in the silicon, both being characterized by very small dielectric relaxation times (typically of the order of picoseconds or smaller), can respond essentially instantaneously to the applied voltage. For the MOS structure, however, the situation is more complicated; the charges cannot respond as quickly to a changing voltage, and the associated capacitance can depend on the frequency of the changing applied voltage. The various cases we consider in the following discussion are summarized in Table 8.4.

Consider first that the MOS system is biased with a steady dc bias V_G that causes the silicon surface to be accumulated. For the p-type silicon sample shown in Figure 8.2, this corresponds to a negative applied voltage and results in a charge configuration like that sketched in Figure 8.4. Electrostatic forces from the gate voltage pull the excess holes at the silicon surface very close to the oxide. When a small ac voltage v_G is superposed on the dc bias V_G, it causes small variations, $-dQ$ and $+dQ$, in the charges stored on the metal gate and at the silicon surface, respectively. If the system is now connected to an instrument that measures the small-signal capacitance associated with these variations, the measured capacitance is nearly that of the oxide itself because the spatial extent of the modulated charge in the silicon is small compared to the oxide thickness. The more the surface is accumulated, the thinner is the accumulation layer; hence, the charge $-dQ$ in the metal gate and the charge $+dQ$ in the silicon are separated by a distance approaching x_{ox}, and the capacitance asymptotically approaches the capacitance associated with the oxide. Thus, the capacitance per unit area C in accumulation approaches

$$C_{ox} = \frac{\epsilon_{ox}}{x_{ox}} \tag{8.4.2}$$

where x_{ox} is the oxide thickness. When the gate voltage approaches the flat-band voltage, the surface accumulation vanishes, and the capacitance decreases as the Debye length at the surface increases. An exact analysis for the capacitance in this bias range requires solving Poisson's equation with free electrons, free holes, and dopant atoms all contributing to the total space charge near the surface [1]. We will not attempt to present this exact analysis; it is important, however, to define a quantity C_{FB} as the capacitance measured when $V_G = V_{FB}$. As shown in reference [3], C_{FB} can be expressed as

$$C_{FB} = \frac{1}{1/C_{ox} + L_D/\epsilon_s} \tag{8.4.3}$$

where L_D, the extrinsic Debye length, was defined in Equation 4.2.14.

When the gate voltage becomes more positive than the flat-band voltage, holes are repelled from the surface of the silicon and the system moves into depletion. Under this

TABLE 8.4 Charge Conditions in the MOS System

Bias Condition	Charge Equilibrium	Charge Distribution	Equivalent Circuit	Comment
Low frequency (LF)	dc equilibrium ac equilibrium	$x_d = x_{dmax}$; SiO_2; δQ_s	C_{ox}	$\delta Q_s = \delta Q_n$ ac coupling only to surface charge
High frequency (HF)	dc equilibrium ac nonequilibrium	$x_d = x_{dmax}$; δQ_d; SiO_2	C_{ox} ϵ/x_{dmax}	$\delta Q_s = \delta Q_d$ ac coupling only to bulk charge
Deep depletion (DD)	dc nonequilibrium ac nonequilibrium	$x_d > x_{dmax}$; δQ_d; SiO_2	C_{ox} ϵ/x_d	both ac and dc coupling to Q_d $x_d > x_{dmax}$

condition, relatively straightforward electrostatic analysis shows that the overall capacitance C corresponds to the capacitance of a series connection of the oxide capacitance and the capacitance C_s across the surface depletion region (Problem 8.3).

$$C = \frac{1}{1/C_{ox} + 1/C_s} = \frac{1}{x_{ox}/\epsilon_{ox} + x_d/\epsilon_s} \tag{8.4.4}$$

where x_d is the width of the surface depletion layer, which depends upon the gate bias as well as the doping and the oxide properties. We can intuitively obtain Equation 8.4.4 without a detailed derivation by considering the position of the charges $\pm dQ$ in the gate and in the silicon. Physically, they are located on the metal electrode and at the neutral edge of the depletion region in the silicon, as if they are on the two plates of a parallel-plate capacitor separated by the oxide with thickness x_{ox} plus the depletion width x_d in the silicon. The resulting capacitance C is a series combination of the oxide capacitance C_{ox} and the depletion capacitance C_s. From Equation 8.4.4, we see that the capacitance of the system decreases as the depletion region widens.

When the gate bias is increased sufficiently to invert the surface, an additional location for charge must be considered to describe the MOS capacitance. Mobile charge can be located in the silicon very close to the oxide–silicon interface; the ability of this inversion-layer charge to change in response to an applied voltage determines the behavior of the capacitance in this region of bias. We recall that the inversion layer at the MOS surface results from the generation of minority carriers. Hence, the population of the inversion layer can change only as fast as carriers can be generated within the depletion region near the surface. This limitation causes the measured capacitance to be a function of the frequency ω of the ac signal used to measure the small-signal capacitance of the system.

The simplest case arises when both the dc gate-bias voltage V_G and the small-signal measuring voltage v_G are changed very slowly so that the silicon can always approach equilibrium and the inversion-layer population can "follow" the applied signal. In strong inversion, the charges on the gate and those in the silicon are then separated only by the gate oxide, and the capacitance of the MOS system approaches C_{ox}. Under these conditions, a plot of measured capacitance versus gate bias follows the curve labeled "LF" (for *low frequency*) in Figure 8.11. The capacitance has a value C_{ox} in the accumulation region of bias and decreases as the surface traverses the depletion region; it then increases toward C_{ox} as the surface becomes inverted.

The results of Problem 8.6 show that a characteristic time τ_{inv} to form an inversion at the surface of an MOS system biased to inversion is of the order of seconds or more and depends on the minority-carrier lifetime τ_0 at the surface ($\tau_{inv} \sim 2N_a\tau_0/n_i$). Therefore, the small-signal measuring voltage must be changed very slowly to observe the low-

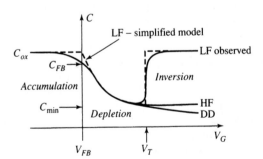

FIGURE 8.11 General behavior of *C-V* curves of an ideal MOS system under different dc bias and ac small-signal conditions. The low-frequency (LF) *C-V* curve corresponding to the simplified model is shown as a dashed line.

frequency C-V curve. We will consider this requirement again when we discuss practical C-V measurement techniques later in this section.

When the dc bias voltage is varied slowly while the ac measuring signal is changed rapidly, the inversion layer can "follow" the dc bias, but not the ac bias. The modulated charge $-dQ$ then corresponds to the movement of holes at the far edge of the depletion region. The capacitance corresponds to the series combination of the oxide capacitance and the depletion-region capacitance, as was true for depletion bias. Because the depletion region reaches a maximum width $x_{d\max}$ (Equation 8.3.6) when the system reaches strong inversion, the measured capacitance approaches a minimum value $C_{\min}$, corresponding to the series connection of the oxide capacitance and the capacitance associated with the maximum depletion width:

$$C_{\min} = \frac{1}{x_{ox}/\epsilon_{ox} + x_{d\max}/\epsilon_s} \tag{8.4.5}$$

The capacitance $C_{\min}$ remains constant at the value given by Equation 8.4.5 as the gate voltage increases further. This *high-frequency* (HF) C-V curve is also shown in Figure 8.11.

At this point we can examine how the C-V curves would behave using the simplified model described in Sec. 8.3. A low-frequency C-V curve given by the model is shown in Figure 8.11 as a dashed line. The simplified model assumes that the transitions between accumulation, depletion, and inversion are abrupt, and that the inversion and accumulation layers are infinitesimally thin. Therefore, the model overestimates the capacitances at both the accumulation-to-depletion and the depletion-to-inversion transitions. The low-frequency C-V curve according to the simplified model gives the inversion charge density Q_n as

$$Q_n = C_{ox}(V_G - V_T) \tag{8.4.6}$$

Equation 8.4.6 is commonly used to calculate the inversion charge in the conducting channel of a MOSFET. This equation overestimates Q_n and thus overestimates the output current of a MOSFET. The overestimation is more significant for lower power-supply voltages (i.e., at small $V_G - V_T$), as in an aggressively scaled MOS technology. We will return to consider Equation 8.4.6 in Chapter 9. Note that the capacitance between the gate and the inversion charge represents the "coupling" of the gate voltage to the current-carrying inversion charge in a MOSFET; this capacitance is desired because higher coupling increases the performance of the transistor.

A final capacitive behavior, called the *deep-depletion* capacitance is sketched in Figure 8.11 as the "DD" curve. It corresponds to the experimental situation in which both the gate bias voltage V_G and the small-signal measuring voltage v_G vary at a faster rate than can be accommodated by generation-recombination processes in the surface depletion region. Because the inversion layer cannot form, the depletion region becomes wider than $x_{d\max}$, as appropriately described by the name *deep depletion*. The capacitance in this case is again given by Equation 8.4.4; however, in deep depletion x_d can exceed $x_{d\max}$, and the capacitance does not reach a minimum.

Practical Considerations in C-V Measurements

C-V Measurement Basics. C-V measurement systems are standard tools in MOS technology both for research and development and for production. In a typical system, the voltage applied to the MOS diode consists of a dc voltage V_G, on which a very small, sinusoidal ac voltage v_G with frequency ω is superposed, as shown in Figure 8.12. To make measurements at different biases, the dc voltage is programmed to step up and down

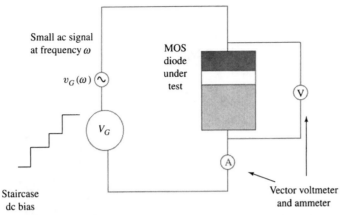

FIGURE 8.12 MOS *C-V* measurement system. The voltmeter and ammeter measure both the magnitude and phase of the voltage across the diode and the current through it.

within a certain desired range. The response is usually measured with a vector ammeter and a vector voltmeter, which determine both the magnitude and the phase of the current through the diode and the voltage across it, respectively. The diode capacitance C is then extracted from either the measured impedance $Z = v/i$ or the conductance $G = i/v$. The response can be modeled as the response from either a parallel or a series combination of capacitance C and (parasitic) resistance R_p. In the series combination,

$$Z = v/i = R_p + 1/j\omega C \tag{8.4.7}$$

For the parallel combination,

$$G = i/v = 1/R_p + j\omega C \tag{8.4.8}$$

The capacitance C can then be determined.

The series and parallel models obviously produce different values of capacitance and resistance. The parallel model is usually more useful unless the series resistance in the test path is significant compared to the magnitude of the impedance of the capacitance. In general, a higher frequency ac voltage produces a larger current and thus a better signal-to-noise ratio and higher accuracy for the measurement. However, the ac measuring signals undergo delay and phase shifts as they travel through the cables; these become increasingly troublesome as the frequency increases. If not properly considered, parasitic effects from the cables can lead to errors in the interpretation of the results. Some commercial capacitance meters have built-in corrections or compensation, but these are usually effective only for specific measuring configurations (e.g., two-cable or four-cable arrangements) and cable lengths. The instruction manuals for the specific measurement instruments being used usually contain detailed measurement recommendations to minimize any errors resulting from signal delays.

Quasi-Static (Low-Frequency) *C-V* Measurements

Low-frequency *C-V* measurements are useful for technology diagnosis and for process monitoring to extract information about the interface-trap densities discussed in the next section. We described typical low- and high-frequency *C-V* behavior earlier. We noted that the characteristic time to form an inversion layer is seconds or more, making low-frequency

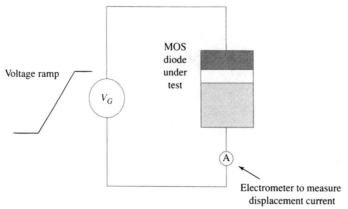

FIGURE 8.13 Quasi-static C-V measurement apparatus used to obtain low-frequency C-V characteristics.

C-V measurements impractical using common ac techniques. Although the electron-hole generation rate can be increased by illuminating the sample to obtain behavior that resembles the low-frequency behavior, illumination may not reduce the time required for an electron to enter or leave a trap. *Quasi-static C-V* measurements provide more useful low-frequency C-V characteristics. The measurement system for a quasi-static C-V measurement is shown in Figure 8.13. Instead of applying a stepped dc voltage with a superposed ac measuring signal, a quasi-static measurement uses a voltage ramp and an electrometer to measure the displacement current. An electrometer is an electronic ammeter with very high sensitivity (resolution of the order of fA) and speed. When the applied voltage is ramped at a rate R_m V/s, the displacement current I in the circuit is related to C and R_m by

$$I = dQ/dt = C \, dV/dt = CR_m \qquad (8.4.9)$$

For a capacitor of the order of 10 pF, a ramp rate $R_m = 0.1$ V/s (suitable for obtaining low-frequency C-V behavior) produces a current of 1 pA, which can easily be measured with high accuracy.

The long time needed to form the inversion layer can lead to problems when making high-frequency C-V measurements because the inversion layer may not reach equilibrium even with a very slowly changing (nearly dc) applied gate voltage. The measurement problems caused by this slow generation can be avoided by first biasing the capacitor into strong inversion under illumination to generate carriers and then turning off the light and measuring the C-V characteristic with the gate voltage stepped from strong inversion toward accumulation.

8.5 NON-IDEAL MOS SYSTEM

Oxide and Interface Charge

The theory presented to this point has not considered an important characteristic of the oxide-silicon system, namely, the influence of charge within the oxide and at the oxide-silicon interface. The presence of these charges is unavoidable in practical systems. To appreciate the importance of oxide charge, we estimate the order-of-magnitude of the charge densities that we considered in our discussion of the MOS system. For example, slightly above the threshold voltage where the system enters the inversion region, the surface density of electrons Q_n/q is of the same order-of-magnitude as the density of dopant atoms (per unit area) within the depletion layer. For homogeneously distributed dopant atoms, the area density is $N_a^{2/3} \approx 10^{10}$ cm^{-2} when N_a is 10^{15} cm^{-3}. Comparing this density to the area density of silicon atoms $(5 \times 10^{22}$ cm$^{-3})^{2/3} = 1.35 \times 10^{15}$ cm^{-2}, we see that a surface-charge density only about 10^{-5} times as large as the silicon atom density can cause the MOS system to depart from its ideal behavior. Fortunately, when they are formed carefully, interfaces between thermally grown, amorphous silicon dioxide and single-crystal silicon can contain charge densities of the order of 10^{10} cm^{-2} or lower. This low charge density is unique and is one of the outstanding characteristics of the silicon/silicon-dioxide system. We will consider specific sources of oxide charge after we discuss its influence on MOS systems.

Theoretical Analysis.   Consider that a density of charge Q_{ox} is located at the plane $x = x_1$ within the oxide as shown in Figure 8.14a. The charges at x_1 induce equal and opposite charges that are divided between the silicon and the metal gate. The closer x_1 is to x_{ox}, the oxide-silicon interface, the greater is the fraction of the induced charge within the silicon. Because this induced charge changes the charge stored in the silicon at thermal equilibrium, it alters the flat-band voltage from the value predicted in the ideal MOS analysis (Equation 8.1.1). The field and potential variations for the zero-bias case (Figure 8.14a) are shown in Figure 8.14c; those for flat-band (Figure 8.14b) are shown in Figure 8.14d.

The size of the shift in the flat-band voltage is readily found by using Gauss' law to obtain the value of gate voltage that causes all of the oxide charge Q_{ox} to be mirrored in the gate electrode so that none is induced in the silicon. This condition is illustrated in Figure 8.14b. From this figure and Gauss' law, we see that the field is constant between the metal (at $x = 0$) and Q_{ox} (at x_1) and zero between x_1 and the silicon surface (at $x = x_{ox}$) for flat-band conditions (Figure 8.14d). Its value $\mathscr{E}_{ox}$ between the gate and x_1 is

$$\mathscr{E}_{ox} = -\frac{Q_{ox}}{\epsilon_{ox}} \qquad 0 < x < x_1 \tag{8.5.1}$$

The gate voltage resulting from the presence of Q_{ox} is the negative integral of $\mathscr{E}_{ox}$ across the oxide. Because it contributes to the flat-band voltage, we call it ΔV_{FB} [i.e., the change in V_{FB} from the ideal MOS case (Equation 8.1.1) because of a planar sheet of oxide charge]. An expression for ΔV_{FB} is

$$\Delta V_{FB} = x_1 \mathscr{E}_{ox} = -\frac{x_1 Q_{ox}}{\epsilon_{ox}} \tag{8.5.2}$$

Equation 8.5.2 can be rewritten by using Equation 8.4.2 to express ΔV_{FB} in terms of C_{ox}, the oxide capacitance per unit area.

$$\Delta V_{FB} = -\frac{Q_{ox} x_1}{C_{ox} x_{ox}} \tag{8.5.3}$$

ΔV_{FB} has its maximum value when Q_{ox} is located at the oxide-silicon interface ($x_1 = x_{ox}$) because the charge induced by Q_{ox} is then contained entirely in the silicon. In contrast, when Q_{ox} is adjacent to the gate ($x_1 = 0$), the induced charge is totally in the metal so the oxide charge has no effect on ΔV_{FB}.

The results given for the sheet of charge at $x = x_1$ can be generalized to describe the shift in flat-band voltage by an arbitrary distribution of oxide charge $\rho(x)$ by superposing and integrating the increments that result from charges distributed throughout the

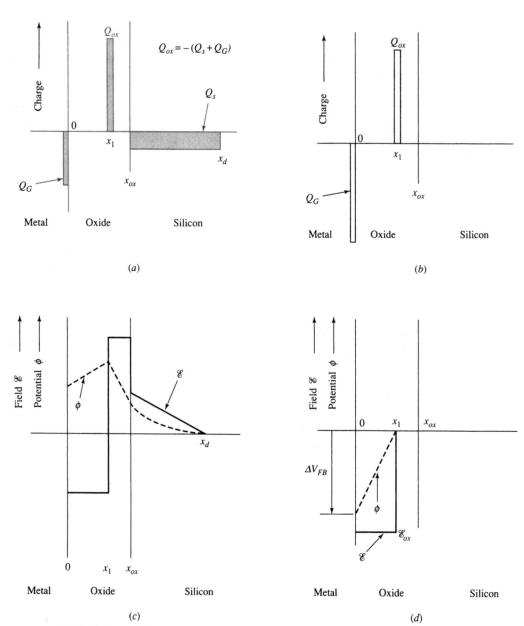

FIGURE 8.14 The effects of a fixed oxide-charge density Q_{ox} on the MOS system.
(a) Charge configuration at zero bias: $Q_{ox} = Q_s + Q_G$; (b) charge at flat band: $Q_{ox} = Q_G$.
(c) field (solid line) and potential (dashed line) at zero bias; (d) field (solid line) and potential
(dashed line) at flat band. The silicon bulk is taken as the reference for potential in (c) and (d).

oxide. The overall result is

$$\Delta V_{FB} = -\frac{1}{C_{ox}} \int_0^{x_{ox}} \frac{x}{x_{ox}} \rho(x)\, dx \qquad (8.5.4)$$

Fixed charge at the oxide-silicon interface is frequently treated separately from
charge incorporated in the oxide itself, even though surface charge can be included in the

formulation of Equation 8.5.4. The fixed interface charge density is designated Q_f, and its contribution to the flat-band voltage is

$$\Delta V_{FB} = -\frac{Q_f}{C_{ox}} \tag{8.5.5}$$

An expression for V_{FB} that includes the effects of differing work functions in the gate and in the silicon, as well as the influence of fixed oxide charge, can be written by combining Equations 8.1.1, 8.5.4, and 8.5.5:

$$V_{FB} = \Phi_{MS} - \frac{Q_f}{C_{ox}} - \frac{1}{C_{ox}} \int_0^{x_{ox}} \frac{x}{x_{ox}} \rho(x)\, dx \tag{8.5.6}$$

Equation 8.5.6 shows that the oxide charge shifts the flat-band voltage from its value in the ideal case. If the oxide charge is stable, the shift in the flat-band voltage produces a corresponding shift in the threshold voltage V_T (Equation 8.3.18). Experimentally, this threshold shift causes the capacitance versus gate-voltage curves to be translated along the V_G axis but does not distort their shape. A typical high-frequency capacitance curve that includes the effect of oxide charge is sketched as the dashed curve in Figure 8.15.

In some cases, oxides and oxide-silicon interfaces contain unstable charges that can be influenced by the applied voltage. In this case, the threshold voltage itself depends on the gate voltage. The capacitance-voltage curve is then distorted, as sketched by the dotted curve in Figure 8.15. To understand this behavior, as well as the influence of fixed charge, we discuss the physical sources of oxide charge.

Origins of Oxide Charge

As is typically done, we consider separately four distinct types of charge in the oxide–silicon system. These are shown in Figure 8.16a, along with the generally accepted names and symbols for the four charge types [4]: Q_f the *fixed interface charge* density, Q_{ot} the *oxide trapped-charge* density, Q_{it} the *interface trapped-charge* density, and Q_m the *mobile charge* density.

We already discussed the fixed interface-charge density Q_f (with corresponding numerical density $N_f = Q_f/q$). This charge is positive and, as shown in Figure 8.16a, located within a very thin (<1 nm) transition layer of nonstoichiometric silicon oxide (labeled SiO_x) at the boundary between the silicon and the stoichiometric SiO_2. The oxide trapped-charge density (Q_{ot}) can be either positive or negative (it is usually negative) and is located in traps distributed throughout the oxide layer. Only a small, often negligible amount

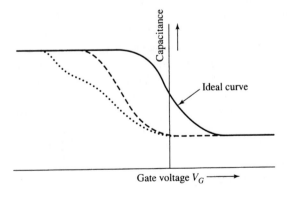

Ideal curve

FIGURE 8.15 Fixed charge in the oxide causes the capacitance–voltage curve to translate along the V_G axis without distortion (dashed curve); charge that is influenced by the gate voltage causes distortion (dotted curve).

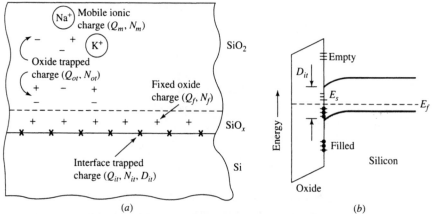

FIGURE 8.16 (a) Four categories of oxide charge in the MOS system. The symbols for the charge densities Q (C cm^{-2}), and state densities N (states cm^{-2}) or D (states cm^{-2} eV^{-1}) have been standardized [4]. (b) Energy levels at the oxide–silicon interface. The interface trapping levels are distributed with density D_{it} (states cm^{-2} eV^{-1}) within the forbidden-gap energies.

of oxide-trapped charge is usually introduced by processing. This charge is fixed except under unusual conditions (discussed in Chapter 10). We defer discussion of the mobile ionic charge until later and consider here the fourth component shown in Figure 8.14a, the interface trapped-charge density Q_{it} residing in trapping levels N_{it}.

The trapping levels N_{it} (traps cm^{-2}) are located at the oxide–silicon interface and, like the traps considered in Chapter 5, have energy levels within the forbidden gap (Figure 8.16b). They are distributed with density D_{it} (traps cm^{-2} eV^{-1}) over energies within the energy gap. Various sources of extra allowed energy levels at a silicon surface were discussed in Sec. 3.5, where we considered metal-semiconductor contacts. That section, which can be usefully reviewed at this time, pointed out that even clean surfaces have extra allowed energy levels different from those in the bulk of a crystal. The inevitable presence of impurities and crystal defects introduced during wafer processing is a source of additional allowed levels. Electrons in these extra levels and ions associated with them both contribute to interface charge.

To relate the behavior of these traps to the distorted capacitance-voltage curve shown in Figure 8.15 (dotted curve), consider an oxide–silicon interface with interface traps at an energy E_s, as shown in Figure 8.16b. If the gate voltage causes the Fermi level at the surface to cross E_s, the charge state of these traps must change. This change of trapped-charge density introduces a voltage-dependent term Q_{it}/C_{ox} into Equation 8.5.6, causing both the flat-band and threshold voltages (Equation 8.3.18) to vary with V_G, and leads to distorted C-V_G curves. The presence of interface-trap densities that approach typical inversion-layer-charge densities (order of 10^{10} cm^{-2} or higher) is generally unacceptable for reliable device operation. High trap densities seemed unavoidable for many years and were a major cause of the 30-year delay between the MOS device concept and its practical realization. However, with modern MOS technology these trap densities can be reduced to tolerable limits. The density of interface-trapping states is typically reduced by annealing the oxidized silicon wafer in hydrogen or forming gas (a mixture of hydrogen and nitrogen).

Fixed charge at the interface (where some is always found) (Q_f) and within the oxide (where fixed charge is less likely) changes the threshold voltage. Therefore, high fixed-charge densities can cause threshold voltages too high for practical circuit operation, especially when the supply voltage is reduced. The fixed charge always present at the interface (Q_f) appears to arise from incomplete silicon-to-silicon bonds. The density of atoms at the surface of a silicon crystal depends on the crystal orientation; and thus Q_f also depends on the orientation of the silicon wafer. Because more bonds are broken in the transition from (111)-oriented silicon to silicon dioxide than in a similar transition for (100)-oriented silicon, Q_f is generally larger when (111)-oriented silicon is used. For this reason, essentially all commercial MOS devices are built on (100)-oriented silicon. The density of fixed interface charge also depends on the high-temperature processing used, especially in the last fabrication steps. Some of the broken bonds can be completed (and Q_f reduced) by annealing at a high temperature.

The mobile charge Q_m shown in Figure 8.16a usually results from alkali-metal ions (mainly sodium and potassium) that are readily absorbed in silicon dioxide. Sodium is especially widely distributed in many metals and chemicals and is easily transmitted to the oxide by human contact. Because the ions are charged, Equation 8.5.4 shows that the distribution of the charges, as well as their total concentration, influences V_{FB}. The alkali ions are sufficiently mobile to drift in the oxide when relatively low voltages are applied. Their mobility increases rapidly with increasing temperature, so the problem of flat-band instability is more serious at higher temperatures. Because the metal ions are positively charged, negative gate voltage causes the ions to migrate to the metal–oxide interface, where they do not affect the flat-band voltage. However, positive voltage moves the ions to the oxide–silicon interface, where their effect is maximum. Consequently, the characteristics of an MOS structure with mobile ions in the oxide are unstable. For threshold voltage stability of about 0.05 V (50 mV), less than 1×10^{11} cm^{-2} mobile ions can be tolerated in the oxide if the oxide thickness is 10 nm. Mobile-ion contamination was a serious problem that delayed development of practical MOS systems. It is avoided by careful processing and by the introduction of impurities that immobilize the alkali ions; chlorine and its compounds, particularly HCl, can be introduced into the thermal oxidation process for silicon to help produce stable MOS structures (although these additions can lead to other problems).

Charge is also introduced into the MOS system by radiation from sources such as energetic electrons incident on the device during its fabrication or by high-energy particles or photons encountered during operation in a space environment or emitted by packaging materials. Both charge in the oxide Q_{ot} and in interface trapping states D_{it} can be altered by irradiation. As an example, high-energy photons (those with energies greater than the approximately 8 or 9 eV bandgap of silicon oxide) generate electron-hole pairs in the oxide, just as light with energy greater than the bandgap of silicon creates electron-hole pairs in the silicon. However, after electron-hole pairs are generated in the oxide, their behavior differs from that of carriers generated in a semiconductor. Because the oxide is thin and contains few free carriers, the probability of recombination is small. Instead, most of the generated electrons are swept out of the oxide by any field across it. However, the oxide typically contains a large number of hole traps, and many holes are immobilized and increase the positive charge in the oxide. This hole trapping shifts the flat-band voltage, as predicted by Equation 8.5.4.

As can be seen in Figure 8.2, electrons can be excited into the oxide by photons of appreciably lower energies than that of the silicon-dioxide bandgap. These electrons can be photoemitted from the metal or the semiconductor if the photon energy exceeds the energy barrier (3.15 eV in the case of aluminum or 3.1 eV if the emission is from the conduction band of the silicon). In practice, appreciable emission from the silicon only occurs when photon energies are sufficient to excite electrons from the silicon valence band where they are more numerous (requiring photons with energies greater than

~4.2 eV). If an oxide is charged by trapped holes, that charge can be reduced by photoemitting electrons from either the metal or the silicon into the oxide. Some of the photoemitted electrons recombine with the trapped holes, thereby reducing the positive oxide charge.

As described in Sec. 4.4, energetic electrons are created in silicon when an avalanching field is present. Thus, avalanche within the silicon provides another means for electrons to gain sufficient energy to surmount the barrier at the oxide interface. This mechanism is especially significant for device applications because it is under electrical control. In Chapter 10, we consider device effects of avalanche injection more fully. For the present discussion, avalanche near the silicon surface can simply be regarded as an alternative to photoemission for providing a supply of energetic electrons that can enter the oxide.

We conclude our description of oxide charge by noting that each type of charge Q_f, Q_{ot}, Q_{it}, and Q_m can affect the flat-band voltage (Equation 8.5.6) and the threshold voltages (Equation 8.3.18). The only charge density mentioned explicitly in these equations is Q_f; the other charge densities are kept low and, if present, included in the integral expression for charge in the oxide in the equation for V_{FB}.

Experimental Determination of Oxide Charge

We just saw that the oxide charge present in the MOS system changes its electrical characteristics. Fixed charge changes the flat-band voltage, and interface traps affect the transition of the silicon surface from depletion to inversion, causing the threshold voltage to change. In the next chapter, we will also see that charged interface traps reduce the mobility of the inversion-layer carriers and thus the drain current of an MOS transistor. To ensure that an MOS technology produces devices with predictable, stable, and reliable characteristics, device and technology engineers need to develop fabrication processes that produce as little oxide charge as possible. Devices produced in an MOS fabrication facility are constantly monitored to track the amount and type of oxide charge they contain. A widely used technique to measure the mobile charge density in the oxide is the "bias-temperature" test, in which a field is applied across the structure at an elevated temperature (e.g., 125°C) to move the mobile ions and change their effect on V_{FB}; the C-V curve is subsequently measured at room temperature. The high-temperature bias is then applied in the opposite direction, followed by another C-V measurement at room temperature. The hysteresis of the C-V curves indicates the amount of mobile oxide charge, and the detailed shape of the curve provides information about trapped charge (Problem 8.14). A mobile charge density equivalent to a few tens of millivolts hysteresis is often the maximum amount allowed.

Two different techniques are commonly used to quantitatively measure the interface trap density, as shown in Figure 8.17. In one technique, the measured high-frequency C-V curve (usually 100 kHz or above) of the MOS capacitor is compared with the corresponding theoretical C-V curve calculated for the same capacitor assuming that the MOS system is ideal (without oxide charge). To obtain the theoretical C-V curve, the exact doping concentration and doping profile and the oxide thickness must be known accurately. These parameters are usually measured using a combination of physical, optical, and electrical techniques during and after fabrication. To illustrate the technique, assume that we have obtained and properly calibrated the experimental and theoretical C-V curves. In the portion of the C-V curve before strong inversion begins, consider two points, one on each curve, at which the capacitance is the same (Figure 8.17a). Note that the experimental C-V curve is measured at high frequency, where the interface states do not respond to the ac signal; therefore, at these two points, the depletion capacitance and the depletion depth into the substrate (and consequently the

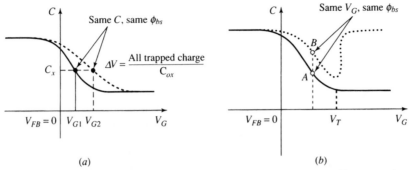

FIGURE 8.17 Measurement of interface-trap density using two different methods: (*a*) ideal and nonideal high-frequency *C-V* characteristics; (*b*) nonideal low-frequency and nonideal high-frequency *C-V* characteristics.

surface potential) are the same. For the measured non-ideal capacitance, however, the interface traps below the Fermi level are filled by electrons, creating a voltage shift between the ideal and experimental points. This voltage shift, $V_{G2} - V_{G1}$, can be written in terms of the interface-state density D_{it} (which depends on energy and, therefore ϕ_s, and has units of cm^{-2}-eV^{-1}):

$$V_{G2} - V_{G1} = \frac{q \int_{\phi_{s1}}^{\phi_{s2}} D_{it}(\phi_s) d\phi_s}{C_{ox}} \qquad (8.5.7)$$

where ϕ_{s1} and ϕ_{s2} are the surface potentials corresponding to V_{G1} and V_{G2}, respectively. The values of the surface potential ϕ_s are known from the ideal *C-V* curve. With such data for each value of capacitance, D_{it} can be numerically calculated as a function of ϕ_s and, thus, the energy distribution of interface traps within the bandgap can be obtained.

With the second technique, the measured high-frequency *C-V* curve is compared to the measured low-frequency *C-V* curve of the same capacitor (Figure 8.17*b*). The low-frequency *C-V* curve is usually obtained using the quasi-static method described earlier. Because the dc condition of the two curves is the same, the same gate bias on the two curves corresponds to the same surface potential in the silicon. For the high-frequency *C-V* measurement the interface traps do not respond, so the equivalent capacitance is simply the series combination of the oxide capacitance C_{ox} and the depletion capacitance $C_{sHF} = \epsilon_s/x_d$ in the silicon, that is,

$$C_{HF} = \frac{1}{1/C_{ox} + 1/C_{sHF}} = \frac{1}{x_{ox}/\epsilon_{ox} + x_d/\epsilon_s} \qquad (8.5.8)$$

At low frequencies, the capacitance in the substrate also includes some contribution from charging and discharging the interface traps.

$$C_{LF} = \frac{1}{1/C_{ox} + 1/C_{sLF}} \qquad (8.5.9)$$

where

$$C_{sLF} = \frac{dQ_s}{d\phi_s} = \frac{d(Q_{it} + Q_d)}{d\phi_s} = qD_{it} + \frac{\epsilon_s}{x_d} \qquad (8.5.10)$$

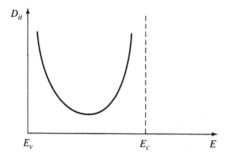

FIGURE 8.18   U-shaped distribution of interface traps within the bandgap of silicon at the oxide–silicon interface.

From Equations 8.5.8, 8.5.9, and 8.5.10, D_{it} can be calculated for the corresponding surface potential ϕ_s. The value of ϕ_s at any gate voltage can be determined using a technique known as Berglund's method [5], described in Problem 8.5. In a practical MOS system, the interface trap density usually has a U-shaped distribution within the bandgap, being higher at the band edges than near midgap, as shown in Figure 8.18.

8.6   SURFACE EFFECTS ON *pn* JUNCTIONS[†]

Several effects important to device operation can occur when a *pn* junction exists in the vicinity of an oxide that is covered by an overlying gate, as in the structure sketched in Figure 8.9. We already saw in the discussion of Figure 8.9 in Sec. 8.3 that the bias applied to the junction can modify the charge Q_n stored in the channel and the charge Q_d stored in the depleted regions of an inverted MOS system. This is an effect of the junction on the MOS system. There is also a strong need for the integrated-circuit designer to understand the effect of the MOS system on the properties of the junction. Most *pn* junctions in devices made by the planar process intersect an oxide-silicon interface. Thus, the properties of the oxide-silicon system can exert a significant influence on the circuit performance of *pn*-junction devices, such as bipolar junction transistors.

Figure 8.19 shows the cross section of a planar n^+p integrated-circuit diode. If this diode is, for example, situated above a second *pn* junction, it might represent the emitter-base junction of a bipolar transistor. In Sec. 5.3, we showed that the generation of electron-hole pairs in the depletion region of a *pn* junction is generally the dominant source for the reverse leakage current of the junction. Similarly, at low forward bias, recombination in the space-charge region is the major current component. Generation and recombination in the space-charge region not only cause deviations from the ideal-diode law but, more seriously (as discussed in Chapter 6), these processes degrade transistor performance by adding components of the base current that are not delivered to the collector. Some special features of these processes in surface space-charge regions deserve our attention.

Consider section A-A' through the diode in Figure 8.19a. The ideal *pn*-junction analysis considered the oxide–silicon interfaces to be free of charge and characterized by flat bands, as sketched in Figure 8.19b. The discussion of the MOS system, however, made clear that the flat-band condition does not correspond to thermal equilibrium. A gate-to-substrate bias equal to V_{FB} must usually be applied to achieve flat-band conditions. If there is no electrode over the oxide at section A-A', then the most significant influence on the surface is that of oxide charges. Oxide charges are almost always positive, causing either depletion or even inversion of *p*-type silicon (Figure 8.19c). Conversely, positive charge in the oxide tends to accumulate *n*-type silicon. (The consequence of these

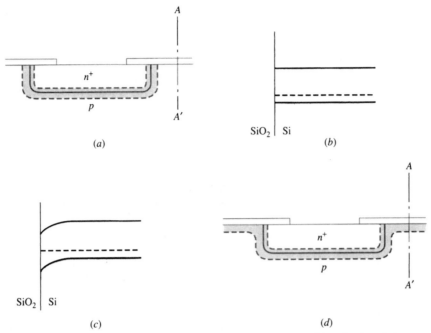

FIGURE 8.19   (*a*) Planar junction diode. (*b*) Ideal analysis considers flat band along section AA'. (*c*) Real oxide-to-*p*-silicon interface is usually depleted because of positive oxide charge. (*d*) Surface depletion region is connected to junction depletion region.

effects on the reliable production of bipolar transistors is considered in Problem 8.13.) Again considering the *p*-region in Figure 8.19, we see that surface depletion in the vicinity of section *A-A'* enlarges the overall junction depletion region because it connects to the depletion region of the diffused junction. If the surface is inverted, there is an extension of the conducting *n*-region along the oxide surface, as well as an enlargement of the depletion region.

One effect of the enlarged depletion region is to provide more volume for the generation of current under reverse bias. Of greater significance, however, is the higher density of generation-recombination sites at the oxide-silicon interface than in the bulk. In addition, the activity of these sites depends on the surface potential at the oxide-silicon surface, as was shown in the Shockley-Hall-Read (SHR) theory for generation and recombination (Sec. 5.2).

Gated-Diode Structure[†]

To consider the effect of the surface potential on the *pn* junction, we introduce the concept of a *gated diode* [6]. This structure is shown in Figure 8.20*a* in which a metal gate overlays both the *p* and *n* regions of a diode. To focus attention on the physical mechanisms in the gated diode, we consider that the bulk of the silicon is at ground potential ($V_B = 0$), and we take the *n*-region to be biased positively to a voltage V_R (reverse bias on the junction).

At gate voltages V_G that are negative with respect to flat band, the surface of the *p*-type region is accumulated, and the depletion region at the surface is small. The depletion region near the surface in the *n*-type region is hardly changed by the gate voltage because it is heavily doped. Therefore, for negative V_G, the leakage current of the

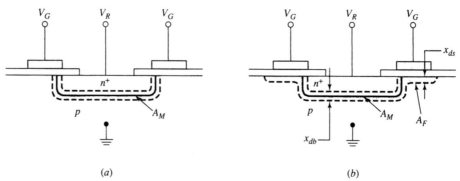

(a)   (b)

FIGURE 8.20 Gated-diode structure showing depletion regions: (a) when the gate voltage is V_{FB} for the p-region, and (b) when the gate causes a depletion region of width x_{ds} at the silicon surface.

diode is approximately determined by the generation of holes and electrons in the fabricated or *metallurgical* junction region. From Equation 5.3.26, this current, which we denote by I_M, is

$$I_M = \frac{qn_i}{2\tau_0} x_i A_M \tag{8.6.1}$$

where τ_0 is the lifetime, A_M is the area of the metallurgical junction, and x_i is the active region for generation as described in Sec. 5.3. In practice, x_i can be taken to be equal to the bulk depletion-layer width x_{db} and therefore varies with V_R in the same way as does x_{db} [often as $(V_R + \phi_i)^{1/2}$]. The current is nearly insensitive to V_G for this bias condition.

If V_G is made more positive than the flat-band voltage, the surface of the p-region under the gate electrode is depleted, as shown in Figure 8.20b. Two components of current now flow, in addition to I_M as given in Equation 8.6.1. First, current results from the generation of carriers in the depletion region induced by the gate. Again from Equation 5.3.26 this component, which we can call I_F (arising from the *field-induced* junction), is

$$I_F = \frac{qn_i}{2\tau_0} x_{ds} A_F \tag{8.6.2}$$

where x_{ds}, the width of the depletion region at the surface, is a function of the applied gate voltage V_G; and A_F, the area of the surface depletion region, is determined by the coverage of the gate electrode over the p-region. The second component of current that arises when the p-region is depleted results from the activity of surface generation sites. This component, which we call I_S, is best described in terms of the parameter called the surface recombination velocity s that we introduced in Sec. 5.2. There, it was shown that the surface recombination velocity is directly proportional to N_{st}, the density of generation-recombination sites at the surface (Equation 5.2.23).* If we consider that the sites have energies near E_i, then (applying Equation 5.2.20 for p_s and $n_s \ll n_i$) we calculate I_S as q times the generation rate

$$I_S = \frac{qn_i s_o A_F}{2} \tag{8.6.3}$$

* The states N_{st} are a subset of the interface trapping-state density N_{it} introduced in Sec. 8.5. The N_{st} states are characterized by energies near the intrinsic Fermi level and by nearly equal rates of interchange with electrons and holes.

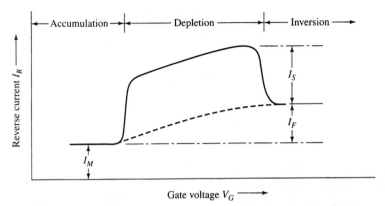

FIGURE 8.21   Reverse current in the gated diode as a function of V_G showing the marked increase in leakage when the surface is depleted. The currents I_S and I_F are given by Equations 8.6.2 and 8.6.3.

where s_o was defined in Equation 5.2.24, $s_o = N_{st}v_{th}\sigma$, v_{th} is the thermal velocity, and σ is the capture cross section associated with the generation-recombination site. The value of s_o is directly proportional to the number of surface generation-recombination centers, so it is strongly influenced by device processing and annealing procedures.

If the gate voltage increases until the surface of the p-type silicon becomes inverted, I_F in Equation 8.6.2 increases and reaches a maximum value when $x_{ds} = x_{dmax}$. Once inversion occurs, however, the surface electron density n_s becomes much greater than the intrinsic density n_i, and the surface recombination velocity decreases markedly from s_o following the prediction of Equation 5.2.23. For typical values (Problem 8.16) I_F is smaller than I_S when the surface is depleted. After inversion I_S (the surface generation component) is negligible and the reverse-bias current becomes the sum of I_M and I_F. Typical measured behavior for reverse leakage current in a gated diode is sketched in Figure 8.21. When the pn junction is forward biased at low voltages, the surface space-charge region also affects the recombination current, but the effect is typically small.

In calculating the behavior of the surface-charge layer, the dependence of the threshold voltage on the reverse bias across the junction must be considered. Quantitatively, the presence of a reverse bias on the junction can be incorporated into the threshold-voltage calculation by using Equation 8.3.18 and letting the channel voltage V_C be equal to the diode reverse-bias voltage V_R while holding the bulk voltage V_B at zero.

In the absence of a gated-diode structure, the surface condition is determined by the oxide charge. In many cases of IC processing for bipolar transistors or MOS isolation structures, the surface charge creates a wide depletion region over the p-region surfaces. In these cases, pn junctions have high leakage currents and are said to be *leaky*. An annealing step is often used to reduce s_o and the oxide charge and usually improves the junction characteristics considerably.

8.7 MOS CAPACITORS AND CHARGE-COUPLED DEVICES

The most straightforward device use of the oxide–silicon system is to make high-quality, precisely controlled capacitors. One example of this application is shown in Figure 8.22a, which is an enlarged view of an integrated circuit whose function is the conversion of

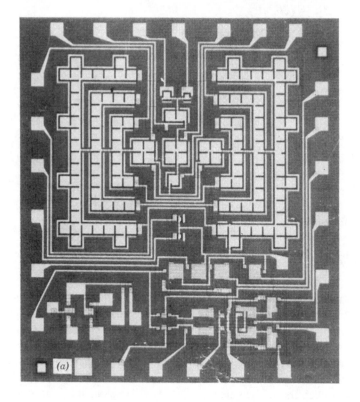

(a)

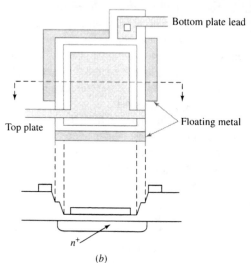

(b)

FIGURE 8.22 (a) Integrated circuit for the conversion of analog signals to a digital representation [7]. The circuit makes use of many precision MOS capacitors having the structure shown in (b).

analog signals to digital representation [7]. The A/D conversion is accomplished by sequential comparison of the signal to fractions of a reference voltage. The reference voltage is divided by comparison with voltages on an array of capacitors having capacitances that are successively reduced by factors of two.

The capacitors in the A/D conversion circuit are MOS devices visible as square structures in Figure 8.22a. For accurate conversion of analog signals, precise control over the ratios of the capacitance values is necessary. This control is successfully achieved with

MOS capacitors having the structure shown in Figure 8.22b. The floating metal strips in Figure 8.22b serve only to assure reliable precision during etch steps used to fabricate the capacitors and do not have any circuit function [7].

MOS Memories

In the A/D converter, an accurate value of capacitance is obtained by biasing the capacitor in the accumulation region so that the MOS capacitance corresponds to the oxide capacitance, which can be accurately controlled.

When the capacitor is biased so that inversion can be obtained, a completely different function is achieved. Consider the n-channel MOS structure shown in Figure 8.23, biased with a positive gate voltage $V_G > V_T$. In steady state, an inversion layer of electrons forms under the gate electrode (Figure 8.23a), and the inversion layer is separated from the p-type substrate by a depletion layer. Part of the charge needed to compensate the gate charge is provided by negative mobile electrons and part is provided by uncompensated acceptors in the depletion layer, corresponding to the curve HF in Figure 8.11. However, we saw in Sec. 8.4 that, if the capacitor is isolated, a significant time is needed to generate thermally the electrons that form the inversion layer. After the bias is initially applied, the capacitor is in deep depletion (Figure 8.23b). All the charge needed to compensate the gate charge is provided by uncompensated acceptors in the depletion layer, corresponding to the curve DD in Figure 8.11. As electrons are generated, the fraction of charge provided by mobile electrons in the inversion layer increases and the capacitor is in a state intermediate between deep depletion and inversion. Consequently, we now have a structure that can contain a variable amount of mobile charge even with the same gate voltage, and we can use the stored mobile charge as an electrical signal.

Because the stored mobile charge can represent information, the MOS capacitor is widely used in memories that store computer data. The channel is eventually filled by thermally generated electrons and reaches inversion, so such a memory cell is called *dynamic*. MOS memory cells are typically organized in regular arrays containing rows and columns, with connections in the two perpendicular directions to access an individual cell. In the typical organization of the memory, any cell in the array can be accessed at any time with equal ease, so the memory organization is called *random*; hence, the name *dynamic, random-access memory* or DRAM.

The memory cell is typically used to store digital data, so it has only two states of interest, represented by mobile charge creating an inversion layer or the absence of mobile charge placing the capacitor in deep depletion. The amount of charge is sensed by other components to determine if a "0" or a "1" is stored. For density and economy, the circuitry that senses and amplifies the charge (the *sense amplifier* or *sense amp*) is placed at the end of each column of the array. The cell is first connected to a *bit line*

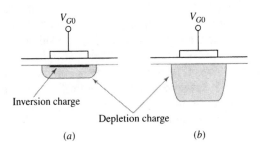

FIGURE 8.23 MOS capacitor with a fixed gate voltage V_{G0} in: (a) inversion, and (b) deep depletion.

(the vertical column of the array) and then detected by the sense amplifier. When the stored charge is connected to the bit line, it is diluted by the capacitance of the bit line. The voltage appearing at the sense amplifier is only $V_{cell} \times C_{cell}/(C_{cell} + C_{bitline})$. As the memory array becomes larger, the bit-line capacitance increases. To retain an adequate ratio $C_{cell}/(C_{cell} + C_{bitline})$ to allow sensing the state of the cell, the capacitance of a dynamic memory remains relatively large and does not scale with the minimum dimensions of the technology used to build it. The typical cell capacitor was only reduced in size moderately (by a factor of ~2 from 50 pF to ~25 pF) as the minimum feature size decreased from 2 μm to 250 nm. To allow more memory cells to be placed on a chip, the projected area of each cell (the *footprint*) must decrease in size by using the area more effectively.

In the early 1970s, a typical gate oxide was ~200 nm thick. As gate-oxide thickness scaled to 4 nm, the capacitance per unit area increased by 50×, so the required area for the somewhat lower capacitance decreased by ~2 × 50 = 100×. However, as the dielectric thickness decreased by 50×, so also did the minimum linear feature on the surface of the chip (to satisfy the scaling laws discussed in Sec. 2.1 and Chapter 9). Consequently, the minimum area of a device feature decreased by 2500×—much more than the 100× reduction in the capacitor area. The area needed for the capacitor becomes inappropriately large if only the gate-oxide thickness scales, and other techniques are needed to reduce the area on the substrate significantly while the capacitance decreases only slowly.

The capacitance varies as $\epsilon_r \epsilon_0 A/x_{ox}$, where x_{ox} is the thickness of the dielectric and ϵ_r is its relative permittivity—often represented by the symbol κ or K. Several of the factors in the expression for the capacitance can be changed to increase the ratio of the capacitor value to the feature size. (1) The thickness of the dielectric can be decreased as technology is improved. Its lower limit is set by current leakage and by the ability to form suitably sized areas of uniform, defect-free dielectric (the *yield*). A thickness approaching 2–3 nm appears to be the limit for pure SiO_2. (2) A dielectric with a higher permittivity can be used. Silicon nitride, with a relative permittivity of ~7, is sometimes used in combination with silicon dioxide (ϵ_r ~4); higher permittivity materials, such as tantalum pentoxide (Ta_2O_5 for which ϵ_r ~22) and more complex materials such as alloys of barium and strontium titanate ($Ba_{1-x} Sr_x TiO_3$) with a relative permittivity of several hundred, are being considered. (3) The projected area on the surface of the chip can be used more effectively by extending the cell in the third dimension or by using an irregular, rough surface to increase the surface area (and thereby the capacitance).

In some cases, a well or *trench* is etched in the substrate. The sides of the trench are oxidized to form a gate dielectric and the trench is filled with a conducting counterelectrode, as shown in Figure 8.24a. The effective area of the *trench capacitor* is approximately $4F \times D$, where F is the *feature size* on the surface (smallest linear surface dimension) and D is the effective depth of the trench. The inversion layer where charge is stored now extends laterally out from the vertical dielectric into the adjacent regions of the substrate. Although the effective area is significantly increased, this technique soon encounters limits. When adjacent capacitors are closely spaced, the depletion regions surrounding the capacitors in the lateral direction can interact by "punching through" the neutral region separating adjacent capacitors (Figure 8.24b). The minimum separation between adjacent capacitors is $2x_d$ where x_d is the maximum extent of the depletion region, essentially that corresponding to deep depletion (i.e., no mobile charge stored on the capacitor). An alternative technique places the counterelectrode, which is held at a fixed

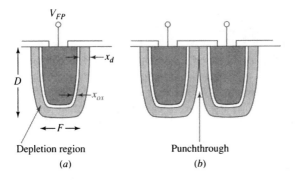

FIGURE 8.24 (*a*) Cross section of a trench capacitor showing the feature size *F* and well depth *D*. (*b*) Interaction between two adjacent trench capacitors when two depletion regions touch.

voltage and common to all the cells, outside the well. The voltage-varying charge-storage region is then inside the trench; this region is heavily doped so that its capacitance does not vary significantly with voltage.

Charge-Coupled Devices

The MOS memory capacitor described above is used to store digital data, so only two states are needed. Analog information can also be stored on an MOS capacitor. An MOS capacitor biased as we discussed above, but operating anywhere between deep depletion and inversion is the basis of the *charge-coupled device* (*CCD*) [8] which combines analog memory with the ability to move charge from one physical location to another. Charge transfer is achieved by building an array of capacitors that are spaced closely enough to one another so that the mobile charge in the inversion layer (the channel) of one MOS capacitor can be transferred to the channel of an adjacent capacitor when the electrodes are appropriately biased.

The CCD is widely used in photo imagers to sense incoming illumination. The ability to form a dense array of sensors allows high-resolution imaging. For this application, a field of CCD cells or *picture elements* (*pixels*) is biased into deep depletion and exposed to a focused image for a time interval. The incoming photons generate electron-hole pairs in the silicon at a rate proportional to the brightness of the image. Within each cell, the electric field in the depletion region separates the electrons and holes, sending the electrons toward the oxide-silicon interface where they are stored. The channel under each gate becomes charged to a level that represents the brightness of the image at its location. As photons continue to illuminate the pixel, the surface potential eventually approaches its inversion value, and the amount of stored charge saturates. Therefore, the cell is only exposed for a limited period of time. After the charge is generated, the analog information in the cell is moved along the array into sensing circuitry built on the edge of the CCD imaging array. This design is used for integrated-circuit still cameras and television cameras.

Because of thermal generation, there is a lower limit for the speed of CCD operation. The charge corresponding to the optical signal must be sensed and transferred out of the array at a rate fast compared to the background charge generation at a depleted silicon surface. No cell of the CCD can be held for a long time in a deep-depletion condition, and noise is added because of thermal generation over the period during which the signal traverses the CCD array.

These constraints impose very severe requirements on the quality of the oxide-silicon system. As we saw from the discussion of surface generation rates, the requirements imposed can be met if the density of generation-recombination sites at the surface is kept very low. Manufacturers of CCD circuits routinely hold these state densities to values in the low 10^9 cm^{-2} range, which results in surface-generation currents that are in the range of 1 to 10 nA cm^{-2}. A small amount of charge is involved in many CCD circuits. Because CCD gates are typically a few $(\mu m)^2$ in area, values of detected charge range from roughly 10 electrons to about 10^7 electrons. By comparison, the thermal generation current of 1 nA cm^{-2} corresponds to roughly 60 electrons s^{-1} under a 1 $(\mu m)^2$ gate.

EXAMPLE *Charge-Coupled Devices*

We want to design a charge-coupled device as an image sensor with square gates 5 μm on a side functioning as picture elements (*pixels*). The detectable charge threshold is 2500 electrons per pixel, and the charge on each pixel is accessed and reset to zero every 10 ms. The inversion-layer charge density for the CCD at thermal equilibrium is 10^{13} electrons cm^{-2}.

If the thermal (unilluminated) generation of electrons is described by exponential time behavior (as derived in Problem 8.6), determine the required minority-carrier lifetime τ_0 in 12 Ω-cm p-type silicon such that less than 5% of the detectable threshold charge is created by thermal generation.

Solution At thermal equilibrium, on each gate there are $10^{13} \times (5 \times 10^{-4})^2 = 2.5 \times 10^6$ electrons. The detectable charge threshold is 2500 electrons and the permissible number of thermally generated electrons is 2500 times $0.05 = 125$ on each gate.

Hence $2.5 \times 10^6 [1 - \exp(-t/\tau_a)] = 125$, which implies $t/\tau_a = 5 \times 10^{-5}$. For a generation time $t = 10^{-2}$ s, the minimum surface-generation lifetime τ_a that meets the requirements is

$$\tau_a = \frac{10^{-2}}{5 \times 10^{-5}} = 2 \times 10^2 \text{ s}.$$

From the analysis in Problem 8.6,

$$\tau_0 = \frac{n_i}{2N_a} \times \tau_a$$

From Figure 1.15, $N_a = 10^{15}$ cm^{-3} for 12 Ω-cm Si; therefore, the minimum acceptable lifetime is

$$\tau_0 = \frac{1.45 \times 10^{10}}{2 \times 10^{15}} \times 2 \times 10^2 = 1.45 \times 10^{-3} \text{ s} = 1.45 \text{ ms}$$

The results of this problem emphasize that the characteristic time τ_a for surface generation in a CCD is much longer (200 s) than the minority-carrier lifetime τ_0 (1.45 ms). ∎

To discuss the charge transfer within a CCD, we assume that the charge generation and transfer times are short compared with the time necessary for thermal generation of an inversion layer. We consider the MOS system shown in Figure 8.25 in which three capacitors are situated side by side on a silicon surface. Assume that the voltage V_2 applied to the middle gate is higher than the voltage applied to either of the side gates, and that V_2 is greater than the MOS threshold voltage. If electrons are introduced into the surface region, they reside in the channel under the middle gate. Applying a more positive voltage V_3 to the right-hand gate causes the electrons to be transferred to the channel beneath it (Figure 8.25b). The voltage on the middle gate can now be decreased to V_1, and then the voltage on the right-hand gate can be reduced to V_2. The net result is a shift of the channel charge one stage to the right.

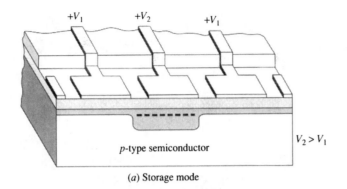

$+V_1$ $+V_2$ $+V_1$

$V_2 > V_1$

p-type semiconductor

(*a*) Storage mode

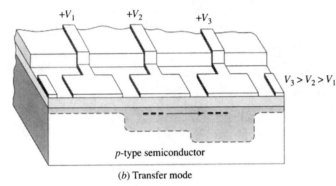

$+V_1$ $+V_2$ $+V_3$

$V_3 > V_2 > V_1$

p-type semiconductor

(*b*) Transfer mode

FIGURE 8.25 Basic transfer mechanism in a CCD. (*a*) In the storage mode, charge is held under the center gate, which has a channel beneath it. (*b*) Application of $V_3 > V_2$ on the right-hand gate causes transfer of charge to the right.

A key factor in the production of a CCD is the spacing of adjacent gates. The gates must be close enough to one another to permit the fringing fields to allow the charge to be transferred when desired. Although systems have been built utilizing only metal gates, employing silicon gates singly or in two or three levels is useful, especially for implementing more elaborate clocking schemes than the one we described here.

A portion of a high-density, frame-transfer, imaging CCD is shown in Figure 8.26. This CCD uses three layers of polysilicon and an *n*-type substrate with an implanted *p*-well. The peripheral circuits to operate the CCD are implemented in CMOS technology (described in Chapter 9).

In addition to imagers, another analog application of a CCD is its use as a delay line. By accessing the signal after differing delay periods, various useful signal-processing tricks, such as convolution (which involves multiplication of a time-varying signal by a delayed version of itself) can be easily implemented with CCDs.

CCDs can also be used to store digital data. Because only two states ("0" and "1") are stored, the requirements on noise and charge generation are somewhat relaxed for the same number of stages. However, a desired bit of information (and all bits in the same column) must be clocked to the sensing circuitry at the edge of the chip to be read. The variable, sometimes long, time periods needed to access an arbitrary bit of information, make this type of memory less flexible than a random-access memory.

From this discussion, we see that the CCD is a very useful device for selected applications, especially imaging. It has also been used for signal processing and digital memories. The ability to form a CCD with very low thermal-carrier generation at the oxide–silicon interface evolved from a detailed understanding of the technology and electronics of the MOS system.

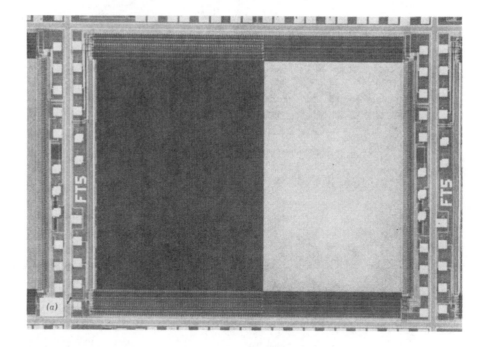

FIGURE 8.26 (*a*) A high-density frame-transfer CCD. The sensor array is on the left side of the chip and the light-shielded storage area is on the right. This CCD has 588 lines of 604 pixels each. The chip area is 38.22 mm². (*b*) Peripheral CMOS circuitry for the CCD. (© 1984, IEEE) [11].

SUMMARY

The electronic behavior of the oxide-silicon system is profitably studied by considering the effects of voltages applied to a metal-oxide-silicon (MOS) structure. An important parameter of the MOS structure is the *flat-band voltage* V_{FB}. The significance of V_{FB} becomes clear when an energy-band diagram is constructed for the MOS system. When a voltage V_{FB} is applied between the metal and the substrate, there is no field or charge at the surface of the silicon. The flat-band voltage is determined by several parameters. For ideal (charge-free) oxides and interfaces, V_{FB} depends only on the work-function difference between the metal and the semiconductor. Practical structures, however, contain charges both in the oxide and at the oxide-silicon interface, so the flat-band voltage also depends on these charges. If the quantity of charge in the oxide is itself changed by the applied voltage, the MOS characteristics are unstable. Generally, variable oxide-charge densities must be kept below a few times 10^{10} cm^{-2} to produce acceptable MOS devices. Alkali metals, particularly sodium, are especially troublesome oxide impurities because they can be moved by the field applied across the MOS oxide.

In many cases there are practical advantages to making the "metal" electrode in an MOS device from doped silicon. The silicon is usually formed over amorphous silicon dioxide by chemical vapor deposition. It is composed of many small crystallites and is therefore known as *polycrystalline silicon.*

A voltage applied across an MOS capacitor between the metal or silicon gate and the substrate can accumulate or deplete the silicon surface of bulk majority carriers. It can also bias the silicon to inversion, in which case a *pn* junction is induced near the silicon surface. The charge controlled by the gate is then distributed, partly as fixed ions in the depletion layer, and partly as free carriers in the inversion layer. The three conditions—*accumulation, depletion,* and *inversion*—can be sensed by measuring the MOS capacitance. In the inversion region, the measured MOS capacitance depends on the frequency of the applied voltage because of the time required to generate the carriers that create the surface inversion layer. For practical MOS systems with *interface traps*, the measured MOS capacitance in the depletion region can also depend on frequency because of the time constants associated with charging and discharging the traps.

When a surface is inverted, the inversion layer is often called a *channel*. If the channel exists adjacent to a *pn* junction, the induced *pn* junction can be biased through the diffused *pn* junction. This method of biasing can alter the distribution of gate-induced charge between the depletion layer, where the charge is fixed, and the inversion layer, where it is mobile. The space-charge layer at the silicon surface can affect the characteristics of *pn* junctions. The activity of *surface states* depends on the surface potential, as demonstrated by the behavior of *gated diodes*.

An important direct application of the MOS system is for the production of integrated-circuit capacitors, often used to store charge in *dynamic, random-access memories* and in analog-to-digital signal conversion. Arrays of closely spaced MOS capacitors are used to form *charge-coupled devices* (CCDs), which can be used for optical imaging.

REFERENCES

1. R. H. KINGSTON and S. F. NEUSTADTER, *J. Appl. Phys.* **26**, 718 (1955).

2. C. E. YOUNG, *J. Appl. Phys.* **32**, 329 (1961).

3. S. M. SZE, *Physics of Semiconductor Devices, 2nd Edition*, Wiley-Interscience, New York, 1981, p. 372.

4. B. E. DEAL, *J. Electrochem. Soc.* **127**, 979 (1980).

5. C. N. BERGLUND, *IEEE Trans. Electron Devices*, **ED-13**, 701 (1966).

6. A. S. GROVE and D. J. FITZGERALD, *Solid-State Electron.* **9**, 783 (1966).

7. J. L. MCCREARY and P. R. GRAY, *IEEE J. Solid-State Circuits*, **SC-10**, 371 (1975). Reprinted by permission.

8. W. S. BOYLE and G. E. SMITH, *Bell Sys. Tech. J.* **49**, 587 (1970).

9. C. JUND and R. POIRER, *Solid-State Electron.* **9**, 315 (1966).

10. N. TERANISHI, A. KOHNO, Y. ISIHARA, E. ODA, and K. ARAI, *IEEE Trans. Electr. Devices*, ED-31, 1829 (1984).

11. A. J. P. THEUWISSEN, C. H. L. WEIJTENS, L. J. M. ESSER, J. N. G. COX, H. T. A. R. DUYVELAR, and W. C. KEUR, *Tech. Digest IEEE Int. Electr. Devices Mtg.* (Dec. 1984), p. 40.

PROBLEMS

8.1* Sketch the energy-band diagrams (i) at thermal equilibrium and (ii) at flat band for ideal MOS systems made with aluminum gates (a) to 1 Ω-cm n-type silicon, and (b) to 1 Ω-cm p-type silicon.

8.2 Repeat the sketches required in Problem 8.1 for an MOS system with a polycrystalline silicon gate. Assume that the silicon gate has a band structure similar to single-crystal silicon but that (i) the gate over n-type silicon is doped with acceptors until it is just at the edge of degeneracy and (ii) the gate over p-type silicon is doped with donors until it is just at the edge of degeneracy. (These conditions correspond to usual silicon-gate technology for reasons to be described in Chapter 9.)

8.3 Prove that the small-signal capacitance of an MOS capacitor C, biased into depletion, is given by Equation 8.4.4. That is, show that C is equal to the capacitance of a series connection of two capacitors: (1) a capacitor made with one plate in the bulk of the silicon and the other plate at the oxide-silicon interface, and (2) a capacitor that has its plates separated by the oxide. (*Hint. Use Gauss' law to express the charge* $\Delta Q = \epsilon_{ox}\Delta\mathscr{E}_{ox}$. *Then, show that the voltage across the capacitor is* $\Delta V = \Delta\mathscr{E}_{ox}x_{ox} + \Delta\mathscr{E}_{ox}\epsilon_{ox}x_d/\epsilon_s$ *and evaluate* $C = \Delta Q/\Delta V$.)

8.4 Take $V_{FB} = -0.5$ V and use Equation 8.4.4 to show the behavior of the overall capacitance C for an MOS system in the depletion region. Sketch a plot of C/C_{ox} versus V_G. Consider that the silicon oxide is 100 nm thick and the silicon is p-type with 1 Ω-cm resistivity. Locate the flat-band capacitance C_{FB} using Equation 8.4.3.

8.5 The value of ϕ_s (the surface potential) is frequently needed for experimental studies of MOS systems.

(a) By using the results of Problem 8.3, show that when the gate voltage V_G is changed on an MOS capacitor biased in the depletion region, it is possible to find the corresponding change in ϕ_s by using the measured capacitance of the MOS system. The

change in ϕ_s can be calculated from the relationship

$$\phi_s(V_{G2}) - \phi_s(V_{G1}) = \int_{V_{G1}}^{V_{G2}}\left(1 - \frac{C}{C_{ox}}\right)dV_G$$

This technique is known as *Berglund's method* after its originator [5]. It can be used conveniently if V_{G1} is taken to be V_{FB}, at which point C is given by Equation 8.4.3.

(b) If V_{G1} is taken as V_{FB}, sketch a low-frequency MOS capacitance curve for p-type silicon (normalized to C_{ox}) and indicate (by shading) an area on the curve equal to $\Delta\phi_s$.

8.6† Consider that an MOS system on p-type silicon is biased to deep depletion by the sudden deposition of a total charge Q_G on the gate at $t = 0$. Carrier generation in the space-charge region at the silicon surface results in a charging current for the channel charge Q_n as described in the discussion of Equation 5.3.26. This allows one to write

$$\frac{dQ_n}{dt} = -\frac{qn_i(x_d - x_{df})}{2\tau_0}$$

where x_d is the (time dependent) depletion-region-width at the surface and τ_0 is the electron lifetime as given in Equation 5.2.14. The quantity x_{df} is the space-charge region width at thermal equilibrium; that is, when $x_d = x_{df}$, channel charging by generation goes to zero.

(a) Show that a differential equation for Q_n is

$$Q_n + \left(\frac{2\tau_0 N_a}{n_i}\right)\left(\frac{dQ_n}{dt}\right) = -[Q_G - qN_ax_{df}]$$

(b) Solve this equation subject to $Q_n(t = 0) = 0$ and thus show that the characteristic time to form the surface inversion layer is of the order of $2N_a\tau_0/n_i$ [10].

8.7 Sketch capacitance-voltage curves of the MOS structures shown in Figures P8.7a, P8.7b, and P8.7c. The capacitance is the small-signal value normalized to that of the oxide and measured at 100 kHz. In all cases, the gate dc bias is varied slowly. Show (by using dotted curves) what effect an increase in positive Q_f would have on the C-V_G curves. Label each

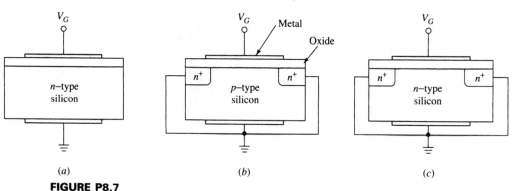

(a)   (b)   (c)

FIGURE P8.7

region on the curves (accumulation, depletion, and inversion). Assume that the substrate resistivity is of the order of 10 Ω-cm in each case and make your sketches qualitatively correct.

8.8 Sketch the curves as described in (a) through (d) below for an MOS capacitor on an n-type substrate that has been biased to inversion. Consider that $V_{FB} = -2$ V mainly because of the presence of fixed oxide charge Q_f. The sketches should show (a) the band diagram, (b) all charge in the system, (c) the electric field, and (d) the potential. (Use the silicon bulk as the reference for potential.)

8.9 Construct a table similar to Table 8.1 to represent the surface-charge conditions for n-type silicon.

8.10 Using the formulas in Sec. 8.3, prove that for $n_s = 10 N_a$, ϕ_s is only 58 mV greater than $-\phi_p$.

8.11[†] Consider the dependence on $(V_C - V_B)$ of the expressions for Q_n (Equation 8.3.16) and V_T (Equation 8.3.18) in order to sketch a qualitative family of (low frequency) curves for C/C_{ox} versus V_G as $(V_C - V_B)$ is varied. This dependence was studied by Grove and Fitzgerald [6].

8.12[*] Find the threshold voltage (a) in 1 Ω-cm p-type silicon and (b) in 1 Ω-cm n-type silicon. The MOS system for each case is characterized by: (i) aluminum gate for which $q\Phi_M = 4.1$ eV, (ii) 100 nm silicon dioxide, (iii) the oxide is free of charge except for a surface density $(Q_f/q) = 5 \times 10^{10}$ cm^{-2}. The channel is not biased except from the gate $(V_C = V_B = 0)$.

8.13[†] Consider the effects of oxide charge on the surfaces of n and p regions as described in Sec. 8.5. Apply these results to the high-resistivity collector region in a double-diffused bipolar transistor. In particular, use sketches and develop arguments that show why these effects make it harder to manufacture reproducible and stable double-diffused pnp bipolar transistors than to produce npn bipolar transistors.

8.14[†] In practical MOS systems, measurements of capacitance versus voltage sometimes show hysteresis effects; that is, the C-V_G curves look like the sketch in Figure P8.14. The sketch refers to measurements made when V_G is swept with a very low frequency triangular wave (~ 1 Hz) and the ac measurement frequency is of the order of 1 kHz or higher. The sense of the hysteresis on such a curve can be observed experimentally to be either counterclockwise, as shown in the sketch, or else clockwise. The hysteresis sense allows one to differentiate between the two most common causes of non-ideal behavior. (a) Show this by considering the following nonideal effects: (i) field-aided movement of positive ions in the insulator and (ii) trapping of free carriers from the channel in traps at the oxide-silicon interface Q_{it}. (b) Using qualitative reasoning, prepare a table with sketches of the expected C-V_G plots for n-

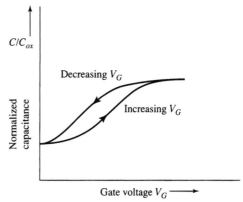

FIGURE P8.14

and p-type substrates; on each sketch indicate the sense of the hysteresis (i.e., clockwise or counterclockwise).

8.15[*] Compare the maximum capacitance that can be achieved in an area 100×100 μm^2 by using either an MOS capacitor or a reverse-biased pn-junction diode. Assume an oxide breakdown strength of 8×10^6 V cm^{-1}, a 5 V operating voltage, and a safety factor of two (i.e., design the MOS oxide for 10 V). The pn junction is built by diffusing boron into n-type silicon doped to 10^{16} cm^{-3}.

8.16[†*] Calculate the area density of surface states that would lead the surface generation rate I_S (Equation 8.6.3) of a fully depleted surface to equal twice the generation rate in the surface depletion region I_F (Equation 8.6.2). Consider the states to be characterized by a capture cross section of 10^{-15} cm^2 and the thermal velocity to be 10^7 cm s^{-1}. Assume that the surface depletion region is 1 μm in width and that the time constant τ_0 is 1 μs.

8.17 The two neighboring MOS capacitors C_1 and C_2 shown in Figure P8.17a have identical C-V characteristics (plotted in Figure P8.17b using the depletion approximation) when measured between gate and bulk. Sketch the capacitance as a function of voltage when it is measured between the two gates with the bulk terminal floating. Identify important features in your sketch.

8.18 The capacitor C_A shown in Figure P8.18a is an ideal MOS capacitor; there are no interface traps at the Si–SiO$_2$ interface and there is also no oxide charge. Capacitors C_B and C_C have the same dimensions and substrate-conductivity type and doping as C_A, but have an oxide charge density of 10^{11}/cm^2 and interface-trap energy distribution as shown in Figure P8.18b. The C-V curve for capacitor C_A under equilibrium conditions is shown. You can assume that $x_{ox} = 100$ nm and $N_a = 10^{15}$ cm^{-3}.

(a) For capacitor C_B sketch the high-frequency C-V curve (at which traps cannot respond to the ac signal)

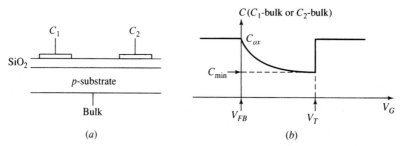

(a)

(b)

FIGURE P8.17

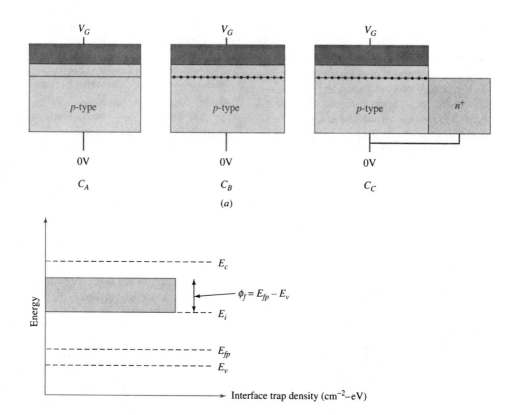

(a)

Interface trap distribution for capacitors C_B and C_C.

(b)

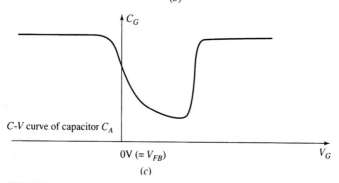

C-V curve of capacitor C_A

0V (= V_{FB})

(c)

FIGURE P8.18

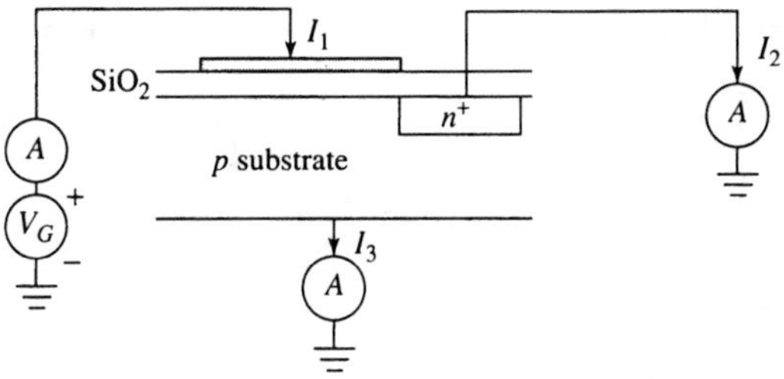

FIGURE P8.19

and low-frequency C-V curve (at which all traps can respond to the ac signal). Superimpose them on top of the C-V curve of capacitor C_A shown in Figure P8.18c. Point out the important features in your sketches. Assume that the MOS system is in equilibrium under the dc bias.

(b) Do the same for capacitor C_C.

8.19 An MOS gated diode is set up for C-V measurements as shown in Figure P8.19. The gate voltage V_G is ramped from positive to negative values at a very slow rate of R_m V/s, and the terminal currents I_1, I_2, and I_3 are monitored by three very sensitive electrometers. Sketch I_1, I_2, and I_3 as functions of V_G on the same graph. Explain important features in your sketch.

TABLE 8.3 Formulas for the Oxide-Silicon System

p-type substrate (n-channel)		n-type substrate (p-channel)
Flat-band voltage (Equation 8.4.6)		
$$V_{FB} = \Phi_{MS} - \frac{Q_f}{C_{ox}} - \frac{1}{C_{ox}}\int_0^{x_{ox}} \frac{x}{x_{ox}}\rho(x)\,dx$$		
Bulk potential (Equation 4.2.9)		
$$\phi_p = -\frac{kT}{q}\ln\left(\frac{N_a}{n_i}\right)$$		$$\phi_n = \frac{kT}{q}\ln\left(\frac{N_d}{n_i}\right)$$
Surface potential for strong inversion (Table 3.1)		
Thermal equilibrium $\phi_s = |\phi_p|$		$\phi_s = -|\phi_n|$
$\phi_s - \phi_p = 2|\phi_p|$		$\phi_s - \phi_n = -2|\phi_n|$
With bias $(V_C - V_B) = V_{CB}$		
$\phi_s = |\phi_p| + V_{CB}$		$\phi_s = -|\phi_n| - |V_{CB}|$
Maximum depletion width, x_{dmax} (Equation 8.3.6)		
Thermal equilibrium		
$$\sqrt{\frac{4\epsilon_s|\phi_p|}{qN_a}}$$		$$\sqrt{\frac{4\epsilon_s|\phi_n|}{qN_d}}$$
With bias V_{CB} (Equation 8.3.8)		
$$\sqrt{\frac{2\epsilon_s(2|\phi_p| + V_{CB})}{qN_a}}$$		$$\sqrt{\frac{2\epsilon_s(2|\phi_n| + |V_{CB}|)}{qN_d}}$$
Work-function difference, Φ_{MS}		
$\Phi_M - (X + E_g/2q + |\phi_p|)$		$\Phi_M - (X + E_g/2q - |\phi_n|)$
Threshold voltage V_T (arbitrary reference)		(Equation 8.3.18)
$V_{FB} + V_C + 2|\phi_p|$		$V_{FB} + V_C - 2|\phi_n|$
$+\dfrac{1}{C_{ox}}\sqrt{2\epsilon_s qN_a(2|\phi_p| + V_C - V_B)}$		$-\dfrac{1}{C_{ox}}\sqrt{2\epsilon_s qN_d(2|\phi_n| + V_B - V_C)}$

CHAPTER 9

MOS FIELD-EFFECT TRANSISTORS I: PHYSICAL EFFECTS AND MODELS

The concept of the metal-oxide-semiconductor field-effect transistor (MOSFET) was actually developed well before the invention of the bipolar transistor. In the early 1930s patents were issued for devices that resemble the modern silicon MOSFET, but which were made from combinations of materials not including silicon.* The lack of a complete understanding of the physics of insulator-semiconductor systems and poor control of insulator-semiconductor interfaces being investigated at that time made practical use of these inventions impossible. Nonetheless, the patent

---

* Early patents include J.E. Lilienfeld, U.S. Patents 1,745,175 (Jan. 28, 1930), 1,877,140 (Sept. 13, 1932), 1,900,018 (Mar. 7, 1933); and O. Heil, British Patent 439,457 (Dec. 6, 1935).

literature shows continuing efforts over the following decades to make the type of device that is now called an MOS transistor.

In his classic contribution to the development of solid-state electronics, "Electrons and Holes in Semiconductors," [1] William Shockley describes the quest at ATT Bell Labs in the 1940s to build a device consisting of a parallel-plate capacitor in which a layer of semiconductor forms one plate arranged with a corresponding metal plate in close proximity to it. "With the metal plate positive, then the additional charge on the semiconductor will be represented by an increased number of electrons" which "should be free to move and should contribute to the conductivity of the semiconductor." The Bell Labs effort to make this device succeeded instead in the invention of the bipolar transistor in 1948 and only later did it produce operating field-effect transistors.

The first MOSFETs suitable for commercial use did not appear until the 1960s. In the four decades since that time, MOSFETs have displaced bipolar transistors to become the most extensively used active solid-state devices. MOSFETs have several advantages over bipolar transistors for digital circuits, an application area that grew exponentially over this same time period. MOSFETs are built with basically simpler fabrication technologies than are bipolar transistors and it is easier to build them in dense arrays. They also typically consume less power than do bipolar integrated circuits, especially at lower frequencies. Because of their simpler fabrication, higher density, and lower power, MOSFETs are widely used in memory circuits, totally displacing bipolar memories. These same advantages also led to the dominance of MOSFETs in logic circuits, especially high-speed microprocessors such as that shown on the cover of this book and in Figure 2.2*e*. The extremely high-volume production of MOSFET integrated circuits has, in turn, financed enormous research programs to improve MOSFET performance; the ongoing research continues to reduce the device size, allowing more devices on a chip, and increasing the frequency response. Over a span of 30 years, the minimum feature size available in MOS technologies has been reduced by a factor of 200—from about 20 μm to below 0.1 μm—and the area density of devices has increased more than 40,000 times.

The term *technology scaling* refers to the process of reducing the sizes of both active devices and passive elements in order to improve both packing density and circuit speed. In 1974, Dennard et al. [2] presented an influential systematic study of the impact of technology scaling on circuit performance. The authors of this study proposed *constant-field scaling* rules, in which dimensions are changed in such a manner that the internal electric fields within the transistor remain (as close as is possible) unchanged in the devices. A summary of the constant-field scaling rules is given in Table 9.1.

Instead of being strictly observed, the scaling rules have functioned more as a guide to size reduction in integrated circuits. In fact, the scaling of both passive and active devices has been influenced more by fabrication-technology limitations and device functionality than by the application of algebraic scaling factors. In Table 9.2 we list some of the past and projected scaling trends covering seven generations of MOS technology as defined by the International Technology Roadmap for Semiconductor (ITRS) [3]. The table shows that both the widths and thicknesses of the metal lines interconnecting devices have been scaled less aggressively than have the gate lengths L of the MOSFETs because

TABLE 9.1 Scaling Rules for Constant-Field Scaling (© 1974 IEEE [2]).

Physical Parameters	Scaling Factor
Surface Dimensions, L	$1/K$
Vertical Dimensions, x_{ox}, x_j	$1/K$
Impurity Concentrations	K
Currents, Voltages	$1/K$
Current Density	K
Capacitance (per unit area)	K
Transconductance	1
Circuit Delay Time	$1/K$
Power Dissipation	$1/K^2$
Power Density	1
Power-Delay Product	$1/K^3$

of the difficulty of patterning small-dimension, highly reflective metal lines on nonplanar surfaces, and also because of the limit placed on the current density by electromigration (described in Sec. 2.7). Scaling rules of the type formulated by Dennard et al. are inherently of limited direct use because effects that can be neglected in larger devices become first significant, and then dominant as device dimensions continue to shrink. With the sub-micrometer-dimension channel lengths that now characterize MOSFETS, device performance is completely dominated by high-electric-field effects that are inconsequential in larger-dimension devices. Our discussion of MOSFETs in this chapter will need to develop both basic- and short-channel physical models. The basic model considered for longer-channel transistors allows us to develop analytical models that provide physical intuition; the basic models can then be augmented to account for additional effects that must be considered to obtain an understanding of the operation of short-channel transistors.

TABLE 9.2 Past and Predicted Technology Scaling Trends for MOS Technology, from *International Technology Roadmap for Semiconductors* (*1997–2001 editions*) [3]

Year of Production	1997	1999	2001	2003	2006	2010	2016
Min. dim. L (μm)	0.25	0.18	0.13	0.1	0.07	0.045	0.022
DRAM density (Gbits/cm^2)	0.18	0.38	0.42	0.91	1.85	4.75	28.85
Logic V_{DD} (V)	2.5–1.8	1.8–1.5	1.2	1.0	0.9	0.6	0.4
Equivalent x_{ox} (nm)	4–5	1.9–2.5	2.3	2.0	1.9	1.2	0.9
Junction depth x_j (nm)	50–100	45–70	30–60	26–52	20–40	15–30	10–20
Local wire pitch (nm)	600	500	350	245	130	105	30
Metal aspect ratio	1.8	2	1.6*	1.6	1.7	1.8	2.0
max. I_{Dsat} NMOS	600	750	900	900	900	900	900
(μA/μm) PMOS	280	350	420	420	420	420	420

* Switching to copper   Manufacturable solution yet to be found.

We begin, in Sec. 9.1 by describing the basic operation of a MOSFET and developing first-order models for the device. An improved model is given in Sec. 9.2; this model includes mobility degradation and velocity saturation, the two most significant second-order effects. Section 9.3 discusses CMOS circuits and circuit-design methodology and develops techniques to determine needed parameters. Future directions for MOSFET design are considered in Sec. 9.4. High electric fields in the MOSFET channel regions are sufficiently important in device design that we devote Chapter 10 to their discussion. Although much of MOSFET device analysis becomes very complicated mathematically if treated in detail, we try to convey a physical picture, instead of presenting rigorous mathematics, wherever possible.

9.1 BASIC MOSFET BEHAVIOR

The two types of MOS transistors are n-channel MOSFETs (in which the conducting carriers are electrons) and p-channel MOSFETs (where the conducting carriers are holes). The two types of MOSFETs are sometimes called NMOSFETs and PMOSFETs, respectively. n-channel MOSFETs are built in p-type silicon substrates so that reverse-biased pn junctions isolate the conducting channels of nearby devices, and p-channel MOSFETs are built in n-type silicon. For n-channel MOSFETs, positive gate voltages that are sufficiently large create a conducting channel; for p-channel MOSFETs, gate voltages that are negative and of sufficient magnitude produce a conducting channel. We present our analysis in terms of the n-channel MOSFET because of its greater importance. The theory developed also applies to p-channel MOSFETs, with suitable changes in the signs of appropriate parameters.

The n-type source and drain regions of the basic MOSFET shown in Figure 9.1 are separated by a lateral distance known as the channel length L. The channel length extends along the y-axis in the figure, while the direction into the silicon (perpendicular to the oxide) is conventionally designated to be the x-direction as in Chapter 8. The source-and-drain regions are electrically disconnected unless there is an n-type inversion layer at the surface to provide a conducting channel between them. When the surface is

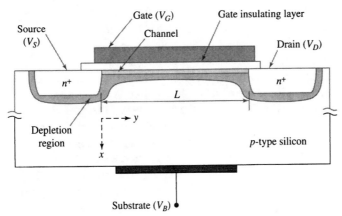

FIGURE 9.1 Basic elements of an n-channel MOSFET. The source-to-drain spacing (*channel length*) is L, and the device width (in the z direction perpendicular to the plane of the paper) is W.

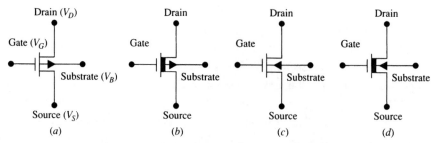

FIGURE 9.2   Electrical symbols for MOSFETs: (a) p-channel enhancement, (b) p-channel depletion, (c) n-channel enhancement, (d) n-channel depletion devices.

inverted and a voltage is applied between the source and drain junctions, carriers can enter the channel at the source and leave at the drain, resulting in current flow from drain to source in n-channel MOSFETs and from source to drain in p-channel MOSFETs. MOSFETs can be fabricated in which the channel is inverted when the gate-to-source voltage is zero. The drain current in this type of MOSFET can be reduced by changing the gate-to-source voltage and hence it is called a *depletion-mode* MOSFET. MOSFETs in which the channel region is not inverted at $V_{GS} = 0$ are called *enhancement-mode* MOSFETs; enhancement-mode MOSFETs are far more frequently used in circuits than are depletion-mode devices.

Voltages at four terminals affect the electronics in a MOSFET: V_G—the gate voltage, V_S—the source voltage, V_D—the drain voltage, and V_B—the bulk voltage (the terms *substrate* and *body* are often used instead of *bulk*). The symbols that have been adopted for MOSFETs are sketched in Figure 9.2a and b for p-channel enhancement and depletion devices, and in Figure 9.2c and d for n-channel enhancement and depletion devices. In the symbols in Figure 9.2, a diode symbol drawn from the source to the substrate indicates the MOSFET type although the diode indication is often omitted if the transistor type is unambiguous from the context. A thickened line along the channel indicates a depletion-mode MOSFET.

To develop the basic theory of the MOSFET, we initially consider the case with the source and the bulk at the same voltage ($V_{SB} = 0$). In a later section, we relax this condition and investigate the influence of a bias between the source and the bulk regions.

In circuit applications, MOSFETs typically function either as voltage-controlled resistors in *analog* circuits or as ON/OFF switches in *digital* circuits. As discussed in Chapter 8, when the voltage applied to the gate of an n-channel MOSFET is lower than the threshold voltage V_T, the surface region of the substrate between the source and drain is either accumulated (many holes are present) or depleted of mobile carriers. In either case, virtually no conduction is possible between the n-type source and drain regions and the MOSFET is in the OFF state. When the gate voltage is increased beyond V_T, the surface region beneath the gate becomes inverted, forming a channel that contains mobile electrons that can carry current in the channel between the drain and the source. Increasing the gate bias strongly increases the density of channel electrons and reduces the source-drain resistance, bringing the MOSFET to its ON state. The threshold voltage V_T is thus conveniently taken as the transition voltage at which the MOSFET changes from OFF to ON when it is used as a switch. This picture builds on our discussion of the MOS capacitor in Chapter 8 where we saw that a voltage on the gate controlled the density of mobile carriers in an inversion layer within the silicon.

Strong Inversion Region

In strong inversion (when the transistor is ON) the channel electrons move mainly by drift, and the current, taken to be positive when it flows into the device, is

$$I_D = WQ_n(y)v(y) \tag{9.1.1}$$

where $-Q_n$ is the inversion charge per unit area at a position y in the channel, and $v(y)$ is the velocity of carriers at that position. Both I_D and $v(y)$ are positive when directed along the positive y-direction. With small y-directed fields (at low drain voltage V_D), the drift velocity $v(y) = -\mu_n \mathcal{E}(y)$ where μ_n is the mobility of the carriers in the channel and $\mathcal{E}(y) = -\partial V(y)/\partial y$. Equation 9.1.1 can be rewritten

$$I_D = WQ_n(y)\mu_n \partial V(y)/\partial y \tag{9.1.2}$$

If we assume (1) that μ_n is constant, (2) that both the channel inversion- and bulk charges are controlled by the vertical field only (*the gradual-channel approximation*) so that the one-dimensional theory introduced in Chapter 8 can be applied in the x- (vertical) direction to calculate the inversion-charge density, and (3) that V_T is not a function of the position y along the channel, we can write for $Q_n(y)$

$$Q_n(y) = -C_{ox}[V_G - V_T - V(y)] \tag{9.1.3}$$

After substituting Equation 9.1.3 into Equation 9.1.2 and integrating along the channel from the source ($y = 0$), which is taken as the voltage reference ($V_S = 0$), to the drain ($y = L$) where $V(y) = V_D$, we have

$$\int_0^L I_D dy = \mu_n W C_{ox} \int_0^{V_D} [V_G - V_T - V(y)]dV \tag{9.1.4}$$

which can be solved to obtain an equation for the drain current that is sometimes called the *long-channel MOSFET equation*:

$$I_D = \mu_n C_{ox} \frac{W}{L}\left[\left(V_G - V_T - \frac{1}{2}V_D\right)V_D\right] \tag{9.1.5}$$

(Some of the more important MOSFET equations we derive are summarized in Table 9.4 at the end of this chapter.) If we consider Equation 9.1.5 with $V_G > V_T$ and increase V_D from zero while V_G stays fixed, we see that drain current initially increases linearly with increasing drain voltage, but that the slope decreases when $V_D/2$ becomes appreciable compared to the value of $(V_G - V_T)$, as shown in Figure 9.3. Equation 9.1.5 predicts that the slope decreases to zero and eventually becomes negative at sufficiently high V_D. This nonphysical

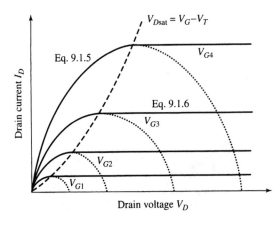

FIGURE 9.3 Drain current as a function of drain voltage at varying gate voltages. The dotted portions of the curves for $V_D > V_{Dsat}$, which are predicted by Equation 9.1.5, are not physical because in this region Equation 9.1.3 predicts an electron density less than zero, and Equation 9.1.6 must be used. The gate voltage increases from bottom to top, and V_S and V_B are both taken to be zero.

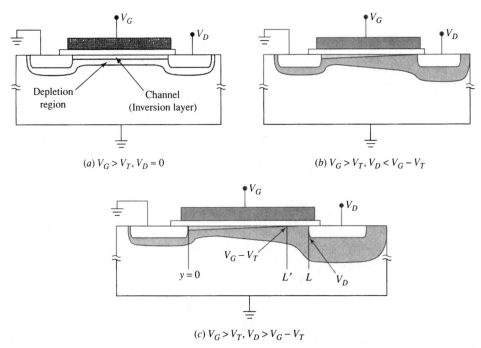

(a) $V_G > V_T$, $V_D \approx 0$

(b) $V_G > V_T$, $V_D < V_G - V_T$

(c) $V_G > V_T$, $V_D > V_G - V_T$

FIGURE 9.4   MOSFET cross sections showing bias effects on the depletion regions: (a) drain voltage small, depletion region nearly uniform along the channel, (b) drain voltage large enough to cause significant variation in depletion-region thickness, (c) drain voltage exceeding V_{Dsat}, channel extends only to $L' < L$; the channel voltage at L' equals V_{Dsat}, which is $(V_G - V_T)$.

negative resistance is predicted because Equation 9.1.5 depends on Equation 9.1.3, and Equation 9.1.3 is only valid when Q_n is negative [i.e., $V(y) < (V_G - V_T)$]. $V(y)$ has its maximum value V_D at the drain; hence Equation 9.1.5 is only valid when $V_D \leq (V_G - V_T)$. The limit of validity occurs when the mobile-electron density Q_n in the channel at $y = L$ decreases to zero (a condition described as *pinch-off*). The channel and inversion regions with different drain voltages including pinch-off are illustrated in Figure 9.4. When the drain voltage exceeds the pinch-off voltage (Figure 9.4c), the conducting channel becomes separated from the doped region of the drain and a voltage drop equal to $[V_D - (V_G - V_T)]$ exists between the drain and L', the location in the channel where the inversion charge becomes zero (the *pinch-off point*). The MOSFET is then operating in the *saturation region* with $V_D > V_{Dsat}$ (where $V_{Dsat} = V_G - V_T$), and the drain current according to our simple model no longer increases with increasing drain voltage. The current in the saturation region can readily be calculated by substituting $V_D = V_G - V_T$ into Equation 9.1.5 to obtain

$$I_{Dsat} = \mu_n C_{ox} \frac{W}{2L} (V_G - V_T)^2 \tag{9.1.6}$$

Combining Equations 9.1.5 and 9.1.6, we obtain expressions that describe the set of I_D versus V_D curves shown in Figure 9.3. The current increases according to Equation 9.1.5 until it reaches I_{Dsat} at $V_D = V_{Dsat}$ (which first occurs when the pinch-off point $L' = L$) and then remains constant at the value determined by Equation 9.1.6 as L' moves toward the source diffusion. This is the "classical" I versus V behavior, which gives a reasonably close prediction to the behavior of MOSFETs made with long channels (≥ 10 μm), but which must be modified considerably for MOSFETs having today's much shorter channels (≤ 1 μm).

At this point, we consider some simple physical implications of the equations derived using the simple MOSFET model. Equation 9.1.5 can be regrouped as

$$I_D = WC_{ox}\left(V_G - V_T - \frac{1}{2}V_D\right)\mu_n\frac{V_D}{L} \tag{9.1.7}$$

In Equation 9.1.7, the term $C_{ox}(V_G - V_T - V_D/2)$ can be interpreted as the amount of inversion charge half way along the channel (at $y = L/2$) if the voltage varied linearly along the channel. The field in this case would be V_D/L, so Equations 9.1.5 and 9.1.7 represent the current that would flow in a channel containing an "average" charge density that drifts under the influence of a constant field. In the pinch-off region (when $Q_n = 0$ at the drain), our simple model predicts that the current flow is determined by the rate at which carriers are delivered to the pinch-off point. Thinking about the MOSFET at pinch-off reveals a troubling aspect of the simple model. Because current must be constant along the channel length ($0 \leq y \leq L'$) while the density of mobile carriers Q_n carrying that current decreases (by Equation 9.1.3), the mobile-charge velocity $v(y)$ must increase continuously along the channel to keep I_D constant (by Equation 9.1.1). Finally, the velocity needs to become infinite at the pinch-off point where Q_n becomes zero. For the long-channel MOSFET, this physical impossibility does not invalidate the model because the drain current is not much affected by events at and beyond the pinch-off point; the drain current is determined by the delivery rate of electrons along the inverted channel between the source and the pinch-off point. An easily visualized physical model provides a useful analogy to this idealized MOSFET behavior.

Water Analogy. Physically, the operation of the MOSFET can be understood in terms of a water analogy. The mobile carriers correspond to water droplets. The source and drain are deep reservoirs who\se relative elevation difference is analogous to the source-to-drain voltage difference. The channel region is like a canal with a depth that depends on the local value of the gate-to-channel voltage as sketched in Figure 9.5.

If the drain and source are held at the same voltage, the water surface is level through the source, canal, and drain in our analogy (Figure 9.5a). When a drain-to-source voltage is imposed, the surface of the drain reservoir is lowered, causing a flow of water along the canal from the source to the drain. The flow speeds up as the elevation difference (analogous to V_D) increases. Because the rate at which water flows past any cross section of the canal is constant, the water velocity increases as the depth of water in the canal decreases toward the drain reservoir. At first, the flow through the canal depends both on its dimensions (as controlled by the gate) and on the elevation difference between the source and drain (Figure 9.5b). At the condition analogous to channel pinch-off, the flow becomes entirely limited by the flow capacity of the canal. If the drain reservoir is lowered further, its surface becomes abruptly disconnected from the water surface at the drain end of the canal. In this condition, the flow into the drain resembles the free fall of water over a waterfall (Figure 9.5c). The rate of flow is equal to the delivery rate to the lip of the waterfall and independent of the total drop over the cataract [which is analogous to ($V_D - V_{Dsat}$)]. Similarly, in the MOSFET, to first order it is unnecessary to consider the transport mechanisms of carriers once they leave the channel to "fall" through the energy drop in the high-field, space-charge region near the drain.

Channel-Length Modulation

The basic MOSFET theory that leads to Equation 9.1.5 allowed sufficiently accurate design of MOSFETs when they were first produced and is called the *long-channel MOSFET*

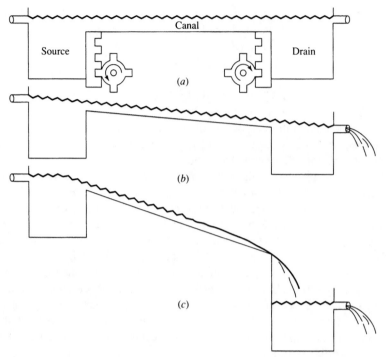

FIGURE 9.5 Water analogy for a MOSFET. (*a*) When the source and drain are level, there is no flow ($V_{DS} = 0$). The water depth in the canal can be varied by the gear and track (V_{GS}). (*b*) When the drain is lower than the source, water flows along the canal. (*c*) The flow is limited by the channel capacity; lowering the drain further only increases the height of the waterfall at its edge.

theory because it provides reasonable accuracy for MOSFETs with long channels; it is especially useful when the fields causing the channel carriers to drift are low enough to justify the assumption of constant mobility along the channel.

Although we have, thus far, only considered the dc theory for MOSFETs, it should be evident that ac behavior (and switching) depends strongly on the length of the channel L, improving directly as L is reduced. A smaller channel length L also increases the drain current for given bias conditions (Equation 9.1.5). For these reasons, reducing L has been a design objective from the earliest days of MOS design. However, as MOSFET channel lengths decreased, improvements to the long-channel theory became essential. With a shorter channel the drain current does not stay constant (at I_{Dsat} as predicted by Equation 9.1.6), but rather increases significantly when the drain voltage exceeds V_{Dsat}. Most of this behavior results from *channel-length modulation*.

Channel-length modulation can be directly incorporated into the analysis as a correction to the long-channel equations by noting that the current in Equation 9.1.6 is inversely proportional to the channel length, which decreases as the drain voltage increases beyond V_{Dsat}. We denote the reduction in channel length by the variable $\Delta L = L - L'$, and include the effect of channel-length modulation by writing

$$I_D(V_D > V_{Dsat}) = I_{Dsat}\left[\frac{L}{L - \Delta L}\right] \approx I_{Dsat}\left[1 + \frac{\Delta L}{L}\right] \tag{9.1.8}$$

where ΔL, the length of the pinch-off region between L' and L in Figure 9.4c, increases as $(V_D - V_{Dsat})$ increases. The approximate form of the equation is adequate in the usual case, where $\Delta L \ll L$ and two terms in a Taylor's series expansion express the functional behavior accurately. Circuit designers need to express currents in terms of voltage, however. A voltage dependence could be derived from Equation 9.1.8 by substituting the expression for ΔL obtained by solving Poisson's equation under the assumption that the charge density is N_a in the pinch-off region:

$$\Delta L = \left[\frac{2\epsilon_s}{qN_a}(V_D - V_{Dsat}) \right]^{1/2} \tag{9.1.9}$$

In practice, this approach is inaccurate (both in its approximation that free charge can be neglected and in treating the problem only one-dimensionally) and cumbersome. For these reasons, MOSFET channel-length modulation is often treated analogously to the bipolar Early effect by writing

$$I_{Dsat} = \frac{\mu W C_{ox}}{2L}(V_G - V_T)^2\left(1 + \frac{V_D}{V_A}\right) \tag{9.1.10}$$

where V_A is the MOSFET analogue of the BJT Early voltage (Section 7.1).

Body-Bias Effect

Thus far, we have considered the MOSFET under the condition that the source and substrate (also called the *body* or sometimes the *bulk*) are held at the same voltage ($V_S = V_B$). Between the source and the body is a *pn* junction that does not conduct appreciable current as long as it is unbiased or under a reverse bias. Although little current flows between the source and the substrate, the reverse bias on this junction does significantly influence the current flowing between the source and the drain. This influence of V_{SB} on I_D is generally described as the *body-bias* or *substrate-bias effect*. Reverse bias on the source-to-substrate junction reduces the amount of charge in the channel for a given gate-to-source bias; alternatively stated, it increases the gate voltage needed to induce a given number of mobile carriers in the channel. That is, it increases the threshold voltage.

We can qualitatively understand the effect of substrate bias by considering the three-dimensional diagrams shown in Figure 9.6, which compare the band structures at zero substrate bias and at a negative substrate bias $V_B = -1$ V for different gate voltages corresponding to flat-band, $2|\phi_p|$, and inversion. When a substrate bias is applied, the greater amount of band bending required to invert the surface increases the threshold voltage as seen by comparing Figures 9.4d and 9.4f.

In Chapter 8, we considered the substrate-bias effect and showed in Equation 8.3.18 that the substrate bias modifies the depletion-charge term in the expression for the threshold voltage

$$\Delta V_T = \frac{\sqrt{2\epsilon_s q N_a}}{C_{ox}}\left(\sqrt{2|\phi_p| + |V_{SB}|} - \sqrt{2|\phi_p|} \right)$$

$$= \gamma\left(\sqrt{2|\phi_p| + |V_{SB}|} - \sqrt{2|\phi_p|} \right) \tag{9.1.11}$$

where we define a *body-effect* parameter γ

$$\gamma = \frac{\sqrt{2\epsilon_s q N_a}}{C_{ox}} \tag{9.1.12}$$

with units of $\sqrt{V}$.

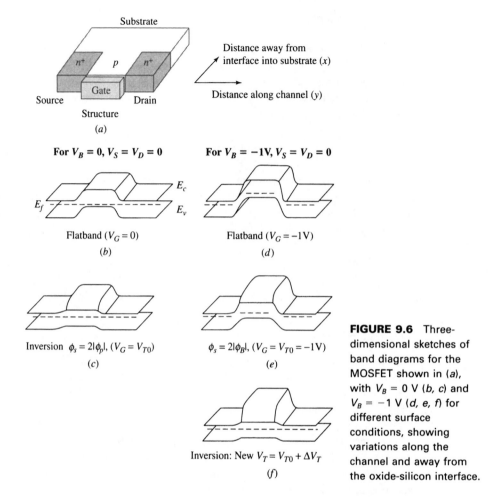

FIGURE 9.6 Three-dimensional sketches of band diagrams for the MOSFET shown in (a), with $V_B = 0$ V (b, c) and $V_B = -1$ V (d, e, f) for different surface conditions, showing variations along the channel and away from the oxide-silicon interface.

EXAMPLE *Parameters for a depletion-mode MOSFET*

Assume that a depletion-mode n-channel MOSFET can be described by Equations 9.1.5 and 9.1.6 if a negative value is used for the threshold voltage. The substrate doping is 1.63×10^{15} cm^{-3}, and the body-effect parameter $\gamma = 0.5$ V$^{1/2}$. The MOSFET is connected in the circuit shown, and a current equal to 30 μA is measured with the supply voltage $V_{SS} = 0$. When V_{SS} is increased to 1 V, the current is reduced to 23.1 μA.

Calculate the threshold voltage $V_T(0)$ when the source voltage $V_S = 0$ V, and also find the prefactor $\mu_n C_{ox} W/L$ in Equations 9.1.5 and 9.1.6.

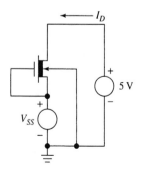

Solution Because $V_{DS} \geq 4$ V for both measurements and $V_{GS} = 0$ V in the circuit, we assume that $V_{DS} > (V_{GS} - V_T)$ so that Equation 9.1.6 can be applied to both measurements. We will check the calculated value of V_T later to see that this assumption is true. Because the source voltage changes, we must consider the body effect using Equation 9.1.11, and we therefore need a value for $|\phi_p|$. From the given dopant concentration in the substrate, we calculate $|\phi_p| = 0.3$ V (Equation 4.2.9b).

At $V_{SS} = 0$ V, the source is at 0 V and

$$30 = \mu_n C_{ox} \frac{W}{2L} [0 - V_T(0)]^2$$

At $V_{SS} = 1$ V, the source is at 1 V and, from Equation 9.1.11,

$$\Delta V_T = 0.5 \left[\sqrt{0.6 + 1} - \sqrt{0.6} \right] = 0.245 \text{ V}$$

Because the body effect makes the threshold voltage less negative for this n-channel MOSFET, $V_T(1) = V_T(0) + 0.245$. Using the measured current at $V_{SS} = 1$ V, we have

$$23.1 = \mu_n C_{ox} \frac{W}{2L} \left[0 - (V_T(0) + 0.245) \right]^2$$

Hence,

$$\left[\frac{I_D(0)}{I_D(1)} \right]^{1/2} = 1.14 = \frac{-V_T(0)}{-V_T(0) - 0.245}$$

From this equation, we calculate $V_T(0) = -2$ V. Using either measured current value, we have $\mu_n C_{ox} W/L = 15$ µA V^{-2}. The calculated value of the threshold voltge is consistent with our assumption that the MOSFET is in saturation because $V_{DS} > (0 - V_T)$ for either of the measured conditions. ∎

In practice, the effect on V_T of a source-substrate bias predicted by Equation 9.1.11 is observed if the channel length of the MOSFET is considerably greater than the depletion width of the reverse-biased source-substrate np junction. For shorter channel lengths, the one-dimensional analysis used to solve Poisson's equation (which led to Equation 8.3.18) loses validity. A two-dimensional theory for the space-charge configuration is then needed to obtain an accurate expression for the threshold voltage.

Bulk-Charge Effect

Our derivation of the drain current for the MOSFET that led to Equations 9.1.5 and 9.1.6 was based on the assumption that the charge Q_d due to ionized acceptors in the depletion region near the surface is constant along the channel. However, because the channel voltage increases from the source to the drain, the width of the depletion region, and therefore this charge (which is sometimes called Q_B, the *bulk* charge) also increases along the channel direction y, as is evident in Figure 9.4b. An expression for $Q_d(y)$ is

$$Q_d(y) = q N_a x_d = \sqrt{2\epsilon_s q N_a [2|\phi_p| + V(y) - V_B]} \tag{9.1.13}$$

The variation in bulk charge along the channel means that the threshold voltage and, therefore, also Q_n depend on y. Bringing this dependence into Equation 9.1.3, we have

$$Q_n(y) = C_{ox} [V_G - V_{FB} - 2|\phi_p| - V(y)] + \sqrt{2\epsilon_s q N_a [2|\phi_p| + V(y) - V_B]} \tag{9.1.14}$$

After substituting Equation 9.1.14 into Equation 9.1.3 and integrating the current-transport Equation 9.1.2 along the channel from source ($y = 0$) to drain ($y = L$), we obtain the long-channel drain-current equation corrected for the bulk charge variation along the channel

$$
I_D = \mu_n \frac{W}{L} \left\{ C_{ox}\left(V_G - V_{FB} - 2|\phi_p| - \frac{1}{2}V_{DS} \right) V_{DS} \right.
$$
$$
\left. - \frac{2}{3}\sqrt{2\epsilon_s q N_a}\left[(2|\phi_p| + V_D - V_B)^{3/2} - (2|\phi_p| + V_S - V_B)^{3/2} \right] \right\} \quad (9.1.15)
$$

When $V_S = V_B = 0$, Equation 9.1.15 reduces to

$$
I_D = \mu_n \frac{W}{L} \left\{ C_{ox}\left(V_G - V_{FB} - 2|\phi_p| - \frac{1}{2}V_{DS} \right) V_{DS} \right.
$$
$$
\left. - \frac{2}{3}\sqrt{2\epsilon_s q N_a}\left[(2|\phi_p| + V_D)^{3/2} - (2|\phi_p|)^{3/2} \right] \right\} \quad (9.1.16)
$$

The mixed square-and-3/2-power law dependences for I_D on V_D makes Equation 9.1.15 clumsy to use. As a result, designers typically represent the bulk-charge effect in a simpler way. Instead of using Equation 9.1.15, the simpler equations are modified by the introduction of a fitting parameter α (called the *bulk-charge* factor) so that Equation 9.1.5 becomes

$$
I_D = WC_{ox}\left(V_G - V_T - \frac{\alpha}{2}V_{DS} \right)\mu_n \frac{V_{DS}}{L} \quad (9.1.17)
$$

The bulk-charge factor α is greater than one (a typical value is 1.5) and therefore the drain current is reduced from the value predicted without the bulk-charge effect. Using Equation 9.1.17, we can calculate the derivative of I_D with respect to V_D and set it equal to zero to obtain an expression for V_{Dsat}

$$
V_{Dsat} = \frac{V_G - V_T}{\alpha} \quad (9.1.18)
$$

When $V_D = V_{Dsat}$, the MOSFET enters the current-saturation region and, in place of Equation 9.1.6, we use Equation 9.1.18 to write

$$
I_{Dsat} = \frac{\mu_n}{2} \frac{W}{\alpha L} C_{ox}(V_G - V_T)^2 \quad (9.1.19)
$$

Threshold-Voltage Adjustment by Ion Implantation

In most CMOS integrated-circuit applications, precisely controlling the threshold voltage V_T of the MOSFETs is critical. As we saw in Chapter 8, controlling the threshold voltage depends on the ability to adjust the local carrier densities near the surface. The introduction of MOS ICs, and particularly CMOS ICs was delayed for many years by the inability to achieve this control; robust threshold-voltage control was achieved in the 1980s by using precise ion implantation. Employing ion implantation (described in Chapter 2), fabrication engineers are able to produce MOSFETs with designed concentrations and positions of dopant atoms near the surface. Ion implantation makes use of a pure beam of ionized dopant atoms that are accelerated in a strong electric field to energies of the order of 100 keV and allowed to strike the wafer surface. The ions penetrate into the silicon (or the oxide or resist film) to a depth that is usually a small fraction of a micrometer. The ions that stop in the oxide are generally electrically inert. The ions implanted into the silicon become active donors or acceptors when the wafer is annealed at a moderate temperature. With properly limited annealing temperatures and times, the junction profiles obtained in previous diffusion steps are not greatly altered by the ion-implantation process.

The implantation can be used either to increase or to decrease (by compensation) the net dopant concentration at the silicon surface. In an *n*-channel MOS process, a heavy implant of boron is used to increase the surface dopant density and, thereby, the threshold voltage in the *field regions* outside the areas of the chip containing the active transistor channels. Using a lighter implant dose, the threshold voltage can be adjusted in the channel regions to inhibit subsurface punchthrough (as will be explained in Sec. 9.2).

Figure 9.7 shows the distribution of dopant atoms from a typical implant as a function of depth into the wafer. After the implant, the atoms are distributed with a Gaussian profile (solid line in the figure). The implant concentration $N_i(x)$ is

$$N_i(x) = \frac{N'}{\sqrt{2\pi}\Delta R_p} \exp\left[-\frac{(x - R_p)^2}{2(\Delta R_p)^2} \right]$$ (9.1.20)

where N' is the number of implanted atoms per unit area (the *dose*), R_p is the average distance an implanted atom penetrates into the solid (the *range*), and ΔR_p is the rms or standard deviation that describes the "width" of the distribution. Both the range and the width of the implant depend on the energy of the implanting ion beam, as well as on the implanted species. Values for R_p and ΔR_p for the most common dopants in silicon are given in Figures 2.16, 2.17, and 2.18.

After the activating anneal, the implanted distribution becomes slightly broader, as shown by the dashed curve in Figure 9.7. Calculating the effect of this implant on the

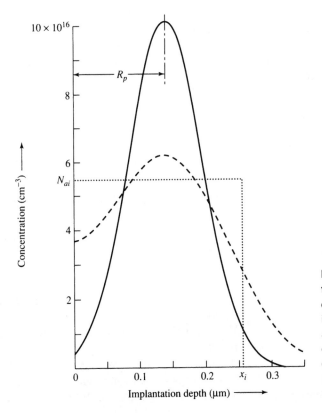

FIGURE 9.7 Distribution of the implanted dopant (solid curve) immediately after the implant, and after activating anneal and diffusions (dashed curve). The dotted "box" curve shows the distribution used to calculate the threshold voltage.

MOSFET threshold voltage is greatly simplified by approximating the actual distribution by the "box" distribution (dotted lines in Figure 9.7). In this approximation, the implanted dopant is assumed to have a constant density N_{ai} from the surface to a depth x_i. The total dose N' per unit area is related to N_{ai} and x_i by the equation $N' = N_{ai} x_i$. The dopant concentration in the region nearest the surface is $(N_{ai} + N_a)$ where N_a is the background acceptor concentration in the substrate. For $x > x_i$, the effective acceptor concentration is N_a.

To calculate V_T with the adjustment implant, two different cases must be considered. In the first case, the effective depth of the implant is greater than the width of the depletion region at threshold. The threshold voltage is then given by Equation 8.3.18, with N_a replaced by $(N_{ai} + N_a)$. However, this type of impurity profile has several drawbacks. The increased doping near the substrate edge of the depletion region increases the substrate capacitance and lowers the breakdown voltage in the channel region, as well as causing greater sensitivity of the threshold voltage to the source-substrate bias (i.e., a more troublesome *body effect*).

In the second case the implanted ions are all contained within the surface space-charge region. Calculating the threshold voltage in this case is more complicated, but it follows the same basic pattern that we used previously. The first step is to find a solution for Poisson's equation when the space charge changes abruptly, as in the approximate "box" distribution. When the solution is found, and the depletion approximation is used, an expression for the depletion-layer width x_d can be written in terms of the surface potential ϕ_s:

$$x_d = \sqrt{\frac{2\epsilon_s}{qN_a}(\phi_s + |\phi_p|) - x_i^2 \frac{N_{ai}}{N_a}} \tag{9.1.21}$$

Real solutions of Equation 9.1.21 are only possible if the second term in the square root is smaller than the first, consistent with our assumption that the highly doped, implanted region is contained within the depletion region. To find the threshold voltage, ϕ_s in Equation 9.1.21 must be set to the inversion value appropriate to the heavier doped surface region: $\phi_{ps} = (kT/q) \ln [(N_{ai} + N_a)/n_i]$. This leads to an expression for $x_{d\max}$ that can be used to find the depletion-charge density Q_d at inversion.

$$Q_d = -qN_{ai}x_i - qN_a x_{d\max}$$
$$= -qN_{ai}x_i - \sqrt{2qN_a\epsilon_s(|\phi_{ps}| + |\phi_p| + V_{SB}) - q^2 x_i^2 N_a N_{ai}} \tag{9.1.22}$$

Equation 9.1.22 can easily be expressed in terms of the implanted dose by using the expression $N_{ai} = N'/x_i$ where N' is the density of atoms per unit area *penetrating into the silicon*. As in Sec. 8.3, the gate voltage is then related to Q_d, and the threshold voltage can be written as

$$V_T = V_{FB} + V_S + |\phi_p| + |\phi_{ps}| + \frac{qN'}{C_{ox}} + \frac{1}{C_{ox}}\sqrt{2qN_a\epsilon_s(|\phi_{ps}| + |\phi_p| + V_{SB}) - q^2 x_i N_a N'} \tag{9.1.23}$$

As expressed in Equation 9.1.23, the threshold voltage has an arbitrary reference point. The zero for potential can be taken at any electrode.

By comparing Equation 9.1.23 to Equation 8.3.18, we see that the nonuniform doping from the ion implantation affects the threshold voltage V_T in three ways: (1) The voltage drop across the surface-depletion region at inversion is $(|\phi_{ps}| + |\phi_p| + V_{SB})$ instead of $(2|\phi_p| + V_{SB})$; (2) the threshold voltage depends linearly on dose through the term qN'/C_{ox}; and (3) the square-root expression for the depletion-charge term is

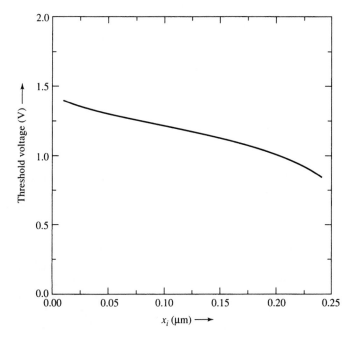

FIGURE 9.8 Dependence of the threshold voltage on the implantation depth x_i as predicted by Equation 9.1.23 for a MOSFET with an 87-nm-thick gate oxide, $Q_f/q = 10^{11}$ cm^{-2}, $N' = 3.5 \times 10^{11}$ cm^{-2}, and $N_a = 2 \times 10^{15}$ cm^{-3}. Both V_S and V_B are assumed to be zero.

altered principally by the added term $-q^2 x_i N_a N'$. Of these three changes, the first depends on the dose only logarithmically and is not of much consequence. The second change, which depends linearly on the dose N', is the dominant effect. It enters the expression for V_T as a direct addition to the term representing fixed surface charge Q_f as is evident from the expression for the flat-band voltage (without distributed oxide charge) $V_{FB} = \Phi_{MS} - Q_f/C_{ox}$. The third change in the threshold-voltage expression is the only one that depends on x_i, the implantation depth. The dependence is not strong, as we can see by inspecting Figure 9.8 which shows the threshold voltage as a function of implantation depth. The values of V_T plotted in Figure 9.8 were calculated from Equation 9.1.23 for the process parameters given in the figure caption. Because the dependence on x_i is weak, a reasonable estimate of V_T can be obtained by disregarding the effect of the implant on the square-root term, and this is often done in practice. For the MOSFET of Figure 9.8 without implantation, the threshold voltage is nominally -0.19 V, too close to zero for reliable fabrication of n-channel MOSFETs, emphasizing the importance of the ion-implantation process in producing useful n-channel transistors.

The effectiveness both of ion implantation and of body bias in adjusting the threshold voltage of n-channel MOSFETs is illustrated in Figure 9.9. The data points on the figure represent the threshold voltages measured as the body bias V_{SB} is varied for a series of n-channel MOSFETs implanted with varying doses of the p-type dopant boron. The solid curves on the figure were calculated using Equation 9.1.23 with V_S taken as the reference. The x-axis in Figure 9.9 corresponds to the magnitude of the square root in Equation 9.1.23, so the curves are linear. The substrate material is (100)-oriented silicon doped with 1.2×10^{16} boron atoms cm^{-3}. The MOSFETs have an oxide thickness of 100 nm, an oxide-charge density $Q_f/q = 8 \times 10^{10}$ cm^{-2}, and $\Phi_{MS} = -0.92$ V. From the figure, we see that, as the implanted dose increases, a higher body bias V_{SB} is needed to cause the depletion region to penetrate completely through the implanted silicon, a condition we assumed in our analysis.

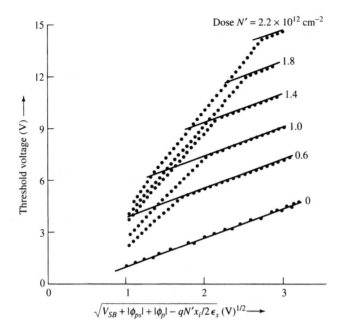

FIGURE 9.9 Dependence of the threshold voltage (referred to the source electrode) for ion-implanted n-channel MOSFETs. The dots represent experimental values measured as V_{SB} is varied. The solid lines are plotted from Equation 9.1.23. The properties of the MOSFETs are given in the text. (*Courtesy of Hewlett-Packard Company*).

Depletion-Mode MOSFETs

In depletion-mode MOSFETs, a conducting channel exists between the source and the drain when the gate and source are at the same potential ($V_{GS} = 0$). The zero-bias channel is especially useful if a MOSFET serves as a load element in a digital inverter stage because it permits fast switching, a maximum voltage swing for a given supply voltage, and well-defined waveforms to represent binary ones and zeros. Because the switching transistor in an inverter circuit should be an enhancement device, using a depletion-mode load requires fabricating MOSFETs with two different threshold voltages in the same IC. Selective ion implantation allows enhancement-switch, depletion-load (E/D) MOSFET circuits.

Looked at most simply, a depletion-mode MOSFET is created by shifting the threshold voltage sufficiently to change its sign. For example, the positive threshold voltage of an n-channel enhancement-mode MOSFET decreases in magnitude and ultimately changes to a negative value as the donor-implant dose is continuously increased. Considering only the threshold-voltage change, we model drain current in the depletion MOSFET by merely changing the sign of V_T in the equations derived earlier in this chapter. This model accurately represents the transistor provided that the donors implanted into the silicon are located in a very thin sheet near the Si–SiO$_2$ interface. Under these conditions, the implant has the same effect as modifying the fixed interface-charge density Q_f. For n-channel MOSFETs with implanted donors, the change in V_{Tn} is negative:

$$\Delta V_{Tn} = -qN'/C_{ox} \qquad (9.1.24)$$

For p-channel MOSFETs, implanted acceptors cause V_{Tp} to increase from its initial negative value:

$$\Delta V_{Tp} = +qN'/C_{ox} \qquad (9.1.25)$$

Although Equations 9.1.24 and 9.1.25 are approximate, modeling depletion-mode MOSFETs in terms of only a threshold-voltage shift is often adequate for circuit design, especially when they are used in the less-critical load elements of the circuit.

Subthreshold Conduction

Our first-order view of the MOSFET as a device in which the gate voltage must reach V_T before any drain current can flow provides a very useful picture for many MOSFET applications. There are, however, important applications in which even very low currents are important. For example, a MOSFET is usually used to access the storage capacitor of the dynamic RAM cell discussed in Sec. 8.7. Even a small current flowing through the MOSFET allows the storage capacitor to discharge, destroying the stored information. In these cases, we need to reexamine our assumption of an abrupt channel turn-off when V_G is reduced to V_T. The small drain current that flows when $V_G < V_T$ is called the *subthreshold current*.

Our discussion in Chapter 8 showed that the mobile-charge density Q_n at the MOS surface changes exponentially as the surface potential ϕ_s changes. Therefore, instead of decreasing to zero and remaining at zero when the gate voltage decreases below the threshold voltage, the mobile charge density and consequently the drain current enter a region where both decrease exponentially. We can use the theory of Chapter 8 to find expressions for currents in this region, but these expressions are too cumbersome for routine circuit design. Fortunately, the functional form of subthreshold current I_D can be derived intuitively by making appropriate approximations.

In the subthreshold bias region, the drain voltage drops almost entirely across the reverse-biased, drain-substrate depletion region. As a result, the drift-current component is negligible. On the other hand, the gradient of the free-carrier density along the channel can be relatively large. Consequently, subthreshold current is carried predominantly by diffusion, analogously to current in the base of a uniformly doped bipolar transistor in which the current depends on the gradient of the carrier density across the base. The applied gate voltage causes greater band bending ϕ_s at the surface which reduces the size of the barrier $q\phi_B$ to electron transport from the heavily doped source region to the channel, as illustrated in the band diagram of Figure 9.10. Electrons are injected from the source (which acts like a bipolar emitter region) into the p-type surface region (which acts under these conditions like the base region of a BJT). Most of these emitted electrons are collected at the drain (which now behaves like a BJT collector region). Because of the effect of the barrier $q\phi_B$, at gate voltages approximately 0.2 V below V_T, the drain current I_D begins to vary exponentially with gate voltage V_G, and the subthreshold drain current can be written as

$$I_D \approx I_{D0} \exp\left(\frac{qV_{GS}}{nkT}\right) \qquad (9.1.26)$$

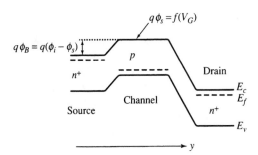

FIGURE 9.10 Band diagram along the channel region of an n-channel MOSFET under bias, indicating that the barrier $q\phi_B$ at the source depends on the gate voltage.

The variation of subthreshold drain current with V_{GS} is linear on a semilog plot down to a very low-current "noise" region, in agreement with Equation 9.1.26. Despite the analogous physics, there are major differences between the behavior of a MOSFET in the subthreshold region and that of a bipolar transistor. First, the injection occurring in MOSFET subthreshold conduction is very localized (at the surface). Second, the analogous "base-emitter bias V_{BE}" is not applied from an external connection as is the case in a BJT, but is instead the difference between the surface band-bending ϕ_s at the source end of the channel and the source voltage V_S.

We express the relationship between the gate voltage and the surface potential by using a parameter η that we define as

$$\eta = \frac{d\phi_s}{dV_{GS}} = \frac{1}{1 + (dV_{ox}/d\phi_s)} = \frac{1}{1 + (C_d/C_{ox})} \qquad (9.1.27)$$

where C_d is the surface depletion-layer capacitance (ϵ_s/x_d). Comparing Equation 9.1.27 with Equation 9.1.26, we see that the "efficiency" factor η relating ϕ_s and V_{GS} equals $1/n$, so the subthreshold current increases exponentially as η increases. Measurements of subthreshold currents, such as those plotted in Figure 9.11, can be used to obtain a value of the slope in the straight-line region of a semilog plot. We denote the reciprocal of the slope by the parameter S, the *inverse subthreshold slope* (often simply called the *subthreshold slope*), which is usually specified in units of mV of applied gate voltage V_{GS} per decade of subthreshold current change. An equation for S is

$$S = \frac{1}{\eta}\frac{kT}{q}\ln 10 = n\frac{kT}{q}\ln 10 \approx 60\, n\,\frac{mV}{decade} \qquad \text{at } 25°C \qquad (9.1.28)$$

where

$$n = 1 + \frac{C_d}{C_{ox}} \qquad (9.1.29)$$

for an ideal interface (free of surface trapping). If interface traps are present, an additional term is included in Equation 9.1.29 to account for the slower variation of surface potential with gate voltage discussed in Sec. 8.5.

At room temperature (25°C), the range of S for a MOSFET in a typical modern technology is between 70 and 120 mV/decade of drain current. The value of S is important to the designer because it determines the gate bias needed to assure an OFF condition

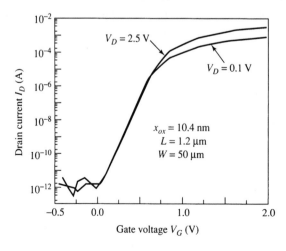

$x_{ox} = 10.4$ nm
$L = 1.2$ μm
$W = 50$ μm

$V_D = 2.5$ V

$V_D = 0.1$ V

Drain current I_D (A)

Gate voltage V_G (V)

FIGURE 9.11 Measured subthreshold characteristics of an MOS transistor with a 1.2 μm channel length. The inverse slope of the straight-line portion of this semilogarithmic plot is called the drain-current subthreshold slope S (measured in mV/decade of drain current).

in a MOSFET switch. For example, a criterion might be that an OFF transistor biased at $V_G = 0$ should not pass more than 10^{-5} (0.001%) times the drain current that flows when $V_G = V_T$. Meeting this requirement places a lower limit on V_T for a given value of S.

EXAMPLE *Subthreshold Current*

An n-channel MOSFET process is to be built in a p-type region doped with 8×10^{16} cm^{-3} acceptors using an n^+ polysilicon gate. The oxide thickness is 15 nm, and the gate length is 0.8 μm. What is the ratio of the leakage current that flows at $V_G = V_T$ to that at $V_G = 0$?

Solution   For an n^+ polysilicon gate, n-channel MOSFET doped with $N_a = 8 \times 10^{16}$ acceptors cm^{-3}, the flat-band voltage and ϕ_p are estimated from Table 8.1 and the discussion in Sec. 8.3 to be -0.9 V and 0.36 V, respectively.

The gate-oxide capacitance C_{ox} is 2.3×10^{-7} F cm^{-2}, and the threshold voltage is calculated to be 0.41 V from Equation 8.3.18.

We calculate $x_{d\max} = 1.28 \times 10^{-5}$ cm $= 0.128$ μm from Equation 8.3.6, and therefore $C_d = 8.2 \times 10^{-8}$ F cm^{-2}.

From Equation 9.1.29, n $= 1.36$, and $S = 82$ mV/decade.

For a threshold voltage of 0.41 V, the current is $(0.41/0.082)$ decades lower at $V_G = 0$ than at V_T. That is, the current at $V_G = V_T$ is 10^5 times that at $V_G = 0$.   ∎

In MOS digital design, requirements for a specific ON/OFF current ratio often limit the minimum allowable value of the threshold voltage that can be used. The minimum threshold voltage, in turn, places a lower limit on the required power-supply voltage and therefore the required power for the circuit. To meet the required performance of some circuits, more elaborate processing is used to fabricate two types of n-channel MOSFETS having different threshold voltages. In these cases, most of the transistors have a higher threshold voltage to limit their OFF current and the resulting standby power dissipation, while a smaller number that must provide maximum current when turned on (usually for high-speed operation) are designed to have a lower threshold voltage. The higher OFF current of these MOSFETs is a penalty for their higher performance. To reduce the leakage current of these transistors, a substrate bias can be applied to increase their threshold voltage when the portion of the circuit in which they are located is not active.

There are two intrinsic limits on the speed of response of a MOSFET. First, as in all current amplifiers, a basic limit is set by the time for charge transport along the channel: that is, the transit-time limitation (described for the bipolar transistor in Sec. 7.3). The second limit is the time required to charge capacitors inherent in the device structures (capacitors shown in the small-signal model). In practical applications, a third (often troublesome) limit on speed arises from the need to charge the unavoidable parasitic capacitors that are not inherent in the device itself. Limitations arising from these effects are usually modeled by adding circuit elements at nodes external to the device itself; we will consider the parasitic elements later in this chapter.

Analysis of the speed limitation of the intrinsic MOSFET itself depends on the operating region in which it is biased. We only consider the saturation region, which dominates for most applications. We start by finding an approximate solution for the field $\mathscr{E}(y)$ along the channel, which is straightforward in the long-channel case. From Equation 9.1.4, we have

$$\int_0^y I_D dy = \mu_n W C_{ox} \int_0^V [V_G - V_T - V(y)] dV \qquad (9.1.30)$$

Performing the integrations in Equation 9.1.30 and solving for $V(y)$, we have

$$V(y) = (V_G - V_T) - \sqrt{(V_G - V_T)^2 - \frac{2I_D y}{\mu_n W C_{ox}}} \qquad (9.1.31)$$

In Equations 9.1.30 and 9.1.31, we have taken the threshold voltage to be independent of y; that is, the variation of depletion-layer charge Q_d with y is not considered. To find the electric field, we differentiate Equation 9.1.31 to find $\mathscr{E}_y = -dV/dy$:

$$\mathscr{E}_y = -\frac{(V_G - V_T)}{2L} \frac{1}{\sqrt{1 - y/L}} \qquad (9.1.32)$$

The transit time T_{tr} along the channel is then found directly by using Equation 9.1.32 in the expression

$$T_{tr} = \int_0^L \frac{1}{v_y} \, dy = -\int_0^L \frac{1}{\mu_n \mathscr{E}_y(y)} \, dy \qquad (9.1.33)$$

which leads to

$$T_{tr} = \frac{4}{3} \frac{L^2}{\mu_n (V_G - V_T)} \qquad (9.1.34)$$

Although this transit-time analysis contains many approximations and does not consider velocity saturation in the channel, the predicted transit time provides a "ball-park" value showing that the transit time is usually appreciably shorter than the fastest switching times obtained in MOSFET circuits. We conclude that the speed of response of real MOSFETs is governed not by the channel transit time, but rather by the time needed to charge capacitances associated with the device and the elements to which it is connected in a circuit. For this reason, the calculation of MOSFET circuit transients is carried out assuming that currents in the MOSFET are governed by the static equations for the device.

Small-Signal Circuit Model

A circuit model to represent the most important effects in the MOSFET is shown in Figure 9.12. Of the four capacitors connected to the gate, only two (C_{GS} and C_{GD}) are intrinsic to the MOSFET. These capacitors represent the flux linkages to the channel charge

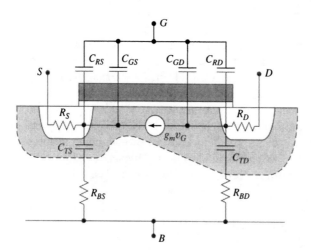

FIGURE 9.12 Small-signal equivalent circuit of a MOSFET.

that allow the basic operation of the MOSFET and are the "good" capacitances. Their values depend on the bias. If V_{DS} is small, each is equal to $C_{ox}WL/2$. When the MOSFET is saturated, C_{GS} becomes $(2/3)C_{ox}WL$, and C_{GD} approaches zero, indicating that few electric flux lines link the gate to the drain. The speed limitations associated with charging C_{GS} and C_{GD} are fundamentally related to the transit time of charge along the channel.

Capacitors C_{RS} and C_{RD} between the gate and the source and between the gate and the drain, respectively, are parasitic elements that result from overlap of the gate with the source and drain diffusions. The two capacitors connected between the substrate and the source and between the substrate and the drain (C_{TS} and C_{TD}) are depletion-region capacitances at the reverse-biased *pn* junctions in these regions. The resistors R_S and R_D represent the series resistances from the external electrodes to the MOSFET channel. Resistors R_{BS} and R_{BD} account for ohmic resistance between the edges of the depletion regions and the contact to the substrate.

The *transconductance* g_m relates the output and the input of the transistor because the output (drain) current typically varies in response to a changing input (gate) voltage. We find g_m by differentiating the drain current given by Equation 9.1.5 or 9.1.6 with respect to V_G

$$g_m \equiv \partial I_D / \partial V_G \tag{9.1.35}$$

For $V_D < V_{D\text{sat}}$, we use Equation 9.1.5 to find

$$g_m = \mu_n C_{ox}(W/L)V_{DS} \tag{9.1.36}$$

In this region of operation, g_m increases linearly with increasing drain voltage, but is independent of gate voltage. The transconductance is also proportional to the ratio of the channel width to length, the mobility, and the oxide capacitance per unit area because an increase in any of these quantities increases the output current for a given change in gate-to-source voltage. When $V_D > V_{D\text{sat}}$ (from Equations 9.1.35 and 9.1.6) the transconductance is, to first order,

$$g_{m\text{sat}} = \mu_n C_{ox}(W/L)\,(V_G - V_T) = \frac{2I_{D\text{sat}}}{(V_G - V_T)} \tag{9.1.37}$$

In the saturated region, $g_{m\text{sat}}$ is independent of V_D, but depends linearly on V_G.

9.2 IMPROVED MODELS FOR SHORT-CHANNEL MOSFETs

Limitations of the Long-Channel Analysis

The channel lengths of the first commercial MOSFETs were more than 20 μm. These MOSFETs are fairly well modeled by the long-channel theory with its successive approximations and the resulting nonphysical consequences (such as infinite carrier velocities near the drain). As we found in our analysis and illustrated with the water analogy, the long-channel theory predicts drain currents to increase with increasing drain voltages (when the applied gate voltage is higher than the threshold voltage) as mobile electrons in the channel move faster in the increasing field along the channel. According to this theory, sufficiently high drain voltages lead to a "pinch-off" condition, in which the channel mobile-carrier density becomes zero near the drain. The channel current must, however, be constant and therefore the pinch-off condition requires an infinite carrier velocity to maintain a constant current with a vanishing carrier density. Despite this physical impossibility, the basic assumptions stated above apply fairly well when channel lengths are greater than about 10 μm, much larger than those in most MOSFETs now being fabricated. In these smaller MOSFETs, other physical effects need to be considered to understand device behavior; these effects are qualitatively discussed in the following sections.

Short-Channel Effects

The term *short-channel effects* is a bit fuzzy, sometimes referring to secondary effects such as mobility degradation and velocity saturation, both of which also occur in long-channel devices. A stricter definition of *short-channel effects* limits this term to behavior observed only in short-channel devices. We can differentiate between the terms *short* and *long* by comparing the channel length L with the thickness of the drain space-charge region. In short-channel MOSFETs the two lengths are the same order of magnitude. In this context, a short-channel effect refers to the decrease in V_T that occurs when L decreases and the drain bias increases. The reduction in V_T results from the combination of three effects: (a) *source/drain charge sharing*, (b) *drain-induced barrier lowering*, and (c) *subsurface punchthrough*. In the following paragraphs, we consider these three effects separately in the order listed.

Source/Drain Charge Sharing. Figure 9.13 represents a short-channel MOSFET biased with $V_S = V_B$, V_D small (close to V_S), and V_G biased at the threshold voltage. At the threshold voltage, we know from Chapter 8 that the fixed substrate-depletion charge Q_d in the device is $-qN_a x_{dmax}WL$, where W is the MOSFET width in the third (z) dimension and L is the channel length. This charge is indicated in the cross section of Figure 9.13 [28] by the cross-hatched region under the gate. Where the shaded region overlaps the cross-hatched regions, we see two pieces of the rectangular-charge region that are also parts of the depletion regions of the source- and the drain-*pn* junctions. Therefore, no gate voltage is required to deplete the mobile charge from these regions. For long-channel MOSFETs the total volume of these two depletion regions, which are roughly triangular in cross section, is insignificant compared to the rectangular depletion region, but they clearly become increasingly important as the channel length is reduced. Because some of the bulk charge in short-channel MOSFETs is supplied by the source and drain depletion regions, the amount that must be induced by the gate decreases, consequently reducing the magnitude of the gate voltage needed to invert the silicon surface (i.e., the threshold voltage of the MOSFET), as indicated in Figure 9.14. This short-channel effect is more serious for MOSFETs with thicker oxides, in which the coupling of the gate to the channel is weaker so that the ratio of the oxide capacitance to the source or drain junction capacitance is lower. From a rigorous solution of Poisson's equation, we can show that V_T varies

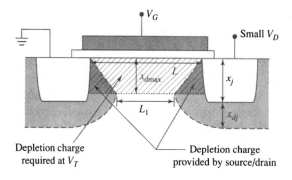

Depletion charge required at V_T

Depletion charge provided by source/drain

FIGURE 9.13 The depletion charge at threshold. The cross hatched rectangle represents the depletion charge Q_d required to achieve threshold. Part of this charge is coupled to the source and drain; therefore less charge needs to be induced by the gate voltage V_G, reducing the gate voltage needed to achieve threshold (© 1983 IEEE [28]).

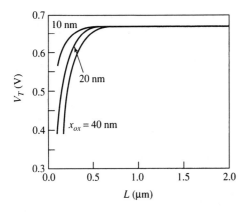

FIGURE 9.14 Simulated values of the threshold voltage versus channel lengths in MOSFETs having different gate-oxide thicknesses.

exponentially with L when L is sufficiently small. The derivation is mathematically complicated [4], and we omit it here. However, we can gain a semi-quantitative feeling for short-channel effects from an approximate geometrical analysis [28] discussed in the following example.

EXAMPLE *Charge Sharing*

Consider the idealized cross section of a short-channel transistor shown in the accompanying figure. The maximum depletion width away from the source and drain junctions is x_{dmax}, r_j is the junction radius, and r_2 is the radial distance to the corner of the trapezoid shown in the figure. W is the width of the channel and N_a is the dopant density.

Considering the reduced depletion-region charge that needs to be induced by the gate electrode, derive an approximate expression for the threshold voltage for small values of V_{DS}.

Solution For small values of V_{DS}, we consider the charge induced by V_G to be approximately contained in a volume whose cross section is the trapezoid of width x_{dmax} and length varying from L at the surface to L_1 at the substrate side of the depletion region. The cross-sectional area is shown cross-hatched on the figure. If this charge is called Q_{d1}, then

$$Q_{d1} = qx_{dmax}WN_a\frac{L + L_1}{2} \tag{1}$$

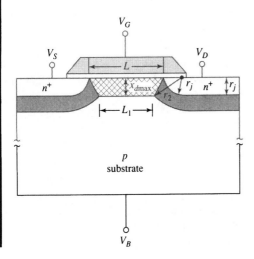

The charge Q_{d1} in Equation 1 is approximately the depletion-layer charge that must be induced by the gate to bring the channel to the threshold condition. If the channel is long so that the space-charge regions at the source and drain are much smaller than L_1, then L_1 approaches L. In that case, from Equation 1, Q_{d1} equals $Q_d = q\, x_{d\max}\, N_a\, WL$, as was assumed in the first-order theory of Sec. 9.1. For shorter channels, L_1 becomes appreciably less than L, and Q_{d1} is therefore less than Q_d, as expected from our qualitative arguments.

For a useful theory, L_1 must be related to the geometry of the MOSFET. This can be done approximately by assuming that when $V_G = V_T$, the depletion layer is $x_{d\max}$ units wide both in the x-direction (perpendicular to the Si–SiO$_2$ interface) and along the radius of the diffused source and drain junctions. With this approximation $r_2 = r_j + x_{d\max}$. From the geometry of the structure, we obtain

$$f \equiv \frac{Q_{d1}}{Q_d} = 1 - \frac{r_j}{L}\left(\sqrt{1 + \frac{2x_{d\max}}{r_j}} - 1\right) \tag{2}$$

The parameter f is therefore a function of the MOSFET geometry. The expression for the threshold voltage is written directly from Equation 8.3.18:

$$V_T = V_{FB} + 2|\phi_p| + V_S - \frac{fQ_d}{C_{ox}}$$
$$= V_{FB} + 2|\phi_p| + V_S + \frac{f}{C_{ox}}\sqrt{2\epsilon_s qN_a(2|\phi_p| + V_S - V_B)} \tag{3}$$

Despite the approximate nature of its derivation, Equation 3 is useful in predicting trends in the behavior of V_T for MOSFETs with short channel lengths.

The analysis leading to Equation 3 did not consider the difference between the space-charge dimensions at the source and at the drain and, therefore, represents V_T for small values of V_{DS}. Because V_{DS} is typically much larger than the source-substrate bias V_{SB}, V_T is sensitive to V_D in short-channel MOSFETs. More elaborate geometric analyses include this effect. ∎

If the width of a MOSFET decreases until it becomes comparable to the space-charge-region width $x_{d\max}$ near the drain, the threshold voltage becomes dependent on W. The detailed dependence of the device behavior on this *narrow-width effect* depends strongly on the specific isolation technology employed in the MOS fabrication process. In raised field-oxide or semi-recessed LOCOS isolation structures (Figure 9.15), two effects cause the threshold voltage V_T to increase as W decreases.

First, because of the fringing field at the edge of the depletion region in the width direction, some of the gate-induced space charge lies outside the channel region. As a consequence, more gate voltage must be applied to induce a conducting channel under the gate, and the threshold voltage increases. The second effect arises from the higher dopant concentrations in the isolation regions than in the channel. In n-channel MOSFETs, boron channel-stop dopant atoms are usually implanted into the isolation regions between devices to prevent inversion there. During continued processing, these atoms can diffuse

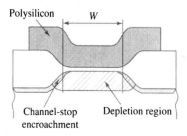

Polysilicon   W

Channel-stop   Depletion region
encroachment

FIGURE 9.15 Schematic diagram of the narrow-channel effects in semi-recessed LOCOS isolation technology.

under the edges of the gate, increasing the dopant concentration near the edges of the channel. In *p*-channel MOSFETs, phosphorus pileup at the silicon surface during field oxidation automatically produces higher concentrations of *n*-type, channel-stop dopant atoms in the field regions. In either case, the doping at the edges of the channel is higher than in the center, tending to increase the threshold voltage. In practice, dopant encroachment is more important than fringing fields, especially in devices with heavy channel-stop implants. An analytical expression that models both effects in semi-recessed MOSFETs is given in reference [5]. For more precise calculations, however, numerical analysis is required.

In newer processes, fully recessed-LOCOS or trench-isolation structures are often used, and a third width-dependent effect can *decrease* V_T. Because this change is in the opposite direction, this effect is sometimes called the *inverse narrow-width effect*. Figure 9.16 depicts the edge of the channel, the trench field-oxide, and the gate for a trench-isolated transistor [6]. When the gate is biased, the field lines from the overlapping region are concentrated at the edge of the channel. Thus, an inversion layer forms at lower voltages at the edges of the channel than at the center, and less applied gate voltage is required to invert the channel across its entire width. An analytical expression for the narrow-width effect in an ideal trench-isolation structure is found in [7]. However, the inverse-narrow-width effect is sensitive to several other factors [8], including (1) the doping concentration at the sidewalls of the silicon, which can be affected by redistribution of the channel-stop dopant, by dopant segregation during field-oxide growth, and by intentional doping of the sidewalls by such processes as large-tilt-angle implantation; (2) trench isolation spacing; and (3) the shape of the corner region at the edge of the gate. Hence, precise calculation of the inverse-narrow-width effect can only be done by numerical simulation (assuming that all the structural parameters are known).

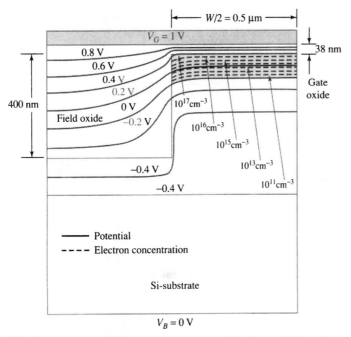

FIGURE 9.16 Two-dimensional contours of equipotentials and electron concentrations in an *n*-channel MOSFET with fully recessed trench isolation (© 1982 IEEE [6]).

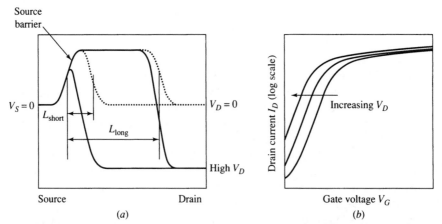

FIGURE 9.17 (*a*) Conduction band of a long-channel and a short-channel MOSFET at $V_D = 0$ V (dotted lines) and high V_D (solid lines). The source barrier of the short-channel device is reduced at high drain voltage. (*b*) Subthreshold conduction characteristics of a MOSFET with drain-induced barrier lowering (DIBL), measured at three different drain voltages.

Drain-Induced Barrrier Lowering. The second short-channel effect, drain-induced barrier lowering (DIBL), refers to the influence of the drain voltage on ϕ_s, the barrier to electron flow at the *np* junction near the oxide surface at the source. The density of electrons entering the channel from the source increases exponentially when the barrier to electron flow from the source into the channel is lowered linearly. In the long-channel theory, only the gate-to-drain voltage V_{GS} lowers the source barrier. However, with short channel lengths, a sufficiently high drain voltage can "pull down" the electron barrier, as sketched in Figure 9.17a, causing the observed threshold voltage to depend on V_D as well as on V_G. The subthreshold current is most sensitive to DIBL, and when barrier lowering occurs, one can most easily detect it by measuring the dependence of subthreshold current on V_D. In contrast, subthreshold current is independent of the drain voltage in long-channel devices. As seen in Figure 9.17b, DIBL can significantly increase the subthreshold current, degrading the behavior of circuits containing short-channel MOSFETs [9].

Subsurface Punchthrough. Like drain-induced barrier lowering, subsurface punchthrough refers to the influence of the drain voltage on the source *np*-junction electron barrier. The distinction between DIBL and subsurface punchthrough is that the latter occurs in the substrate away from the surface. The physics of subsurface punchthrough is essentially equivalent to the punchthrough effect we described in our discussion of IC diodes in Sec. 5.6. In short-channel *n*-channel MOSFETs, the surface *p*-region is more heavily doped than the bulk, making the junction-depletion region wider below the surface than in the channel region. Consequently, it is possible for the drain-substrate depletion region to reach the source-substrate depletion region at a sufficiently high drain bias. As shown in Figure 9.18, this effect can provide a punchthrough region where increasing drain bias lowers the source-substrate barrier, leading to undesirable, typically destructive, drain currents. Figure 9.19 sketches the unacceptable MOSFET $I_D - V_D$ and subthreshold-region characteristics of a device exhibiting subsurface punchthrough.

Both, DIBL and subsurface punchthrough degrade MOSFET operation in integrated circuits, most notably by leading to unacceptably high and variable leakage currents in normally OFF devices. These problems become more severe as channel dimensions decrease

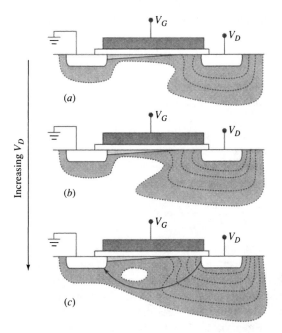

FIGURE 9.18 Cross sections of a short-channel MOSFET with its source contact grounded, gate voltage V_G above the threshold voltage V_T and held constant as the drain voltage V_D is successively increased from (a) to (c). The space-charge region is shaded and equipotentials are shown as dotted lines. At punchthrough, indicated in (c), a new subsurface path for drain current is shown by a solid arrow.

and the density of devices increases. Designers typically seek ways to minimize these effects by increasing the substrate doping as much as possible, by thinning the silicon dioxide as much as feasible, and by reducing the depth of the source and drain junctions.

Mobility Degradation

The drain current in a MOSFET is carried by a mobile-charge density Q_n that moves in a region near the surface under the influence of varying fields. The electrons are scattered by collisions with the interface as well as with charged acceptor sites and thermal phonons [10]. An exact theory for the influence of these effects on carrier mobility is not feasible, but an approximate treatment [11] that leads to a simplified mobility formulation uses an effective x-directed electric field $\mathscr{E}_{eff}$ near the surface to write a simple field-dependent expression for carrier mobility. Experiments have validated this formulation for mobility for both holes and electrons and for gate oxides as thin as 4 nm [12, 13]. The effective field

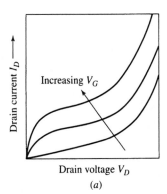

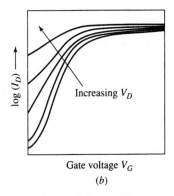

FIGURE 9.19 (a) $I_D - V_D$ characteristics and (b) subthreshold conduction characteristics of a MOSFET with punchthrough.

$\mathscr{E}_{eff}$ is the average x-directed field experienced by all the mobile carriers at the MOS surface

$$\mathscr{E}_{eff} = \int_0^\infty n(x)\mathscr{E}_x(x)dx \bigg/ \int_0^\infty n(x)dx \tag{9.2.1}$$

We derive a simplified and useful expression for $\mathscr{E}_{eff}$ using the relation in calculus for the integral of a product of functions, and (as in Sec. 8.3) definitions for the total charge density at the surface Q_s which is the sum of the charge stored on the depleted acceptors Q_d in the surface space-charge region and the mobile-electron density Q_n in the region. As in Equation 8.3.14, $Q_s = Q_n + Q_d = -\epsilon_s\mathscr{E}_{s0}$, where $\mathscr{E}_{s0}$ is the field at the oxide–silicon interface. The mobile-electron density Q_n is, in turn, the integral of $-qn(x)$ over all x, $Q_n = -q\int_0^\infty n(x)dx$. From these definitions, we can write Equation 9.2.1 as

$$\mathscr{E}_{eff} = -\frac{1}{\epsilon_s}\left(Q_d + \frac{Q_n}{2}\right) \tag{9.2.2}$$

A physical picture can help us understand Equation 9.2.2. To develop this picture, we consider Figure 9.20 which shows an electron moving in the MOS surface-space-charge region. Electric field lines in this region terminate either on mobile electrons or on ionized acceptors. Because most of the mobile electrons are close to the oxide–silicon interface, essentially all of them are influenced by the field lines that terminate on all of the acceptors, as indicated by the term Q_d in Equation 9.2.2. Only half the charge represented by the mobile electrons $Q_n/2$ appears in Equation 9.2.2 because field lines that terminate on mobile charges near the interface do not affect carriers that are moving below them. The "average" electron within the spatial distribution is located with half the carriers above and half below so that only half the electron density contributes to the effective field. Thus $\mathscr{E}_{eff}$ is determined by the sum of Q_d and half of Q_n, as shown in Equation 9.2.2.

Equation 9.2.2 can be manipulated further to bring it into a more useful form. From Sec. 8.3 and Equation 8.3.19 we have $Q_n = -C_{ox}(V_G - V_T)$ and $Q_d = -C_{ox}(V_T - V_{FB} - 2|\phi_p|)$. To simplify further, we note that, as MOSFET dimensions have been repeatedly decreased, the sum $(V_{FB} + 2|\phi_p|)$ has remained fairly constant at a value we represent by an empirical constant V_Z, where V_Z is roughly 0.5 V. Using the equations for Q_n and Q_d as well as the definition of V_Z and recognizing that $\epsilon_s/\epsilon_{ox} = 3$, we can write a useful form of Equation 9.2.2

$$\mathscr{E}_{eff} = \frac{(V_G - V_T)}{6x_{ox}} + \frac{(V_T + V_Z)}{3x_{ox}} \tag{9.2.3}$$

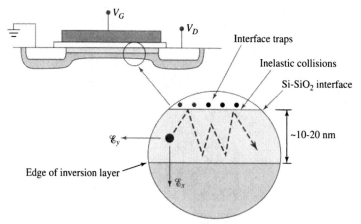

FIGURE 9.20 Motion of inversion electrons under the influence of lateral electric field $\mathscr{E}_y$ and vertical electric field $\mathscr{E}_x$.

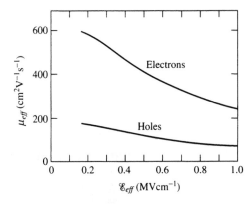

FIGURE 9.21 Effective mobility μ_{eff} for electrons and holes in MOSFET channels as a function of the x-directed, effective surface field $\mathscr{E}_{eff}$ (© 1986 IEEE [13]).

Figure 9.21 shows values for the effective mobilities of electrons and holes measured in n- and p-channel MOSFETs as functions of the effective x-directed surface field $\mathscr{E}_{eff}$ (Equation 9.2.3) [13]. An empirical equation to represent this mobility degradation in equations for the MOSFET is

$$\mu_{eff} = \frac{\mu_0}{1 + (\mathscr{E}_{eff}/\mathscr{E}_0)^\nu} \qquad (9.2.4)$$

where μ_0, $\mathscr{E}_0$, and ν are all fitting parameters. Values of the parameters fitting a number of MOSFETs are given in Table 9.3 [13]. Equation 9.2.4 holds over the wide range of conditions where the mobility is limited by phonon scattering; however, it fails when scattering is dominated by other mechanisms such as Coulomb scattering at low temperatures (e.g. 77 K) [14].

Although mobility degradation affects long-channel transistors, it is a more serious limitation in short-channel devices. The x-directed gate oxide fields are higher in short-channel transistors because the power-supply voltage is not scaled as much as suggested by constant-field scaling when the gate-oxide thickness is reduced.

Velocity Saturation

Our discussion of mobile-carrier velocities in solids in Sec. 1.2 pointed out that, at lower fields, drift velocity is linearly proportional to the applied field (constant mobility) in the direction of carrier drift (the y-direction for the MOSFET channel). When high fields are applied, carrier velocities approach a limiting value (Figure 1.18). In MOSFET channels where interface scattering also limits transport, the low-field mobility is smaller, but similar velocity saturation is observed. Measured data [15], [16], [17] indicate that the saturation velocity of an electron in a surface channel is between 6 and 10×10^6 cm s^{-1} and that of a hole is between 4 and 8×10^6 cm s^{-1}. These values are slightly lower than limiting velocities in the bulk, which are shown in Figure 1.18. Measured surface velocities can be fit by a convenient mathematical expression that is piecewise continuous with a

TABLE 9.3 Parameters for effective mobility
© 1986 IEEE [13]

	Unit	Electron (surface)	Hole (surface)	Hole (subsurface)
μ_0	cm^2/V-s	670	160	290
$\mathscr{E}_0$	MV/cm	0.67	0.7	0.35
ν		1.6	1.0	1.0

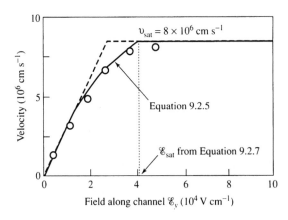

FIGURE 9.22 Measured electron velocity at different lateral electric fields $\mathscr{E}_y$ as fit to Equation (9.2.5) (© 1984 IEEE [18]).

discontinuity in the derivative at a tangential (y-directed) field $\mathscr{E}_{sat}$, above which the velocity remains constant [18]. At lower fields, the velocity is fit by an equation of the form

$$v = \frac{\mu_{eff}\mathscr{E}}{(1 + \mathscr{E}/\mathscr{E}_{sat})} \qquad \mathscr{E} < \mathscr{E}_{sat} \tag{9.2.5}$$

while for fields greater than the saturation field $\mathscr{E}_{sat}$ the velocity becomes constant

$$v = v_{sat} \qquad \mathscr{E} > \mathscr{E}_{sat} \tag{9.2.6}$$

The parameter $\mathscr{E}_{sat}$ is determined by solving for the field at which $v = v_{sat}$ as predicted by Equation 9.2.5 so that the v–$\mathscr{E}$ relation is continuous. Hence,

$$\mathscr{E}_{sat} = \frac{2v_{sat}}{\mu_{eff}} \tag{9.2.7}$$

The fit of Equations 9.2.5 and 9.2.6 to measured data is shown in Figure 9.22. Despite the many simplifications and the empirical nature of Equations 9.2.5 and 9.2.6, these equations have proven useful for predicting drain currents in short-channel MOSFETs.

Drain Current in Short-Channel MOSFETs

If we consider the variation of mobility with lateral field expressed in Equation 9.2.5, we can rewrite the MOSFET drain current (Equation 9.1.1) as

$$I_D = C_{ox}W[V_G - V_T - V(y)]\frac{\mu_{eff}\mathscr{E}(y)}{1 + [\mathscr{E}(y)/\mathscr{E}_{sat}]} \tag{9.2.8}$$

where $\mathscr{E}(y) = -\partial V/\partial y \approx -dV/dy$. Substituting for $\mathscr{E}(y)$, we obtain a differential equation in V that can be integrated from the source ($V = 0$) to the drain ($V = V_D$). If we treat V_T as constant (ignoring the variation of Q_d along the channel), we obtain

$$I_D = \mu_{eff}C_{ox}\frac{W}{L}\left(V_G - V_T - \frac{V_D}{2}\right)V_D\frac{1}{1 + (V_D/\mathscr{E}_{sat}L)} \tag{9.2.9}$$

which converges to Equation 9.1.5 for $\mathscr{E}_{sat} \gg V_D/L$ and $\mu_{eff} = \mu_n$. The derivation of Equation 9.2.9 treated μ_{eff} as a constant along the channel although Equation 9.2.4 shows that the mobility does depend on $\mathscr{E}_{eff}$, the effective surface field, which clearly varies along y. Despite this discrepancy and the further approximation that V_T does not vary along y (which ignores the varying bulk charge), Equation 9.2.9 is sufficiently accurate and simple to be very useful.

Drain-Current Saturation. Because of the carrier-velocity limitation, we expect the drain current to become saturated when the electrons arrive at the drain with their limiting velocity v_{sat}. If we designate the drain voltage at which this occurs as $V_{D\text{sat}}$, the electron charge per unit area at the drain Q_{nD} can be written as $Q_{nD} = -WC_{ox}(V_G - V_T - V_{D\text{sat}})$. Because this charge is moving at v_{sat}, the resulting drain current is given by

$$I_{D\text{sat}} = WC_{ox}(V_G - V_T - V_{D\text{sat}})v_{\text{sat}} \qquad (9.2.10)$$

The drain saturation voltage ($V_{D\text{sat}}$) is calculated by inserting Equation 9.2.10 into Equation 9.2.9 to obtain

$$V_{D\text{sat}} = \frac{\mathscr{E}_{\text{sat}}L(V_G - V_T)}{\mathscr{E}_{\text{sat}}L + (V_G - V_T)} \qquad (9.2.11)$$

As expected, the value predicted for $V_{D\text{sat}}$ by Equation 9.2.11 converges to $(V_G - V_T)$ when $\mathscr{E}_{\text{sat}}$ becomes very large so that velocity saturation is not important.

In the following example, we investigate use of the short-channel model.

EXAMPLE   *Short-Channel Model*

Use the short-channel model to find values for $V_{D\text{sat}}$ and $I_{D\text{sat}}$ for a MOSFET with the following parameters: $x_{ox} = 20$ nm, $W = 50$ μm, $L = 0.5$ μm, $V_T = 0.7$ V when its source is at zero potential, and with bias voltages $V_G = 3$ V and $V_D = 1.5$ V. Compare these results to the predictions of the long-channel model.

Solution   From Equation 9.2.3 with $V_Z = 0.5$ V,

$$\mathscr{E}_{\text{eff}} = \frac{(V_G - V_T)}{6x_{ox}} + \frac{(V_T + V_Z)}{3x_{ox}} = 3.92 \times 10^5 \text{ Vcm}^{-1}$$

From Equation 9.2.4 and Table 9.3

$$\mu_{\text{eff}} = \frac{\mu_0}{1 + (\mathscr{E}_{\text{eff}}/\mathscr{E}_0)^{\nu}} = 470 \text{ cm}^2 \text{ V}^{-1}\text{s}^{-1}$$

Taking $v_{\text{sat}} = 8 \times 10^6$ cm s^{-1}, we calculate $\mathscr{E}_{\text{sat}} = 2\dfrac{v_{\text{sat}}}{\mu_{\text{eff}}} = 3.4 \times 10^4$ Vcm^{-1}

From Equation 9.2.11, $V_{D\text{sat}} = \dfrac{\mathscr{E}_{\text{sat}}L(V_G - V_T)}{\mathscr{E}_{\text{sat}}L + (V_G - V_T)} = 0.98$ V

Because the applied drain voltage is 1.5 V ($>V_{D\text{sat}}$), the MOSFET is in the saturated-drain-current region; hence, using Equation 9.2.10,

$$I_{D\text{sat}} = WC_{ox}(V_G - V_T - V_{D\text{sat}})v_{\text{sat}} = 9 \text{ mA}$$

To compare these results to those from the long-channel theory, we first calculate $V_{D\text{sat}} = V_G - V_T = 2.3$ V so that the MOSFET is operating below the saturation region. From Equation 9.1.5 we find

$$I_D = \mu_n C_{ox}\frac{W}{L}\left[\left(V_G - V_T - \frac{1}{2}V_D\right)V_D\right] \quad \text{or} \quad I_D = 24 \text{ mA}$$

using a typical low-field mobility value of 600 cm^{2}V^{-1}s^{-1}. The very different results obtained using these two models show the degree to which the MOSFET behavior is influenced by short-channel effects. ■

Transconductance.   When MOSFETs are used as amplifiers, the *transconductance* in the saturation region $g_{m\text{sat}} = \partial I_{D\text{sat}}/\partial V_G$ is an important parameter. For the long-channel theory, we found $g_{m\text{sat}}$ by differentiating Equation 9.1.6 with respect to V_G (Equation 9.1.37, which is repeated here):

$$g_{m\text{sat}} = \frac{2I_{D\text{sat}}}{(V_G - V_T)} \qquad (9.2.12)$$

To include velocity saturation, $g_{m\text{sat}}$ is calculated from Equation 9.2.10 to obtain

$$g_{m\text{sat}} = W v_{\text{sat}} C_{ox} \left[1 - \frac{\partial V_{D\text{sat}}}{\partial V_G} \right] \tag{9.2.13}$$

Examination of Equations 9.2.11 and 9.2.7 shows that $V_{D\text{sat}}$ depends weakly on V_G, allowing us to derive an equation for $g_{m\text{sat}}$ in short-channel MOSFETs

$$g_{m\text{sat}} = W v_{\text{sat}} C_{ox} \frac{(V_G - V_T)(V_G - V_T + 2\mathscr{E}_{\text{sat}} L)}{(V_G - V_T + \mathscr{E}_{\text{sat}} L)^2} \tag{9.2.14}$$

Equations 9.2.9, 9.2.10, 9.2.11, and 9.2.14 form a basic and useful set of equations describing short-channel MOSFET operation.

MOSFET Scaling and the Short-Channel Model

Although the short-channel MOSFET representation we presented is relatively simple in its form, the insight it provides has proven useful in the continuing development of MOSFET integrated circuits; for example, as a guide for scaling MOSFETs. In the following discussion, we consider some effects of scaling as expected from the short-channel representation of the MOSFET. These considerations will help us to develop a physical picture for short-channel effects.

Drain Current in the Nonsaturated Region. By comparing Equation 9.1.5 to Equation 9.2.9, we see that short-channel effects only minimally modify the behavior of drain current in the nonsaturated region. In this lower drain-voltage region, the major degradation of the drain current is caused by mobility reduction, which becomes more significant as V_G increases (Equations 9.2.3 and 9.2.4).

Drain-Saturation Voltage. As we see from Equation 9.2.11, the drain-saturation voltage $V_{D\text{sat}}$ is sensitive both to the effective gate bias ($V_G - V_T$) and to the product of gate length L and $\mathscr{E}_{\text{sat}}$. When $\mathscr{E}_{\text{sat}} L \gg (V_G - V_T)$, $V_{D\text{sat}}$ converges to the long-channel value ($V_G - V_T$). When the MOSFET is scaled to smaller dimensions, L decreases and $V_{D\text{sat}}$ tends toward the value $V_{D\text{sat}} = \mathscr{E}_{\text{sat}} L$. Scaling of the MOSFET surface dimensions is accompanied by a reduction in the gate-oxide thickness x_{ox} and a corresponding increase in the oxide capacitance C_{ox} and the field at the surface. These effects reduce the surface mobility μ_{eff} (Equation 9.2.7) and increase $\mathscr{E}_{\text{sat}}$ and thereby $V_{D\text{sat}}$.

Drain-Saturation Current. In the long-channel theory the saturated drain current depends quadratically on the effective gate voltage [$I_{D\text{sat}} \propto (V_G - V_T)^2$ (Equation 9.1.6)] while in the short-channel theory the dependence is linear [$I_{D\text{sat}} \propto (V_G - V_T - V_{D\text{sat}})$ (Equation 9.2.10)]. The strikingly different behavior predicted by these two theories is sketched in Figure 9.23. At first, Equation 9.2.10 may seem peculiar because the channel length L does not appear explicitly even though the equation arose from our consideration of channel-shortening effects on MOSFET behavior. A changing L affects $I_{D\text{sat}}$ only through its influence on the term $V_{D\text{sat}}$ in Equation 9.2.10. As L decreases, $V_{D\text{sat}}$ also decreases (Equation 9.2.11); from Equation 9.2.10 we see that this decrease of $V_{D\text{sat}}$ tends to increase $I_{D\text{sat}}$. Physically, this increase of $I_{D\text{sat}}$ is a consequence of the growing fraction of the mobile charge Q_n that is moving at v_{sat} when L decreases. If L becomes vanishingly small, then from Equation 9.1.11, $V_{D\text{sat}}$ also tends toward zero, but $I_{D\text{sat}}$ from Equation 9.2.10 is limited to

$$I_{D\text{max}} = v_{\text{sat}} W C_{ox} (V_G - V_T) \tag{9.2.15}$$

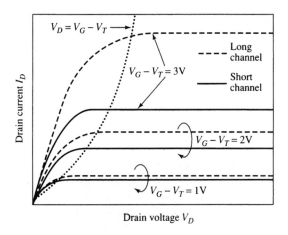

FIGURE 9.23 Drain-current characteristics of MOSFETs as predicted by the long-channel theory (dashed) and by the short-channel theory (solid).

In contrast to this result, the long-channel theory predicts (Equation 9.1.6) that I_{Dsat} grows toward infinity as channel length is reduced to zero ($I_{Dsat} \propto L^{-1}$). Figure 9.24 compares these two results.

The current I_{Dmax} calculated in Equation 9.2.15 is a useful benchmark to be considered when MOSFETs are scaled to smaller dimensions. We can define an *ideality factor* K_I that compares I_{Dsat} as measured for a scaled MOSFET to the limiting value I_{Dmax}. Using Equations 9.2.10 and 9.2.15, we obtain an expression for K_I

$$K_I = \frac{I_{Dsat}}{I_{Dmax}} = \frac{(V_G - V_T - V_{Dsat})}{(V_G - V_T)} \tag{9.2.16}$$

It is clear that K_I is always less than unity; not apparent is that K_I has different values for different MOSFET processes. As an example, consider a MOSFET with $x_{ox} = 40$ nm and $L = 1$ μm that is observed to saturate at $V_{Dsat} = 1.3$ V when $(V_G - V_T) = 4.3$ V. The ideality factor K_I is therefore $(4.3 - 1.3)/4.3 = 0.72$, indicating that reducing the channel length further can at best increase the device driving current by roughly 40%. The factor K_I given by Equation 9.2.16 is plotted against L for two different oxide thicknesses (7 and 20 nm) in Figure 9.25. Figure 9.25 shows that K_I for the thicker-oxide MOSFET varies more gradually with L than it does for the thinner-oxide MOSFET. Thus, reducing the size of the MOSFET more favorably affects the drain current of the device with the thinner oxide, and this improvement becomes stronger as the channel length decreases. The

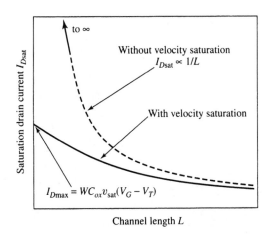

FIGURE 9.24 Predicted MOSFET saturation drain current as a function of channel length without velocity saturation (dashed) and with velocity saturation (solid).

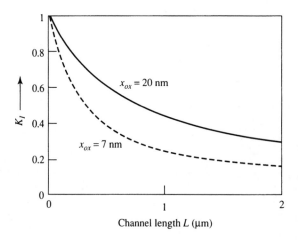

FIGURE 9.25 The calculated drain-current ideality factor K_I versus channel length L for two different gate-oxide thicknesses.

increased gate capacitance of the thin-oxide MOSFET tends to dominate all other effects, and MOSFET design emphasizes minimizing x_{ox}, thereby maximizing C_{ox}. Measured data for very short-channel MOSFETs, presented in Figure 9.26 [19], show the marked improvement obtained by decreasing x_{ox}, supporting this conclusion. Ultra-thin oxides, however, pose major reliability and yield problems, requiring a compromise on the best choice for x_{ox} that balances these competing factors.

Speed of Short-Channel MOSFETs.
The speed of short-channel MOSFETs can be derived in a manner similar to that we used to obtain Equation 9.1.34, but the resulting theoretical expression is overly complicated for effective use as an intuitive guide. In short-channel MOSFETs, the velocity-saturation region between L' and L in Figure 9.4c is often a large fraction of the channel length. We defer finding a solution for this length until Chapter 10, opting at this point to present a more qualitative discussion of MOSFET switching speed.

The maximum switching speed of an electronic device can be obtained by measuring the amplifying behavior of that device as a function of frequency and recording the cut-off frequency f_T at which the gain decreases to one. The expression for f_T in a MOSFET, as derived from the small-signal model of Sec. 9.1, is

$$f_T = \frac{g_{msat}}{2\pi C_{in}} \tag{9.2.17}$$

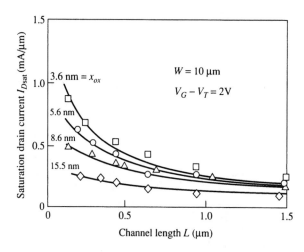

FIGURE 9.26 Measured and calculated saturation drain currents I_{Dsat} for an array of deep submicrometer (short-channel) n-channel MOSFETs with four different gate-oxide thicknesses (© 1987 IEEE [19]).

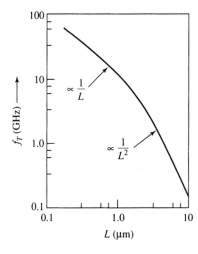

FIGURE 9.27 Intrinsic unity-current-gain "cutoff" frequency f_T of n-channel MOSFETs as a function of the channel length L, calculated using Equations 9.2.14 and 9.2.17.

where $g_{m\mathrm{sat}}$ is the MOSFET transconductance in the saturation region, and C_{in} is the capacitance at the input node.

For a long-channel MOSFET, $g_{m\mathrm{sat}}$ is given by Equation 9.1.37 and C_{in} is dominated by C_{GS} [which is equal to $(2/3)C_{ox}$ in the saturation region]. Hence, f_T for a long-channel MOSFET can be written

$$f_T = \frac{3\mu_n(V_G - V_T)}{4\pi L^2} \tag{9.2.18}$$

In Equation 9.1.34, we derived an expression for the transit time T_{tr} in the long-channel MOSFET that can be compared to Equation 9.2.18 to find $f_T = 1/\pi T_{tr}$.

For a short-channel MOSFET, the transconductance is given by Equation 9.2.14, and Equation 9.2.17 predicts

$$f_T = \frac{3v_{\mathrm{sat}}(V_G - V_T)(V_G - V_T + 2\mathscr{E}_{\mathrm{sat}}L)}{4\pi L(V_G - V_T + \mathscr{E}_{\mathrm{sat}}L)^2} \tag{9.2.19}$$

In the short-channel limit, $\mathscr{E}_{\mathrm{sat}}L \ll (V_G - V_T)$, and Equation 9.2.19 can be simplified to

$$f_T = \frac{3v_{\mathrm{sat}}}{4\pi L} \tag{9.2.20}$$

A comparison of Equations 9.2.18 and 9.2.20 highlights the strong influence of MOSFET size reduction on the basic behavior of the device; the change in the dependence of f_T on channel length (from L^{-2} to L^{-1}) is especially notable, as shown in Figure 9.27. The values of f_T derived in this section are relevant to the MOSFET structure itself and thus represent the highest frequencies at which the unloaded device has unity gain. In essentially all applications, the actual value of f_T (or the switching speed) is governed by the time to charge capacitances connected to the device nodes and is smaller than f_T of the intrinsic device.

9.3 DEVICES: COMPLEMENTARY MOSFETs—CMOS

Early in the development of MOS ICs, it was realized that when digital circuits are built with p- and n-channel MOSFETs connected in series, they have low "standby" (dc) power dissipation [20]. Circuits of this type are called *complementary* MOS transistor circuits

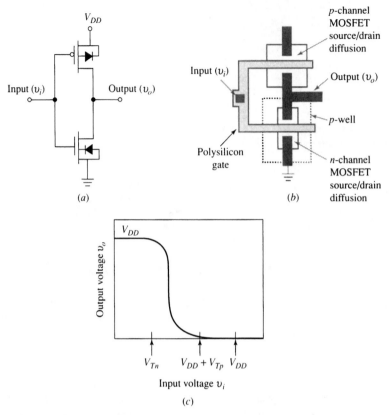

FIGURE 9.28   (a) CMOS inverter circuit, (b) layout of a p–well CMOS inverter, (c) voltage-transfer characteristics of a CMOS inverter.

or simply CMOS circuits. The basic building blocks for CMOS circuits are CMOS inverters which deliver output signals that are the binary complement of their input signals. Figure 9.28 shows the circuit schematic, layout, and voltage-transfer characteristic for a CMOS inverter cell. In the inverter, the two MOSFETs are connected in series (p-channel drain to n-channel drain), and their gates are tied together.

To understand the operation of this inverter, assume that the input voltage is lower than the n-channel MOSFET threshold voltage, and sufficiently negative with respect to the bulk of the p-channel MOSFET; the n-channel MOSFET is then OFF (non-conducting), and the p-channel MOSFET is ON (conducting). Under these conditions, the p-channel MOSFET provides a conducting path from the output terminal to the V_{DD} power supply voltage while the n-channel MOSFET isolates the output terminal from ground. Because the output terminal is typically tied to the inputs of other inverter circuits, which draw no dc current, the output voltage (at the drain of the p-channel MOSFET) is in its "high" state ($v_o \approx V_{DD}$). If the input voltage now increases, the p-channel MOSFET turns off, and when the input voltage becomes larger than the threshold voltage of the n-channel transistor, its channel is turned on, pulling the output voltage toward ground. Thus, under dc conditions one of the two MOSFETs is always turned off, and there is no path to carry current from the positive voltage supply to ground except for junction leakage. For this reason, the voltage-transfer characteristic is very abrupt, and almost all the power dissipation in CMOS circuits takes place during switching transients when current must flow to charge and discharge the input and output nodes. The steep, well-defined voltage-transfer characteristic of the CMOS inverter is very desirable in digital designs, as is the noise immunity that

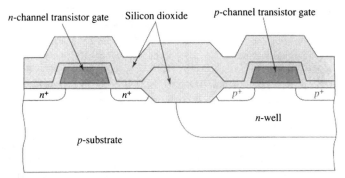

n-channel transistor gate   Silicon dioxide p-channel transistor gate

n^+ n^+ p^+ p^+

n-well

p-substrate

FIGURE 9.29 Cross section of complementary MOSFETs for an n–well CMOS inverter.

results from the low impedance between the logic signal and either the supply voltage or the ground terminal. The combination of these advantages has lead to the dominant position of CMOS compared to other IC technologies.

CMOS Design Considerations

Because the basic idea in a CMOS process is to provide both n- and p-channel MOSFETs in the same IC, one of the transistors must be placed in a region of different conductivity type than that of the substrate. This region is usually formed by dopant-ion implantation and diffusion and is called a *well*. If the substrate is n-type, the p-channel MOSFETs are made directly in it; p-type regions (p–*wells*, sometimes called *p-tubs*) must be formed wherever n-channel MOSFETs are to be placed. Alternatively, CMOS can be built on p-type wafers by forming n–wells.

Both n–well and p–well realizations have advantages and drawbacks. Whichever MOSFET is built in the well is formed in compensated silicon, which (because of the higher total dopant density) has a lower channel-carrier mobility than does a transistor formed in a substrate with the same net dopant density. Because nearly equal drive currents are desired in the n-channel and p-channel MOSFETs, a p–well process (with n-channel transistors in the p–well) is usually preferable because electron mobility is higher than hole mobility. However, modern circuits may be built with sections that are not made of CMOS inverters and which typically contain n-channel MOSFETs so that the overall IC contains many more n-channel MOSFETs than p-channel MOSFETs. In such cases, an n–well process is favored. A cross-sectional view of MOSFETs made in an n–well process is shown in Figure 9.29.

In the design of a CMOS process, an important consideration is the necessary depth of the well region needed to avoid punchthrough from the bottom of the drain to the substrate below. The following example illustrates this consideration.

EXAMPLE *CMOS Well-Depth Design*

An n–well CMOS process is designed for circuit operation at $V_{DD} = 1.5$ V. The starting wafers are p-type with $N_a = 5 \times 10^{14}$ cm^{-3}. The n-wells are to have an average dopant density $N_d = 3 \times 10^{15}$ cm^{-3}. The p-channel MOSFET sources and drains are to have junction depths $x_j = 0.8$ μm and an average dopant density $N_a = 10^{18}$ cm^{-3}. What is the minimum n–well depth that will avoid vertical punchthrough to the substrate?

Solution Vertical punchthrough occurs in a path through the two back-to-back pn junctions from the p-channel MOSFET source, biased at V_{DD} (1.5 V), to the grounded substrate (see Figure 9.29).

The source to n–well junction is essentially one-sided with a built-in voltage $\phi_i \approx 0.78$ V. From Table 4.1 or Equation 4.3.1, the depletion-layer width extends into the n–well 0.58 μm. The np junction to the substrate has a built-in voltage $\phi_i \approx 0.58$ V, and the total depletion width at 1.5 V bias is found from Equation 4.3.1 to be 2.51 μm. Applying Equation 4.2.6, we see that one seventh of the depletion width (0.36 μm) is in the n–well. The n–well must therefore be thick enough to accommodate the depth of the drain junction (0.8 μm), as well as the total 0.94 μm (0.58 + 0.36) depleted width to avoid vertical punchthrough from the source to the substrate. Thus, the minimum well depth is 1.74 μm. Good engineering design makes it advisable to increase this dimension in order to allow a reasonable safety factor.

An additional consideration is that when the p-channel MOSFET is in the OFF-state, its drain is essentially at ground potential. In this condition, the depletion layer from well to drain is wider than that from well to source because of the added V_{DD} voltage drop at the drain junction. However, even if the depletion regions between drain-well and well-substrate touch one another, no high punchthrough currents flow because the drain and substrate are both at ground potential. Despite this, having depletion regions touch is not good design because it may cause remote sections of the well to become pinched off. Carrying out an analysis similar to that above shows that a well depth of 2.16 μm is needed to assure that there are charge-neutral regions throughout the well under all bias conditions. With some added margin for safety, a reasonable design might make the well 2.5 μm deep. ∎

In modern CMOS technologies designed for very high performance CMOS, more elaborate and expensive fabrication technologies are employed to obtain the highest possible performance from both the n- and p-channel MOSFETs. Twin- or, sometimes, even triple-well technologies have been used. The fabrication steps for a twin-well CMOS process are indicated in Figure 9.30.

MOSFET Parameters and Their Extraction

Circuit simulation is an essential tool in the design of integrated circuits. Accurate simulation depends on accurate models for all circuit elements, and much work has been done to develop these models. Of the many MOSFET models proposed, we will focus our discussion on a model developed at the University of California, Berkeley, called BSIM [21]. BSIM3v3 (Berkeley Short-channel IGFET Model 3 version 3) was selected as a standard by the Compact Model Council [22], an organization that has members from many leading companies in the semiconductor industry. The BSIM3v3 model is valid for MOSFETs with gate lengths as short as 0.1 μm and therefore includes many complicated physical effects. The physical framework used in the BSIM3v3 model was described in Sec. 9.2. We discuss here the technique used to determine important device parameters needed as input to the BSIM3v3 model. The parameter-extraction technique is very useful because it links circuit design and process technology. It starts with the production of CMOS prototype samples at an IC fabrication facility, such as a foundry. These samples are then measured to extract the BSIM3v3 parameters without reference to the fabrication process used to make the CMOS samples. There are alternate approaches that use detailed device modeling of known or simulated device structures to compute device parameters but we do not discuss them here.

Global Optimization. The broad objective of parameter extraction is to find a set of parameters such that the error between the experimental data and the model prediction is minimized over the entire range of the available data. We represent the error by a parameter ϵ, defined as the square root of the sum shown in Equation 9.3.1

$$\epsilon^2 = \sum_i \left[D_i | (V^i_{G,D,S,B}) - M_i(p_1, p_2, \ldots, p_n) | (V^i_{G,D,S,B}) \right]^2 \qquad (9.3.1)$$

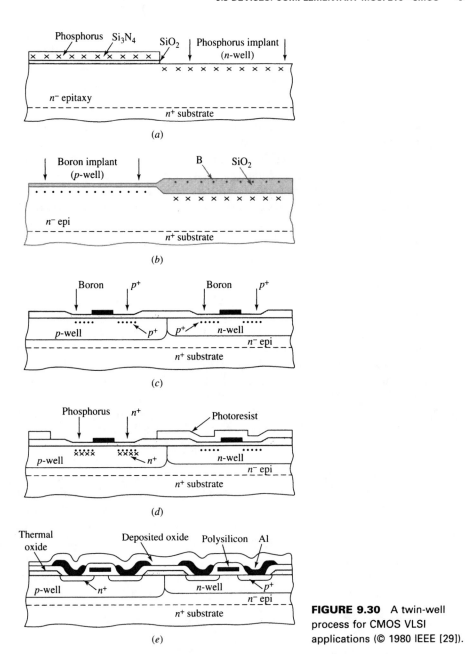

FIGURE 9.30 A twin-well process for CMOS VLSI applications (© 1980 IEEE [29]).

where D_i is the ith measured value at bias voltages ($V^i_{G,D,S,B}$) and M_i is the prediction by the model at the same bias point with $\{p_1, p_2, \ldots, p_n\}$ as parameters. The goal is to find a set of parameters $\{p_1, p_2, \ldots, p_n\}$ such that ϵ and, therefore, the sum in Equation 9.3.1 is minimized.

Global optimization treats the modeling purely as a minimization problem and uses computer analysis to find one set of parameters that best fits all the experimental data. Because this method is really empirical, any particular parameters extracted using it may not be near the value one would estimate by carrying through focused physical modeling. In some cases, totally unphysical implications can be drawn from the parameters, but they are still very useful in the regions of operation over which they were extracted. However, using a parameter set in bias regions outside that for which it was derived can lead to large errors. Successful use of the ϵ-minimization scheme therefore requires extensive measurements over as broad a range of biases as will be encountered by ICs to be designed with the devices being investigated.

Local Optimization. *Locally optimized modeling* refers to a technique in which a selected parameter is extracted in the operating region over which it dominates device behavior. Parameters that are locally optimized may not yield models that track experimental data accurately over all operating regions, but they are closely related to physical processes, in contrast to the global parameters described in the previous section. Because of its strong connection to physical processes, locally optimized modeling is explained in greater detail in this section. The techniques used to extract several important BSIM3v3 model parameters—oxide thickness x_{ox}, low-field threshold voltage V_T, and effective mobility μ_{eff}—are used as examples.

Gate-Oxide Thickness. The circuit used to extract the gate-oxide thickness x_{ox} is shown in Figure 9.31a. In the circuit, the MOSFET source and drain are connected together at the ground terminal, and the gate is biased positively. The body is left floating and the capacitance is measured between the gate and the source/drain connection as the gate voltage is slowly increased (quasi-dc). A typical measurement in this configuration (which is similar to that used with the gated diode described in Sec. 8.6) shows an abrupt transition when $V_G = V_T$, as shown in Figure 9.31b. The change in capacitance is equal to $C_{ox}WL$, where W and L are known from the MOSFET layout; therefore, first C_{ox} and then $x_{ox} = \epsilon_{ox}/C_{ox}$ can be extracted. To minimize errors in this method, it is usual to perform the measurements on large MOSFETs, typically having W and L greater than 10 μm so that their values are known with high precision.

Low-Field Threshold Voltage. A circuit schematic showing the technique used to extract the threshold voltage V_T is shown in Figure 9.32a. The drain current is measured with the drain voltage at a small, constant value (typically ~50 mV) while the gate voltage V_G is increased. Values from a series of these measurements, with three different substrate voltages V_B, are plotted in Figure 9.32b. A straight line tangent to the measured

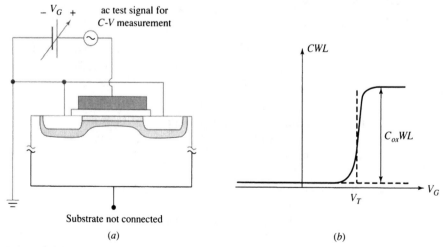

(a)   (b)

FIGURE 9.31 (a) Measurement apparatus, and (b) C-V characteristics used to obtain gate capacitance and gate-oxide thickness.

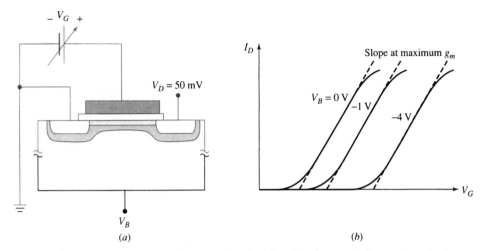

FIGURE 9.32 (a) Measurement apparatus and (b) $I_D - V_G$ characteristics used to obtain threshold voltage V_T at low gate voltages. The threshold voltages at the three values of substrate bias V_B are found at the $I_D = 0$ intercepts of the straight-line segments.

data at the point of maximum slope is extended to the V_G axis to determine V_T. Theory that justifies this technique follows from Equation 9.2.9, which for very small values of V_D [$V_D \ll 2(V_G - V_T)$] can be approximated as

$$I_D \approx \mu_{eff} C_{ox} \frac{W}{L} (V_G - V_T) V_D \qquad (9.3.2)$$

Equation 9.3.2 shows that I_D is linearly proportional to V_G for $V_G > V_T$ and zero at $V_G = V_T$ as assumed in Figure 9.31b. The decreasing slope observed on each curve at higher values of V_G indicates the mobility reduction that occurs at higher gate fields.

Effective Mobility. The circuit of Figure 9.32a is also used to extract the effective mobility as a function of gate voltage V_G. At very small drain voltages V_D, the gate-to-source and gate-to-drain voltages are approximately the same and the channel and bulk charge each have an almost constant value from $y = 0$ to $y = L$. Therefore, the mobility is constant along the length of the channel. Using Equation 9.3.2, we can solve for the effective mobility:

$$\mu_{eff} \approx \frac{I_D L}{C_{ox} W (V_G - V_T) V_D} \qquad (9.3.3)$$

Obtaining an accurate value for μ_{eff} using this method depends on knowing precise values for all terms in Equation 9.3.3. A specific problem can arise, for example, in a MOSFET in which unknown resistance is in series with the channel (at the source and drain contacts or in the diffused regions forming the source and drain). If series resistance is present, the drain voltage can differ from the applied bias. Using large MOSFETs reduces this source of error, just as it helps improve the precision in measurements of C_{ox}.

EXAMPLE *Parameter Extraction*

To extract the threshold voltage and mobility of the n-channel MOSFETs obtained from a foundry, test devices are measured by applying two different gate voltages with the drain voltage kept at 0.05 V. The measured drain currents are shown in the following table.

	$V_G = 1\text{ V}$	$V_G = 2\text{ V}$
$V_D = 0.05\text{ V}$	$I_D = 14\ \mu A$	$I_D = 34\ \mu A$

We assume that mobility degradation is not important and that the source/drain resistance is negligibly small compared to the channel resistance. Use the measurements to obtain the low-field threshold voltage and the mobility in the MOSFET if the known parameters describing it are $x_{ox} = 45$ nm, $W = 10$ μm, and $L = 1.0$ μm.

Solution At very small drain voltages V_D, Equation 9.2.9 can be used. From the values given in the table,

$$14 \text{ μA} \approx \mu_{eff} C_{ox} \frac{W}{L}(1 - V_T)V_D \tag{1}$$

$$34 \text{ μA} \approx \mu_{eff} C_{ox} \frac{W}{L}(2 - V_T)V_D \tag{2}$$

Dividing Equation (2) by Equation (1), we have $\dfrac{2 - V_T}{1 - V_T} \approx \dfrac{34}{14} \Rightarrow V_T = 0.3$ V

Knowing V_T, we can use Equation 9.2.9 to find

$$\mu_{eff} \approx \frac{I_D L}{W C_{ox}(V_G - V_T)V_D}$$

$$= \frac{14 \times 10^{-6} \times 1 \times 10^{-4}}{10 \times 10^{-4} \times \left(\dfrac{3.9 \times 8.85 \times 10^{-14}}{45 \times 10^{-7}}\right)(1 - 0.3)0.05}$$

$$= 521 \text{ cm}^2\text{V}^{-1}\text{s}^{-1}$$ ∎

CMOS Latch-up[†]

A requirement in designing CMOS circuits is to avoid a condition known as *latch-up*, in which regenerative bipolar-transistor action causes a clamped, low-resistance path between the power supply and ground. To understand the basic latch-up phenomena, consider the *p*–well CMOS structure shown in Figure 9.33. Superimposed on the MOS cross sections shown in Figure 9.33 are unwanted or *parasitic npn* and *pnp* bipolar transistors. These

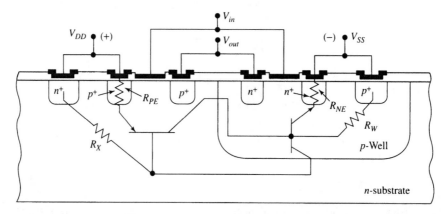

FIGURE 9.33 Cross section of a *p*–well CMOS inverter. The parasitic *pnp* and *npn* bipolar transistors are indicated along with associated substrate resistor R_x and well resistor R_W. The two resistors R_{PE} and R_{NE} represent contact and diffused-region resistance in the emitters.

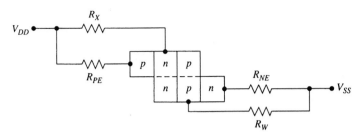

FIGURE 9.34 Circuit and schematic representation of the cross-coupled parasitic *npn* and *pnp* transistors in *p*–well CMOS.

transistors are cross connected so that the base-collector junctions are common. From the resulting bipolar equivalent circuit shown in Figure 9.34, we see that, under active bias, the *pnp* collector delivers current to the *npn* base, and the *npn* collector delivers current to the *pnp* base. If these bipolar transistors have even moderate current gains (βs), this interconnection can easily lead both devices to saturate so that the supply voltages become connected across a low resistance in series with two voltage drops: (1) the voltage across a saturated base-collector junction V_{CEsat}, and (2) the voltage across a saturated base-emitter junction V_{BEsat}.*

Under normal CMOS operating conditions, the base-emitter junctions for both bipolar transistors are reverse-biased, which makes latch-up impossible. A successful circuit design must, however, preclude latch-up under any conditions that might be experienced by the circuit. To understand the ways in which latch-up can be initiated, we refer to Figure 9.35, in which the cross-connected bipolar pair is redrawn and two elements—a capacitor C_{PS} and a current source I_0—are added in parallel across the base-collector junctions. The capacitance C_{PS} is much larger than that of a typical base-collector junction because this capacitor represents the large junction between the *p*–well and the substrate. The current source I_0 normally models only junction leakage and is very small in magnitude. Several mechanisms, however, can cause I_0 to increase markedly.

Among the possible sources for current through I_0 are (1) minority carriers injected into the substrate by transient forward bias on *pn* junctions (typically in input or output circuits); (2) photogeneration by ionizing radiation; and (3) impact ionization by hot carriers. The large capacitor C_{PS} can also deliver currents when voltage transients occur, especially during the power-up phase of circuit operation. Any of these sources of current can turn on one or both of the bipolar devices. Thereafter, latch-up occurs if the gain of the cross-coupled bipolar pair is sufficient and if the V_{DD} power supply can deliver enough current.

Latch-up Models. A simple expression that reveals conditions on device gain that can lead to latch-up is obtained by simplifying Figure 9.35 by omitting I_0 and C_{PS} and

---

* Merged *pnp* and *npn* transistors in which the *pnp* base is driven by the *npn* collector and vice versa are useful and important power-handling switches. These switches are frequently called *silicon controlled rectifiers*, abbreviated SCRs. Because SCR switching has been the subject of much research, latch-up in CMOS is often described as an SCR effect.

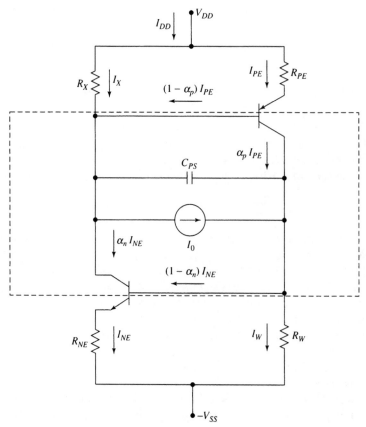

FIGURE 9.35 Latch-up equivalent circuit including well-to-substrate capacitor C_{PS} and parasitic current source I_0. The dashed lines surround all elements connected between the well and substrate nodes.

considering R_{PE} and R_{NE} to be negligible (Figure 9.36). Under these conditions, the current driving the base of the *pnp* transistor is equal to the base current of the *npn* transistor times β_n reduced by the divider action of the input resistance of the *pnp* transistor base and the substrate resistor R_X, which is in parallel with it. On a small-signal basis this is $\beta_n R_X/(r_{\pi pnp} + R_X)$, where $r_{\pi pnp}$ is the reciprocal of δg_m given in Equation 7.5.3. An analogous expression with the well resistor R_W in place of R_X applies for the base drive to the *npn* transistor [23]. Thus, the overall loop gain G_L for the cross-coupled pair is

$$G_L = \beta_n \times \frac{R_X}{r_{\pi pnp} + R_X} \times \beta_p \times \frac{R_W}{r_{\pi npn} + R_W} \tag{9.3.4}$$

For a latched condition, the loop gain must equal one; conversely, latch-up is not possible if this loop gain is less than one. This simple calculation shows that a design that avoids latch-up must reduce the bipolar transistor gains (βs) and also make R_X and R_W as small as possible. These basic guidelines underlie virtually all of the ways that have been explored to design CMOS that is free of latch-up.

To prevent latch-up in very dense CMOS circuits, more complicated structures may be needed in place of basic "bulk CMOS" design, in which a single well is diffused into the substrate as shown in Figure 9.29. The most common method used to

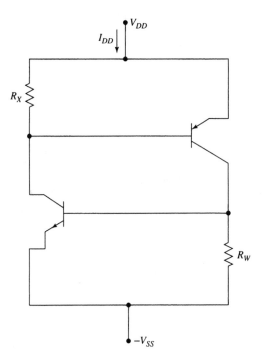

FIGURE 9.36 Simplified model for analysis of gain requirements for CMOS latch-up.

avoid latch-up is to place the transistors in an epitaxial layer grown on a highly doped substrate to reduce the substrate resistance. The minority-carrier lifetime is low in the heavily doped substrate, so the gains of the parasitic bipolar transistors are also reduced. Using an epitaxial structure reduces the tendency for latch-up, even for deep sub-micrometer CMOS circuits. The lifetime of minority carriers can also be reduced by irradiating the entire structure, thereby decreasing the bipolar transistor gains. However, reducing the lifetime in the source-drain junction regions increases the leakage current, which is contrary to the low-standby-power advantage of CMOS. Other techniques use closed patterns with heavily doped guard rings on the surface to clamp voltages at sensitive locations or to collect minority carriers before they can reach the well-to-substrate junctions.

EXAMPLE   *Latch-up in CMOS*

Use the circuit in Figure 9.35 to calculate the power-supply current I_{DD} as a function of the current I_W in the well, the current I_X in the substrate, and the well-substrate current source I_0 at the well junction. Assume that both transistors are active, and find conditions on the transistor common base gains α that cause I_{DD} to become unbounded. Assume that any voltage changes occur slowly.

Solution   Because the voltage changes occur slowly, the capacitor C_{PS} need not be considered. By applying Kirchhoff's Current Law to the circuit of Figure 9.35, we have

$$I_{DD} = I_X + I_{PE}$$
$$I_{DD} = I_W + I_{NE}$$
$$I_W = \alpha_p I_{PE} - (1 - \alpha_n)I_{NE} + I_0$$
$$I_X = \alpha_n I_{NE} - (1 - \alpha_p)I_{PE} + I_0$$

Eliminating I_{PE} and I_{NE}, we can write

$$I_{DD} = I_X + \frac{(I_W - I_0)}{\alpha_p} + \frac{(1 - \alpha_n)}{\alpha_p} \times \left[\frac{(I_X - I_0)}{\alpha_n} + \frac{(1 - \alpha_p)}{\alpha_n} \times (I_{DD} - I_X) \right]$$

Solving this expression for I_{DD}, we have

$$I_{DD} = \frac{I_0 - \alpha_p I_X - \alpha_n I_W}{1 - (\alpha_n + \alpha_p)}$$

From this expression we see that I_{DD} tends toward infinity (and the circuit becomes latched) when the sum $(\alpha_n + \alpha_p)$ approaches unity. This condition on the transistor αs can be compared to the latch-up constraint expressed through Equation 9.3.4. Equation 9.3.4 was obtained by applying the small-signal equivalent circuit for bipolar transistors, and thus represents a condition on the circuit gain during the transient build-up to the latched condition. In this example, steady state is considered, and the r_π resistor-divider terms are not relevant. If these terms are not considered, Equation 9.3.4 simplifies to the condition $\beta_n \times \beta_p = 1$, or

$$\left(\frac{\alpha_n}{1 - \alpha_n} \right) \left(\frac{\alpha_p}{1 - \alpha_p} \right) = 1$$

which reduces to $(\alpha_n + \alpha_p = 1)$ as concluded above. ∎

9.4 LOOKING AHEAD

The minimum surface dimension in a MOSFET process is a key benchmark for the density of devices that can be built in a process. This dimension, typically the MOSFET channel length, has decreased continually for more than 35 years as subsequent new generations of MOSFET circuits were built with ever-increasing numbers of active devices. As briefly discussed in Chapter 2, this trend is most frequently described in terms of *Moore's Law*, which was first stated by Gordon Moore in the early 1970s. Moore's Law predicts that the number of transistors in an integrated circuit will double through advances in technology and design every 18–24 months. Basic intuition tells us that, as with any rapidly developing field, a limit will eventually be reached. In the early 1980s, the minimum possible surface dimension was predicted to be approximately 0.5 μm. Ten years later, the predicted minimum dimension had decreased to about 0.1 μm and now, in the early 21st century, a frequently quoted limit is roughly 25 nm. Examples of device parameters in the past and predictions into the future for IC technology scaling are shown in Table 9.2.

Scaling Goals

The main goals of scaling are (1) to reduce transistor size, (2) to increase current drive per unit width, (3) to reduce power-supply voltage, and (4) to reduce overall load capacitance. Progress towards these goals helps to increase the degree of functionality that can be integrated into a given silicon area using scaled devices. All of these goals help to improve performance, reduce power consumption, and provide a more economical product. The constraints that need to be faced when considering methods for scaling the size of IC devices include (1) retaining adequately low leakage current in the OFF state; (2) minimizing short-channel effects (as much as possible); and (3) maintaining circuit reliability. Some common parameters such as ON-current (I_{on}), OFF-current (I_{off}), subthreshold slope (S), input impedance (Z_{in}), transconductance (g_m), and breakdown voltage (BV) are used to quantify the performance of a scaled MOSFET.

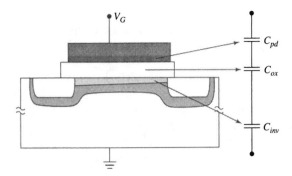

FIGURE 9.37 Cross-sectional sketch to show that the MOSFET gate capacitance C_G is made up of three capacitances in series: the oxide capacitance C_{ox}, a capacitance due to a depletion layer in the polycrystalline-silicon gate material C_{pd}, and a capacitance due to the finite thickness of the conducting channel C_{inv}.

Gate Coupling

Reducing the gate-oxide thickness helps increase the desired coupling of the gate voltage to the channel charge, and every new generation of CMOS technology uses a thinner gate oxide. However, this scaling of oxide thickness is limited. When the oxide becomes extremely thin, direct tunneling of electrons through it causes significant gate current to flow and reduces the input impedance. The acceptable limit for gate-oxide scaling appears to be about 2.5 nm. In addition to the limitation resulting from direct tunneling, the scaling of gate oxide below 2.5 nm cannot effectively increase the gate coupling because of carrier depletion in the polysilicon-gate electrode and the finite thickness of the inversion layer (channel) in the silicon, as illustrated in Figure 9.37.

In our formulation of MOSFET theory, we made several assumptions related to the gate-to-channel coupling. (1) We assumed that the polysilicon gate is metal-like and that no portion of the polysilicon can be depleted of mobile carriers; if the region of the polysilicon near the gate oxide becomes depleted, this depleted region acts as a dielectric, increasing the effective thickness of the gate insulator and reducing the coupling of the gate voltage to the channel in the underlying silicon. (2) We also assumed that the inversion layer in the silicon containing the mobile carriers is infinitesimally thin and located at the surface of the silicon; the finite thickness of the inversion layer [30] reduces the coupling between the gate voltage and the conducting carriers in the channel and weakens gate control. The assumptions we made are very good when the gate oxide is thick, but can produce substantial errors when the gate oxide becomes very thin (e.g., ≤ 5 nm). The combination of polysilicon depletion and inversion-layer capacitance can reduce MOSFET current drive significantly. To increase gate coupling, some new approaches are being investigated. High dielectric constant (high permittivity or high-κ) gate dielectrics are being actively explored, as are gates made of metal silicides. Although these new materials and technologies have not yet become part of production processes, most engineers believe that their use will be necessary in future CMOS technology generations.

The reason for using high-permittivity insulators is obvious. The gate-oxide capacitance per unit area is given by ϵ_{ox}/x_{ox}. Increasing ϵ_{ox}, rather than continuing to reduce x_{ox}, avoids the problem of direct tunneling. Difficulties encountered in using high-permittivity dielectrics in the past were mainly poor uniformity of the thicknesses of the deposited films and poor interface quality. However, recent reports have shown that fairly good interfaces can be achieved. Some of the high-permittivity materials being investigated are Al_2O_3 ($\epsilon_r = 9.5$), ZrO_2 ($\epsilon_r = 20 - 40$), HfO_2 ($\epsilon_r = 20 - 30$), TiO_2 ($\epsilon_r = 80$), and Ta_2O_5 ($\epsilon_r = 25$).

Along with the use of a high-permittivity material to increase the gate-dielectric capacitance, metallic or metal-silicide gate electrodes can be used to reduce the effect of polysilicon depletion. For n-channel MOSFETs the metal used should have a work function that positions its Fermi level at approximately the same energy as the conduction-band edge of silicon (as does n^+ polysilicon); for p-channel MOSFETs the metal Fermi level should be close to the valence-band edge of silicon (as obtained with p^+ polysilicon). Hence, achieving proper band alignment for CMOS requires using two different metals, significantly increasing process complexity. A compromise is to use a metal having a work function close to the midgap of silicon for both types of MOSFETs at the cost of having somewhat higher threshold voltages. Some common metals that have been investigated include double layers of W and TiN, and triple layers composed of Ta, TaN, and NiSi. In general, metal-gate processes are not compatible with polycrystalline-silicon-gate CMOS processes [24], so replacing polysilicon with metal for the gate electrodes will require extensive process development.

Velocity Overshoot

In Sec. 9.2, we discussed the effect of velocity saturation assuming that the mean-free path of the carriers is short compared to the channel length. The carriers then lose excess energy to the lattice and reach a limiting or *saturation* velocity, as discussed in Chapter 1. For extremely short channel lengths, free carriers can move at velocities greater than the saturation value. This effect, called *velocity overshoot*, can lead to currents higher than predicted in our previous discussion.

We can understand velocity overshoot by recalling our discussion of free-carrier mobility in Chapter 1 where we pointed out that the energy gained by a free carrier as it travels in an applied electric field is transferred to the lattice through collisions. At low fields the average velocity in the field direction increases linearly with the magnitude of the field; at relatively high fields, more efficient energy-transfer mechanisms become important, effectively limiting the average velocity. As a result, the field-directed velocity saturates at v_{sat}. The average distance the carrier travels between collisions is λ. Velocity overshoot occurs when the channel length becomes comparable to or even shorter than this mean-free path λ. In this case, the likelihood of energy loss from collisions with the lattice decreases. Many electrons can travel through the entire channel without losing energy to the lattice and the average carrier velocity can exceed v_{sat}, increasing the drain current.

Theoretical calculations suggest that velocity overshoot will eventually become important. However, the carrier mean-free path is short, so velocity overshoot is only expected in the shortest devices. Behavior observed in very-short-channel transistors has been explained in terms of velocity overshoot [25–27]; however, uncertainties in dimensions and other parameters in these experimental devices make a velocity-overshoot hypothesis difficult to confirm.

Trends. CMOS technology is the dominant integrated-circuit manufacturing technology and MOSFETs are the most widely used devices in modern integrated circuits. For the foreseeable future, CMOS appears certain to remain the mainstream IC technology, and the reduction of device dimensions will continue. It is realistic to expect device performance to increase by continuing efforts in technology scaling for the next decade; however, fundamental materials and processing changes appear likely to be needed.

Device scaling has allowed transistors with channels shorter than 25 nm to be built in research laboratories. However, the optimization of MOSFETs with gate lengths shorter than 100 nm usually involves the use of complicated structures and less well-known materials. The increasing complexity makes scaling unlikely to continue to yield the previous rates of improvement in circuit performance and density. More complex and unconventional approaches now being explored by researchers throughout the world include the use of high-permittivity (high-κ) gate materials and metal gate electrodes.

SUMMARY

In a metal-oxide-silicon field-effect transistor (MOSFET), the conductance in the *channel* between the *source* and *drain* electrodes can be modulated by a voltage applied to the *gate* electrode. The MOSFET is most often used as a gate-controlled current switch. It can also function as an amplifier because high power in the source-drain circuit can be controlled by low power supplied to the gate-source circuit. A MOSFET in which the channel can carry current between the source and the drain when the gate and source are at the same potential ($V_{GS} = 0$) is called a *depletion-mode* MOSFET. If a finite value of gate voltage is necessary to induce a conducting channel, the MOSFET is called an *enhancement-mode* transistor.

The dependence of drain current on drain voltage in a MOSFET (the *output characteristic*) can be divided into two regions: At low drain biases, carriers in the channel move by drift along a continuous conducting path extending from the source to the drain. At higher drain biases, the conducting path in the channel does not reach the drain so that, near the drain, the current must flow through a high-field, space-charge region containing few mobile carriers. First-order analysis suggests that the drain current saturates when high drain biases are applied; therefore this region of operation is called the *current-saturation* region.

Simple, analytical models for the drain current that assume the presence of a conducting path along the entire length of the channel are only accurate for drain voltages lower than $V_{D\text{sat}}$, the voltage at which the free charge in the channel becomes zero near the drain. For $V_D > V_{D\text{sat}}$, the current is assumed to be independent of V_D and limited by the flow along the channel from the source to the *pinch-off point* where the conducting channel ends. Because the pinch-off point moves toward the source as the drain-source voltage increases, the length of the conducting channel decreases. This *channel-length modulation* causes the drain saturation current $I_{D\text{sat}}$ to increase as the drain-source voltage increases above $V_{D\text{sat}}$. The *body effect* is the variation of threshold voltage that occurs when the bias between the source and the substrate is changed. The body effect becomes more important as the substrate doping increases, and it varies inversely with the oxide capacitance.

The *transconductance* g_m of a MOSFET increases linearly with increasing drain voltage until saturation occurs, but (to first order) does not depend on gate voltage below saturation. Once saturation is reached, g_m becomes a linear function of gate voltage, but is independent of drain voltage. The speed of response of a MOSFET is generally determined, not by the channel transit time, but by the charging and discharging times of capacitances present in the device. A circuit model for the MOSFET consists of elements representing the dc circuit equations for the device, together with associated capacitances and resistances.

The basic MOSFET theory contains many approximations that are useful in providing a set of equations for designers. An important simplification comes from the assumption that there are no free carriers in the channel if the magnitude of the gate voltage is less than the magnitude of the threshold voltage. The free-carrier densities do not, however, change abruptly at $V_G = V_T$; instead, they decrease exponentially as the gate voltage decreases below the threshold voltage. The channel free charge present when $|V_G| < |V_T|$ is responsible for *subthreshold* current. Subthreshold current is transported by *diffusion*, unlike the *drift current* that dominates when $|V_G| > |V_T|$.

A number of the approximations in the basic MOSFET theory become increasingly weak as device dimensions are reduced or voltages increase. One particular limiting assumption in the basic MOSFET theory is that free charge moves along the channel by drift with a constant mobility. Although drift is the dominant transport mechanism when $|V_G| > |V_T|$, mobility is not constant in the channel. Instead, the mobility depends on both the local field $\mathscr{E}_y$ along the channel and the field $\mathscr{E}_x$ normal to the Si–SiO$_2$ interface. Even in the basic long-channel theory $\mathscr{E}_y$ becomes very large near the drain and therefore assuming that mobility remains constant over the entire channel length is often questionable.

For short channels, the drift velocity saturates at low voltages, and the velocity is nearly constant over most of the channel length. Velocity saturation in the channel reduces the current below that predicted for a constant mobility. Because of velocity saturation, $I_{D\text{sat}}$ varies more slowly than $(V_G - V_T)^2$, and $V_{D\text{sat}}$ is smaller than $(V_G - V_T)$. The threshold voltage depends on both the channel length and the channel width in MOSFETs with small dimensions. As the channel length decreases, the magnitude of the threshold voltage tends to decrease because of the influence of the source and drain space-charge regions on the charge in the channel. In addition to this *charge sharing*, the characteristics can be degraded by *drain-induced barrier lowering* and *subsurface punchthrough*. If the channel is made narrow, the threshold voltage can increase because of fringing fields or extra dopant atoms, or it can decrease because of field concentration by the isolation structure.

Threshold-voltage adjustment using ion implantation is a key step in the processing of *complementary MOS (CMOS)* circuits. In CMOS ICs, both *p*-channel and *n*-channel MOSFETs are fabricated on the same IC chip, demanding the creation of a *well* of opposite conductivity type to the substrate. Both *p*–well and *n*–well technologies exist, and there are advantages and drawbacks to each. With CMOS, digital inverter circuits can be built that consume little dc power because only small leakage currents flow between the supply voltage and ground when the circuit is not switching. This feature of CMOS is especially important as the density of transistors increases. However, the power dissipated during switching is a significant problem in dense ICs operating at high frequencies. A special problem in CMOS design is avoiding *latch-up*, in which regenerative bipolar action drives the parasitic bipolar transistors present in most CMOS processes into saturation.

Scaling rules aid the design of MOSFETs as device dimensions decrease. One set of scaling rules attempts to keep fields in the MOSFET constant as the dimensions are reduced. However, practical considerations limit this *constant-field-scaling* technique, so that it is not used in the form originally proposed.

Predicting the benefits and drawbacks of device scaling requires a coherent and accurate, physics-based, short-channel, MOS transistor model. The most important scaling effects to include when modeling the drain current are mobility degradation due to the oxide field and velocity saturation. A transistor behaves more like a long-channel device if it has a thinner gate oxide to increase the coupling between the gate and the channel carriers. Because the drain current is roughly proportional to the gate capacitance, reducing the gate-oxide thickness is especially beneficial for deep sub-micrometer technology. For short-channel transistors, the *cutoff frequency* f_T increases only linearly as the channel length decreases, instead of quadratically, as in a long-channel transistor.

REFERENCES

1. WILLIAM SHOCKLEY, *Electrons and Holes in Semiconductors*, D. Van Nostrand Co., Princeton, NJ, 1950, p. 29.
2. R. H. DENNARD, F. H. GAENSSLEN, H. N. YU, V. L. RIDEOUT, E. BASSOUS, and A. R. LeBLANC, *IEEE J. Solid-State Circuits*, SC-9, 256–68, 1974.
3. International Technology Roadmap for Semiconductors (1997–2001 editions), http://public.itrs.net
4. Z. H. LIU, C. HU, J. H. HUANG, T. Y. CHAN, M. C. JENG, P. K. KO, and Y. C. CHENG, *IEEE Trans. Electron Devices*, ED-40, 86–95, January 1993.
5. L. A. AKERS, M. M. E. BEGUWALA, and F. Z. CUSTODE, *IEEE Trans. Electron Devices*, ED-28, 1490–1495, Dec. 1981.
6. J. Y. CHEN and R. C. HENDERSON, *IEEE IEDM Technical Digest*, 1982, p. 233–36.
7. L. A. AKERS, *IEEE Electron Dev. Lett.*, EDL-7, 419–21, July 1986.
8. K. OHE, S. ODANAKA, K. MARIYAMA, T. HORI, and G. FUSE, *IEEE Trans. Electron Devices*, ED-36, 1110–1116, June 1989.
9. R. R. TROUTMAN, *IEEE J. Solid-State Circuits*, SC-14, 383–91, 1979.
10. Y. C. CHENG and E. A. SULLIVAN, *J. Appl. Phys.* 45, 187, 1974.
11. A. G. SABNIS and J. T. CLEMENS, *Tech. Dig.*, Int. Electron Devices Meeting, 1979, pp. 18–21.
12. S. C. SUN and J. D. PLUMMER, *IEEE Trans. Electron Devices*, ED-27, 1497–1508, 1980.
13. M. S. LIANG, J. Y. CHOI, P. K. KO, and C. HU, *IEEE Trans. Electron Devices*, ED-33, 409, 1986.
14. K. K. HUNG, *Electrical characterization of the Si–SiO₂ interface for thin oxides*. Doctoral thesis, Dept of Electrical Engineering, Hong Kong University, 1987.
15. F. FANG and X. FOWLER, *J. Appl. Phys.* 41, 1825, 1969.
16. R. COEN and R. S. MULLER, *Solid-State Electron.* 23, 35–40, 1980.
17. J. A. COOPER and D. F. NELSON, *IEEE Electron Device Lett.* EDL-2, 169–73, 1983.
18. C. G. SODINI, P. K. KO, and J. L. MOLL, *IEEE Trans. Electron Devices*, ED-31, 1386, 1984.
19. M. C. JENG, J. CHUNG, A. WU, J. MOON, T. Y. CHAN, G. MAY, P. K. KO, and C. HU, *IEEE IEDM Tech. Digest*, 1987, p. 710.
20. F. M. WANLASS and C. T. SAH, *IEEE Int. Solid-State Circuits Conf.*, Philadelphia, PA (Feb. 1963).
21. B. SHEU, D. L. SCHARFETTER, P. K. KO, and M. C. JENG, *IEEE Journal of Solid-State Circuits*, SC-22, p. 558–66, April 1987.
22. The Compact Modelling Council: http://www.eigroup.org/cmc
23. K. W. TERRILL, *CMOS Latch-up Modeling and Prevention*, Doctoral thesis, Department of EECS, University of California, Berkeley, Dec. 1985.

24. A. Chatterjee et al., *IEEE IEDM Tech. Digest*, 1997, pp. 821–824.
25. T. Kobayashi and K. Saito, *IEEE Trans. Electron Devices* 32, 788–92 April 1985.
26. G. G. Shahidi, D. A. Antoniadis, and H. I. Smith, *IEEE Electron Device Letters*, 9, 94–96, Feb. 1988.
27. F. Assaderaghi, P. K. Ko, and C. Hu, *IEEE Electron Device Letters*, 14, 484–86, October 1993.
28. R. J. C. Chwang, et al., *IEEE J. Solid-State Circuits*, SC-18, p. 457, October 1983.
29. Adapted from L. C. Parillo et al., *Tech Digest*, 1980 IEEE Intl. Electron Devices Mtg. (Dec. 1980), p. 752.
30. T. I. Kamins and R. S. Muller, *Solid-State Electron.* 10, 423 (1967).

PROBLEMS

9.1 Construct a table showing the threshold voltage V_T (taking $V_S = V_B = 0$) as a function of dopant concentration for both n- and p-channel MOSFETs. Take values of the substrate doping N_a and N_d to be 10^{15}, 10^{16}, 10^{17}, and 10^{18} cm^{-3} and assume that there is a surface density of fixed positive charge $Q_f/q = 10^{11}$ cm^{-2} at the oxide-silicon interface in all cases. The silicon dioxide is 20 nm-thick and all gates are made of n^+ polysilicon with its Fermi level at the conduction-band edge. Indicate on the table whether the MOSFET is a depletion-mode or an enhancement-mode device.

9.2 Assume that under pinch-off conditions, the pinch-off region starts at the point with voltage V_{Dsat} in the channel and extends to the drain. Assuming the pinch-off region is completely depleted of mobile carriers and that the drain is much more heavily doped than the channel, derive Equation 9.1.9.

9.3* An n-channel MOSFET for which $W/L = 5$, the gate-oxide thickness is 20 nm, and the mobility is constant with a value $\mu_n = 600$ cm^2 V^{-1}s^{-1} is to be used as a controlled resistor.
(a) Calculate the free-electron density in the channel Q_n/q that is required for the MOSFET resistance to be 500 Ω between the source and the drain at low values of V_{DS}.
(b) Using the long-channel theory, calculate the gate voltage in excess of the threshold voltage ($V_G - V_T$) needed to produce the desired resistance under the conditions of part (a).

9.4* An n-channel MOSFET with $L = 0.8$ μm, $x_{ox} = 15$ nm, and $V_T = 0.7$ V is biased with $V_{GS} - V_T = 3$ V and $V_{DS} = 2$ V.
(a) Using the long-channel theory, determine the velocity of the channel carriers near the source and drain. Use $\mu_n = 670$ cm^2 V^{-1}s^{-1}.
(b) Repeat part (a) using the refined model given in Sec. 9.2. Use $v_{sat} = 8 \times 10^6$ cm s^{-1}.

9.5* Consider the effect of body bias on an n-channel MOSFET connected as in the depletion-mode

example in Sec. 9.1 (page 436). Calculate the current I_D with $V_{SS} = 2, 3, 4$, and 4.5 V. (Note that the MOSFET is *not* saturated for all of these voltages.)

9.6 Repeat Problem 9.5 with $V_{SS} = 6, 8$, and 10 V. (Note that the role of source and drain are interchanged under these bias conditions and that $V_{GD} = 0$ V.)

9.7* A series of measurements made on an n-channel MOSFET are given in the accompanying table.

V_{GS} (V)	V_{DS} (V)	V_{SB} (V)	I_D (μA)
3	4	0	120
3	6	0	130
3	4	4	76.8
4	4	0	270

Choose parameters $[V_T(0), k = \mu_n C_{ox} W/L, \gamma$, and $V_A]$ to represent the MOSFET if it is modeled using Equations 9.1.10 and 9.1.11. Assume that $2|\phi_p| = 0.6$ V (note that ϕ_p is only a weak function of N_a).

9.8 (a) Compare I_{Dsat} calculated using Equations 9.1.6 and 9.1.16 for an n-channel MOSFET for which the oxide thickness is 50 nm, the substrate doping $N_a = 2 \times 10^{15}$ cm^{-3}, the flat-band voltage $V_{FB} = -0.2$ V, and $\mu_n W/L = 5 \times 10^3$ cm^2/V-s. Consider $V_{GS} = 5.5, 4.5, 3.5$, and 2.5 V.
(b) For I_{Dsat} calculated using Equations 9.1.16 in part (a), find the value of α such that Equation 9.1.19 gives the same value as Equation 9.1.16 at $V_{GS} = 3.5$ V. Using this value of α, compare I_{Dsat} obtained from Equations 9.1.16 and 9.1.19 for the other values of V_{GS} given in part (a).

9.9 Derive Equation 9.1.21 for the depletion-layer width for the approximate "box" dopant profile obtained from the ion-implantation process used to adjust the threshold voltage discussed in Sec. 9.1.

9.10 Carry through the steps indicated in the text to obtain Equations 9.1.22 and 9.1.23 for the depletion-charge density at inversion and the threshold voltage, respectively, in an ion-implanted MOSFET.

9.11 Consider a MOSFET that is biased in strong inversion in the linear region.

(a) Find an expression for the diffusion current at any point y along the channel.

(b) Find an expression for the drift current at any point y.

(c) Compare the expressions derived in (a) and (b). Which current component dominates? [Hint: you may need to simplify and also use Einstein's relationship $D_n/\mu_n = kT/q$].

9.12 In the circuit shown in Figure P9.12, the MOSFET is described by Equations 9.1.5 and 9.1.6 with $\mu_n C_{ox} W/L = 25$ μA/V^2, $V_T = 1$ V, and $W/L = 2$. The voltage V_i is varied form 0 to 4 V.

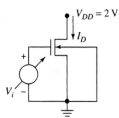

FIGURE P9.12

(a) Make a plot of $\sqrt{I_D}$ as a function of V_i, showing any break points on the curve.

(b) Plot the MOSFET transconductance using a solid line.

(c) On the plot of part (b), use a dotted line to indicate the output conductance ($g_D = \partial I_D/\partial V_{DS}$).

9.13 Obtain an expression for Q_n in the saturation region from Equation 9.1.3. From this expression, show that the small-signal capacitance between the gate and the source under saturation conditions is

$$C_{GS} \equiv \left|\frac{\partial Q_n}{\partial V_{GS}}\right| = \frac{2}{3}C_{ox}WL$$

9.14 Check the scaling rules in Table 9.1 to see if the field remains constant as each of the variables is scaled.

9.15 Show that the refined I-V model (Equations 9.2.9, 9.2.10, 9.2.11, and 9.2.14) developed in Sec. 9.2 converges to the classical long-channel model (Equations 9.1.5, 9.1.6, and 9.1.31) for long-channels L or thick gate oxides x_{ox}.

9.16* Company A developed a submicrometer CMOS technology with the following specifications: $x_{ox} = 15$ nm, $x_j = 0.2$ μm, and $V_T = 0.7$ V for the n-channel MOSFET with an n^+ polysilicon gate and $V_T = -0.7$ V for the p-channel MOSFET with a p^+ polysilicon

gate. Estimate I_{Dsat} and $g_{msat}(= dI_{Dsat}/dV_G)$ for the n-channel transistor and for the p-channel transistor. For both devices $L_{eff} = 0.5$ μm and $W = 100$ μm. Do the calculation both for $V_{DD} = 3.3$ V and for $V_{DD} = 5.0$ V. What are the percentage improvements in the values of I_{Dsat} if n-channel and p-channel transistors can be made with $L_{eff} = 0.02$ μm (with no punchthrough problems)?

9.17 Consider the relative improvement in I_{Dsat} in the following two cases:

a) The saturation velocity v_{sat} is increased by a factor of 2.

b) The effective mobility μ_{eff} is increased by a factor of 2 (while v_{sat} remains unchanged).

Plot $I_{Dsat}(new)/I_{Dsat}(old)$ as a function of V_G for the two cases showing calculations. Use the following device parameters: $x_{ox} = 20$ nm, $L = 1.0$ μm, $v_{sat}(old) = 10^7$ cm s^{-1}, $\mu_{eff}(old) = 500$ cm^2/V-s, and $V_T = 0$ V. Assume that μ_{eff} does not depend on V_G.

9.18 Let x be the coordinate normal to the surface in an MOS transistor and y be the coordinate from the source to drain. The inversion charge is given by $Q_n(y) = C_{ox}[V_G - V_T - V(y)]$ for $V_D < V_{Dsat}$.

(a) Find an expression for $V(y)$ and the lateral electric field $\mathcal{E}_y(y)$ in the inversion layer at $x = 0$.

(b) From the expressions derived in part (a), obtain equations for $V(y)$ and $\mathcal{E}_y(y)$ when the MOSFET is at the edge of saturation.

(c) Using the results of (a) and (b) (irrespective of their validity), sketch $\mathcal{E}_y(y)$ as a function of y over the range $y = 0$ to $y = L$ for $V_G > V_T$ and $V_D = 0.5 \, V_{Dsat}$. On the same graph, plot the behavior when V_D is changed to V_{Dsat}.

9.19 Consider the CMOS inverter shown in Figure 9.28a.

(a) Copy the voltage-transfer characteristics (VTC) sketched in Figure 9.28c and indicate on your copy the state of each MOSFET as v_i is changed. For example, with v_i near zero the p-channel MOSFET is ohmic and the n-channel MOSFET is cut off. Indicate all points on the VTC where a MOSFET changes its conduction state.

(b) Calculate the voltage at all points indicated in part (a) if both MOSFETs are characterized by Equations 9.1.5 and 9.1.6 with the following parameters. For the n-channel MOSFET: $\mu_n C_{ox} W/L = 40$ μA/V^2 and $V_T = 1$ V. For the p-channel MOSFET: $\mu_p C_{ox} W/L = 35$ μA/V^2 and $V_T = -1$ V. For both MOSFETs, take $\gamma = 1/V_A = 0$. The supply voltage $V_{DD} = 5$ V.

9.20 The circuit shown in Figure P9.20 is an enhancement-load inverter. Consider that both transistors are described by Equations 9.1.5 and 9.1.6 with

$\mu_n C_{ox} W/L = 40 \times 10^{-6}$ A/V^2 and $V_T = 2$ V. Take the supply voltage $V_{DD} = 8$ V and ignore the body effect and channel-length modulation (that is, take γ and $1/V_A$ to be zero). Note that the output voltage v_o cannot exceed $(V_{DD} - V_T)$ because no current can flow in the upper (load) device unless v_o is below this value.

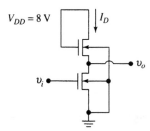

$V_{DD} = 8$ V

v_o

v_i

FIGURE P9.20

(a) Construct a set of output characteristics (I_D versus V_{DS}) for the lower MOSFET with gate voltages equal to 0, 2, 4, and 6 V as the drain voltage varies from 0 to 8 V.

(b) On the characteristics of part (a), plot the *load line* for the circuit; that is, draw a curve connecting the values of v_o (which is V_{DS} for the lower MOSFET) that correspond to each of the input gate voltages.

(c) Repeat part (b) if the upper transistor is replaced by a 20 kΩ resistor.

9.21* The circuit shown in Figure P9.21 is a depletion-load inverter. Consider that the lower MOSFET is described by Equations 9.1.5 and 9.1.6 with $\mu_n C_{ox} W/L = 50$ μA/V^2 and $V_T = 1$ V. Also use Equations 9.1.5 and 9.1.6 to describe the upper (depletion-mode) MOSFET, but take $\mu_n C_{ox} W/L = 10$ μA/V^2 and $V_T = -3$ V. In contrast to the inverter in Problem 9.20, the output voltage for this inverter can reach the supply voltage $V_{DD} = 5$ V because the load MOSFET conducts when $V_{GS} = 0$ V.

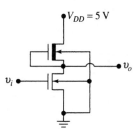

$V_{DD} = 5$ V

v_o

v_i

FIGURE P9.21

(a) If the input $v_i = 5$ V, calculate the output voltage v_o. (In this part, do not consider the body effect or channel-length modulation.)

(b) For this part, assume that the body-effect parameter γ is 0.4 V$^{1/2}$ and $|\phi_p| = 0.3$ V. Calculate the

threshold voltage of the load device when v_o is at its maximum value.

(c) Use the value of γ given in part (b), and repeat the calculation called for in part (a).

9.22† Use Figure 9.36 to prove that the current I_{DD} that the voltage source V_{DD} must supply to keep the CMOS circuit in a latched condition is

$$I_{DD} = \frac{(V_{BE}/R_W)\beta_n(\beta_p + 1) + (V_{BE}/R_X)\beta_p(\beta_n + 1)}{(\beta_n\beta_n - 1)}$$

(This example shows that a way to keep a CMOS circuit from latching-up is to limit the current supplied to it.)

9.23 Consider the simple geometrical model given in the example describing charge sharing in Sec. 9.2 (page 449). Derive Equation (2) for the fractional reduction f of the depletion region charge and Equation (3) for the threshold voltage.

9.24† (a) Consider Poisson's equation (Equation 4.1.10) in the channel region of a MOSFET. Using the condition that there is charge neutrality in the substrate, show that the impurity doping terms $N_d - N_a$ can be written $n_i[\exp u_B - \exp(-u_B)] = 2n_i \sinh u_B$, where u represents the potential normalized to (kT/q) and u_B is the substrate potential ($u_B = -q\phi_p/kT$) in a p-type substrate.

(b) Using the form derived in part (a), show that Poisson's equation can be written in terms of u as

$$\frac{d^2u}{dx^2} = \frac{1}{L_{Di}^2}[\sinh(u) - \sinh(u_B)]$$

where L_{Di} is the intrinsic Debye length

$$L_{Di} = \left(\frac{\epsilon_s kT}{2q^2 n_i}\right)^{1/2}$$

(c) Calculate L_{Di} at 300 K.

9.25† (a) Solve the form of Poisson's equation derived in Problem 9.24 using an integrating factor equal to $2du/dx$ which multiplies both sides of the equation so that the left-hand side becomes a perfect differential for

$$\left[\frac{du}{dx}\right]^2 = \mathscr{E}^2$$

(b) Show that the field in the silicon at the silicon-silicon dioxide interface $\mathscr{E}_s$ where $u = u_s$ is

$$\mathscr{E}_s = \pm \frac{kT}{qL_{Di}} F_s(u_s, u_B)$$

where the positive sign is used if $(u_B - u_s) > 0$ and the negative sign is used if $(u_B - u_s) < 0$. The function F_s is defined as the positive root in the following expression

$$F_s \equiv \sqrt{2}[(u_B - u_s)\sinh u_B - (\cosh u_B - \cosh u_s)]^{1/2}$$

(c) Use a band-diagram sketch to show that $(u_B - u_s) > 0$ corresponds to accumulation of the surface and $(u_B - u_s) < 0$ occurs in depletion and inversion for a p-type substrate.

9.26† The charge Q_s at the silicon surface is related to $\mathscr{E}_s$ by $Q_s = \epsilon_s \mathscr{E}_s$, hence

$$Q_s = \pm \frac{\epsilon_s kT}{qL_{Di}} F_s$$

where the sign choice is made as in Problem 9.25. Consider a substrate doped with $N_a = 10^{15}$ cm^{-3} at

300 K. Sketch a semilogarithmic plot of $|Q_s|$ as the surface potential u_s is changed from accumulation ($u_B - u_s$ positive) through flat-band ($u_s = u_B$) to inversion ($u_s > -u_B$). Make the range of u_s be from -20 to $+20$. Identify the flat-band voltage and conventional threshold voltage on your plot. [Problems 9.24 to 9.26 provide the basis for an exact treatment of subthreshold currents because Poisson's equation fully specifies the surface space change.]

TABLE 9.4 MOSFET Equations

Basic Electrostatic Equations

n-channel	p-channel
Depletion charge density at threshold (Equation 8.3.9)	
$Q_d = -qN_a x_{dmax} = -\sqrt{2\epsilon_s qN_a(2|\phi_p| + |V_{SB}|)}$	$Q_d = +qN_a x_{dmax} = +\sqrt{2\epsilon_s qN_a(2|\phi_p| + |V_{SB}|)}$
Flatband voltage (Equation 8.5.6)	
$V_{FB} = \Phi_{MS} - \dfrac{Q_f}{C_{ox}} - \dfrac{1}{C_{ox}} \displaystyle\int_0^{x_{ox}} \dfrac{x\rho(x)}{x_{ox}} dx$	
Threshold voltage (Equation 8.3.18)	
$V_T = V_{FB} + V_S + 2|\phi_p| + \dfrac{|Q_d|}{C_{ox}}$	$V_T = V_{FB} - V_S - 2|\phi_n| - \dfrac{|Q_d|}{C_{ox}}$
Threshold-voltage shift with body bias (Equation 9.1.11)	
$\Delta V_T = \dfrac{\sqrt{2\epsilon_s qN_a}}{C_{ox}}\left(\sqrt{2|\phi_p| + |V_{SB}|} - \sqrt{2|\phi_p|}\right)$	$\Delta V_T = -\dfrac{\sqrt{2\epsilon_s qN_a}}{C_{ox}}\left(\sqrt{2|\phi_p| + |V_{SB}|} - \sqrt{2|\phi_p|}\right)$
Long-Channel Current-Voltage Equations	
Linear-region drain current (Equation 9.1.5)	
$I_D = \mu_n C_{ox} \dfrac{W}{L}\left[(V_{GS} - V_T)V_{DS} - \dfrac{V_{DS}^2}{2}\right]$	$I_D = -\mu_p C_{ox} \dfrac{W}{L}\left[(V_{GS} - V_T)V_{DS} - \dfrac{V_{DS}^2}{2}\right]$
Long-channel saturation drain voltage	
$V_{Dsat} = (V_{GS} - V_T)$	
Saturation-region drain current (Equation 9.1.6)	
$I_{Dsat} = \mu_n C_{ox} \dfrac{W}{2L}(V_{GS} - V_T)^2$	$I_{Dsat} = -\mu_p C_{ox} \dfrac{W}{2L}(V_{GS} - V_T)^2$
Channel-length modulation (Equation 9.1.10)	
$I_{Dsat} = \mu_n C_{ox} \dfrac{W}{2L}(V_{GS} - V_T)^2\left(1 + \dfrac{V_{DS}}{V_A}\right)$	$I_{Dsat} = -\mu_p C_{ox} \dfrac{W}{2L}(V_{GS} - V_T)^2\left(1 - \dfrac{V_{DS}}{V_A}\right)$
Saturation-region transconductance (Equation 9.1.37)	
$g_{msat} = \mu_n C_{ox} \dfrac{W}{L}(V_{GS} - V_T) = \dfrac{2I_{Dsat}}{(V_{GS} - V_T)}$	$g_{msat} = -\mu_p C_{ox} \dfrac{W}{L}(V_{GS} - V_T) = -\dfrac{2I_{Dsat}}{(V_{GS} - V_T)}$

Long-Channel Current-Voltage Equations with Substrate Charge

Linear-region drain current (Equation 9.1.17)

$$I_D = \mu_n C_{ox} \frac{W}{L}\left[(V_{GS} - V_T)V_{DS} - \frac{\alpha V_{DS}^2}{2}\right] \qquad I_D = -\mu_p C_{ox} \frac{W}{L}\left[(V_{GS} - V_T)V_{DS} - \frac{\alpha V_{DS}^2}{2}\right]$$

Saturation drain voltage (Equation 9.1.18)
$$V_{Dsat} = \frac{(V_{GS} - V_T)}{\alpha}$$

Saturation-region drain current (Equation 9.1.19)

$$I_{Dsat} = \mu_n C_{ox} \frac{W}{2\alpha L}(V_{GS} - V_T)^2 \qquad I_{Dsat} = -\mu_p C_{ox} \frac{W}{2\alpha L}(V_{GS} - V_T)^2$$

Short-Channel Current-Voltage Equations

Effective vertical field (Equation 9.2.3)

$$\mathscr{E}_{eff} = \frac{(V_{GS} - V_T)}{6x_{ox}} + \frac{(V_T + V_Z)}{3x_{ox}} \qquad \mathscr{E}_{eff} = \frac{(V_{GS} - V_T)}{6x_{ox}} + \frac{(V_T - V_Z)}{3x_{ox}}$$

Effective mobility (Equation 9.2.4)

$$\mu_{eff} = \frac{\mu_0}{1 + (\mathscr{E}_{eff}/\mathscr{E}_0)^\nu} \qquad \mu_{eff} = \frac{\mu_0}{1 + (-\mathscr{E}_{eff}/\mathscr{E}_0)^\nu}$$

Saturation $\mathscr{E}$-field (Equation 9.2.7)
$$\mathscr{E}_{sat} = 2v_{sat}/\mu_{eff} \qquad \mathscr{E}_{sat} = -2v_{sat}/\mu_{eff}$$

Linear-region drain current (Equation 9.2.9)

$$I_D = \mu_{eff} C_{ox} \frac{W}{L}\left[(V_{GS} - V_T) - \frac{V_{DS}}{2}\right]\frac{V_{DS}}{1 + (V_{DS}/\mathscr{E}_{sat}L)} \qquad I_D = -\mu_{eff} C_{ox} \frac{W}{L}\left[(V_{GS} - V_T) - \frac{V_{DS}}{2}\right]\frac{V_{DS}}{1 + (V_{DS}/\mathscr{E}_{sat}L)}$$

Saturation drain voltage (Equation 9.2.11)
$$V_{Dsat} = \frac{(V_{GS} - V_T)\mathscr{E}_{sat}L}{V_{GS} - V_T + \mathscr{E}_{sat}L}$$

Saturation-region drain current (Equation 9.2.10)
$$I_{Dsat} = WC_{ox}v_{sat}(V_{GS} - V_T - V_{Dsat}) \qquad I_{Dsat} = -WC_{ox}v_{sat}(V_{GS} - V_T - V_{Dsat})$$

CHAPTER 10

MOS FIELD-EFFECT TRANSISTORS II: HIGH-FIELD EFFECTS

In **Chapter** 9, we described the basic theory of MOSFET operation and derived equations for the currents flowing under normal biases. We first considered long-channel transistors where the *gradual-channel approximation* allows straightforward analysis and the corresponding physical understanding of the device operation. After we obtained the basic equations, we introduced additional effects arising from velocity saturation, mobility degradation, and the effects of short and narrow channels. Although the theory developed in Chapter 9 describes the behavior of MOSFETs under many conditions, the approximations made in that theory limit its accuracy in describing a number of important physical effects occurring in the velocity-saturation region. In particular, high-field effects that limit conventional MOSFET operation, but which can enable the operation of specialized devices, have not been fully considered.

In this chapter, we investigate how the high fields in today's MOSFETs can cause carriers to move into regions outside the normal conducting channel and we focus our attention on the following effects: (1) reduced output resistance caused by channel-length modulation; (2) current flowing from the channel to the substrate contact; and (3) current flowing from the channel through the gate oxide to the gate electrode. We discuss modifications of the MOSFET structure that can help avoid the limits imposed by these effects. Scaling the MOSFET dimensions typically leads to markedly higher electric fields in the device, with the peak field occurring near the drain junction. Because of the energy they gain from the high electric fields, carriers in the channel can initiate effects in surface regions near the drain junction that cause irreversible degradation of the MOSFET performance. The accumulated damage can, over time, result in device failure.

Because these serious limits to MOSFET scaling and performance depend on the magnitudes of the fields in the device, especially their maximum values, we start our discussion by determining the electric field and its spatial distribution more accurately than we did in Chapter 9. We use a "pseudo-two-dimensional" approach to solve Poisson's equation in the velocity-saturation region to gain insight into the relevant physics. With this understanding of the fields in the MOSFET, we can consider the magnitude and importance of both substrate current and gate current. We then discuss the long-term degradation that results from these undesirable currents, noting that their physical effects differ for $n-$ and $p-$channel MOSFETs. Finally, we see how high-field charge-injection into the oxide can be used advantageously to build nonvolatile MOS memory devices that retain information even when they are disconnected from the power supply.

10.1 ELECTRIC FIELDS IN THE VELOCITY-SATURATION REGION

When the drain voltage applied to a short-channel MOSFET exceeds the saturation drain voltage V_{Dsat}, carriers travel at their limiting or *saturation* velocity through a portion of the channel region near the drain; we call this physical portion of the transistor the *velocity-saturation region* (VSR). As V_D increases, the length of this region increases, and the length of the "normal" channel (in which the *gradual-channel analysis* used in Chapter 9 is applicable) continues to shorten. Because the voltage V_{Dsat} at the end of this channel is dropped over a decreasing length, the average field along the channel, and hence the average carrier velocity and the drain current I_D, continue to increase as V_D increases beyond V_{Dsat}. In integrated circuits, MOSFETs very frequently operate with carriers flowing at saturated velocities; therefore, it is essential to develop a reliable model for this condition. A central feature needed for this model is an understanding of the physics that controls channel-length modulation under the condition that mobile carriers reach saturated velocities at the end of the channel.

The basic MOS descriptions presented in Chapters 8 and 9 are based on the assumption that all mobile charges in the channel region of the MOSFET provide locations at which an x-directed field can terminate. This assumption is equivalent to assuming that we can approximate Poisson's equation as being one-dimensional. In the gradual-channel approximation, the y-directed field acts only to move charge that is induced by the x-directed field. We saw in Chapter 9 how the resulting gradual-channel model produces physical inconsistency when V_D increases to $V_{Dsat} = (V_G - V_T)$ and Q_n decreases to zero. If we revisit our basic assumptions above, we note another inconsistency that becomes evident at $V_{Dsat} = (V_G - V_T)$; for drain voltages of this magnitude and larger, a one-dimensional (only x-dependent) form for Poisson's equation cannot reasonably be justified. For $V_D > V_{Dsat}$ and increasing, x-directed electric-field lines actually change direction in the *pinch-off region* we defined in Chapter 9. Some field lines emanate from ionized donors in the space-charge region at the drain junction, travel through the gate oxide, and terminate on the gate terminal. The drain current in the pinch-off region is pulled away from the oxide surface because of this field. To obtain values for the drain current under these conditions, it is essential to work with the proper assumptions—most especially to formulate Poisson's equation as a two-dimensional differential equation. The general form for Poisson's equation in two dimensions with the charge density $\rho(x, y)$ represented as the sum of depleted-acceptor density $-qN_a$ and the mobile-charge density $-qn$ is

$$\frac{\partial^2 V(x, y)}{\partial x^2} + \frac{\partial^2 V(x, y)}{\partial y^2} = \frac{q[N_a(x, y) + n(x, y)]}{\epsilon_s} \tag{10.1.1}$$

where x is the direction away from the oxide-silicon interface, and y is the direction parallel to it. To solve Equation 10.1.1, we must first obtain expressions that specify the variation of the electron density n in the region of interest. Typically, analytical solutions cannot be obtained and numerical techniques, such as those provided by the software programs CADDET [1], MINIMOS [2], or PISCES [3] are necessary. To understand the physical picture, however, it is preferable to examine this problem using more classical methods; therefore, we develop an approximate solution for Equation 10.1.1, using what we call a "pseudo two-dimensional model" to make clear that it, too, is really an approximation.

Pseudo Two-Dimensional Model

The basis for the original pseudo two-dimensional model was presented by El Mansy and Boothroyd in 1977 [7] and, because of its utility, the model was subsequently enhanced [8], [9] to include the effects of velocity saturation and junction depth. The model has been used to gain an understanding of a number of hot-carrier effects [10–13]. We develop the essentials of the pseudo two-dimensional model in this section.

We want to apply the psuedo two-dimensional model to the velocity-saturation region shown in Figure 10.1, which shows a Gaussian box that encloses a portion of the space-charge region under the gate in the velocity-saturation region near the drain. The electric field emanating from the Gaussian box integrated over all its surfaces is proportional

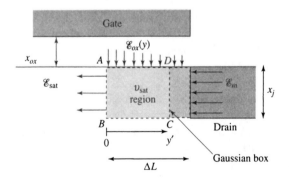

FIGURE 10.1 Cross-sectional geometry in the velocity-saturation region near the drain used in the pseudo two-dimensional approximation to find the electric field.

to the amount of charge enclosed in the box, and the total charge in the saturation region is composed of the sum of the depletion charge density qN_a and the mobile charge density qn. The Gaussian box is bounded, on one side, by a plane in yz (labeled AB) extending from the point adjacent to the oxide at which $\mathscr{E} = \mathscr{E}_{sat}$ down into the substrate to the depth of the drain junction x_j and, on the other side, by a plane in yz (labeled CD) located at a variable position y'. We call the field in the oxide adjacent to the oxide-silicon interface $\mathscr{E}_{ox}(y)$. The plane AB marks the beginning of the velocity-saturation region, and the plane CD corresponds to a position y' within the velocity-saturation region. In the analysis we assume: (1) that all carriers in the velocity-saturation region travel at their saturation velocity (by definition); (2) that the drain junction is an abrupt pn junction; (3) that the heavily doped drain region is perfectly conducting; and (4) that no electrons carrying the drain current reach depths greater than the drain-junction depth x_j. Assumption (2) limits our analysis to arsenic-doped n^+/p and boron-doped p^+/n junctions, which both have abrupt metallurgical transitions. However, the analysis can readily be extended to apply to graded junctions such as those incorporated in *lightly doped drain* (LDD) structures (to be discussed in Sec. 10.4) [14], [15].

The coordinate system in Figure 10.1 is chosen such that $y' = 0$ at point A and $y' = \Delta L$ at the edge of the drain. We apply Gauss' Law to the box $ABCD$:

$$-\mathscr{E}_{sat}x_j + \mathscr{E}(y')x_j + \frac{\epsilon_{ox}}{\epsilon_s}\int_0^{y'}\mathscr{E}_{ox}(y)\,dy = \frac{q}{\epsilon_s}x_j y'(N_a + n) \tag{10.1.2}$$

We made two approximations in writing Equation 10.1.2: (1) any field lines that cross boundary BC contribute negligibly compared to the other surface integrals; this is reasonable because the field lines in the velocity-saturation region that originate near the bottom plane of the junction are approximately horizontal; and (2) the y-directed field $\mathscr{E}_y$ is independent of x (a good approximation for large gate voltage V_G but only approximately true at low V_G). Differentiating Equation 10.1.2 with respect to y', we obtain

$$x_j\frac{d\mathscr{E}(y')}{dy'} + \frac{\epsilon_{ox}}{\epsilon_s}\mathscr{E}_{ox}(y') = \frac{q}{\epsilon_s}x_j(N_a + n) \tag{10.1.3}$$

At the interface between the oxide and silicon, the oxide field $\mathscr{E}_{ox}$ is given by

$$\mathscr{E}_{ox} = \frac{V_G - V_{FB} - 2|\phi_p| - V(y')}{x_{ox}} \tag{10.1.4}$$

Substituting Equation 10.1.4 into Equation 10.1.3, we obtain

$$x_j\frac{d\mathscr{E}(y')}{dy'} + \frac{\epsilon_{ox}}{\epsilon_s}\frac{1}{x_{ox}}[V_G - V_{FB} - 2|\phi_p| - V(y')] = \frac{q}{\epsilon_s}x_j(N_a + n) \tag{10.1.5}$$

Because

$$\frac{V_G - V_{FB} - 2|\phi_p| - V_{D\,sat}}{x_{ox}} = \mathscr{E}_{ox}(y' = 0) = \frac{q}{\epsilon_{ox}}x_j(N_a + n)$$

Equation 10.1.5 can be simplified to

$$\epsilon_s x_j\frac{d\mathscr{E}(y')}{dy'} = C_{ox}[V(y') - V_{D\,sat}] \tag{10.1.6}$$

or

$$\frac{d\mathscr{E}(y')}{dy'} = \frac{[V(y') - V_{D\,sat}]}{\ell^2} \tag{10.1.7}$$

where

$$\ell^2 = \frac{\epsilon_s}{\epsilon_{ox}} x_{ox} x_j \tag{10.1.8}$$

is a parameter with the units of length squared that is related to the MOSFET dimensions.

It is helpful to consider a physical interpretation of Equation 10.1.6. At the channel position we have taken for the origin of y' (point A of the Gaussian box) the voltage $V(y' = 0) = V_{D\text{sat}}$. The drain current consists of mobile-charge Q_n traveling at the saturated velocity v_{sat}. We assume that at this point all electric-field lines to Q_n emanate from the gate (i.e., they are linked to $\mathcal{E}_{ox}$). Moving toward the drain from the point $y' = 0$, $\mathcal{E}_{ox}$ must decrease as the voltage in the semiconductor increases toward V_D. Because I_D must be continuous in this velocity-saturated region, Q_n is also constant. If we consider our Gaussian box, these conditions can only be met if the field extending from the drain into the velocity-saturation region increases by the same amount that the field from the oxide decreases. In terms of Poisson's equation the charge released by the decreasing oxide field must be taken up by an increasing y-directed field. In Equation 10.1.6, the right-hand side is the amount of charge released by the oxide field as a result of the increase in channel voltage equal to $V(y') - V_{D\text{sat}}$, and the left-hand side is the corresponding increase of the channel-field gradient that supports this charge. The fractions of the charge controlled by the oxide field and by the drain field are illustrated in Figure 10.2.

With the boundary condition $\mathcal{E}(0) = \mathcal{E}_{\text{sat}}$ and $V(0) = V_{D\text{sat}}$, we can solve Equation 10.1.7 to obtain

$$\mathcal{E}(y') = \mathcal{E}_{\text{sat}} \cosh(y'/\ell) \tag{10.1.9}$$

and

$$V(y') = V_{D\text{sat}} + \ell \mathcal{E}_{\text{sat}} \sinh(y'/\ell) \tag{10.1.10}$$

At the drain end of the channel where the field is maximum

$$\mathcal{E}_m = \mathcal{E}(y' = \Delta L) = \mathcal{E}_{\text{sat}} \cosh(\Delta L/\ell) \tag{10.1.11}$$

and

$$V_D = V_{D\text{sat}} + \ell \mathcal{E}_{\text{sat}} \sinh(\Delta L/\ell) \tag{10.1.12}$$

Equations 10.1.11 and 10.1.12 can be combined to find the length of the velocity-saturation region

$$\Delta L = \ell \ln \left\{ \frac{[(V_D - V_{D\text{sat}})/\ell] + \mathcal{E}_m}{\mathcal{E}_{\text{sat}}} \right\} \tag{10.1.13}$$

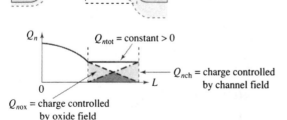

FIGURE 10.2 Fractions of the channel charge controlled by the oxide field and by the drain field found using the pseudo two-dimensional approximation.

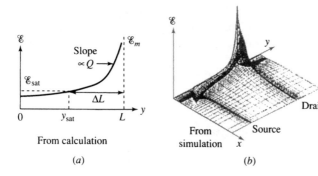

FIGURE 10.3 Electric field in the silicon near the oxide-silicon interface calculated using (*a*) the pseudo two-dimensional approximation and (*b*) two-dimensional numerical simulation. Note the steep increase of the field near the surface in the velocity-saturation region ΔL near the drain.

and the maximum electric field in the channel

$$\mathscr{E}_m = \left[\frac{(V_D - V_{D\text{sat}})^2}{\ell^2} + \mathscr{E}_{\text{sat}}^2 \right]^{1/2} \tag{10.1.14}$$

Important physical insight can be gained from Equations 10.1.9 to 10.1.14. Equation 10.1.9 indicates that the channel field increases almost exponentially toward the drain. As noted earlier, this steep increase in channel field is necessary to support the charge released by the oxide field. The field $\mathscr{E}(y')$ in the channel calculated using Equation 10.1.9 is plotted in Figure 10.3 to show similarities with the numerically simulated two-dimensional results. The parameter ℓ is adjusted to give approximately the same maximum channel field in Figures 10.3*a* and 10.3*b*.

To explore the magnitude of the maximum channel field in a short-channel MOS transistor, we substitute some realistic values for x_{ox} and x_j into Equations 10.1.10 and 10.1.14. With $x_{ox} = 25$ nm and $x_j = 0.2$ μm, which are typical for an *n*–channel MOS transistor in a 1 μm CMOS technology, $1/\ell = 8 \times 10^4$ V/cm (from Equation 10.1.8). At low $(V_G - V_T)$, $V_{D\text{sat}}$ is small, and we neglect it here. Using $\mathscr{E}_{\text{sat}} = 5 \times 10^4$ V/cm and $V_D = 5$V in Equation 10.1.14, we find the maximum field $\mathscr{E}_m = 4.0 \times 10^5$ V/cm, a high field that causes several hot-carrier effects. Because $\mathscr{E}_m$ is proportional to $1/(x_{ox}^{1/2} x_j^{1/2})$, continued scaling of the MOSFET to smaller dimensions makes these high-field effects more severe. Reducing the supply voltage below 3.3V moderates the high-field effects somewhat, but hot-carrier-resistant structures are still needed to improve device reliability, as we will discuss in Sec. 10.4.

The maximum channel field $\mathscr{E}_m$ depends explicitly on $(V_D - V_{D\text{sat}})$. However, the MOSFET channel length influences $\mathscr{E}_m$ only indirectly through $V_{D\text{sat}}$. For a drain voltage V_D more than one or two volts larger than $V_{D\text{sat}}$, $(V_D - V_{D\text{sat}})/\ell \gg \mathscr{E}_{\text{sat}}$, so that the maximum field is approximately proportional to $(V_D - V_{D\text{sat}})$ and can be expressed as

$$\mathscr{E}_m \approx \frac{(V_D - V_{D\text{sat}})}{\ell} \quad \text{for} \quad (V_D - V_{D\text{sat}}) > 1\text{V} \tag{10.1.15}$$

Extensive two-dimensional numerical analysis has confirmed the basic forms of Equations 10.1.9 and 10.1.13, suggesting that the pseudo two-dimensional analysis, while approximate, preserves the essence of the governing physics. Note that the substrate doping N_a does not appear explicitly in any of the equations. Both two-dimensional numerical simulation and experimental results show that N_a has little effect on ℓ. Most of the influence of N_a is through $V_{D\text{sat}}$, which is minor for short-channel devices. Two-dimensional simulation shows an approximate functional dependence for the parameter ℓ that has a reduced dependence on x_{ox} compared to that derived in Equation 10.1.8:

$$\ell = 0.22 x_j^{1/2} x_{ox}^{1/3} \tag{10.1.16}$$

(where ℓ, x_j, and x_{ox} are in cm) over wide ranges of x_{ox}, x_j, and N_a, and this form is widely used. The numerical simulations indicate the slightly weaker (1/3-power dependence) on x_{ox} instead of the square-root dependence predicted by the pseudo two-dimensional model.

EXAMPLE *Maximum Electric Field*

Sketch the maximum lateral field $\mathscr{E}_m$ as a function of V_D.

Solution $\mathscr{E}_m$ has different functional forms depending on whether V_D is larger or smaller than V_{Dsat}. For $V_D < V_{Dsat}$, the device is in the linear region. From Equation 9.2.8, we have

$$I_D = C_{ox}W[V_G - V_T - V(y)]\frac{\mu_{eff}\mathscr{E}(y)}{1 + [\mathscr{E}(y)/\mathscr{E}_{sat}]}$$

Expressing the field as the negative derivative of voltage and integrating to an arbitrary location y, we have

$$I_D y = \mu WC_{ox}\left(V_G - V_T - \frac{V(y)}{2}\right)V(y)\frac{1}{1 + (V_D/\mathscr{E}_{sat}\, y)}$$

By rearranging terms, we can solve for $V(y)$:

$$V(y) = (V_G - V_T) - \sqrt{(V_G - V_T)^2 - \frac{2[1 + (V_D/\mathscr{E}_{sat}\, y)]I_D y}{\mu WC_{ox}}}$$

The field $\mathscr{E}(y)$ is

$$\mathscr{E}(y) = -\frac{dV(y)}{dy} = \frac{I_D}{\mu WC_{ox}\sqrt{(V_G - V_T)^2 - \frac{2[1 + (V_D/\mathscr{E}_{sat}\, y)]I_D y}{\mu WC_{ox}}}}$$

$\mathscr{E}_m$ at the drain ($y = L$) is

$$\mathscr{E}_m(V_D) = \frac{\frac{\mu WC_{ox}}{L}\left(V_G - V_T - \frac{V_D}{2}\right)V_D\frac{1}{1 + (V_D/\mathscr{E}_{sat}L)}}{\sqrt{[\mu WC_{ox}(V_G - V_T)]^2 - 2(\mu WC_{ox})^2\left(V_G - V_T - \frac{V_D}{2}\right)V_D}}$$

For $V_D > V_{Dsat}$, $\mathscr{E}_m$ is given by Equation 10.1.14 and can be approximated by Equation 10.1.15. Hence, a sketch of the field can be drawn.

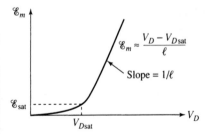

Using the expression for ΔL in Equation 10.1.13, we can derive both the drain current in the saturation region and the equivalent output resistance R_{out}. The drain current for $V_D > V_{Dsat}$ is

$$I_D = I_{Dsat}\left[\frac{V_G - V_T + \mathscr{E}_{sat}L}{V_G - V_T + \mathscr{E}_{sat}(L - \Delta L)}\right] \tag{10.1.17}$$

The equivalent output resistance R_{out} (the inverse slope of the $I_D - V_D$ characteristic) in the saturation region can be found by differentiating Equation 10.1.17 with respect to V_D. After some mathematical manipulation, we obtain the approximate expression for R_{out}

$$R_{out} = \frac{1}{I_{Dsat}} \left(\frac{V_D - V_{Dsat}}{\ell} \right) \left(L + \frac{V_G - V_T}{\mathscr{E}_{sat}} \right) \tag{10.1.18}$$

The finite output resistance predicted by Equation 10.1.18 degrades circuit performance. Ideally, we want the drain current to be delivered from a current source that depends only on the gate voltage; that is to be independent of the load. Equation 10.1.18 predicts that R_{out} is proportional to $1/I_{Dsat}$. This functional dependence is very close to what is observed, except for MOSFETs exhibiting severe drain leakage or punchthrough. Measured experimental results are shown in Figures 10.4a and 10.4b, with channel length L and oxide

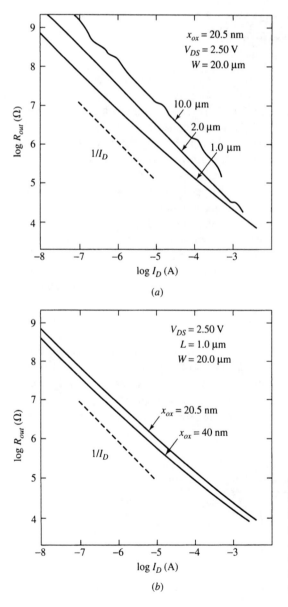

FIGURE 10.4 Experimental measurements of output resistance R_{out} in the saturation region as a function of drain current with: (a) channel length L, (b) oxide thickness x_{ox} as a parameter. The nearly reciprocal dependence of output resistance on drain current is evident. [16] (© 1987 IEEE).

thickness x_{ox}, respectively, as parameters [16]. These data verify the L and $1/x_{ox}^n$ dependences (with $1/3 < n < 1/2$). The dependence on junction depth x_j through ℓ (Equation 10.1.16) cannot be verified easily through experiments because R_{out} depends on both x_j and the junction profile, and it is difficult to fabricate devices with changing x_j values while maintaining the same junction profiles. Two-dimensional analyses, however, tend to confirm the dependence on ℓ inferred from Equations 10.1.16 and 10.1.18.

In this section we showed that the pseudo two-dimensional model method provides a useful analysis technique to determine the approximate values of the electric field in the critical velocity-saturation region of the transistor. In the following sections, we first consider some detrimental consequences of the high electric fields on device behavior, and then we show that we can also make use of the high electric fields in short-channel MOSFETs to obtain specialized devices.

10.2 SUBSTRATE CURRENT

Hot-Carrier Effects

Some of the most serious problems created by the continued scaling of MOSFETs well below 1 μm (generally described as scaling into the deep submicrometer regime) are caused by *hot-carrier effects*. A voltage difference of one volt across 0.5 μm, for example, imposes an average field of 2×10^4 V/cm, sufficiently high to saturate the velocity of both holes and electrons. In such fields, carriers are described as "hot" and their high energies make them capable of physical effects that can degrade the MOSFETs. Figure 10.5 illustrates the major parasitic-current components (typically undesirable) that are created through hot-carrier effects. The substrate current, if not somehow limited, can cause local potential fluctuations or electron injection into the substrate, and induce "snap-back" device breakdown (decreasing voltage with increasing current) as well as the CMOS latch-up discussed in Sec. 9.3. The gate current caused by electron injection into the oxide can degrade the oxide, but it also provides a mechanism for programming nonvolatile devices, such as those we discuss in Sec. 10.5. Because of the detrimental effects of hot carriers and the opportunities to make use of them in some devices, an understanding of hot-carrier effects is very important for proper device design.

Electrons in the inversion layer reach their saturation velocity when the electric field exceeds about 3 to 5×10^4 V cm^{-1}. Hot-carrier effects at fields lower than this are typically insignificant so we only need to consider carriers in the velocity-saturation region. The maximum channel field $\mathscr{E}_m$ is the most significant parameter to monitor when considering the influence of hot-carrier effects. The expressions for $\mathscr{E}_m$ were derived in Sec. 10.1 as Equations 10.1.14 and 10.1.15. The magnitude of $\mathscr{E}_m$ is mainly determined by $(V_D - V_{Dsat})$ and ℓ. From the theory we developed, we see that hot-carrier effects are

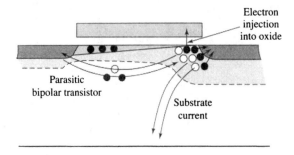

FIGURE 10.5 Detrimental effects resulting from channel hot carriers in a MOSFET; carriers can be injected into the gate oxide; current can flow through the substrate; and a parasitic bipolar transistor can become forward-active biased.

more significant at higher drain voltages V_D, shorter channel lengths L (because of smaller V_{Dsat}), thinner oxides (smaller ℓ), and shallower junction depths (smaller ℓ). The major-current components affected by hot carriers are the substrate and gate currents. Each of these components is discussed in the following sections.

Substrate-Current Model

If electrons in the channel of a MOSFET acquire more than about 1.5 eV of kinetic energy, they can cause impact ionization when they collide with the lattice and can generate electron-hole pairs. Electrons produced by impact ionization are either attracted to the drain because of the high potential there (adding to the drain current), or they can be injected into the oxide (if they possess sufficient energy). The generated holes, on the other hand, are attracted to the substrate (a location with a lower potential), creating a parasitic substrate current I_{sub}. Although this mode of carrier generation can lead to avalanche breakdown, the generated holes move rapidly into a lower field region and avalanching does not always occur. To estimate the magnitude of the substrate current in an n–channel MOSFET correctly, we need to calculate the number of holes generated by impact ionization in the velocity-saturation region of the channel by the high electric fields there.

The number of impact-ionization events produced by one carrier per unit length is calculated using the impact-ionization coefficient, which varies with the electric field as $A_i \exp(-B_i/\mathscr{E})$, as discussed in the derivation of Equation 4.4.13. The ionization parameter B_i is roughly twice as large for impact ionization by holes as for impact ionization by electrons, making hole-induced generation less important. If we assume that all the holes travel to the substrate, the substrate current in an n–channel MOSFET due to impact ionization by the drain current can be expressed as [17]

$$I_{sub} = \int_0^L I_D A_i \exp[-B_i/\mathscr{E}(y)]\,dy \tag{10.2.1}$$

Replacing dy with $(dy/d\mathscr{E})d\mathscr{E}$ or $-\mathscr{E}^2(dy/d\mathscr{E})d(1/\mathscr{E})$ in Equation 10.2.1, we have

$$I_{sub} = \int_{\mathscr{E}_s}^{\mathscr{E}_m} -I_D A_i \exp[-B_i/\mathscr{E}(y)]\,\mathscr{E}^2(y)\frac{dy}{d\mathscr{E}}d\left(\frac{1}{\mathscr{E}}\right) \tag{10.2.2}$$

where $\mathscr{E}_s$ is the electric field at the source end of the channel. Using Equation 10.1.9, and approximating $\cosh(y/\ell)$ by $e^{y/\ell}/2$, we have

$$\frac{d\mathscr{E}}{dy} \approx \mathscr{E}_{sat}\frac{e^{y/\ell}}{2\ell} = \frac{\mathscr{E}}{\ell} \tag{10.2.3}$$

so that

$$\mathscr{E}^2(y)\frac{dy}{d\mathscr{E}} \approx \ell\mathscr{E} \tag{10.2.4}$$

Because the exponential term in Equation 10.2.2 has a pronounced maximum at $\mathscr{E} = \mathscr{E}_m$, we evaluate it at $\mathscr{E} = \mathscr{E}_m$ and let it be constant over the region so we can remove it from the integral. After this simplification, Equation 10.2.4 can be solved for I_{sub}

$$I_{sub} = A_i \ell \mathscr{E}_m I_D \int_{\mathscr{E}_s}^{\mathscr{E}_m} \exp[-B_i/\mathscr{E}(y)]\,d\left(\frac{1}{\mathscr{E}}\right)$$

$$= \frac{A_i}{B_i}\ell\mathscr{E}_m I_D \exp(-B_i/\mathscr{E})\bigg|_{\mathscr{E}_s}^{\mathscr{E}_m} \approx \frac{A_i}{B_i}\ell\mathscr{E}_m I_D \exp(-B_i/\mathscr{E}_m) \tag{10.2.5}$$

which can be approximated by

$$I_{sub} \approx \frac{A_i}{B_i}(V_D - V_{Dsat})I_D \exp\left(-\frac{\ell B_i}{V_D - V_{Dsat}}\right) \tag{10.2.6}$$

The form of Equation 10.2.5 agrees with our earlier statement that the substrate current primarily depends on the difference between V_D and V_{Dsat}, and on the value of the parameter ℓ. Experimental results show that B_i is approximately 1.7×10^6 V/cm and A_i/B_i is about 1.2 V^{-1}.

Before continuing our discussion of substrate current, a comment about the magnitude of the quantities of interest is worthwhile. Electrons gain energy from an applied electric field through the integrated coulombic force on them multiplied by the distance over which the force is applied between electron collisions. The amount of energy gained from the field is an integral over the distance between collisions, and we are not justified in calculating the energy of hot electrons by an averaging method such as taking the product of the mean-free path times the size of the local field. We need to consider this fact because of some of our derivations. For example, the expression we used for the impact-ionization coefficient (Equation 4.4.13) depends very strongly on the electric field. When the field varies rapidly with position, the energy of the electron can be less than the energy we estimate from the local electric field and the mean-free-path length. If we use the electric field at the collision point in the expression for the ionization coefficient, we overestimate the impact-ionization rate. The mathematical complexity of a detailed analysis requires numerical, rather than analytical, methods. However, numerical methods often limit physical insight, so an approximate analysis is valuable.

We can obtain a useful representation of I_{sub} by superimposing lines of constant I_{sub}/I_D on a plot of the $I_D - V_D$ characteristics. The lines of constant I_{sub}/I_D are parallel to each other, as shown in Figure 10.6. Each line represents not only constant I_{sub}/I_D, but also constant $\mathscr{E}_m$ (from Equation 10.2.5) and, therefore, constant $V_D - V_{Dsat}$ (from Equation 10.2.6). Figure 10.6 confirms that the lines of constant I_{sub}/I_D are parallel to the line representing V_{Dsat} [18].

Figure 10.7 is a plot of (I_{sub}/I_D) as a function of $1/(V_D - V_{Dsat})$, in a form suggested by Equation 10.2.6, for transistors with channel lengths of 0.95, 1.45, and 2.7μm and for gate voltages of 0.9 and 1.5 V. All data points for the different gate voltages and the different channel lengths fall on a single straight line. The linear relationship supports the

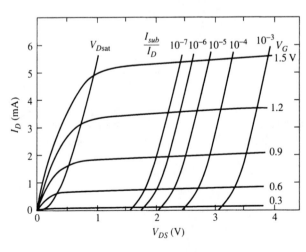

FIGURE 10.6 Lines of constant I_{sub}/I_D superimposed on the I_D versus V_D characteristics of an n–channel MOSFET with $x_{ox} = 15.2$ nm and $L = 1.45$ μm [18] (© 1984 IEEE).

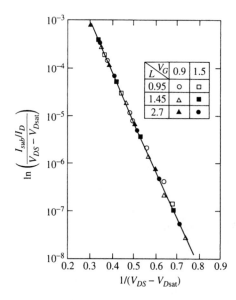

FIGURE 10.7 Experimental measurements of (I_{sub}/I_D) as a function of $1/(V_D - V_{Dsat})$ for transistors with channel lengths of 0.95, 1.45, and 2.7 μm and gate voltages of 0.9 and 1.5 V. The form of the vertical axis is suggested by Equation 10.2.6 [18] (© 1984 IEEE).

validity of the assumptions used to derive Equation 10.2.6 and confirms our assertion that the important dependences of substrate current on gate bias and channel length are contained in the V_{Dsat} term. In deriving Equation 10.2.6 from Equation 10.2.5, we assumed $\mathscr{E}_{sat}$ to be much less than $\mathscr{E}_m$ and neglected it. Experimental data show that the linear relationship in Figure 10.7 is valid for ratios of I_{sub}/I_D as small as 10^{-6}, below which $\mathscr{E}_m$ is no longer much larger than $\mathscr{E}_{sat}$. Note that the slope of the line in Figure 10.7 is $-\ell B_i$ from which the value of B_i can be obtained experimentally. From Equation 10.2.6, the intercept of the line with the vertical axis gives the value of A_i/B_i. Equation 10.2.6 is also valid for p–channel MOSFETs with $B_i = 3.7 \times 10^6$ V/cm and $A_i/B_i = 2.2$.

While the plot in Figure 10.7 is useful for verifying Equation 10.2.6, it does not directly give the dependence of I_{sub} on the terminal voltages. Figure 10.8, which is a plot of I_{sub} versus V_G for five different values of V_D, relates the physics to the applied voltages more directly [19]. This figure shows that I_{sub} reaches a maximum at a gate voltage V_G between V_T and $V_D/2$. This behavior can be physically understood using the following

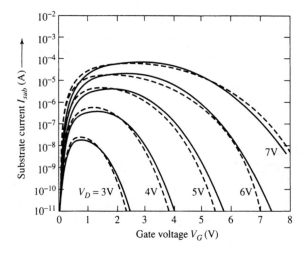

FIGURE 10.8 Substrate current I_{sub} as a function of gate voltage V_G for different drain voltages V_D; solid lines are from measurements, dashed curves are calculated [19] (© 1989 Academic Press).

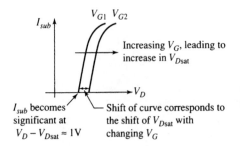

FIGURE 10.9 Substrate current I_{sub} as a function of drain voltage V_D for two different gate voltages V_{G1} and $V_{G2} > V_{G1}$; the curve shifts to the right because V_{Dsat} increases.

argument. First, at low gate voltages, to initiate impact ionization some inversion electrons must be present in the velocity-saturation region and these electrons carry a drain current I_D. Equation 10.2.6 shows the proportionality between I_{sub} and I_D. Increasing the gate voltage, therefore, tends to increase I_{sub} through the drain-current increase. Increasing the gate voltage, however, increases the value of V_{Dsat}, which, as indicated by Equation 10.1.14, reduces the size of the maximum field $\mathscr{E}_m$. Equation 10.2.5 shows that reducing $\mathscr{E}_m$ lowers the substrate current I_{sub}. The combined effect of these two competing trends results in the successive maxima observed in the plots of substrate current I_{sub} versus gate voltage V_G shown in Figure 10.8.

The dependence of substrate current I_{sub} on drain voltage V_D is sketched in Figure 10.9 with V_G as a parameter. A significant substrate current I_{sub} is first observed at $(V_D - V_{Dsat}) \approx 1$ V (similar to the condition used to obtain Equation 10.1.14 from Equation 10.1.13); the substrate current increases linearly with increasing drain voltage. A typical slope of the curves in Figure 10.9 corresponds to approximately one decade increase of substrate current for each 0.5 V increase in drain voltage. Higher gate voltage V_G causes higher V_{Dsat}, leading to a shift in the I_{sub} curve along the V_D axis by an amount equal to the change in V_{Dsat} as seen in Figure 10.9. This behavior is reasonable because I_{sub} does not depend directly on V_D, but rather depends on the difference between V_D and V_{Dsat}.

Effect of Substrate Current on Drain Current

As I_{sub} flows into the bulk region of the MOSFET, a voltage develops across the resistive substrate, as indicated in Figure 10.10. This voltage drop forward-biases the neutral region under the channel, causing a *reverse-body-bias effect* that reduces the MOSFET threshold voltage and increases its drain current. The substrate current I_{sub} depends on the drain voltage V_D, so this phenomenon is another mechanism by which the drain voltage modulates the saturation-region drain current and thus reduces the output resistance.

Because we have considered a number of effects, it is worthwhile to summarize important factors that affect the saturation drain current (and thereby R_{out}): (1) channel-length

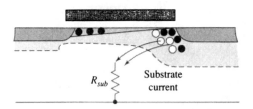

FIGURE 10.10 Substrate current flowing through the resistive substrate to the substrate contact leads to a difference between the applied substrate bias and the voltage in the neutral region immediately under the channel. The effective substrate bias and, therefore, the threshold voltage of the transistor are consequently changed.

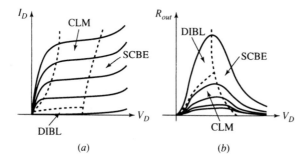

FIGURE 10.11 (*a*) Increase of drain current, and (*b*) reduction of output resistance resulting from channel-length modulation (CLM), drain-induced barrier lowering (DIBL), and substrate-current-induced body effect (SCBE), showing bias regions in which each effect dominates.

modulation (CLM), (2) drain-induced-barrier lowering (DIBL), and (3) substrate-current-induced body effect (SCBE) [20]. Our earlier consideration about channel-length modulation predicts R_{out} to be proportional to $(V_D - V_{Dsat})$, as indicated in Equation 10.1.18, leading us to conclude that R_{out} increases with increasing V_D. However, experimentally R_{out} is found to reach a maximum at a drain voltage V_D somewhat above V_{Dsat} and then to decrease rapidly to a very low value at high V_D, as shown in Figure 10.11. This rapid decrease of R_{out} occurs when the hot-carrier-induced substrate current discussed above becomes dominant. In addition to reducing R_{out}, drain-induced barrier lowering (discussed in Sec. 9.2) also decreases the threshold voltage V_T of a transistor biased in strong inversion as the drain voltage V_D increases. MOSFETs can, however, be designed to minimize this threshold-voltage reduction. The relative contributions of channel-length modulation, drain-induced barrier lowering, and substrate-current biasing to the increase of the saturation drain-current and the reduction of the output resistance depend strongly on the bias conditions as shown in Figure 10.11. Accurate modeling of the degradation of the output resistance by the substrate current is complex [20], and we do not consider it quantitatively.

When the substrate current in a MOSFET becomes large and the voltage induced beneath the channel increases beyond the turn-on voltage of the source-body junction, this junction acts like the emitter-base junction of a bipolar transistor. The neutral region under the channel is the base, and the drain is the collector of this parasitic *npn* bipolar transistor, shown in Figure 10.12. With a forward bias on the source-body junction, this parasitic transistor becomes active, causing electrons to flow from the source into the body region, further increasing the voltage there and reducing the threshold voltage of the MOSFET. The channel current increases, as shown in Figure 10.13, and this increase leads to more impact ionization (higher I_{sub}) and eventually to MOSFET breakdown. Because of these effects, the breakdown voltage of a MOSFET depends strongly on the magnitude of the substrate current and on the effective substrate resistance. Figure 10.13 shows the frequently observed *C*-shaped breakdown curve (dashed line), implying that the lowest breakdown voltage occurs at an intermediate value of gate voltage. The critical gate voltage is related to the magnitude of I_{sub}, which reaches its maximum value at a gate voltage V_G between V_T and $V_D/2$ as shown in Figure 10.8.

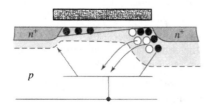

FIGURE 10.12 When the source-body junction of the MOS transistor becomes forward biased, it can act as the emitting junction of a parasitic bipolar transistor in which the drain acts as the collector.

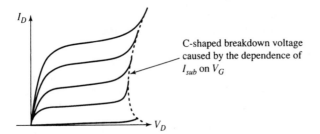

C-shaped breakdown voltage caused by the dependence of I_{sub} on V_G

FIGURE 10.13 Parasitic bipolar-transistor action below the MOSFET channel can lead to breakdown with characteristic C-shaped behavior.

10.3 GATE CURRENT

In our discussion up to this point, we considered current flowing in various regions of the silicon. Now, we extend our discussion to consider current flowing into and through the gate oxide, which we have so far assumed to be a perfect insulator. *Gate current* is caused by carriers flowing from the channel to the gate electrode through the oxide. The most important conduction mechanism involves injection of energetic (*hot*) electrons into the oxide. The term *hot-carrier* injection implies that the injected carriers are no longer in equilibrium with the lattice at the point of injection; that is, they possess far more kinetic energy than the energy corresponding to the ambient lattice temperature. The carriers gain energy as they are accelerated by the high fields in the channel until they have enough energy to be thermally injected over the barrier at the oxide-silicon interface shown in Figure 8.2. Hot-carrier injection differs from *tunnel injection* in which the carriers traveling through a very thin oxide are in thermal equilibrium with the silicon lattice (i.e., they are "cold").

Lucky-Electron Model

Although both substrate current and gate current are caused by hot-carrier effects, the bias-voltage dependences of these two currents differ substantially. The observed gate current of an *n*–channel MOSFET is plotted together with the substrate current in Figure 10.14 [21]. As shown in this figure, the gate current is much smaller than the substrate current and is largest when the gate voltage and drain voltage are approximately equal. For hot electrons in the channel to reach the gate, two conditions must be satisfied: (1) The hot electrons must gain sufficient kinetic energy from the channel field to surmount the potential barrier between the bulk silicon and the gate oxide [11], and (2) the hot electrons must be redirected toward the oxide–silicon interface [22]. The necessary conditions are illustrated in Figure 10.15: Electrons gain energy as they move in the channel from point *A* to point *B*, and are redirected at point *B* toward the oxide (point *C*). As a hot electron is redirected toward the interface, it must not lose too much energy in collisions so that it is sufficiently energetic to surmount the potential barrier at the oxide–silicon interface. Once the electron is emitted from the silicon over the potential barrier into the oxide, it is swept toward the gate electrode by the aiding field in the oxide, provided that the gate is at a higher potential than the surface channel in the silicon. Because the relevant processes are statistically independent, the overall probability for electron acceleration and injection is the product of the probability that each individual event takes place.

For a hot electron to surmount the Si–SiO$_2$ potential barrier of height $q\phi_B = 3.1$ eV (Figure 8.2), its kinetic energy must be greater than $q\phi_B$. If we assume the accelerating electric field ($\mathcal{E}_y$) in the channel to be constant, the hot electron must travel a distance d equal to $(\phi_B/\mathcal{E}_y)$ to acquire this kinetic energy. The probability that a channel electron travels this distance or longer without any collision can be written as $\exp(-d/\lambda)$, where λ is the scattering mean-free-path of the hot electron. Hence, we can take $\exp[-\phi_B/(\mathcal{E}_y\lambda)]$ as

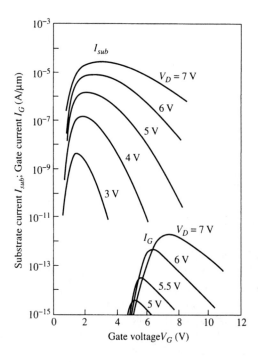

FIGURE 10.14 Measured substrate and gate currents (A/μm) in an *n*–channel MOSFET versus gate voltage V_G for different drain voltages V_D, showing that I_{sub} is larger and reaches maxima at lower V_G than does I_G [21] (© 1982 IEEE).

the probability that an electron acquires a kinetic energy greater than the Si–SiO$_2$ potential barrier. This approach to calculating the gate current is sometimes called the *lucky-electron model* [11]. The probability P of injection and collection illustrated in Figure 10.15 can be found as a function of the oxide field $\mathscr{E}_{ox}$. The gate current I_G can then be derived by considering the product of the number of carriers (represented by the drain current), the probability that a carrier gains enough energy (expressed as $\exp\left[-\phi_B/(\mathscr{E}_y\lambda)\right]$), and the probability P of injection and collection. Integrating this product over the channel, we have

$$I_G = I_D \int_0^L \exp\left(-\frac{\phi_B}{\mathscr{E}_y\lambda}\right) P(\mathscr{E}_{ox})\,dy \tag{10.3.1}$$

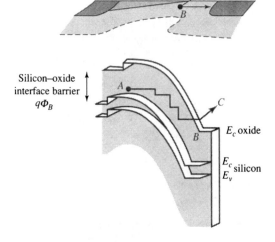

FIGURE 10.15 For gate current to flow, electrons must traverse the silicon dioxide; (1) the carriers must gain adequate energy to escape the silicon as they move from point *A* to point *B* on an energy diagram, (2) they must be redirected (at point *B*) toward the oxide–silicon interface where they enter the oxide conduction band and ultimately reach the conducting-gate region (point *C*).

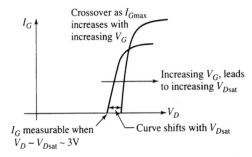

FIGURE 10.16 Gate current I_G as a function of drain voltage V_D, showing steep increase and then saturation of gate current with increasing drain voltage.

which can be approximated as

$$I_G \approx C\, I_D \exp\!\left(-\frac{\phi_B}{\lambda \mathscr{E}_m}\right) \tag{10.3.2}$$

where C is approximately 2×10^{-3} when $V_G > V_D$.

In the following qualitative discussion we frequently compare the magnitudes of the gate and the drain voltages. To simplify the discussion, we assume the threshold voltage V_T to be zero (alternatively, we can think of the effective gate voltage as the "overdrive" above V_T) and we take the source voltage as our reference.

As shown in Figure 10.16, at a fixed gate voltage V_G, the gate current I_G increases with increasing drain voltage until $V_D = V_G$ because of an increasing maximum field $\mathscr{E}_m$. At higher drain voltages ($V_D > V_G$), the gate current remains constant or decreases slightly because the field in the gate oxide near the drain changes direction. The sketches in Figure 10.17 illustrate the channel and oxide fields in different ranges of gate voltage and drain voltage. When V_D is smaller than V_G (Figure 10.17a), the oxide field at the drain end of the channel (where the y-directed field is maximum) is directed from the gate to the channel ($\mathscr{E}_{ox} > 0$), favoring the collection of injected electrons by the gate. Increasing V_D increases the drain current and thereby the number of electrons having sufficient energy to surmount the Si–SiO$_2$ barrier. Because the gate current is limited by the number of electrons that gain adequate energy from the channel field, gate currents in this case are called *channel-field-limited* gate currents. When V_D equals V_G (Figure 10.17b), the oxide field is zero at the drain end of the channel. When $V_D > V_G$ (Figure 10.17c), the oxide field at the drain opposes the collection of the hot electrons ($\mathscr{E}_{ox} < 0$), retarding even those carriers having adequate energy to surmount the barrier, so gate currents in this range are called *oxide-field-limited* gate currents. The gate current saturates or decreases when $V_D > V_G$, as shown in Figure 10.16.

At high gate voltages, $\mathscr{E}_m$ also decreases, lowering the gate current as shown in Figure 10.14. This reduction in gate current is analogous to the decrease of substrate current I_{sub} with increasing V_G that we discussed previously.

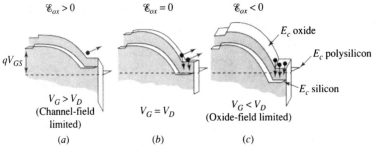

FIGURE 10.17 Schematic representation of gate and channel fields important for hot-electron injection for three cases of electric field at the drain.

EXAMPLE   *MOSFET Degradation Time*

In a short-channel MOSFET with $x_{ox} = 30$ nm, a gate current of 1 nA is momentarily caused by hot-electron injection. This gate current is estimated to flow through a section of oxide measuring 200 nm $\times$ 10 μm near the drain end of the channel. Assume that a fraction equal to 10^{-6} of the injected electrons becomes trapped at an average distance of 0.1 x_{ox} from the Si-SiO$_2$ interface. Calculate the time needed for gate-current flow (under the conditions stated) to change the threshold voltage by 100 mV in the region where the injection is taking place.

Solution   The gate current of 1 nA is carried by a current density across the injection area of

$$10^{-9}/2 \times 10^{-8} = 5 \times 10^{-2} \text{ A cm}^{-2}$$

Because 1 in 10^6 electrons become trapped, the rate of charge trapping in the SiO$_2$ is $5 \times 10^{-2} \times 10^{-6} = 5 \times 10^{-8}$ Coulombs s^{-1} cm^{-2}. From Equation 8.5.4, trapped charge in the oxide shifts the flat-band voltage and thereby the threshold voltage (Equation 8.3.18). Assuming that the charge is concentrated in a sheet, we calculate ΔV_{FB} from Equation 8.5.3 to be

$$\Delta V_{FB} = \Delta V_T = \left(\frac{1}{C_{ox}}\right)\left(\frac{0.9 x_{ox}}{x_{ox}}\right)\Delta Q_{ot}$$

or

$$\Delta Q_{ot} = \frac{C_{ox}\Delta V_T}{0.9} = \frac{1.15 \times 10^{-7} \times 0.1}{0.9} = 1.28 \times 10^{-8} \text{ C cm}^{-2}$$

The time required to trap this quantity of charge is

$$t = \frac{\Delta Q_{ot}}{J_{ot}} = \frac{1.28 \times 10^{-8}}{5 \times 10^{-8}} = 0.256 \text{ s} \qquad \blacksquare$$

Carrier Injection at Low Gate Voltages

The lucky-electron-model discussion presented in the previous paragraphs explains the carrier injection that leads to gate current at high gate and drain voltages—conditions where gate currents can be most troublesome. This model considers carrier acceleration directly by the field in the channel to create *channel hot electrons* (CHE) that do not result from impact-ionization processes. At lower gate voltages, other mechanisms leading to gate current become important. The behavior of gate current as a function of gate voltage is illustrated over a wide range of bias conditions in Figure 10.18 [23]. The data in this figure show that the electron current through the gate has a maximum at an intermediate gate voltage in addition to the one at high gate voltages caused by channel hot electrons and predicted by the lucky-electron model. At very low gate voltages, gate currents can also flow in the reverse direction, indicating the injection of holes into the oxide. These two additional components result from impact ionization. Because impact ionization also causes substrate current I_{sub}, these additional components of gate current are most serious at bias conditions that produce the maximum substrate current. Injection of hot holes and hot electrons created by impact ionization is called *drain-avalanche hot-carrier current* (D.A.H.C.).

First, consider the holes that cause gate current at very low gate voltages. These holes are created in the high-field space-charge region near the drain by impact-ionization events taking place there. When the gate voltage is much smaller than the drain voltage (but still greater than the threshold voltage V_T), the field in the oxide near the drain reverses sign. Some of the sufficiently energetic holes are injected into the gate oxide and pass through it to the gate electrode, giving rise to current flowing out of the gate lead.

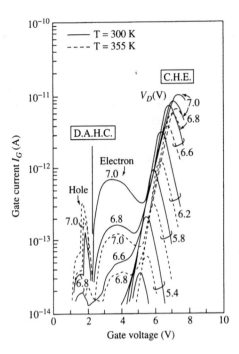

FIGURE 10.18 Magnitude of measured gate current in an *n*–channel MOSFET versus gate voltage at various values of drain voltage. At low V_G, drain-avalanche hot-carrier currents (D.A.H.C) predominate with hot holes causing negative gate current that peaks below $V_G \sim 2V$ and hot electrons causing a maximum in positive gate current for $V_G \sim 3V$. Channel hot electrons (C.H.E) as predicted by the lucky-electron model lead to maxima in gate currents for $V_G > 4V$ [23] (© 1982 IEEE).

This hole injection is the dominant mechanism for gate current at high drain voltages and low gate voltages. Its magnitude is very small and it is difficult to measure, especially when the gate oxide is thicker than 10 nm. In the data of Figure 10.18, hole injection is the largest gate-current component when the gate voltage is much smaller than the drain voltage (e.g., $V_G = 1.2$ V and $V_D = 6$ V). As V_G increases above 2 V (while V_D remains at 6 V), the hot-hole gate current decreases to negligibly small values.

When the gate voltage is increased, injected hot electrons instead of hot holes begin to dominate gate current. Even though the oxide field near the drain opposes collection of the injected electrons by the gate electrode, the excess energy of the injected carriers allows some of them to reach the gate. As the field between the drain and the gate decreases (because the gate voltage increases while the drain voltage remains at a constant, high value), this component of gate current becomes especially important because of the increasing probability that hot electrons injected into the oxide conduction band reach the gate electrode before being forced back into the substrate by the opposing oxide field. Thus, in the intermediate gate-voltage regime, the gate current initially increases because of electrons created by impact ionization. However, further increases of the gate voltage reduce the lateral electric field in the channel near the drain so that hot electrons generated by impact ionization no longer contribute significantly to the gate current. Because gate currents in typical MOSFETs are so small (below pA as seen in Figure 10.18), experimental verification of the theoretical description we have given for drain-avalanche hot-carrier injection is not entirely complete. At higher gate voltages, more studies have been done which validate that hot-electron injection due to channel *lucky electrons* predominates.

Gate Current in *p*–Channel MOSFETs

Hot-carrier effects for holes differ from those found for electrons, and measurable channel hot-hole injection is not observed in *p*–channel MOSFETs. The major differences between *n*– and *p*–channel transistors are consequences of the larger barrier height for hole injection into the oxide (Figures 8.1 and 8.2) and the shorter mean-free path of holes than of electrons in silicon. The dominant gate current in p–channel MOSFETs arises from *electron* injection into the oxide, as in n–channel MOSFETs. The injected electrons are generated by impact ionization and are subsequently accelerated by the channel field. The injection mechanism is analogous to that of the hole current in n–channel MOSFETs shown in Figure 10.18 at low gate voltages. Gate current in p–channel MOSFETs can even be larger than in n–channel MOSFETs. Comparing the substrate current I_{sub} to the gate current I_G of p–channel MOS-FETs indicates that both I_{sub} and I_G have maxima at low magnitudes of gate voltage [24].

10.4 DEVICE DEGRADATION

Degradation Mechanisms in n–Channel MOSFETs

The hot-carrier effects described above cause degradation that accumulates with time. For example, Figure 10.19 shows the degradation of the measured output characteristics of a MOSFET. This degraded performance can be interpreted as resulting from an increase of the threshold voltage of the MOSFET. This increase of threshold voltage, in turn, leads to lower drain currents and a resulting decrease of the MOSFET switching speed. Ulti-mately, circuits made from such degraded devices may cease to operate at the required speed, causing failure of the system into which they are built.

Hot-carrier-induced gate currents cause this degradation by two mechanisms: (1) charge trapping in the oxide, and (2) generation of interface states. Both mechanisms cause negative charge to build up either in the gate oxide and/or at the oxide–silicon interface. Charge continues to accumulate, primarily near the drain of the MOSFET, whenever cir-cuit conditions cause hot-carrier injection. The effect of this charge accumulation on the observed device behavior is somewhat mitigated because the size of the threshold voltage at the drain is less important than its magnitude near the source end of the channel.

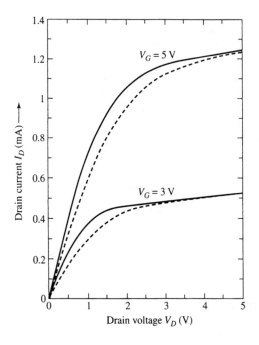

FIGURE 10.19 Current-voltage charac-teristics of an n–channel MOSFET both before (solid curves) and after (dashed curves) hot-carrier stressing ($W/L = 8/2$ μm/μm, $x_{ox} = 28$ nm, $\Delta V_T = 220$ mV) [19] (© 1989 Academic Press).

However, if the source- and drain electrodes are reversed, the degraded region is then near the source where changes in threshold voltage are most important. Although assessing the relative importance of oxide-charge trapping and interface-state generation on the observed changes of the threshold voltage is difficult, experimental results suggest that interface-state generation is most important. If oxide-charge trapping were the dominant effect, MOSFET degradation should be most severe under bias conditions that cause the highest values of gate current. However, the degradation rate is experimentally found to be highest when V_G is approximately $V_D/2$, a bias that produces the highest substrate current and not the maximum gate current. This observed voltage dependence of the degradation indicates that it is predominantly caused by the generation of interface states (see the discussion in Sec. 8.5) [26]. The exact mechanism for the interface-state generation is still the subject of research, but a reasonable model has been presented based on the creation of interface-charge-trapping states when colliding hot electrons transfer sufficient energy to break chemical bonds (probably those between silicon atoms and hydrogen).

Many of the bonds between silicon atoms at the interface are weakened or broken by disorder in the lattice structure in the transition region where the crystalline silicon lattice links to the amorphous oxide. As one of the final steps in an IC fabrication process, the wafer is generally annealed in a hydrogen-containing ambient; the hydrogen usually bonds with those silicon atoms that are not tightly bonded to other silicon atoms or to oxygen atoms. The resulting Si — H bond is weaker and more easily broken than a Si — Si bond, and is, therefore, a likely location for trapped-interface charge to accumulate when hot carriers traverse this region.

Characterizing *n*–Channel MOSFET Degradation

The severity of device degradation is commonly characterized by measuring the change of threshold voltage (ΔV_T) or the percentage change in drain current ($\Delta I_D/I_D$) in the MOSFET "linear" region at low V_D and high V_G (e.g., $V_D = 0.1$ V, $V_G = V_{DD}$). In some cases, degradation of the transconductance (g_m) or the inverse subthreshold slope (S) is used as a measure of MOSFET degradation.

In the following paragraphs we discuss how to determine if a MOSFET can be operated reliably under the desired circuit-operating conditions. The requirement is usually specified in terms of allowable degradation resulting from hot-carrier damage during a certain length of operating time. A commonly used requirement is that the maximum permissible change in MOSFET threshold voltage ΔV_T should be 10 mV; this specification roughly corresponds to a 3% degradation in the MOSFET drain current, which might be the maximum change that a circuit can tolerate and still function.

To design a degradation test, we first identify a quantity that is correlated with the stress degradation and develop a method of measuring this accumulated-stress "monitor." A possible monitor quantity is the substrate current I_{sub}. The value of the quantity monitored during the test and the elapsed time indicate the degree of accumulated stress in the MOSFET. Another quantity, such as the change of threshold voltage, determines "failure" of the device.

The bias conditions for the stress test can be the same as those expected to be used during operation of the final circuit or else the test can be performed in an "overstress" state, for example by using a higher value of the system drain power-supply voltage V_{DD}. The gate voltage V_G is adjusted to produce the maximum substrate current I_{sub} so that the degradation is measured under the worst-case conditions. The device is then operated with these bias voltages for an extended period of time with I_{sub} being continuously monitored. The test is periodically interrupted to measure the device parameter selected to indicate failure (e.g., ΔV_T or ΔI_D). The stress test continues until the failure criterion is reached.

The time τ to reach failure is then called the "hot-carrier-device lifetime," "the MOSFET time-to-failure," or simply the "MOS lifetime."

Accelerated Testing of Device Lifetime

Devices are often designed to operate reliably under normal operating conditions for at least 10 years. Having to wait for 10 years to verify the reliability of a device is obviously not practical. Instead, we need to find ways to accelerate the stress test so that we can obtain the time-to-failure τ in a reasonable time. As we just pointed out, one technique is to carry out the stress test at higher biases than are normally used during system operation; typically the drain voltage is increased to accelerate the life test. The monitor quantity (e.g., I_{sub}) is correspondingly higher than observed under normal bias, so τ is shorter.

The time-to-failure τ and I_{sub} are related through a power-law function of the general form [19]

$$\tau \approx K_1\left(\frac{I_{sub}}{I_D}\right)^{-m} \tag{10.4.1}$$

where K_1 and m are determined empirically. The MOS lifetime τ is first measured at several high values of bias voltage. The measured high-stress lifetimes are then extrapolated using Equation 10.4.1 to predict τ under normal operating conditions. Because I_D is proportional to the channel width W, it is useful to evaluate the reliability of an MOS technology by measuring MOSFETs having various channel widths and using an alternative expression to express τ as a function of W [19]:

$$\tau \approx K_2\left(\frac{I_{sub}}{W}\right)^{-m} \tag{10.4.2}$$

Some typical values for the parameters in Equations 10.4.1 and 10.4.2 are $m = 3$, $K_2 = 10^6$, and $K_1 = 3$, respectively. For τ in seconds, I_{sub}/W has units of mA/μm. Figure 10.20 is a typical plot of τ as a function of I_{sub}, showing the validity of the power-law relation between the two quantities [26].

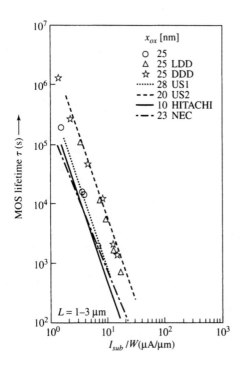

FIGURE 10.20 Device lifetimes τ as a function of substrate current I_{sub} for several different technologies, showing the inverse power-law relation between these quantities [26] (© 1987 IEEE).

EXAMPLE *MOSFET Time-to-Failure*

For a MOSFET with the parameters given below, calculate the time-to-failure τ for operation at $V_{DD} = 5$ V. If the device is to be used in a system that must operate reliably for 10 years, what is the maximum drain-supply voltage that can be used?

Device parameters are $x_{ox} = 20$ nm, $x_j = 0.2$ μm, $V_{Dsat} = 1$ V, $V_T = 0.7$ V, channel length $L = 1$ μm. Use the failure criteria $\Delta I_D / I_D = 10\%$ with $K_1 = 10$ and $m = 3$.

Solution Using Equation 10.1.16,

$$\ell = 0.22 x_{ox}^{1/3} x_j^{1/2} = 0.124 \ \mu m$$

Then from Equation 10.1.15,

$$\mathcal{E}_m \approx \frac{V_D - V_{Dsat}}{\ell} \approx 3.23 \times 10^5 \ \mathrm{Vcm}^{-1}$$

From Equation 10.2.6, and the associated value of A_i / B_i,

$$\frac{I_{sub}}{I_D} = 1.2(V_D - V_{Dsat}) \exp\left(-\frac{B}{\mathcal{E}_m}\right) = 0.025$$

Finally, using Equation 10.4.1,

$$\tau = K_1 \left(\frac{I_{sub}}{I_D}\right)^{-3} = 7.4 \ \text{days}$$

Note that in this problem, L and V_G are not explicitly utilized because V_{Dsat} is specified for the given biasing condition.

For a lifetime of 10 years, $\tau = 3.15 \times 10^8$ s. Therefore, we need

$$\frac{I_{sub}}{I_D} = 3.17 \times 10^{-3} \Rightarrow \mathcal{E}_m < 2.4 \times 10^5 \ \mathrm{Vcm}^{-1}$$

The corresponding value of $V_{DD(max)} \approx 4.22$ V ∎

Structures that Reduce the Drain Field

The preceding example shows that 5 V operation would result in an unacceptably short time-to-failure for the parameters of the 1 μm CMOS technology considered. Reducing the power-supply voltage is one way to increase τ, but reducing V_{DD} lowers the available gate voltage V_G and therefore the maximum drive current I_{Dsat}, degrading circuit speed and system performance. An alternative and usually better solution is to retain the higher power-supply voltage, but to reduce the maximum field $\mathcal{E}_m$ by redesigning the MOSFET circuit so that the excess drain voltage $V_D - V_{Dsat}$ is dropped somewhere other than directly across the velocity-saturated region. This redesign should not affect the current drive appreciably because I_{Dsat} depends only weakly on V_D. A simple method to limit the voltage across the channel is to add a resistor at the drain; however, this method is not very effective because the amount of voltage that can be dropped across a resistor without degrading the circuit performance is small. A more effective means of limiting the maximum field in the channel is to fabricate a lightly doped (n^-) buffer region between the heavily doped drain and the channel within the MOSFET itself [27].

The lightly doped buffer region can be incorporated in MOSFETs by using a number of different structures. The most widely adopted technology to form the buffering n^- region is the lightly doped drain (LDD) technology illustrated in Figure 10.21. To form the lightly doped drain, the gate electrode (usually made from polysilicon) is first patterned

LDD *S/D* implant

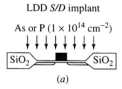

(a)

Deposit SiO$_2$

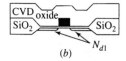

(b)

Heavy *S/D* implant

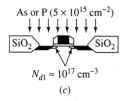

(c)

FIGURE 10.21 Process used to form the lightly doped drain (LDD) structure: (*a*) implant moderate dose of *n*-type species, (*b*) deposit CVD oxide, (*c*) anisotropically etch CVD oxide to form "sidewall spacer" and then implant high dose of *n*-type species.

and etched. Then, a moderate amount of *n*-type dopant (density N_{d1}) is added to the source and drain regions by implantation, using the edge of the polysilicon to position one side of the implanted regions. Afterward, an oxide *sidewall spacer* is created to cover the n^- regions adjacent to the edges of the gate, followed by a heavy n^+ implant to form the normal heavily doped regions of the source and drain electrodes. The n^- region under the spacer remains lightly doped to drop the unwanted, excess drain voltage [28].

In an optimized lightly doped drain structure, the n^- LDD region should be totally depleted, creating a positive space-charge density equal to qN_{d1}. This added charge modifies the total charge in the velocity-saturation region, and Equation 10.1.7 can then be rewritten as

$$\frac{d\mathscr{E}(y')}{dy'} = \frac{[V(y') - V_{D\text{sat}}]}{\ell^2} - \frac{qN_{d1}(y')}{\epsilon_s} \tag{10.4.3}$$

As we can see from Equation 10.4.3, the LDD structure decreases the maximum field $\mathscr{E}_m$ in the channel region near the drain. The improvement is seen in the sketch shown in Figure 10.22 of the fields with and without a lightly doped drain region. To effectively drop the excess voltage in the LDD region, control of the doping concentration of the n^- region is critical. If the doping is too low, the series resistance is excessive and limits

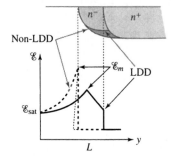

FIGURE 10.22 Electric-field distributions for an *n*–channel MOSFET with (solid lines) and without (dashed lines) a lightly doped drain (LDD) structure.

$I_{D\text{sat}}$; if the doping is too high, the LDD region is not completely depleted, and the fields are not significantly lower than those obtained without the LDD process. A useful design goal is to obtain a constant field in the LDD region. The maximum field $\mathscr{E}_m$ in the channel under this condition is found from the expression

$$\mathscr{E}_m \approx \frac{V_D - V_{D\text{sat}} - \mathscr{E}_m L_{n^-}}{\ell}$$

$$= \frac{V_D - V_{D\text{sat}}}{\ell + L_{n^-}} \tag{10.4.4}$$

where L_{n^-} is the length of the lightly doped drain region. To fabricate MOSFETS meeting this design target, doping in the LDD region must be very precisely controlled.

Another structure that forms a buffering n^- region to reduce the maximum drain field is the *double-diffused-drain (DDD)* structure. To form the double-diffused drain, two different dopant species (phosphorus and arsenic) are co-implanted into the drain-contact region using a medium phosphorus dose (e.g., 1×10^{14} to 1×10^{15} cm^{-2}) and a heavy arsenic dose (e.g., 5×10^{15} cm^{-2}) as shown in Figure 10.23. During subsequent high-temperature annealing steps, the faster-diffusing phosphorus is driven farther under the gate edge than is the arsenic, creating a less-abrupt concentration gradient in the drain junction. [Under usual processing conditions, the source also contains the double diffusion. While the gradual transition is not needed at the source junction to limit the field, added process complexity (i.e., an extra photomasking step at added cost) would be needed to block the second dopant implant from the source region. This extra complexity is typically not justified.] While forming the double-diffused drain is simple, it allows very little flexibility in adjusting the length of the n^- region and creates a deep junction which is detrimental for short-channel devices. Therefore, the DDD structure is not widely used in modern technology.

p–Channel MOSFET Degradation

Hot-carrier effects in *p*–channel MOSFETs are not as serious as those in *n*–channel MOSFETs because the rate of impact ionization by holes is usually one-to-two orders-of-magnitude lower than the rate of ionization by electrons. A lightly doped drain structure is often not necessary for *p*–channel MOSFETs (although it may be used in a CMOS circuit for fabrication-process compatibility with the *n*–channel MOSFET). Also, in contrast to the behavior of *n*–channel MOSFETs after extended operation, drain currents of *p*–channel MOSFETs often increase rather than decrease.

The mechanism responsible for degradation of *p*–channel MOSFETs can be qualitatively explained as follows [29]. At high-channel fields, holes are accelerated sufficiently to cause impact ionization, thereby generating electron-hole pairs. If some of the generated electrons gain sufficient energy from the channel field and are also scattered in the correct direction they can be injected into the gate oxide. Some of these electrons become trapped in the oxide where they accumulate especially near the drain (Sec. 10.3).

The major impact of hot-electron trapping in *p*–channel MOSFETs is the negative-charge buildup in the oxide near the drain. This charge eventually inverts the *n*–type silicon surface near the drain, extending the p^+ drain region and effectively decreasing the channel length. The shorter effective channel length increases the drain current, an effect that is

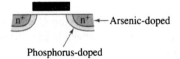
Arsenic-doped

Phosphorus-doped

FIGURE 10.23   A less-abrupt junction is obtained in the double-diffused-drain (DDD) structure by implanting two dopant species that diffuse at different rates.

generally less harmful to circuit performance than a decrease of drain current. However, shortening of the channel length can lead to higher leakage current or punchthrough (Sec. 9.2). Channel shortening also increases the maximum channel field, making the mean-time-to-failure of p–channel MOSFETs more sensitive to channel-length scaling.

10.5 DEVICES: MOS NONVOLATILE MEMORY STRUCTURES

Thus far, we have focused on problems caused by hot-electron effects in MOSFETs—chiefly their influence on the degradation of performance. There are, however, important commercial devices that depend on mechanisms associated with hot carriers for their basic operation. The chief use of these devices is in the area of nonvolatile memory—that is, memory elements that retain information even without applied bias voltages. Many different types of nonvolatile memories exist including floating-gate memories that store charges on an electrically isolated electrode, metal-nitride-oxide-silicon (MNOS) devices which operate by trapping charge near an interface, and ferroelectric random-access (FRAM) memories.

We focus our attention on floating-gate memory structures that depend on the high-field effects we have been discussing in this chapter. We consider several different types of memories in two categories: *erasable, programmable, read-only memories* (EPROMs or UV EPROMs) that are written electrically and erased by ultraviolet light, and *electrically erasable, programmable* ROMs (EEPROMs or E²PROMs) that are both programmed and erased electrically. In conventional E²PROMs, any cell (bit) can be accessed at random for reading or writing. FLASH E²PROMs employ the same basic structure for the memory element, but the arrays are organized differently from those of conventional E²PROMs. To achieve high density in FLASH E²PROMs, an individual memory device cannot be accessed directly. A FLASH memory is erased or cleared in groups of devices (called "blocks"); information is subsequently written sequentially into groups of bits within the block by changing the state of some of the bits to that opposite the erased state.

The structure of a floating-gate memory is shown in Figure 10.24a. This device is essentially an n–channel MOSFET with two gates—a control gate and a floating gate. The control gate is connected to an external circuit, but the floating gate has no external connection. In our discussion of the MOS structure in Chapter 8, we noted that an MOS system usually reaches a charge state corresponding to thermal equilibrium if enough time is allowed because of explicit or parasitic paths for charge transport. The floating gate eventually reaches a charge condition determined by the work functions of the materials in the system. However, because of the high barrier height of 3.1 eV for transfer of an electron from the conduction band of the floating gate (usually polysilicon) to the conduction band of the surrounding oxide and the high resistance of the oxide, any excess

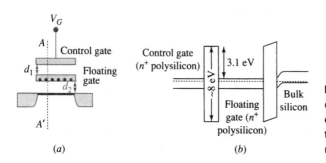

(a)   (b)

FIGURE 10.24 (a) Structure of a floating-gate memory element. (b) Band diagram through section AA' of the memory element.

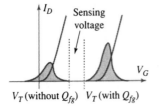

FIGURE 10.25   Drain current vs. gate voltage for a floating-gate memory cell. The shaded regions represent the variations in threshold voltage for measurements without any stored gate charge ($Q_{fg} = 0$), and for negative charge $-|Q_{fg}|$ stored on the gate.

electrons in the floating gate escape only very slowly. The device can remain in a *metastable* state of nonthermal equilibrium for decades; it can therefore retain charge for a useful period of time and be used as a memory element. The floating-gate element shown in Figure 10.24 contains two insulating films. One (having thickness d_2) is a thermally grown oxide above the channel region in the substrate, and the other (having thickness d_1, typically substantially larger than d_2) is a high-quality deposited oxide. The overlying control gate in Figure 10.24 is usually a second layer of deposited polysilicon.

The threshold voltage for the gate electrode to induce mobile charge in the substrate channel depends on the amount of charge $-|Q_{fg}|$ stored on the floating gate. An equation for this threshold voltage is

$$V_T = V_{FB} + 2|\phi_p| + \frac{|Q_d|}{\epsilon_{ox}}(d_1 + d_2) + \frac{|Q_{fg}|}{\epsilon_{ox}}d_1 \qquad (10.5.1)$$

The band diagram along section A-A' in Figure 10.24a is shown in Figure 10.24b. Figure 10.25 is a sketch showing that negative charge $-|Q_{fg}|$ stored in the floating gate increases the memory threshold voltage and thereby displaces a measured plot of I_D versus V_G to more positive gate voltages. Typical process variations lead to a distribution of values for V_T, as shown on the figure; for clarity, the corresponding variations in I_D with threshold voltage are not shown on Figure 10.25. Figure 10.25 illustrates the two memory states that can be stored by the cell based on the presence or absence of the stored floating-gate charge Q_{fg}. A gate voltage with a value between the threshold voltages of the two states therefore causes the cell to conduct unless its floating gate has been charged with a sufficient amount of negative charge $-|Q_{fg}|$. Because the two overlying electrodes make this floating-gate memory structure different from the basic MOSFET, calculating the drain current is complicated. Quantitatively estimating the drain current requires more than considering that Q_{fg} alters the threshold and then using the modified value of the threshold voltage from Equation 10.5.1 in the basic MOSFET equations. To treat the device accurately, the voltage on the floating gate must first be calculated using the capacitive network shown in Figure 10.26, and then using this voltage as the gate voltage of an n–channel MOSFET with the thin oxide d_2. The threshold voltage used to calculate the drain current is the voltage on the floating gate that induces a channel in the underlying silicon. Finding the drain current quantitatively is not trivial; however, in a nonvolatile memory device, we primarily want to know whether the device is conducting or not and are less concerned with predicting the exact value of the drain current.

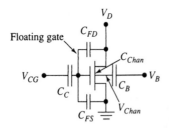

FIGURE 10.26   Capacitor model used to calculate the floating-gate voltage.

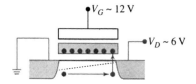

$V_G \sim 12$ V

$V_D \sim 6$ V

FIGURE 10.27 Schematic diagram for hot-electron programming of a floating-gate memory cell.

Programming Floating-Gate Memory Cells

While reading (i.e., determining the information state that is stored on the floating-gate memory cell) is straightforward, writing (transferring electrons either onto or off of the floating gate) is more challenging. Writing, or adding electrons to the floating gate, is usually called *programming*; the process of removing excess electrons from the floating gate is called *erasing*. Taken together, *reading* or *sensing*, *writing* and *erasing* are the primary operations of a floating-gate memory. Unlike reading, both programming and erasing typically require much higher voltages than the power supply voltage V_{DD} connected to the chip. These higher voltages are usually generated by on-chip, *charge-pump* circuits that increase the voltage beyond V_{DD} [30].

The most important mechanism for programming a floating-gate memory array, such as a FLASH E²PROM, is the channel hot-electron injection described in Sec. 10.3. A sketch illustrating the process of electron injection near the drain end of the channel is shown in Figure 10.27. The programming current is described by the lucky-electron model and given in Equation 10.3.1. Design of the drain-side injection, channel-hot-electron (CHE) cell involves several tradeoffs. For efficient programming, large gate currents are desirable, so the lateral electric field should be maximized by decreasing the channel length. However, we need a minimum channel length to prevent a device on a nearby, unselected row of a memory array from punching through when programming voltages are applied. Several factors limit the efficiency of the floating-gate memory device. First, hot electrons are only a minor component of the channel current; most channel electrons are not sufficiently energetic to surmount the barrier at the interface. Therefore, much of the power supplied during programming is wasted. In addition, as discussed in Sec. 10.3, there are conflicting requirements on the optimum voltage for programming. To have the high channel field needed for efficient hot-electron creation, the floating gate should be at a low voltage. However, with a low floating-gate voltage, the field is primarily directed from the channel to the floating gate, repelling hot electrons from the floating gate; thus, a high floating-gate voltage is needed for efficient collection. Because of the conflicting requirements, the typical injection efficiency (the ratio of injected electrons to total channel electrons) is only about 10^{-6}.

EXAMPLE   *Flash EPROM Programming Current*

An *n*–channel FLASH EPROM cell is programmed by hot-carrier gate current. The device parameters are $V_T = 0.7$ V, $x_j = 0.2$ μm, $W = 100$ μm, $L = 0.5$ μm, and $(V_{FB} + 2|\phi_p|) = -0.2$ V. For simplicity, treat the device as a MOSFET with an extra capacitor in series with the normal gate capacitor. That is, assume that the floating-gate voltage is given by the voltage divider composed of C_{GS} and the control-to-floating-gate capacitor. Assume that the floating gate does not carry any charge initially and estimate the programming current immediately after the MOSFET is biased by applied voltages $V_{GS} = 10$ V and $V_{DS} = 5$ V.

$x_{ox} = 10$ nm

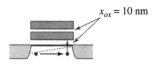

Solution   From Equation 10.5.1, V_T measured from the control gate is

$$V_T = V_{FB} + 2|\phi_p| + \frac{|Q_d|}{\epsilon_{ox}}(d_1 + d_2) + \frac{|Q_{fg}|}{\epsilon_{ox}}d_1 = 0.7 \text{ V}$$

with the values given above for $V_{FB} + 2|\phi_p|$, d_1, d_2, and Q_{fg}, we have

$$\frac{|Q_d|d_1}{\epsilon_{ox}} = 0.45 \text{ V}$$

Therefore, the threshold voltage V_{TF} measured from the floating gate is

$$V_{TF} = V_{FB} + 2|\phi_p| + \frac{|Q_d|d_1}{\epsilon_{ox}} = 0.25 \text{ V}$$

For hot-electron programming, the MOSFET is biased in the saturation region. Using the model of Sec. 9.1, $C_{GS} = (2/3)C_{ox}$. Assuming a simple voltage divider, we find the voltage at the floating gate to be

$$V_{GF} = 10\frac{C_{ox}}{C_{ox} + (2/3)C_{ox}} = 6 \text{ V}$$

We calculate the drain current as in Sec. 9.2:

$$\mathscr{E}_{eff} \approx \frac{V_G - V_T}{6x_{ox}} + \frac{V_T + 0.5}{3x_{ox}} \approx 1.21 \times 10^6 \text{ V cm}^{-1}$$

$$\mu_{eff} \approx \frac{\mu_0}{1 + (\mathscr{E}_{eff}/\mathscr{E}_0)^\nu} = 187 \text{ cm}^2 \text{ V}^{-1}\text{s}^{-1}$$

$$V_{Dsat} = \frac{(V_G - V_T)\mathscr{E}_{sat}L}{(V_G - V_T) + \mathscr{E}_{sat}L} = 2.45 \text{ V}$$

$$\mathscr{E}_{sat} = \frac{2v_{sat}}{\mu_{eff}} = 8.55 \times 10^4 \text{ V cm}^{-1}$$

$$\therefore I_{Dsat} = WC_{ox}(V_G - V_T - V_{Dsat})v_{sat} = 23 \text{ mA}$$

We can also calculate $\mathscr{E}_m$ using

$$\ell \approx 0.22x_{ox}^{1/3} x_j^{1/2} \approx 0.0984 \text{ }\mu\text{m}$$

$$\mathscr{E}_m \approx \frac{V_D - V_{Dsat}}{\ell} \approx 2.59 \times 10^5 \text{ V cm}^{-1}$$

The charging current just after the bias is applied is found by using Equation 10.3.2:

$$I_G \approx CI_{Dsat}\exp\left(-\frac{\phi_B}{q\mathscr{E}_m\lambda}\right) \approx 2 \times 10^{-3} \times 23 \text{ mA} \times \exp\left(\frac{-3.32}{2.59 \times 10^5 \times 7.3 \times 10^{-7}}\right) \approx 1.1 \text{ pA} ■$$

An approach to channel hot-electron injection that significantly improves injection efficiency is *source-side injection*. To use this technique, a small, floating *injection gate* is added adjacent to the source side of the main floating gate, as illustrated schematically in Figure 10.28 [31]. When a high voltage is applied on the control gate, the main floating gate is capacitively biased by the control gate to a high-positive voltage, and the majority of the

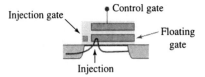

FIGURE 10.28   Schematic diagram of structure used for source-side injection.

channel is highly conductive. However, because of the geometry, the injection gate is capacitively biased only slightly above the threshold voltage. This low effective bias limits the channel current under the injection gate and leads to a high-field region at the end of the injection gate nearest the drain. Electrons flowing in the channel under the injection gate are accelerated in the high electric field and are hot when they reach the drain end of the injection gate and enter the channel region under the floating gate. The high positive voltage applied to the floating gate favors the collection of electrons but has little effect on the fields in the channel under the injection gate. As a result, the conflicting requirements described above are both satisfied. The typical injection efficiency of source-side injection is approximately 10^{-3}.

Erasing Floating-Gate Memory Cells

Erasing refers to the process of removing electrons from the floating gate. In erasable, programmable ROMs (EPROMs), erasing is accomplished by using ultraviolet light to excite the electrons in the floating gate, with all device electrodes grounded. The repulsion between nearby electrons drives most of the electrons out of the floating gate. While this mechanism is simple, it is very slow and all the cells on the chip are simultaneously erased. In addition, the chip generally must be moved from the circuit board to a different location where it can be illuminated rather than erasing it on the circuit board by electrical means. Because of their limited flexibility, UV-erasable PROMs are less popular than other types of programmable memory devices.

　　To permit electrical erasing, a thinner oxide is usually formed to allow Fowler-Nordheim tunneling, the main mechanism employed to erase most E^{2}PROMs. Fowler-Nordheim tunneling is a combination of thermal excitation part way to the top of a barrier and subsequent tunneling through a thin part of the barrier as shown in Figure 10.29a. As indicated in Figure 10.29b, Fowler-Nordheim tunneling current can flow between the floating gate and the source, drain, body, or even the control gate, depending on the bias and the device structure. For example, when a voltage of approximately -10 V is applied to the control gate and about $+5$ V is applied to the p-type substrate or $body$ with the source and drain both floating, a vertical field is created between the floating gate and the holes accumulated in the substrate at the oxide–silicon interface. Electrons can then tunnel from the floating gate to the p-type substrate. Using a negative control-gate bias during erasing reduces the voltage needed for the substrate bias to a value lower than V_{DD}, so the substrate can be connected to the power supply rather than to the output of a charge-pump circuit. This is very important because the tunneling-electron current can be as large as 1 nA/cell. With over 10^6 cells on a chip, the current can be in the mA range and generating this magnitude of on-chip, charge-pump current is difficult. On the other hand, virtually no current flows to the control gate, and a high voltage can be applied from the charge-pump circuit.

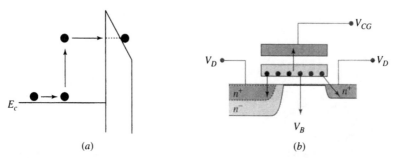

(a)　　　　　　　　　　　　　　　　　　(b)

FIGURE 10.29 (*a*) Fowler-Nordheim tunneling mechanism. (*b*) Various possible Fowler-Nordheim tunneling paths for erasing a memory cell.

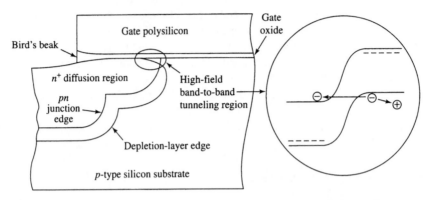

FIGURE 10.30 Cross section showing representative features during erasing by charge transfer from the diffused-source region extending under the edge of the gate [32] (© 2000, John Wiley and Sons, reprinted by permission).

Erasing to the source is another option for removing electrons from the floating gate and this erase mechanism is often used. In this case, a bias of approximately positive 5 V is applied to the source; approximately negative 10 V is applied to the control gate; and the drain is allowed to float. With these biases, the electric field in the vicinity of the source is very nonuniform as illustrated in Figure 10.30. The high field that is needed to induce tunneling between the n^+ diffused source region and the gate depletes a portion of the n^+ source diffusion and causes the depletion region to curve back under the gate. The electric field across the gate oxide is reduced in the vicinity of this depletion region. On the other hand, the very high field in the silicon in this depletion region leads to *band-to-band tunneling (BBT)*. As shown in the inset of Figure 10.30, band-to-band tunneling can occur when the reverse field across the junction is high enough for electrons in the valence band of the n^+ region to tunnel to the empty states in the conduction band of the p-type region. The holes remaining behind in the n^+ region are accelerated by the electric field, potentially degrading the device performance and reliability. To reduce the impact

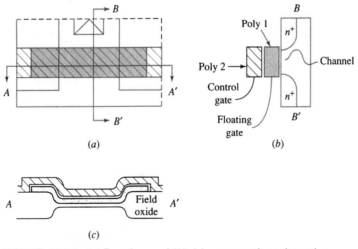

FIGURE 10.31 (a) Top view and (b), (c) cross sections through a "T"-cell floating gate memory [32] (© 2000, John Wiley and Sons, reprinted by permission).

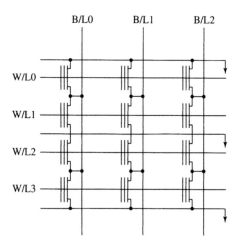

FIGURE 10.32 Schematic diagram of a portion of a T-cell array [32] (© 2000, John Wiley and Sons, reprinted by permission).

of band-to-band tunneling, an n-type region is usually introduced between the n^+ source region and the p-type substrate as indicated in Figure 10.29b; this added region reduces the electric field between the substrate and the source and consequently the amount of band-to-band tunneling.

Floating-Gate Memory Array

To be useful, the nonvolatile memory cells must be assembled into regular arrays to store digital information. There are many variations in the design of nonvolatile cells to meet ever more demanding criteria as MOS technology improves, and it is beyond our reasonable scope to discuss this broad field in any detail.

It is worthwhile, however, to look at the creation of an array using one nonvolatile cell as an example; we consider the "T" cell having the layout and cross sections shown in Figure 10.31. The arrangement of the T-cells into a nonvolatile-memory array is shown in Figure 10.32 where the rows connecting T-cell control gates form the word-lines and the columns connect drains to form the bit lines. The bit-line contacts are shared between adjacent cells and all sources are common. A given cell is programmed if it is situated at the intersection of activated row- and column-lines.

SUMMARY

The typically large electric fields found in submicrometer MOSFETs cause non-negligible *substrate and gate currents* and other effects that are unimportant in long-channel devices. To understand the behavior of the electric field in the channel region of very short-channel MOSFETs, we can use a *pseudo-two-dimensional approximation* to solve Poisson's equation. The pseudo two-dimensional analysis shows that the electric field in the channel increases sharply (nearly exponentially) to a maximum near the drain electrode, where important new physical effects arise. The rapidly varying electric field causes the effective source-drain resistance $(\partial I_D / \partial V_D)^{-1}$ to vary inversely with drain current I_D.

In short n–channel MOSFETs, the electrons carrying the drain current are sometimes sufficiently "heated" by high fields near the drain that impact ionization can occur. The holes generated by the ionizing collisions are accelerated through the depletion region, giving rise to *substrate current* that can disable the MOSFET when the ohmic $I\text{-}R$ drop from the p-region under the source junction to the substrate contact approaches the turn-on voltage of the drain-substrate np junction. The magnitude of the substrate current depends strongly on the maximum electric field $\mathscr{E}_m$ in the channel; $\mathscr{E}_m$ is, in turn, a function of the gate-oxide thickness and the drain-junction depth. To reduce substrate current and increase breakdown voltage, both a thicker oxide and deeper drain junction can be used. However, increasing these dimensions conflicts with the short-channel requirements discussed in Chapter 9; a deeper junction shortens the effective channel length, causes more charge sharing, and also makes the device more susceptible to sub-surface punchthrough. All of these effects degrade the performance of a short-channel MOSFET.

Because of their high energies, electrons near the drain can have sufficient energy to surmount the channel-to-oxide barrier, giving rise to gate current. Some of the electrons traversing the oxide can be trapped, degrading the MOSFET characteristics. The main degradation mechanism in n–channel MOSFETs is *interface-state generation* at the oxide–silicon interface near the drain, and the worst bias condition for degradation is high drain voltage V_D and moderate gate voltage V_G, bias values at which the substrate current I_{sub} is maximum. On the other hand, the most serious degradation mechanism in p–channel MOSFETs is *electron trapping* in the oxide, with the worst bias condition being high drain voltage and lower gate voltage. The degradation of p–channel devices results

in effective channel shortening and increased likelihood of punchthrough. Short-channel MOSFETs with improved reliability and reduced susceptibility to high-field effects are produced by adding *lightly doped* regions near the *drain* end of the channel. The performance of these LDD MOSFETS has made them a preferred short-channel design.

A specialized, yet important class of MOSFET-related devices—*floating-gate, nonvolatile memory elements*—is designed so that charge can be stored on a gate that is "floating" (i.e., not directly connected to external leads). Charge can be stored (or both stored and removed in *erasable* devices) on the floating gate to modify the threshold voltage of a MOSFET built in the underlying silicon. To "read" the binary state of the memory MOSFET, a gate voltage greater than the threshold voltage of the non-charged device, but smaller than the threshold voltage of the charged device, is applied. Floating-gate, nonvolatile memory MOSFETs are typically charged by *hot-electron injection* across the oxide–silicon barrier and discharged by *Fowler-Nordheim electron tunneling.*

REFERENCES

1. T. TOYABE et al., *IEEE Trans. Electron Devices* **ED-25**, 825–834 (1978).

2. S. SELBERHERR, A. SCHUTZ, and H. W. POTZL, *IEEE Trans. Electron Devices* **ED-27**, 1540 (1980).

3. M. PINTO, R. DUTTON, H. IWAI, and C. HEGARTY, *Tech. Digest—IEEE Int'l Elect. Devices Mtg.* (1985), p. 288.

4. G. BAUM and H. BENEKING, *IEEE Trans. Electron Devices* **ED-17**, 481–482 (1972).

5. A. POPA, *IEEE Trans. Electron Devices* **ED-19**, 774–780 (1972).

6. M. FUKUMA and Y. OKUTO, *IEEE Trans. Electron Devices* **ED-27**, 2109–2114 (1980).

7. Y. A. EL MANSY and A. R. BOOTHROYD, *IEEE Trans. Electron Devices* **ED-24**, 254–262 (1977).

8. P. K. KO, R. S. MULLER, and C. HU, *Tech. Digest—IEEE Int'l Elect. Devices Mtg.* (1981), p. 600.

9. P. K. KO, "Hot-electron effects in MOSFETs," Doctoral Thesis. Dept of Electrical Engineering and Computer Science, University of California, Berkeley (1982).

10. S. TAM, P. K. KO, C. HU, and R. S. MULLER, *IEEE Trans. Electron Devices* **ED-29**, 1740 (1982).

11. S. TAM, P. K. KO, and C. HU, *IEEE Trans. Electron Devices* **ED-31**, 1116 (1984).

12. F. C. HSU, P. K. KO, S. TAM, R. S. MULLER, and C. HU, *IEEE Trans. Electron Devices* **ED-29**, 1735 (1982).

13. F. C. HSU, *IEEE Trans. Electron Devices* **ED-30**, 571–576 (1983).

14. K. MAYARAM, K. LEE, and C. HU, *IEEE Trans. Electron Devices* **ED-34**, 1509 (1987).

15. K. W. TERRILL, C. HU, and P. K. KO, *IEEE Electron Device Lett.* **EDL-5**, 440 (1984).

16. C. G. SODINI, P. K. KO, and S. S. WONG, *Tech. Papers, IEEE VLSI Circuits Symp.*, 1987, pp. 57–61.

17. Y. A. EL MANSY and D. M. CAUGHEY, *Tech. Digest—IEEE Int'l Elect. Devices Mtg.* (1975), pp. 31–34.

18. T. Y. CHAN, P. K. KO, and C. HU, *IEEE Electron Device Lett.* **EDL-5**, 505–507 (1984).

19. C. HU, "Hot-Carrier Effects," in *Advanced MOS Device Physics*, N. G. Einspruch and G. Gildenblat, Eds., Vol. 18, VLSI Electronics Microstructure Science, Academic Press, New York, 1989.

20. J. H. HUANG, Z. H. LIU, M. C. JENG, P. K. KO, and C. HU, *Tech. Digest—IEEE Int'l Elect. Devices Mtg.* (1992), pp. 569–572.

21. E. TAKEDA, H. KUME, T. TOYABE, and S. ASAI, *IEEE Trans. Electron Devices* **ED-29**, 611 (April 1982).

22. C. N. BERGLUND and R. J. POWELL, *J. Appl. Phys.* **42**, 573–579 (1971).

23. E. TAKEDA, N. SUZUKI, and T. HAGIWARA, *Tech. Digest—IEEE Int'l Elect. Devices Mtg.* (1982), pp. 396–399.

24. K. K. NG and G. W. TAYLOR, *IEEE Trans. Electron Devices* **ED-30**, 871–876 (August 1983).

25. F. H. HSU and K. Y. CHIU, *Tech. Digest—IEEE Int'l Elect. Devices Mtg.* (1984), pp. 96–99.

26. J. Y. CHOI, P. K. KO, and C. HU, *Digest Tech. Papers, IEEE Symp. VLSI Tech.* 1987, pp. 45–46.

27. S. OGURA et al., *IEEE Trans. Electron Dev.* **ED-27**, 1359 (1980).

28. P. J. Tsang et al., *IEEE Trans. Electron Dev.* **ED-29**, 590 (April 1982).

29. M. Koyanagi, A. Lewis, R. Martin, T. Huang, and J. Chen, *Tech. Digest*—IEEE Int'l Elect. Devices Mtg. (1986), pp. 722–725.

30. T. Tanzawa, T. Tanaka, T. Tanaka, H. Nakamura, H. Oodaira, K. Sakui, M. Momodomi, S. Shiratake, H. Nakano, Y. Oowaki, S. Watanabe, K. Ohuchi, and

F. Masuoka, *Tech. Papers*, IEEE Symp. VLSI Circuits, 1994, pp. 65–66.

31. A. T. Wu, T. Y. Chan, P. K. Ko, and C. Hu, *Tech. Digest*— IEEE Int'l Elect. Devices Mtg. (1986), pp. 584–587.

32. J. Caywood, and G. Derbenwich, "Nonvolatile Memory," Chapter 8 in *ULSI Devices*, Edited by C. Y. Chang and S. M. Sze, Wiley Interscience, 2000.

PROBLEMS

10.1 Show that the output resistance of a MOSFET in saturation, when $(V_D - V_{Dsat})/\ell \gg \mathcal{E}_{sat}$, is approximately given by

$$R_{out} = \frac{dV_D}{dI_D} = \left(L + \frac{V_G - V_T}{\mathcal{E}_{sat}}\right)\left(\frac{\mathcal{E}_m}{I_{Dsat}}\right)$$

10.2 Plot g_{msat}, R_{out}, and $g_{msat} R_{out}$ for an n–channel MOSFET with $L = 1$ μm as a function of I_{Dsat} on a log–log graph. Confine your plot to $V_G - V_T > 0.1$ V. The other device parameters are $x_{ox} = 20$ nm, $x_j = 0.3$ μm, and $W = 100$ μm.

10.3 The ratio I_{sub}/I_D under hot-carrier-induced breakdown usually falls within the range 0.05 to 0.2. For this problem, we assume that it is equal to 0.05 and is device independent.

(a) A 1-μm CMOS technology has the following device and process parameters: $x_{ox} = 20$ nm, $x_j = 0.3$ μm, $V_{Tn} = 0.7$ V, and $V_{Tp} = -0.7$ V. Derive an expression for the hot-electron–induced breakdown voltage V_{BD} of the n–channel MOSFET as a function of L and $V_G - V_T$. Plot log (V_{BD}) as a function of log (L) from $L = 0.1$ μm to $L = 100$ μm for $V_G - V_T = 2$ V. Use $A_i = 2 \times 10^6$ cm^{-1} and $B_i = 1.7 \times 10^6$ V/cm in Equation 10.2.6.

(b) Repeat part (a) for a p–channel MOSFET. Use $A_i = 8 \times 10^6$ cm^{-1} and $B_i = 3.7 \times 10^6$ V/cm.

10.4* Use the following parameters for an n–channel MOSFET: $L_{eff} = 1$ μm, $x_j = 0.3$ μm, $x_{ox} = 20$ nm, and $V_T = 0.7$ V. Based on device degradation results, it was recommended that n–channel MOSFETs should not be operated with $\mathcal{E}_m > 2 \times 10^5$ V/cm.

(a) What is the ratio of I_{sub}/I_D at this value of $\mathcal{E}_m$? Determine the recommended maximum supply voltage V_{DD}. Assume that $V_G = 3$ V, and that this value of V_G produces the maximum substrate current I_{sub}.

(b) What is the breakdown voltage of the device with gate voltage equal to 3 V? Assume that the device breaks down at $\mathcal{E}_m = 4 \times 10^5$ V/cm.

10.5 Assuming that $\mathcal{E}_{sat}L \gg V_G - V_T$, derive the expressions for the magnitude and location of the peak in the ln (I_{sub}) versus V_G plot. To further simplify the calculation, assume that V_D mainly affects Equation

10.2.6 through the exponential term and that the linear term can be neglected; that is, Equation 10.2.6 can be approximated by

$$I_{sub} \approx \frac{A_i}{B_i} I_{Dsat} \exp\left(-\frac{\ell B_i}{V_D - V_{Dsat}}\right)$$

10.6* With short-channel effects and DIBL, it can be shown that $V_T \approx V_{T0} - 2V_D e^{-L/\ell}$ where V_{T0} is the long-channel V_T. It is also known that in the ON-state, $\mathcal{E}_m \approx (V_D - V_{Dsat})/\ell$. Assume that $\mathcal{E}_m$ must be kept below 2×10^5 V/cm for hot-carrier reliability and that V_T must decrease by less than 0.2 V for (a) a 0.2-μm decrease in L, and (b) a 20% decrease in L resulting from imperfect process control. Find the maximum acceptable V_D and the corresponding value of ℓ for $L = 2$ μm, 1.2 μm, 0.6 μm, and 0.3 μm. Evaluate V_{Dsat} for $L - 0.2$ μm and $L - 0.2L$ for the worst-case value of L. To simplify the calculation, assume that all approximations mentioned in Problem 10.5 can be applied in this problem.

10.7* One theory for oxide "wearout" is that hot electrons become trapped in the oxide, increasing the field until ultimately the breakdown value of 10 MV cm^{-1} is reached. Consider the example on page 499 in Sec. 10.3 concerned with threshold-voltage shifting by hot electrons and assume that the hot-electron injection described there occurs continuously. The oxide is initially free of charge, and the voltage across it is maintained at 20 V.

(a) How long does it take for breakdown to occur in the oxide?

(b) Sketch the field and voltage versus distance through the oxide, showing the initial condition and the condition at breakdown.

(c) Repeat parts (a) and (b) for a new oxide in which all parameters are the same except that the trapping sites are concentrated 3 nm away from the gate electrode.

10.8* (a) Show that the number of electrons on a MOSFET gate that can cause an oxide to break down is a function only of the gate area and the oxide permittivity.

(b) If a MOSFET having $L = 6$ μm and $W/L = 5$ is charged by static electricity, how many electrons can be transferred to the gate before the oxide field reaches its breakdown value?

(c) How long does it take to deliver these electrons to the gate if the average current is 1 pA?

10.9 For an LDD MOSFET with the gate overlapping the LDD region, qualitatively sketch $\mathcal{E}_y(y)$ in the LDD region. On the same graph, sketch $\mathcal{E}_y(y)$ if the n-region in the LDD structure is replaced by intrinsic (undoped) silicon. Sketch $\mathcal{E}_y(y)$ for a device with the same structure but without the intrinsic region (with the i-region removed). Identify and explain all critical features. (From these plots, you should be able to observe the trend of the maximum field $\mathcal{E}_m$ as a function of doping.)

10.10 Repeat Problem 10.9 if the gate does not overlap the LDD region.

10.11 If we could fabricate an arbitrary lateral doping profile in the n-region, we might want to have $d\mathcal{E}_y/dy = 0$ there so that the electric field does not increase because of the effects explained in Sec. 10.4. Find the doping profile $N_d(y)$ in the n-region that would produce the constant field along the y-direction. (Convince yourself that this is the optimum case.)

10.12 Calculate $\mathcal{E}_m$ and I_{sub} in a non-LDD MOSFET described by the following parameters: $x_{ox} = 20$ nm, $x_j = 0.2$ μm, $W = 50$ μm, $L = 0.5$ μm, $V_T = 0.7$ V when the source is at zero potential, the gate voltage $V_G = 3$ V, and the drain voltage $V_D = 3$ V. How much are the maximum field $\mathcal{E}_m$ and substrate current I_{sub} reduced by adding a 0.1-μm LDD region having an optimum doping profile?

10.13* For a MOSFET with the parameters given in Problem 10.12, calculate the reduction of drain current resulting from the added LDD region at the same bias conditions as given in Problem 10.12. Assume that the LDD region has a resistance of 1 kΩ/square.

10.14* Consider the FLASH memory cell given in the example on page 509 in Sec. 10.5. Estimate the time required to program the device to a threshold voltage V_T that is 1V above the initial state given in the example. (Note that the programming current changes with time. To simplify calculations, use the "average" programming current found by taking the mean of the initial and final currents.)

10.15 Under the programming condition given in Problem 10.14, derive an expression for the programming current as a function of charge (Q_{fg}) injected onto the floating gate. Determine the amount of charge (Q_{fg}) to be injected onto the floating gate to give the maximum charging current. In this problem, assume $V_{FG} - V_T \ll \mathcal{E}_{sat}L$ to simplify the calculation.

ANSWERS TO SELECTED PROBLEMS

The following are numerical solutions to those problems marked with an asterisk in the text. A solutions manual showing calculations is available (to teaching faculty only) by direct request from the publisher.

CHAPTER 1 **1.2** (a) $E_i - E_f = 0.35$ eV, $n = 2.1 \times 10^4$ cm^{-3}, (b) $E_f - E_i = 0.29$ eV, $p = 2.1 \times 10^5$ cm^{-3}. **1.4** (a) 2.3×10^5 cm^{-3}, (b) 4×10^9 cm^{-3}, (c) 9.5×10^{16} cm^{-3}. **1.6** 160 cm^2 V^{-1} s^{-1}. **1.8** (a) 207 collisions, (b) 162 mV. **1.10** -825.6 A cm^{-2}. **1.14** (a) 1.85 μm (IR), (b) 1.10 μm (IR), (c) 0.87 μm (near IR), (d) 0.14 μm (UV).

CHAPTER 2 **2.1** (a) 1.36×10^{15} cm^{-3}, (b) 2.21×10^{15} cm^{-3}. **2.4** 1.1 hour. **2.7** 0.73 μm. **2.9** (a) 1.79 μm, (b) 1.88 hours. **2.18** 34.5 ns. **2.20** (a) 147 Ω per square, (b) an added *acceptor* density of approximately 6.5×10^{16} cm^{-3} (implant *dose* of 2.6×10^{13} cm^{-2}).

CHAPTER 3 **3.1** (a) 4.094 eV, 4.261 eV, (b) 0.167 V. **3.5** (a) $C = 0.282$ pF, (b) $V_R = -0.813$ V. **3.7** (a) $N_{dmax} = 2.9 \times 10^{17}$ cm^{-3}, (b) $\rho > 0.04$ Ω cm. **3.16** (a) for n-type doping, a Schottky barrier is obtained if $N_d > 7.45 \times 10^{11}$ cm^{-3}; this is smaller than any practical IC doping level, (b) since $\Phi_M - X_S \approx E_g/2q$, any level of p-doping will result in a blocking contact.

CHAPTER 4 **4.1** (a) $\phi_i = 0.72$ V, $x_d = 0.97$ μm, $\mathcal{E}_{max} = 1.48 \times 10^4$ V cm^{-1} (at 0 V), (b) $x_d = 3.73$ μm, $\mathcal{E}_{max} = 5.75 \times 10^4$ V cm^{-1} (at -10 V). **4.3** $\mathcal{E}_{max} = 648$ V cm^{-1}; for abrupt junction $\mathcal{E}_{max} = 1.53 \times 10^4$ V cm^{-1}. **4.8** (a) $N_{do} = 6.1 \times 10^{17}$ cm^{-3}, (b) $a = 1.83 \times 10^{21}$ cm^{-4}, (c) $\mathcal{E}_{max} = 3.4 \times 10^4$ V cm^{-1}. **4.10** (a i) 0.33 eV, (a ii) 0.36 eV, (b i) 0.62 V, (c) total depletion possible at $V_R = 23.5$ V.

CHAPTER 5 **5.1** (a) 7.25×10^{16} cm^{-3} s^{-1}, (b) 2.6×10^{11} cm^{-3} s^{-1}. **5.2** (a) $n = 10^{16}$ cm^{-3}, $p = 10^{13}$ cm^{-3}, (b) 10 ns. **5.9** $\gamma = 1.55 \times 10^{-4}$.

5.11 (a) 0.738 V, (b) $n_p(-x_p) = 1.37 \times 10^{14}$ cm^{-3}, $p_n(x_n) = 6.83 \times 10^{13}$ cm^{-3}, (c) $J_T = 0.472$ A cm^{-2}, (d) currents are equal at 1.35 μm from the physical junction (in p-type region).

5.21

	V_a (V)	r_d (Ω)	Capacitance (pF)
(a)	0.1	2.95×10^{10}	3.36
	0.5	6.14×10^3	6.68
	0.7	2.8	3580
(b)	0	1.38×10^{12}	3.15
	-5	5×10^{95} *	1.18
	-20	10^{346} *	0.63

* other mechanisms actually limit the resistance

CHAPTER 6 **6.1** 9.9×10^{14} cm^{-3}. **6.5** (a) 1.346 μC cm^{-2}, (b) 0.84 μm. **6.8** $\phi_i = 0.91$ V, $K_A = 0.163$ pC V$^{-1/2}$, Q_{VE} (-50 V) $= -1$ pC, Q_{VE} (0.3 V) $= +0.028$ pC. **6.12** $V_{CEsat} = 0.048$ V. **6.17** (a) $\gamma = 0.99722$, (b) $\alpha_T = 0.99993$, (c) $\beta_F = 360$, error is 2.57%.

CHAPTER 7 **7.3** $\dfrac{V_A \text{ (constant doping)}}{V_A \text{ (exponential doping)}} = 0.74$. **7.7** $\beta_F = 1.333$. **7.9** $\beta_F = \beta_o[0.75 + 0.25\sqrt{1 + 4I_F/I_K}]^{-1}$ where I_F is the forward-injected current and $I_K = qD_nA_EN_a/x_B$. **7.23** (a) $Q_F = 24$ pC, (b) $Q_F = 81.1$ pC, $Q_R = 156$ pC, (c) charge ratio $= 9.88$.

7.29

	constant doping	exponential doping
g_m	77.5 mS	77.5 mS
C_D	1.94 pF	1.61 pF
δ	2.5×10^{-4}	2.08×10^{-4}
η	1.5×10^{-4}	2.2×10^{-4}

CHAPTER 8 **8.1** (a) $V_{FB} = -0.17$ V, (b) $V_{FB} = -0.90$ V. **8.12** (a) 1.18 V, (b) -2.02 V. **8.15** $C_{MOS}/C_{junction} = 22.8$. **8.16** $N_{st} = 2 \times 10^{10}$ cm^{-2}.

CHAPTER 9 **9.3** (a) $Q_n/q = 4.17 \times 10^{12}$ cm^{-2} (b) $V_{GS} - V_T = 3.82$ V

9.4 (a) at source, $v = 1.26 \times 10^7$ cm/s
at drain, $v = \infty$
(b) at source, $v = 3.68 \times 10^6$ cm/s
at drain, $v = 8 \times 10^6$ cm/s

9.5 (a) 18.7 μA (b) 15.5 μA (c) 12.2 μA
(d) 7.56 μA **9.7** $V_T(0) = 1$ V, $k = 50$ μA/V^2,
$\gamma = 0.3$ V$^{1/2}$, $V_A = 20$ V

9.16

	V_{DD}(V)	$\mathscr{E}_{eff}$ (MV/cm)	μ_{eff} (cm^2/V-s)	$\mathscr{E}_{sat}$ (V/cm)	V_{Dsat} (V)	I_{Dsat} (mA)	g_{msat} (mA/V)
$L = 0.5$ μm							
NMOS	5	0.744	307.0	5.21×10^4	1.62	49.36	15.8
	3.3	0.556	384.6	4.16×10^4	1.16	26.52	14.8
PMOS	5	0.744	92.78	12.9×10^4	2.58	23.76	8.84
	3.3	0.556	112.0	10.7×10^4	1.75	11.7	7.56
$L = 0.02$ μm							Improvement
NMOS	5	0.744	307.0	5.21×10^4	0.102	77.31	56.6%
	3.3	0.556	384.6	4.16×10^4	0.081	46.39	74.9%
PMOS	5	0.744	92.78	12.9×10^4	0.243	56.04	135.9%
	3.3	0.556	112.0	10.7×10^4	0.198	33.18	183.6%

9.21 (a) $V_o = 0.232$ V (b) $V_{TD} = -2.36$ V
(c) 0.223 V

CHAPTER 10 **10.4** (a) $I_{sub}/I_D = 8 \times 10^{-4}$
(b) $V_{DD(max)} = 4.35$ V (c) $V_D = 7.6$ V

10.6

	2 μm		1.2 μm		0.6 μm		0.3 μm	
	(a)	(b)	(a)	(b)	(a)	(b)	(a)	(b)
ℓ (μm)	0.4	0.36	0.25	0.24	0.12	0.14	0.043	0.081
V_D (V)	9.5	8.6	5.9	5.7	2.9	3.3	1.0	1.9

10.7 (a) 2.56 s (c) 23 s **10.8** (b) 2.72×10^7
electrons (c) 4.35 s **10.13** reduced by 1.04%
10.14 0.9 s

SELECTED LIST OF SYMBOLS

Symbol	Definition	Section	Symbol	Definition	Section
$A_{C,E}$	BJT collector, emitter area	6.3	D_{it}	density of interface traps	
A_i	impact ionization			(per area and energy)	8.5
	parameter	4.4, 10.2	$D_{n,p}$	diffusion coefficient for	
B	magnetic field	1.3		electrons, holes	1.2
B	parabolic rate coefficient	2.3	D_s	surface-state density	3.5
B	characteristic field for		E_1	minimum energy for	
	tunneling	4.4		ionizing collision	4.4
B_i	impact ionization		E_a	acceptor energy	1.1
	parameter	4.4, 10.2	E_a	activation energy	2.6
B/A	linear rate coefficient	2.3	E_f	Fermi-level energy	1.1
BV	breakdown voltage	4.4	$E_{fn,fp}$	quasi-Fermi levels for	
C	small-signal capacitance	3.2		electrons, holes	1.1
C_D	diffusion capacitance in		E_g	band-gap energy	1.1
	hybrid-pi model	7.5	E_i	intrinsic Fermi-level energy	1.1
C_{FB}	flat-band capacitance	8.4	E_o	energy of free electron	3.2
C_{GD}	intrinsic gate-drain		E_{st}	energy of surface states	5.2
	capacitance	9.1	$\mathscr{E}$	electric field	1.2
C_{GS}	intrinsic gate-source		$\mathscr{E}_{s0}$	surface electric field	9.2
	capacitance	9.1	$\mathscr{E}_c$	characteristic field for	
C_{RD}	parasitic gate-drain			channel velocity	1.2
	capacitance	9.1	$\mathscr{E}_{eff}$	average x-directed field	
C_{RS}	parasitic gate-source			in MOS channel	9.2
	capacitance	9.1	$\mathscr{E}_l$	characteristic field for	
C_{TD}	drain depletion-region			velocity limitation	7.1
	capacitance	9.1	$\mathscr{E}_s$	surface electric field	8.2
C_{TS}	source depletion-region		$\mathscr{E}_{ox}$	electric field in oxide	8.2
	capacitance	9.1	$\mathscr{E}_m$	maximum electric field	10.1
C_d	diode diffusion capacitance	5.4	$\mathscr{E}_{max}$	maximum electric field	3.2
C_g	impurity concentration in		$\mathscr{E}_{sat}$	field at which carrier	
	gas stream	2.5		velocity saturates	9.2
C_j	diode junction capacitance	5.4	G_A	Auger generation rate	5.2
$C_{jc,je}$	small-signal capacitance		$G_{n,p}$	generation rate for electrons,	
	at BC, BE junctions	7.1		holes	5.1
C_o	concentration of oxidant		G_{sp}	spontaneous generation rate	5.2
	in gas	2.3	GN	Gummel number	6.2
C_{ox}	oxide capacitance	8.3	I_{0E}	emitter-base saturation	
C_p	peak dopant concentration	2.5		current in BJT	6.4
C_s	impurity concentration at		$I_{C0,CB0}$	collector current	
	solid surface	2.6		(emitter open-circuited)	6.4
C_{sHF}	high-frequency silicon		I_{CE0}	collector current (base	
	surface capacitance	8.5		open-circuited)	6.4
C_{sLF}	low-frequency silicon		I_D	drain current	9.1
	surface capacitance	8.5	I_{DD}	drain power supply current	9.3
C^*	equilibrium gas-phase		I_F	field-induced junction	
	oxidant concentration	2.3		current	8.6
D_{eff}	effective diffusivity	2.5	I_G	gate current	10.3

Symbol	Definition	Section
$I_{KF,KR}$	"knee" currents in Gummel-Poon model	7.7
I_M	metallurgical junction current	8.6
I_S	surface generation current	8.6
I_{pE}	emitter hole current in a BJT	6.2
I_{rB}	base-region recombination current	6.2
I_{Dmax}	maximum value of I_{Dsat}	9.2
I_{Dsat}	MOSFET saturation drain current	4.5, 9.1
I_{sub}	substrate current	10.2
J	current density	1.1
J_0	diode saturation-current density	5.3
J_1	characteristic current for Kirk effect	7.1
J_S	linking-current density (BJT)	6.1
J_g	generation density current	5.3
$J_{n,p}$	electron, hole current density	1.2
J_r	recombination current	5.3
J_t	total current in a diode	5.3
K	scaling factor for MOSFET dimensions	Table 9.1
$K_{1,2}$	failure time parameters	10.4
K_I	ideality factor relating I_{Dsat} to I_{Dmax}	9.2
L	pattern length	2.10
L	FET channel length	4.5, 9.1
L_D	Debye length	3.4
L_{Di}	intrinsic Debye length	4.2
L_{n^-}	length of lightly doped drain (LDD) region	10.4
$L_{p,n}$	diffusion length for holes, electrons	5.3
L'	undepleted length of FET channel	4.5
M	avalanche multiplication factor	4.4
M_i	parameter value from model	9.3
N_{ai}	"box" approximation to implant density	9.1
N_c	effective density of states in the conduction band	1.1
$N_{d,a}$	donor, acceptor atomic density	1.1
N_i	implanted dopant concentration	9.1
N_{st}	area density of surface states	5.2
N_v	effective density of states in the valence band	1.1
N'	area density of dopant (implant dose)	2.5
Q_B	base majority charge	6.1

Symbol	Definition	Section
Q_B	bulk charge (of MOSFET)	9.1
$Q_{F,R}$	forward, reverse charge-control variables	7.3
$Q_{VE,VC}$	emitter, collector stored junction charge	6.3
Q_d	depletion-charge density	8.2
Q_f	fixed interface charge density	8.5
Q_{fg}	charge stored on floating gate	10.5
Q_{it}	interface trapped-charge density	8.5
Q_m	mobile charge density (oxide)	8.5
Q_n	channel free-electron charge	8.2
Q_{nB}	minority charge storage in base (npn BJT)	7.3
Q_{ot}	oxide trapped-charge density	8.5
Q_{ox}	charge in an oxide layer	8.5
Q_{pE}	minority charge storage in emitter (npn BJT)	7.3
$Q_{p,n}$	stored minority charge (diode)	5.4
Q_s	semiconductor space charge	3.2
R_A	Auger recombination rate	5.2
R_B	base spreading resistance	7.2
R_{BD}	parasitic drain resistance	9.1
R_{BS}	parasitic source resistance	9.1
R_D	drain-region resistance (MOSFET)	9.1
R_H	Hall coefficient	1.3
R_S	source-region resistance (MOSFET)	9.1
R_W	well resistance (CMOS)	9.3
R_X	substrate resistance (CMOS)	9.3
R_d	deposition rate	2.6
R_p	implant range	2.5
R_{sp}	spontaneous recombination rate	5.2
R_{out}	output resistance	10.1
S	inverse subthreshold slope	9.1
T	temperature	1.1
T_{tr}	transit time	9.1
U	net rate of recombination	5.2
U_A	net rate of Auger recombination	5.2
U_s	surface recombination rate	5.2
$V_{A,B}$	Early voltage (forward, reverse bias)	7.1
V_B	voltage of MOS transistor body, bulk, or substrate	9.1
V_{BD}	breakdown voltage	9.4
V_C	voltage in MOS channel	8.3
V_D	voltage of MOS transistor drain	9.1
V_{FB}	flat-band voltage	8.1
V_G	voltage of MOS transistor gate	9.1

Symbol	Definition	Section	Symbol	Definition	Section
V_H	Hall-effect voltage	1.3	p_{no}	equilibrium hole density (n-region)	5.3
V_S	voltage of MOS transistor source	9.1	p'	excess free-hole density	5.2
V_T	MOSFET threshold voltage	8.3	q	electronic charge	1.1
V_{Tn}	threshold voltage (n-channel MOSFET)	9.1	$q\phi_B$	energy barrier	3.2
V_{Tp}	threshold voltage (p-channel MOSFET)	9.1	r	capture and emission rates	5.1
V_a	applied voltage	3.2	r_b	base resistor (hybrid-pi model)	7.5
V_t	thermal voltage (kT/q)	7.5	r_j	junction radius	9.2
V_{CEsat}	collector-emitter voltage when BJT is saturated	6.4	s	surface recombination velocity	5.2
V_{Dsat}	saturation drain voltage	9.1	v_d	drift velocity	1.2
V_{Dsat}	drain voltage at edge of saturation	4.5	v_l	limiting carrier velocity	1.2
W	pattern width	2.10	v_{th}	mean thermal velocity of electrons	1.2
Z_{in}	input impedance	9.4	v_{sat}	saturation (maximum) carrier velocity	1.2, 9.2
$e_{n,p}$	emission-rate constant	5.1	x_B	charge-neutral base width (BJT)	5.1
f_D	Fermi-Dirac distribution function	1.1	x_E	width of the emitter charge-neutral region	6.2
f_T	frequency at which β_F equals unity	7.5	x_d	depletion-region width	3.2
$g_\square$	conductance of a square resistor pattern	2.10	x_i	"box" approximation to implant depth	9.1
g_m	transconductance	4.5	x_j	junction depth	2.10
g_{msat}	transconductance in saturation	4.5	$x_{n,p}$	depletion-region edge (n, p region)	4.3
$g(E)$	density of available electron energy states	1.1	x_{ox}	oxide thickness	2.3
h	Planck's constant	1.1	x_{dmax}	maximum width of depletion layer	8.2
k	Boltzmann's constant	1.1	α	absorption coefficient	1.1
k_s	surface reaction-rate coefficient	2.6	α	bulk-charge factor (MOSFET)	9.1
ℓ	MOSFET dimensional parameter	10.1	$\alpha_{F,R}$	forward-, reverse-active common-base current gain	6.4
m	mass	1.1	α_T	base-transport factor (BJT)	6.2
n	electron density	1.1	$\alpha_{n,p}$	ionization coefficient for electrons, holes	4.4
n	diode ideality factor	3.3	$\beta_{F,R}$	forward-, reverse-active common-emitter current gain	6.2
n	subthreshold-slope ideality factor	9.1	γ	emitter efficiency	6.2
n_i	intrinsic free-carrier density	1.1	γ	body-effect parameter (MOSFET)	9.1
n_{ie}	effective intrinsic-carrier density	1.1	δ	defect factor in hybrid-pi model	7.5
n_p'	excess electron density (p region)	5.3	η	efficiency: surface potential to gate voltage	9.1
n_{po}	equilibrium electron density (p-region)	5.3	η	parameter in hybrid-pi model	7.5
n_s, p_s	surface electron, hole density	3.3	ϵ	permittivity	1.1
n'	excess free-electron density	5.2	μ_{eff}	effective mobility in MOS channel	9.2
p	hole density	1.1	$\mu_{n,p}$	mobility of electrons, holes	1.2
p	gas-phase partial pressure	2.3	λ	scattering mean-free path	1.2, 4.4, 10.3
p	parameter in model	9.3	λ_{ph}	optical phonon mean-free path	Table 1.3
$p_{n'}$	excess hole density in n region	5.3			

Symbol	Definition	Section	Symbol	Definition	Section
$\phi(x)$	potential at x	3.3	τ_A	Auger lifetime	5.2
ϕ_{Bn}	barrier to electron injection across junction	4.2	τ_B	base transit time	7.3
ϕ_{Bp}	barrier to hole injection across junction	4.2	$\tau_{BF,BR}$	forward, reverse base charge-control variable	7.4
ϕ_{bs}	band bending between bulk and surface	8.3	$\tau_{F,R}$	forward, reverse charge-control time constants	7.3
ϕ_{fi}	intrinsic Fermi potential	1.1	τ_c	mean time between collisions	1.2
$\phi_{fn,fp}$	quasi-Fermi potential for electrons, holes	1.1	τ_{inv}	time to form inversion layer	8.4
ϕ_i	built-in potential	3.2	$\tau_{n,p}$	excess electron, hole lifetime	5.2
$\phi_{p,n}$	potential in a p-type, n-type region	4.2	τ_{nA}	lifetime with Auger recombination	5.2
ϕ_s	surface potential	9.1	τ_r	dielectric relaxation time	3.6
ρ	resistivity	1.0	$\Gamma_{n,p}$	Auger coefficient for electrons, holes	5.2
ρ	space-charge density	3.4	ΔE_c	conduction-band discontinuity	4.2
σ	conductivity	1.2	ΔE_g	band-gap discontinuity	4.2
$\sigma_{n,p}$	capture cross section (electrons, holes)	5.1	ΔL	Length of pinch-off region	9.1
τ	time to failure	10.4	ΔR_p	implant "width", (straggle)	2.5, 9.1
τ	initial oxide thickness parameter	2.3	$\Phi_{M,S}$	work-function voltage for metal, semiconductor	8.1
τ_0	minority-carrier lifetime	8.4	Θ	tunneling probability	4.4
			X	electron affinity (voltage)	3.2

INDEX

CPSIA information can be obtained at www.ICGtesting.com
Printed in the USA
BVOW06*1952180114

342278BV00010B/102/P